Key Steel Manufacturing Technologies by Collaborative Innovation

The series were written to address the needs of collaborative innovation and development within the iron and steel industry. They focus on the primary challenges facing the industry in China, including resources, energy, environment, and product performance. Six research directions in key common technologies throughout the entire iron and steel manufacturing process were proposed as follows: beneficiation; smelting; continuous casting; hot rolling; short-process hot rolling; cold rolling.

The series share insights into the intelligent manufacturing, the internationally original processing theory and production equipment. The publication of this series of books will be of great significance to improve the theoretical system of key common technologies in the iron and steel industry.

The series focus on expounding the new generation of ladle bottom powder spraying refining technology and equipment, high-quality continuous casting billet production process and equipment, integrated structure and performance control of smelting, casting and rolling, heat treatment technology and equipment for hot-rolled sheet and strip products with extreme specifications, thin slab endless/semi-endless rolling + non-pickling coating process technology, complete process technology and equipment for high-performance silicon steel via thin strip continuous casting, high-precision flatness and edge thinning control technology and equipment, advanced annealing and coating technology and equipment, new technology of complex refractory iron ore pre-enrichment-suspension roasting-magnetic separation (PSRM), short-process green preparation of super iron concentrate and clean steel base material, intelligent manufacturing of long profiles, intelligent flat products manufacturing and other key and common technologies urgently needed in the steel industry. Most of these key and common technologies are of original innovation of Chinese scientific and technological workers and have been successfully popularized and applied in many Chinese steel enterprises, which have greatly promoted the green transformation and development of the steel industry in China. Most of the technologies have reached the internationally leading level, helped the China's steel industry leap from "Following Suit" to "Taking the Lead", and made an important contribution to achieving the goal of "green technology, intelligent equipment, high-quality products, and service-oriented supply".

This series of books are edited by Guodong Wang, director of Northeastern University's Collaborative Innovation Center of Steel Technology, and member of China Engineering Academy, with a number of academicians in the industry working as consultants, and many authoritative experts and scholars in various fields of the steel industry as co-authors.

This series are supported by the NATIONAL PUBLICATION FOUNDATION.

Cheng Ji · Miaoyong Zhu · Rui Guan

Solidification End Reduction of Continuous Casting Process

冶金工业出版社
Metallurgical Industry Press

Springer

Cheng Ji
Northeastern University
Shenyang, China

Miaoyong Zhu
Northeastern University
Shenyang, China

Rui Guan
University of Science and Technology
Liaoning
Anshan, China

ISSN 2731-9733 ISSN 2731-9741 (electronic)
Key Steel Manufacturing Technologies by Collaborative Innovation
ISBN 978-981-96-8543-1 ISBN 978-981-96-8544-8 (eBook)
https://doi.org/10.1007/978-981-96-8544-8

Jointly published with Metallurgical Industry Press
The print edition is not for sale in China (Mainland). Customers from China (Mainland) please order the print book from: Metallurgical Industry Press.
ISBN of the Co-Publisher's edition: 978-7-5240-0239-0

This Springer imprint is published by the registered company Springer Nature Singapore Pte Ltd.
The registered company address is: 152 Beach Road, #21-01/04 Gateway East, Singapore 189721, Singapore

Foreword

As an important raw material industry, iron and steel industry plays an important role in the "supply side". It strives to provide the society with sufficient high-quality and high-performance iron and steel products with the lowest resource and energy consumption, the lowest environmental and ecological load, and the highest efficiency and labor productivity, so as to meet the needs of social development, national security, and people's life.

In the early stage of reform and opening-up, China's iron and steel industry was in the "following others" stage, mainly relying on the introduction of production lines and technologies from abroad. After more than 40 years of reform, innovation, and development, China has a steel production capacity of more than 1 billion tons, which exceeds more than half of the world's steel output, and the iron and steel industry has developed rapidly. With continuously improved technical level, in the fierce international competition, China's iron and steel industry is not in any single stage of development (i.e., "following others" to "keeping pace" and even "leading the pack"), but rather in all of the three stages concurrently. However, it still faces the following four major problems in the technological development. Firstly, the consumption of iron and steel production resources and energy is huge, pollutant emissions are serious, and the environment is overwhelmed. Therefore, there is an urgent need to realize green technology. Secondly, the stability, uniformity, and consistency of production equipment are poor, and the production efficiency is low. Thus, it is an urgent task for the iron and steel industry to realize the intellectualization of equipment, in-depth perception of information, coordinated and accurate control, intelligent optimization of decision-making and improvement of autonomous learning. Thirdly, insufficient product quality, unbalanced product structure, and insufficient supply capacity of high-performance products and independent innovation products all lead to strong demand for high-quality products. Fourthly, the supply-side development quality of China's iron and steel industry is not high enough and the service is not in place. Consequently, it is necessary to focus on improving the quality and efficiency of development, take supporting the supply-side structural reform as the main line, take improving the quality of the supply system as the main direction, build a service-oriented iron and steel industry, and realize the service-oriented supply.

Having experienced a period of rapid development, China's iron and steel industry has entered a stage of structural adjustment and transformation in recent years. Iron and steel enterprises must change the mode of development, optimize the economic structure, and foster new drives of growth. They should put quality first and give priority to performance, pursue supply-side structural reform as the main task, work hard for better quality, higher efficiency, more robust drivers of economic growth through reform, and raise total factor productivity. They should also endeavor to make China's iron and steel industry a global leader in "green technology, intelligent equipment, high-quality products, and service-oriented supply" and build China's iron and steel industry into a world leading iron and steel industry cluster.

In October 2014, the Collaborative Innovation Center of Steel Technology, with the two metallurgical universities of Northeastern University and Beijing University of Science and Technology as the core, and jointly established by enterprises, research institutes, and other colleges and universities, was officially put into operation after being approved by the Ministry of Education and the Ministry of Finance.

From October 2014 to the end of 2018, Collaborative Innovation Center of Steel Technology has been operating for 4 years. The process and equipment R&D platform focuses on the key common processes and equipment technologies in the iron and steel industry. According to the overall development plan of the platform's top-level design and the tasks and indicators formulated in various research directions, and through the in-depth integration and collaborative innovation of industry, academia, and research, the platform has made great achievements in six research directions, i.e., mining and beneficiation, smelting, hot rolling, short process, cold rolling, informatization, and intellectualization. These achievements are all key common technologies urgently needed by the iron and steel industry. Major achievements are as follows:

- A new generation of ladle bottom powder spraying refining process and equipment technology;
- High-quality continuous casting slab production process and equipment technology;
- Integrated structure and performance control of refining, casting, and rolling;
- Heat treatment process and equipment of hot rolled strip products with extreme specifications;
- Thin slab endless/semi-endless rolling + pickling-free coating process technology;
- Complete set of process technology and equipment for preparing high-performance silicon steel by thin strip casting;
- High-precision flatness and edge thinning control technology and equipment;
- Advanced annealing and coating technology and equipment;
- New technology of complex refractory iron ore preconcentration suspension roasting magnetic separation (PSRM);

- Short process green preparation of super iron concentrate and clean steel base material;
- Intelligent manufacturing of long profile;
- Intelligent manufacturing of flat material.

Most of these key common technologies belong to the original technologies of Chinese scientific and technological workers and have been successfully promoted and applied in China's iron and steel enterprises, greatly promoting the green transformation and development of China's iron and steel industry. Most of the technologies have reached the international leading level as a whole, played an important part in promoting the role of China's iron and steel industry from "following others" to "leading the pack", and made an important contribution to realizing the goal of "green technology, intelligent equipment, high-quality products, and service-oriented supply".

General Secretary Xi Jinping pointed out in his speech at the 2014 Academician Conference of the Chinese Academy of Sciences and the Chinese Academy of Engineering, "We must strengthen overall planning and coordination, vigorously carry out collaborative innovation, concentrate our efforts on major events, and form a strong synergy to promote independent innovation". Looking back on more than two years of refining and application, and four years of arduous research and development, we are carrying out our work under the guidance of this idea. The expectations of the leaders and the workers of steel and iron enterprises for China's original technology have struck a chord in our souls and inspired us to rectify the results of cooperation and innovation and promote them, to make them become a sharp weapon for the technicians of iron and steel enterprises to overcome difficulties and win new victories. Therefore, we have the desire to write a series of books. This series of books will be based on the series of innovative achievements of the Collaborative Innovation Center of Steel Technology, with the whole process, green technology, equipment and engineering, and industrialization as the main line. Combined with the actual engineering projects and examples of high-quality steel produced on the production line of iron and steel industry, this series of books will systematically gather the research progress of industry-university-research collaborative innovation and application, as well as key common technologies, cutting-edge leading technologies, and modern engineering technology innovation, so as to provide reference for enterprise technological reconstruction, transformation and upgrading, high-quality development, and planning the future development blueprint. This idea has won the positive response, and full support and close coordination of the vast number of colleagues in the enterprises. The leaders and editors of the Metallurgical Industry Press specially came to the university to give enthusiastic guidance, put forward suggestions, and discuss specific matters in publishing the series of books.

The needs of the country and the expectations of the iron and steel industry have touched our hearts and inspired us to move forward; the support and guidance of colleagues, publishers, and editors have given us strong confidence. The chiefs and academic backbones of the Collaborative Innovation Center and our intimate friends in enterprises and scientific research units acted immediately, splashed ink, and made great plans. We believe that through the joint efforts of all parties in the industry, academia, and research and the publishing house, we will dedicate a series of books that are visible, valuable, informative, and influential to the students in the steel and iron industry, as a return for the vigorous support from colleagues in enterprises. At the same time, on the occasion of the 70th anniversary of the founding of the People's Republic of China, we would like to present a gift forged with hard work, sweat, innovation, and a pure heart to the 70th birthday of our great motherland.

July 2019

Guodong Wang
Academician
Chinese Academy of Engineering
Beijing, China

Preface

Solidification end reduction technology for continuous casting process, as an important technique for improving the internal quality of billets/blooms/slabs, has been widely applied in the industry. A compressive deformation force near the solidification end in the continuous casting strand could extrude the molten steel, homogenize the solute segregation, and compensate for solidification shrinkage and porosity. The soft reduction technique was first proposed by Nippon Steel Corporation (NSC) of Japan in the late 1970s. In the 1990s, Voestalpine AG (VAI) in Austria first implemented dynamic soft reduction, which involved real-time tracking of the liquid core position in the continuous casting strand and dynamically adjusting the gap between two supporting rollers. However, the behaviors near the solidification end, such as the shell deformation in the two-phase zone, interfacial heat transfer, and macroscopic and microscopic solute transport during the soft reduction process, are extremely complex, and there is currently a lack of systematic theoretical research. Therefore, the reduction process can only be adjusted and optimized through repeated industrial trials. It not only consumes a large amount of human and material resources, but also offers little valuable reference for developing soft reduction processes for new continuous casting machines. Since 2003, Northeastern University (NEU) has been at the vanguard of research into the theory of the soft reduction process and its industrial application in China. The NEU has attained relatively systematic research outcomes, making significant contributions to this field. The methods of numerical simulation, physical simulation, and on-site experiments were employed to study and clarify the metallurgical behaviors during the continuous casting process, such as solidification heat transfer, gap shrinkage, bulging deformation, microstructure evolution, and solute segregation. The NEU has proposed calculation methods for the reduction amount and reduction zone in the dynamic soft reduction process and developed an online control model to overcome the limitations of industrial trials. This technology has been successively applied in more than ten enterprises, including Baosteel, Panzhihua Steel, Xingcheng Special Steel, Xianggang Steel, and Liancheng Steel, which has strongly promoted the iteration and innovation of continuous casting technology.

However, with the increase in the cross-sectional area of continuous casting billets/blooms/slabs, their internal cooling conditions deteriorate remarkably. The microstructure is predominantly characterized by columnar dendrites, while the equiaxed grains grow coarser. The residual molten steel, which is rich in solute-segregating elements among dendrites, has a tendency to flow toward an equilibrium state. This phenomenon exacerbates defects within the billets/blooms/slabs, such as segregation, porosity, and shrinkage. Besides, the dissipative effect of the solidified shell on the reduction amount escalates exponentially as its thickness increases. Conventional soft reduction deformation is no longer adequate to penetrate into the core of large-cross-section continuous casting billets/blooms/slabs. As a result, the effectiveness of the process experiences a significant decline.

To address this practical problem, NEU has further developed the dynamic heavy reduction technology at the solidification end for large-cross-section continuous casting blooms/slabs. The first continuous casting machines equipped with the dynamic heavy reduction function for blooms and wide-thick slabs were successively commissioned and put into production at Panzhihua Steel and Tangshan Steel. It enables the stable mass production for high-strength construction steel with a thickness of 150 mm under a reduction ratio of 1.87:1, and the successful production of large-size bar products such as axle-shaft steel under a reduction ratio of 3.74:1. This technology has been applied and promoted on nearly 20 production lines of domestic and foreign steel enterprises, including Baowu Steel and Ansteel, ensuring the stable production of large-size steel products such as high-strength and high-toughness heavy-duty rails, high-strength steel for construction machinery, high-performance weathering steel, large-size crankshaft steel, and high-grade gear steel.

Since the development and application of heavy reduction technology, there has been frequent discussion among industry experts and technical personnel about whether soft reduction or heavy reduction is better. In our view, both soft reduction and heavy reduction aim to improve the segregation and porosity defects at the center of the blooms/slabs, enhancing the bloom/slab homogeneity and density. The difference lies in the specific cross-sectional size of the blooms/slabs and the requirements for the rolled products, which lead to variations in the reduction amount and the reduction position. Therefore, both are considered solidification end reduction technologies. Over the past 20 years, our research has focused on key scientific questions such as "Where to apply reduction? How much reduction is needed?" as well as the related technical challenges of accurately and reliably implementing the reduction process. To this end, we have written this book to summarize the research and experience of our team in the field of solidification end reduction, particularly our understanding and exploration in recent years. We hope this will serve as a reference for industry experts and technical personnel and invite feedback and constructive criticism, with the aim of jointly advancing the innovative development of reduction technology. The main content of this book includes: Formation and control measures of central segregation and porosity in continuous casting product (Chap. 1); Metallurgical rheological characteristics of continuous casting strand during the reduction process (Chap. 2); Deformation laws and crack risks of continuous casting strand under the interaction of roller pressure, driving force, thermal stress, and other internal and external forces (Chap. 3); The effect of reduction on the closure of central shrinkage porosity in blooms (Chap. 4); Solute transport and segregation improvement under the collaborative control of reduction and electromagnetic stirring (Chap. 5); Recrystallization behavior of austenite during the reduction process and its impact on the refinement of microstructure, as well as the evolution in subsequent transport and rolling processes (Chap. 6); Key technologies for solidification end reduction, core equipment, and their application in enterprises such as Baosteel and Panzhihua Steel (Chap. 7).

We would like to take this opportunity to express our sincere gratitude to our experts and colleagues for their care, encouragement, support, and assistance. We also thank the leaders, experts, and frontline technicians from our partner companies for their trust and selfless help. Our heartfelt thanks go to every member of our team: Yong Chen, Qiyong Lin, Qi Zhao, Wei Guo, Shuyan Zhang, Haiqi Yu, Sen Luo, Yutang Ma, Changzheng Dong, Xueqian Cao, Jianping Tan, Chenhui Wu, Rui Guan, Tianci Chen, Junlu Yao, Zilin Wang, Shimin Deng, Guoliang Li, Qian Zhou, Yinghuan Li, Guorong Wu, Lang Liu, Zijian Wei, Yi Chen, and others. Although your names are not listed as authors, your tireless efforts and hard work have been the foundation of both the successful application of this technology and the completion of this book.

Finally, due to our limited expertise, there may inevitably be oversights, shortcomings, or inaccuracies in the process of writing this book. We sincerely invite readers to provide constructive criticism and suggestions.

Shenyang, China Cheng Ji
Shenyang, China Miaoyong Zhu
Anshan, China Rui Guan

Competing Interests The authors have no competing interests to declare that are relevant to the content of this manuscript.

Contents

Chapter 1
Introduction

1.1 Formation and Control Measures of Central Segregation and Porosity in Continuous Casting Product

During the solidification process of continuous casting products, a variety of external forces, such as thermal solute buoyancy, grain precipitation, and shell deformation, exert an influence on the relative flow of the solute-rich liquid phase and solute-poor solid phase. This promotes the long-distance transfer of solute elements, leading to the formation of macrosegregation defects. Simultaneously, because of factors such as insufficient compensatory shrinkage in the molten steel, numerous voids, and shrinkage cavities occur between dendrites in the final solidification region of the strand, resulting in center porosity and looseness defects. These defects cannot be eliminated during subsequent heating and rolling processes, which leads to discrepancies in the organizational performance and non-destructive testing of rolled materials. Center segregation and porosity, along with cracks and inclusions, are recognized as the three major quality defects in continuous casting products. Figure 1.1 illustrates the genetic defects of center segregation and porosity in continuous casting products during the rolling process.

1.1.1 Mechanisms of Center Segregation and Porosity Formation in Continuous Casting Product

Central segregation and porosity are often associated phenomena. Depending on the specific cross-sectional shape of continuous casting product and casting conditions such as steel type, there are currently mainly theoretical approaches involving thermal

C. Ji et al., *Solidification End Reduction of Continuous Casting Process*, Key Steel Manufacturing Technologies by Collaborative Innovation,
https://doi.org/10.1007/978-981-96-8544-8_1

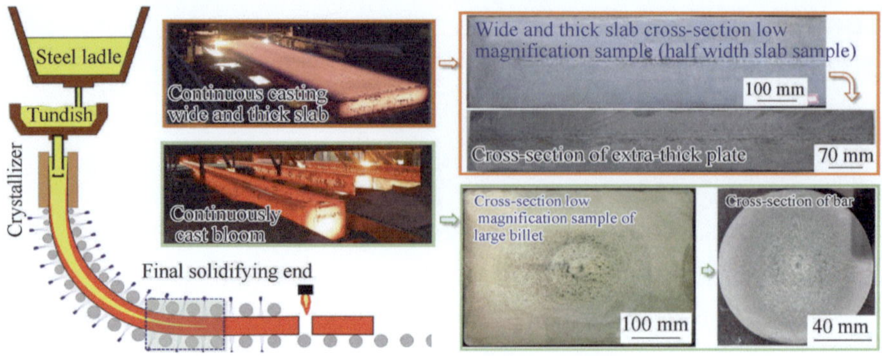

Fig. 1.1 Formation of central segregation and porosity in continuous casting product and their hereditary defects in rolled materials

solute buoyancy-driven liquid-phase flow [1–4], dendrite bridging and solidification shrinkage [5–8], shell bulging deformation [9–17], and thermal shrinkage [18–21].

1.1.1.1 Thermal Solute Buoyancy-Driven Liquid-Phase Flow

During the solidification process of steel, differences in the solubility of solute elements (such as carbon, phosphorus, and sulfur) in the solid and liquid phases cause them to continuously precipitate from the solid phase and become enriched in the liquid phase. Under the influence of thermal buoyancy and solute buoyancy, these solutes are transported with the liquid flow and gradually accumulate in the liquid core. Eventually, the diffusion and mass convection mechanisms ultimately lead to the formation of centerline segregation in the continuous casting product.

Based on this theoretical background, thermal buoyancy, and solute buoyancy are the two principal driving forces for the molten metal convection in the two-phase regions. During the solidification process, the cooling of the molten metal inevitably gives rise to local temperature disparities, which, under the influence of temperature gradients, lead to convection induced by thermal buoyancy. Simultaneously, solute redistribution causes uneven solute distribution, and solute buoyancy under concentration gradients induces convection. The combined action of these two forces gradually promotes the development of solute segregation. In 1985, Huppert and Worster [1] systematically investigated the fluid dynamics behavior under the influence of thermal and solute buoyancy by conducting solidification experiments on sodium nitrate under different boundary conditions. Subsequently, Turner [2] further pointed out the distinct effects of thermal and solute buoyancy in the liquid-phase region. In this region, stable thermal buoyancy and unstable solute buoyancy can induce double-diffusive convection effects. In the mushy zone, where molten metal temperature and concentration are strongly coupled, the buoyancy field is dominated by solute buoyancy. In 1995, Aboutalebi et al. [3] simulated the solidification process of Fe–C alloy round billets during continuous casting, analyzing the behaviors of solidification heat transfer, fluid flow, and solute transport under the influence of thermal and solute buoyancy. They explained the reasons for the formation of centerline segregation due to the continuous migration of solute elements. Professor Jiaquan Zhang and colleagues from the University of Science and Technology Beijing [4] used a continuous medium model to simulate the solidification transport behavior of large blooms during continuous casting. They explored the influence of thermal solute buoyancy on liquid flow and solute segregation, observing the gradual enrichment of solute elements expelled from the solid phase in the liquid phase. However, this theory does not fully consider the impact of the evolution of internal solidification structure on segregation and cannot accurately explain the mechanism behind negative segregation near the centerline of the continuous casting product.

1.1.1.2 Dendrite Bridging and Solidification Shrinkage Theory

When the carbon content exceeds 0.45%, the effect of columnar dendrite bridging becomes more conspicuous due to the solidification characteristics of the strand. Japanese scholars, such as Suzuki et al. [5], have analyzed the relationship between the solidification structure of the strand and centerline segregation, as illustrated in Fig. 1.2. During the unsteady-state casting process, certain columnar dendrites in the continuous casting product grow rapidly, and the accumulated equiaxed crystals precipitate at the columnar dendrite tip, forming dendrite bridging. Pores are formed at the bottom of the dendrite bridging, and solute-enriched liquid flows toward the center, leading to segregation near the dendrite bridging, forming V-shaped segregation. Simultaneously, owing to inadequate or complete lack of replenishment of the liquid phase, shrinkage cavities are formed in the strand center.

In recent years, Japanese scholars like Murao et al. [6] have used numerical simulation methods to dissect the phenomenon of dendrite bridging. They have pointed out that a negative pressure emerges at the bottom of the dendrite bridging, which prompts the flow of the liquid phase with a high solute concentration toward the center. Consequently, positive segregation occurs at the bottom of the dendrite bridging, while negative segregation forms at the top. In other words, the centerline segregation in the continuous casting product is the result of both dendrite bridging and solidification shrinkage caused by the instability of solidification. However, as the cross-section of the casting increases, the contraction of the shell and the development of columnar dendrites will further deteriorate the internal quality of the strand. Sivesson [7] and Gabathuler [8] suggested that in the region below a certain point in the strand ($f_s = 0.2$ and $f_l = 0.63$), the probability of ineffective replenishment of the volume shrinkage in the strand by the molten steel increases, leading to the formation of shrinkage cavities. This theory is also known as the small ingot segregation theory, commonly used to elucidate the formation mechanism of solute segregation in small billets. Currently, with the increase in the cross-section of continuous casting products and the intervention of external factors such as electromagnetic stirring and solidification end reduction, this theory exhibits certain limitations in probing

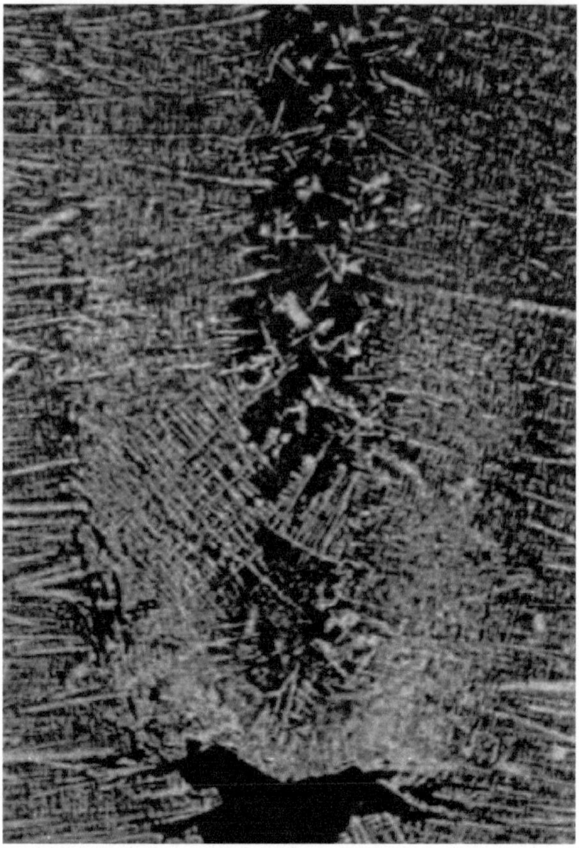

Fig. 1.2 Solidification bridging phenomenon in ingot [5]

the formation mechanism of solute segregation under intricate conditions involving multiple external fields.

1.1.1.3 Shell Bulging Deformation Theory

When the strand traverses the support rollers, the solidifying shell experiences periodic bulging deformations, which expedite the relative movement of solid and liquid phases in the two-phase region. This results in the transport of solute elements, leading to the formation of centerline segregation in the later stages of strand solidification. Currently, numerous researchers [9–17] have dedicated their efforts to studying the impact of bulging deformations in the shell on macrosegregation. In the 1970s, Flemings and Nereo [9] first proposed that the bulging deformation of the shell would enhance the flow of the melt between dendrites, intensifying centerline segregation. Subsequently, Miyazawa and Schwerdtfeger [10] carried out a series of studies on the macrosegregation in slabs caused by steady-state bulging deformations

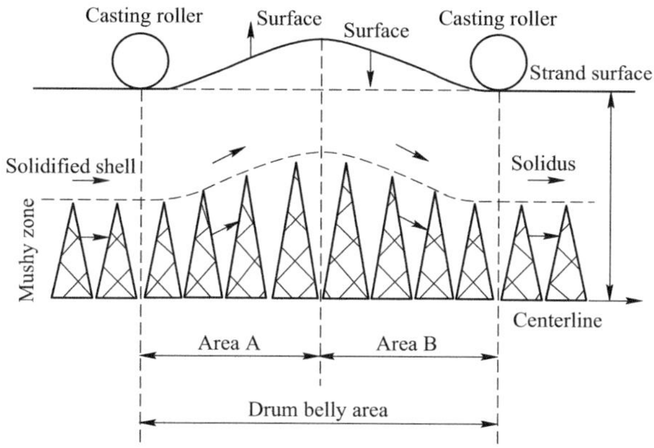

Fig. 1.3 Bulging deformation of the shell between support rollers [14]

of the shell, suggesting that bulging deformations are the main contributors to the formation of centerline segregation and adjacent negative segregation. In 2001, Kajitani et al. [11] employed the morphology curve of steady-state bulging in the shell to calculate the velocity and pressure fields of the liquid phase between dendrites. They revealed the promoting effect of bulging deformations in the shell on centerline segregation. However, they also pointed out that the negative segregation near the centerline is attributable to solidification shrinkage.

By incorporating the multiphase solidification model [12–14], as illustrated in Fig. 1.3, Domitner et al. [15, 16] analyzed the formation mechanism of macrosegregation under the influence of periodic bulging deformations by assuming a cosine curve-shaped shell bulging morphology. They demonstrated that the swelling effect is the primary cause of centerline segregation formation. Nevertheless, the actual shape of the bulging shell between adjacent support rollers is not a cosine curve, and the maximum deflection of the shell bulge between adjacent support rollers, shifting toward the downstream support roller, directly affects the flow of melt and solute segregation behavior within the casting product. Our research team [17] further considered the impact of non-uniform shell bulging solute transport in wide and thick continuous casting slabs. In comparison with the cosine curve-shaped shell bulge morphology, the non-uniform bulging morphology will aggravate centerline segregation defects in the slab.

1.1.1.4 Thermal Shrinkage Theory

In the 1980s, Lesoult and Sella [18] proposed that bulging deformation might not be the primary cause of macrosegregation and introduced the concept of solid-phase thermal shrinkage. According to this theory, during the solidification process, the continuous change in density with decreasing temperature leads to linear or volumetric contraction in the solidifying phase, creating voids. The negative pressure effect of these voids promotes the relative flow between the solute-rich liquid phase and the solid phase at the interdendritic regions. This phenomenon leads to centerline segregation, where the molten steel is drawn into the core of the strand to form centerline segregation, and the unreplenished portions create central looseness and shrinkage cavities. Raihle and Fredriksson [19] simulated the continuous casting process through ingot solidification experiments and developed corresponding mathematical models for analysis. They also ascribed the formation of segregation to the decrease in temperature at the solidification endpoint, causing volumetric contraction and the formation of tubular voids, along with segregation under the negative pressure suction effect. El-Bealy [20, 21] further systematically analyzed the strain and solute segregation behavior between dendrites under different cooling rates. Their results indicated that increasing cooling rates led to increased compressive strain between dendrites while decreasing cooling rates resulted in the conversion to tensile strain within the same strain zone. The study revealed that interdendritic tensile strain causes positive segregation, while compressive strain leads to negative segregation, and the segregation level depends on the strain history and its rate. However, in the central region of continuous casting products, not only positive segregation exists but also negative segregation defects near the center. The singular thermal shrinkage theory alone cannot provide a scientific explanation for adjacent positive and negative segregation defects. Additionally, for the internal defects such as looseness and shrinkage cavities in continuous casting products, relying merely on the contraction theory cannot accurately reveal their formation causes.

1.1.1.5 Temperature Gradient Theory

In 1953, Pellini [22] pointed out the crucial role that temperature gradients play in the formation of pores. In 1981, Niyama et al. [23] extensively discussed this phenomenon through numerous commercial casting experiments, identifying a temperature gradient threshold that could prevent looseness. This threshold depends on the specific shape and size of the strand. In 1982, Niyama et al. [24] experimentally determined the critical temperature gradient, which is inversely proportional to the square root of the cooling rate for each type of casting. This led to the establishment of the well-known Niyama criterion, $N_y = G/\sqrt{v}$, where G is the temperature gradient, and v is the cooling rate. In 1983, Engström et al. [25] studied the differences in temperature gradient at the two solidification fronts during the solidification process, indicating that such differences could result in the formation of gas pores in the continuous casting process. In 2005, Monroe [26] applied the Niyama criterion to predict the microporosity in cast steel due to shrinkage, where lower Niyama values corresponded to higher porosity rates. In 2009, Carlson and Beckermann [27] introduced the concept of solidification characteristics and the critical pressure in the mushy zone, proposing a dimensionless Niyama criterion. This criterion is directly applicable to predicting the volume fraction of shrinkage porosity during the metal solidification process and is currently one of the most widely used models for predicting shrinkage cavities.

1.1.1.6 Dissolved Gas Evaporation Pressure Theory

In 2005, Stefanescu [28] indicated that the formation of closed shrinkage defects is intimately correlated with the nucleation and growth of pores in the mushy zone, which hinges on the level of impurities in the metal and the amount of dissolved gases. In 2011, Stefanescu and Catalina [29] analyzed the physical laws of pore formation based on pressure balance. They posited that shrinkage cavities are formed only when the pressure in the mushy zone compels the liquid to stream out of the final region during solidification. In 2020, Kweon et al. [30] pointed out that shrinkage is the primary factor in the formation of micropores. Additionally, due to the significantly lower solubility of gases in solid metal compared to liquid metal, bubbles may form at the end of solidification, possibly entrapped within the dendritic mushy zone. The main hypothesis of this theory is that if the gas pressure in the mushy zone exceeds the sum of the local pressure and the pressure caused by surface tension, micropores will form. Presently, the theory regarding the pressure exerted by dissipating dissolved gases has emerged as a central aspect in the research on the microscopic formation of shrinkage cavities.

1.1.2 Control Measures for Central Segregation and Porosity in Continuous Casting Product

In the exploration of the mechanisms underlying center segregation and porosity in continuous casting products, the techniques for diminishing or eliminating these issues can be generally classified into three principal categories:

(1) The reduction of non-metallic inclusions and harmful elements in steel is accomplished via representative technologies. For instance, in the clean steel smelting process, refining is carried out to eliminate inclusions or to control their distribution patterns.

The research team led by Dianzhong Li at the Institute of Metal Research, Chinese Academy of Sciences [31], has recognized that the transport behavior of inclusions under buoyancy as one of the contributing factors to the formation of segregation defects. During the solidification process of steel, harmful elements such as phosphorus, sulfur, hydrogen, nitrogen, and oxygen tend to form fine inclusions that disperse, aggregate, and grow. In the low solid fraction mushy zone, inclusions obstructed by dendrite arms undergo flow and transport under buoyancy, effectively disturbing the flow field in the mushy zone, altering the local solute flow pattern, and thus resulting in solute segregation defects. Additionally, under buoyancy, inclusions cause accompanying flows of solute-rich melt, low-melting-point impurities (sulfides), and bubbles, leading to localized temperature reduction, remelting, or erosion of dendrite arms. This further promotes the formation of solute segregation defects. Figure 1.4 illustrates the schematic diagram of inclusion transport under buoyancy between dendrites.

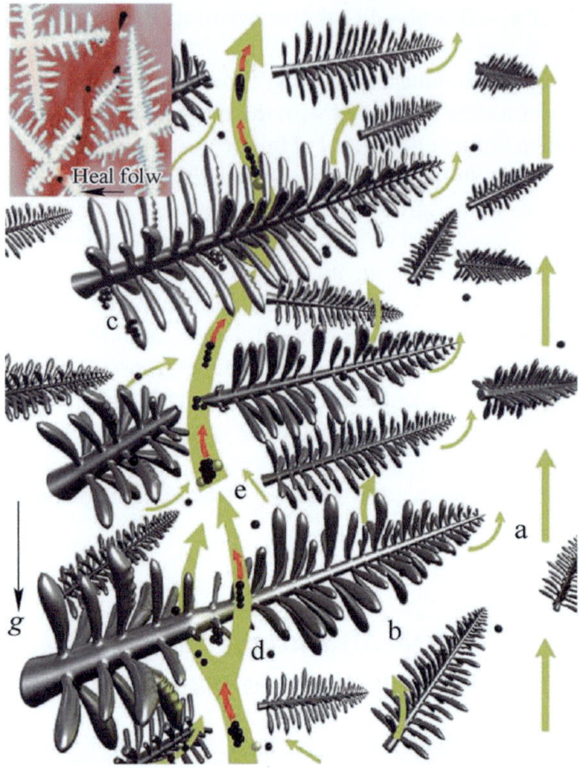

Fig. 1.4 Schematic diagram of buoyancy-driven movement of inclusions between dendrites [31]

Via procedures like hot metal pretreatment and ladle refining, effective desulfurization, and dephosphorization, as well as the removal of non-metallic inclusions in steel, can be achieved. Nevertheless, these approaches cannot wholly expunge harmful elements and are incapable of eradicating inclusions entirely. Additionally, for steels with high sulfur content, high phosphorus content, and higher levels of microalloying elements (prone to the formation of carbonitrides), these methods are insufficient in resolving the segregation problems.

(2) Augmenting the equiaxed crystal ratio of continuous casting products. Typical techniques encompass low superheat casting, electromagnetic stirring within the crystallizer, feeding steel strip/wire into the crystallizer, and magnetic pulse oscillation of the crystallizer.

The equiaxed crystals are formed when the solidification interface migrates freely in three-dimensional space, while columnar dendrites are formed when the solidification interface migrates in the opposite direction of the heat flow in three-dimensional space. The solute distribution in the dendrite trunk of columnar dendrites and equiaxed crystals is relatively uniform. Under the action of redistribution, solute enriches between dendrite arms, leading to localized segregation. If columnar dendrites are well developed, it can easily cause dendrite bridging, blocking the melt shrinkage channels, and resulting in defects such as centerline segregation, looseness, and shrinkage. In contrast, fine and densely distributed equiaxed crystals in the center of the cast product compress the distribution space of solute-rich melt between grains, significantly reducing the risk of defects such as centerline segregation, looseness, and shrinkage.

By implementing low superheat casting [32], electromagnetic stirring in the crystallizer [33], feeding steel strip/wire into the crystallizer [34], and pulse magnetic oscillation [35] in the intermediate ladle, crystallizer, and their adjacent areas (see Fig. 1.5), the equiaxed crystal ratio of continuous casting product can be effectively increased, thereby enhancing the homogeneity of the cast product. Among them, technologies such as induction heating in the intermediate ladle and plasma heating can achieve stable low superheat casting, effectively increasing the number of equiaxed crystals in the liquid core. The technique of feeding the steel strip or wire into the crystallizer can reduce the superheat within the crystallizer by leveraging the heat absorption during the melting of the steel strip or wire. Additionally, it can utilize the vibration of the steel strip/wire to introduce energy and composition fluctuations, increase nucleation sites, and enlarge the nucleation undercooling, aiming to improve the equiaxed crystal rate and alleviate defects such as centerline segregation and porosity. However, the introduction of inclusions is unavoidable during the feeding process, limiting its industrial application. Mold electromagnetic stirring enhances the fluid flow velocity of the molten steel, facilitating rapid heat dissipation in the core of the cast product, increasing nucleation undercooling, and significantly improving the equiaxed crystal rate in the core of the strand, thereby improving internal defects in the cast product. Pulse magnetic oscillation technology induces electromagnetic forces through pulsed current, causing already solidified nuclei to detach from the solidification interface under the flow induced by the pulse wave. This serves the purpose of refining the solidification structure and improving solidification defects. Figure 1.5 illustrates the effective techniques for increasing the equiaxed crystal ratio of continuous casting products.

(3) Improving solute transport and compensating for solidification shrinkage are technologies that represent technologies such as final electromagnetic stirring and solidification end reduction.

The long-distance transport of solute-segregated molten steel in the two-phase zone of continuous casting product is a major cause of macrosegregation. During the transport process, it forms a "V"-shaped segregation defect, which, after gathering at the center of the strand, results in a central segregation defect. Simultaneously, the solidification structure obstructs the feeding channels, causing a decrease in the fluidity of the melt and the formation of porosity and shrinkage defects in the strand. Technologies such as final electromagnetic stirring [36] and pressing down at the end of solidification [17] can regulate the solidification structure at the center of the strand and alter the transport of solute-enriched molten steel in the late stages of solidification, thereby improving center segregation, porosity, looseness, and other metallurgical defects.

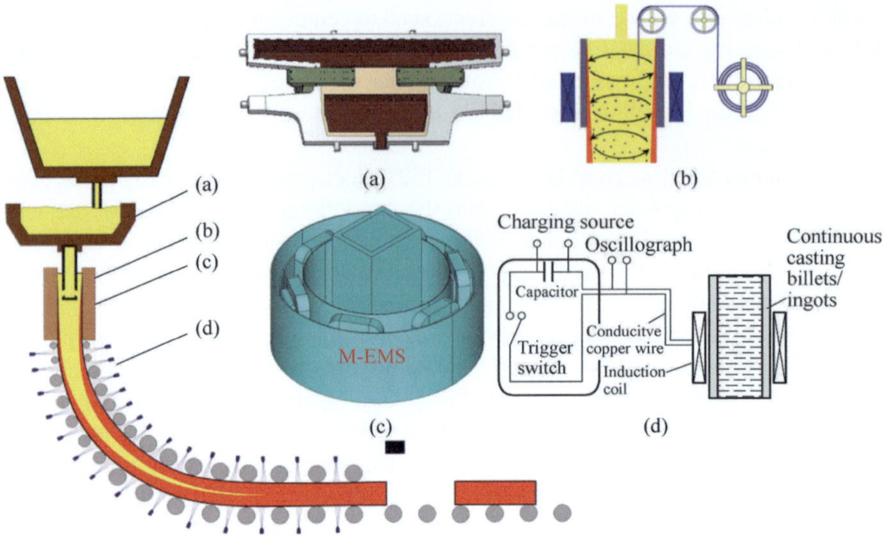

Fig. 1.5 Effective technologies for increasing the equiaxed crystal ratio of continuous casting product. **a** Low superheat casting; **b** feeding steel wire/strip into the crystallizer; **c** mold electromagnetic stirring; **d** pulse magnetic oscillation

The inductive motion of electromagnetic stirring suppresses the formation of columnar dendrites, avoiding the bridging phenomenon of large dendritic or free dendrites in the final solidification zone. Electromagnetic stirring technology can reduce V-shaped segregation, but its stirring position is generally challenging to adjust flexibly based on steel grades, cross-sections, casting speed, etc. Inappropriate stirring positions may cause negative segregation in the bright area. Moreover, when the solid-phase fraction is high, and the cross-section of the strand is large, the stirring effect may be insignificant. Figure 1.6 illustrates the impact of mold electromagnetic stirring on the solidification structure.

Applying pressure (thermal and mechanical stress) near the solidification end of continuous casting products to induce a certain amount of reduction can impede the gathering of molten steel enriched with segregating elements, thereby eliminating central segregation. Simultaneously, it compensates for the solidification shrinkage of continuous casting products to eliminate central looseness [37]. Compared to thermal stress reduction techniques (mainly achieved through strong cooling at the solidification end to reduce the shrinkage of small-section continuous casting products), roller reduction has a broader application scope, lower equipment investment and maintenance costs, and can implement larger reduction deformations. It is the most common method of implementing soft reduction.

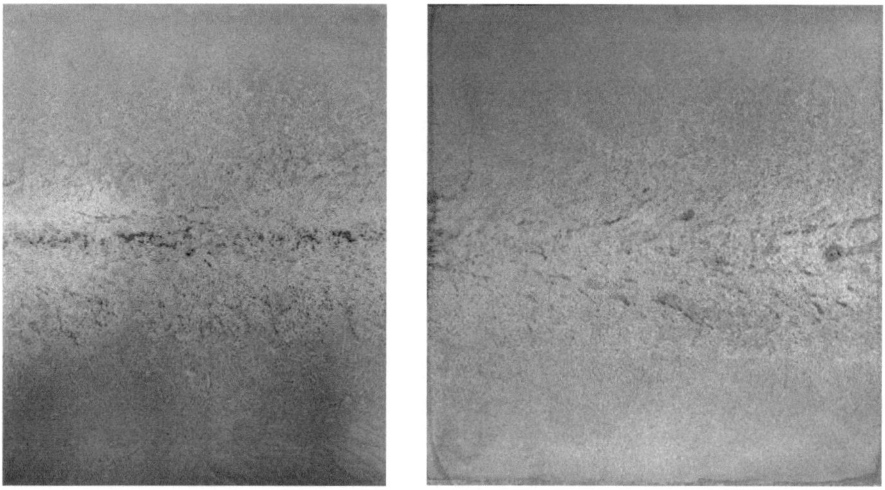

(a) (b)

Fig. 1.6 Mold electromagnetic stirring on solidification structure. **a** Weak electromagnetic stirring; **b** strong electromagnetic stirring

1.2 Solidification End Reduction Process Mechanism and Key Parameters in Continuous Casting Products

Solidification end reduction is a crucial external control technique in continuous casting. Based on different process objectives, applicable casting machine types, and development history, it can be broadly categorized into two techniques: soft reduction and heavy reduction. Soft reduction, initially proposed by Nippon Steel, involves a reduction deformation generally not exceeding 5% of the strand thickness. It is mainly targeted at continuous casting products with conventional section dimensions. The technique involves applying a certain reduction before complete solidification to extrude molten steel enriched with solute-segregating elements, promoting solute homogenization distribution, and simultaneously compensating for solidification shrinkage to improve looseness. Due to its significant effects on improving central segregation and porosity of the strand, soft reduction technology is widely used on over a hundred continuous casting lines globally and has become a standard technology for advanced casting machines.

In recent years, with the continuous improvement in steel quality and specifications, the development of casting machine sections is trending toward larger sizes. For instance, new continuous casting machines for medium-thick plates, with a thickness of 470 mm, have been built in Yingkou. Companies such as Benxi Steel, Shougang, and Xingcheng Special Steel have constructed continuous casting machines for large-section bloom with dimensions exceeding 400 mm × 500 mm. Currently, more than 50 continuous casting machines capable of producing slabs with a thickness greater than or equal to 250 mm have been commissioned in China, along with over 30 continuous casting machines producing bloom with a section greater than or equal to 350 mm × 350 mm. The production of high-end, large-sized steel products using large-section continuous casting products has become one of the main development directions for continuous casting. However, with the widening and thickening of strand sections, the solidification time has increased threefold to fourfold, proportionally enlarging the central segregation and porosity areas. The conventional soft reduction deformation is insufficient to address this issue. Therefore, for practical requirements such as large-section continuous casting products with a significant increase in shell thickness, strict ultrasonic testing requirements for thick plates/special thick plates, and large cross-section continuous casting products, solidification end heavy reduction technology has gradually developed based on the foundation of soft reduction technology. Solidification end heavy reduction is implemented within a certain range before complete solidification, at the solidification endpoint, and within a specified range after complete solidification. The reduction amount can exceed 10% of the strand thickness. On the one hand, it can overcome the limitations of the conventional soft reduction deformation, which cannot fully penetrate the core of a large-section continuous casting product. On the other hand, it fully utilizes the temperature difference of up to 500°C between the inner heat and outer cold at the solidification end of the strand to achieve efficient transmission of the reduction amount to the core of the strand. This aims to achieve substantial improvements in segregation and porosity closure.

From a control perspective, both light and heavy reductions share the same operating principle, namely, implementing a certain reduction deformation within a specific range at the solidification end to enhance the internal quality of the strand. Both can be classified as solidification end reduction technology, requiring answers to the core question of "where to apply pressure and how much". The corresponding two key process control parameters are the reduction zone and the reduction amount.

As shown in Fig. 1.7 [38, 39], the reduction zone refers to the region where the reduction is applied, generally implemented before complete solidification for soft reduction. It is usually represented by the solid fraction (f_s) in the two-phase zone at the solidification end of the continuous casting product, indicating the solid fraction. The reduction amount characterizes how much deformation is applied within the reduction zone. The reduction amount can be further categorized into the reduction amount applied by a single reduction roller or the total reduction amount applied by multiple reduction rollers. Since the continuous casting production process of slab sections is often completed in fan-shaped segments for solidification end reduction, the reduction amount per unit length within a fan-shaped segment remains constant. Therefore, it can also be represented by the reduction rate, with the unit typically being mm/m. Since the reduction effect needs to be transmitted to the core of the strand to achieve improvements in solute segregation and central porosity, the reduction efficiency is commonly used to represent the transmission efficiency of the reduction amount from the surface to the core of the strand. This is calculated as the deformation amount at the core of the strand divided by the surface deformation amount. Additionally, there is the reduction rate parameter, which is the deformation amount squeezed per unit of time calculated based on the casting speed of the strand.

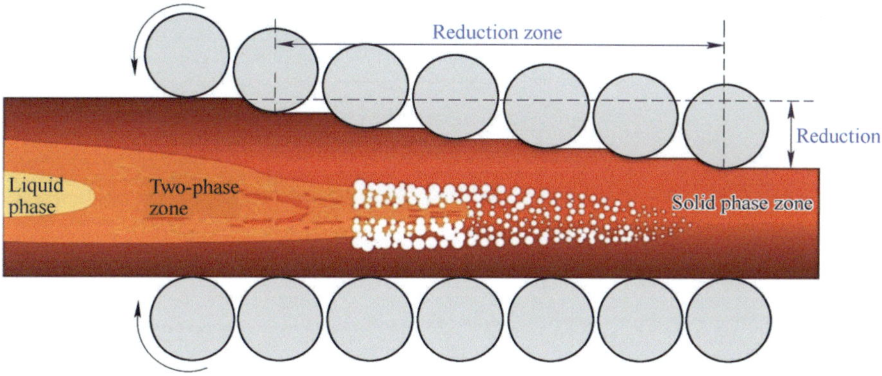

Fig. 1.7 Schematic diagram of solidification end roller reduction

1.2.1 Reduction Zone

The reasonable parameters of the reduction zone are crucial for effectively addressing center segregation and porosity while avoiding the generation of mid-crack defects. Numerous scholars worldwide have conducted extensive scientific research from the perspectives of physical experiments and numerical simulations. In the 1980s, based on the mechanism that hindered solute diffusion between dendrites during the solidification process was the primary cause of solute segregation, as shown in Fig. 1.8, Takahashi and Hagiwara [40, 41] divided the mushy zone into three regions based on different solid-phase fractions. They suggested that the reduction should be implemented in the q_1 zone ($0.3 < f_s < 0.7$) where interdendritic melt can flow. In the p zone with a higher solid-phase fraction ($f_s > 0.7$), interdendritic solute flow is hindered and can only exist in partially solidified areas, forming solute segregation defects. As the solid-phase fraction decreases, the interdendritic solute in the q_1 zone faces greater flow resistance but can flow between the solid-phase framework. In the q_2 zone with a lower solid-phase fraction ($f_s < 0.3$), the flow resistance of interdendritic solute significantly decreases, allowing for thorough mixing with the central liquid phase and effectively inhibiting positive segregation. Ogibayashi et al. [42] also used the solid-phase fraction to partition the reduction zone, suggesting that reduction should start when the solid-phase fraction is 0.1–0.3 and end when it is 0.6–0.9. When the solid-phase fraction exceeds 0.9, the flow of the liquid phase is hindered, rendering the reduction process ineffective. Subsequently, to maximize the benefits of continuous casting end reduction technology in improving the internal quality of the strand, researchers such as Bleck [43], Ito [44], Yim [45], and Qi [46] determined through physical simulations and production tests that the solid-phase fraction at the end of the reduction zone should not exceed 0.96.

Our team, led by Rosen et al. [47], considered the impact of each solute element on the solidification behavior of the strand, suggesting that each element has an optimal reduction zone position corresponding to it. Based on solute segregation

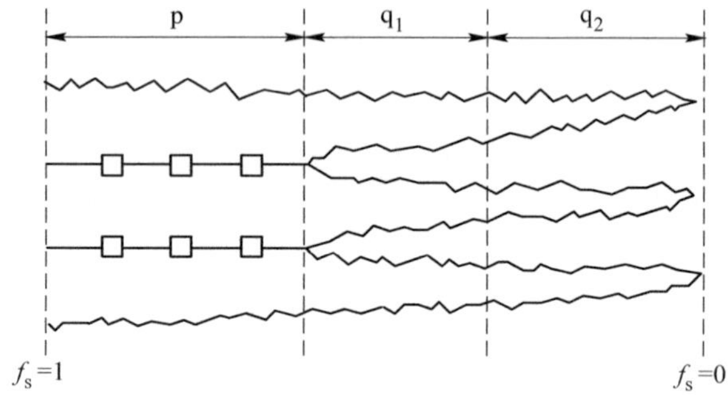

Fig. 1.8 Schematic diagram of two-phase region during solidification process [40]

calculations, a theoretical calculation model for continuous casting end reduction was established. Additionally, for the non-uniform solidification of wide and thick slabs, our team proposed the reduction zone process [48], breaking the limitation of designing the reduction zone solely based on the solidification progress along the centerline of the cast slab. This approach effectively improves segregation defects in the 1/8–1/4 region of the slab width and ensures the synchronized improvement of solute segregation across the entire cross-section.

While researchers have generally reached a consensus in theoretical studies on the reduction zone, namely the application of pressure in the two-phase region of continuous casting product, practical results from production lines still exhibit significant differences, as shown in Table 1.1.

In fact, due to the limitations of simple solidification heat transfer models that cannot account for complex phenomena like solute segregation in the two-phase region, the thermal tracking models of the process control system struggle to accurately predict the solidification process. This is a critical factor leading to the inability

Table 1.1 Reduction zone parameters in domestic and international enterprises

Company	Application range		Reduction zone (f_s)
Henan Jiyuan Iron & Steel (Group)Co., Ltd., China [49]	Medium-thickness slabs		0.5–0.95
Rautaruukki, Finland [50]	Microalloyed steels w[C] = 0.088%	210 mm × 1825 mm 210 mm × (1250–1475) mm	0.15–0.8 0.3–0.9
China Steel Corporation, Taiwan, China [51]	Blooms		0.55–0.75
POSCO, South Korea [52]	S82 square billet 250 mm × 330 mm		0.3–0.7
Hyundai Steel, South Korea	Pipeline steels 250 mm × 2200 mm		0.2–1.0
Bayuquan Steel Branch of Angang, China	Pipeline steels 250 mm × 2000 mm		0.6–0.95

to precisely correlate the reduction zone with the low-magnification quality of the strand.

In recent years, our research team has addressed this challenge by establishing a multiphase solidification model incorporating various external field effects such as electromagnetic stirring during the continuous casting process and solidification end pressing. This model, describing the "liquid-phase columnar dendritic equiaxed crystal" multiphase solidification, enables multiscale simulations of macroscopic molten metal flow, heat transfer, and microscopic grain nucleation and growth. Through systematic analysis of the effects of thermal shrinkage, solidification shrinkage, thermal buoyancy, and solute buoyancy in the two-phase region [53], our team has proposed a novel mechanism that effectively suppresses the accelerated enrichment of solutes during significant deformation pressing in the later stages of continuous casting solidification. This theoretical advancement provides support for the stable production of high-homogeneity strand [17], as discussed in detail in Chap. 5 of this book.

1.2.2 Reduction Amount

The reduction amount must fully compensate for the volume shrinkage of the molten steel during the solidification process in the reduction zone to prevent the flow of solute-enriched molten steel. However, excessive reduction can lead to internal cracking in the strand and damage to the fan-shaped segment pinch rolls. On the other hand, insufficient reduction has minimal impact on central segregation and porosity improvement. Therefore, the reduction amount must meet three requirements: (1) compensate for the solidification shrinkage in the reduction zone, (2) reduce central segregation and central porosity, and (3) avoid internal cracking of the strand within the permissible pressure range of the caster's fan-shaped segment.

Researchers have proposed various methods for calculating the reduction amount [54–58]. Utilizing the principle of mass conservation along the withdrawal direction of the strand, Sugimaru et al. [54] from Nippon Steel initially derived a model for the reduction rate of large blooms and analyzed the factors influencing the reduction rate. Zeze et al. [55], through experimental studies on the segregation and deformation behavior of un-solidified steel ingots under soft reduction, investigated the comprehensive impact of liquid core thickness and reduction amount on V-shaped segregation. As shown in Fig. 1.9, the V-shaped segregation defects decrease continuously with increasing reduction amount. Excessive reduction leads to the formation of a white band. Additionally, the required reduction amount increases with an increase in liquid core thickness. However, when the liquid core thickness reaches a critical value, further increasing the reduction amount cannot improve V-shaped segregation and may worsen internal quality, resulting in internal cracking. Figure 1.10 indicates that an increase in reduction amount does not affect improving V-shaped segregation when the reduction rate is less than 0.02 mm/s. This is because the reduction rate is less than the solidification shrinkage rate, and there is insufficient time to fully

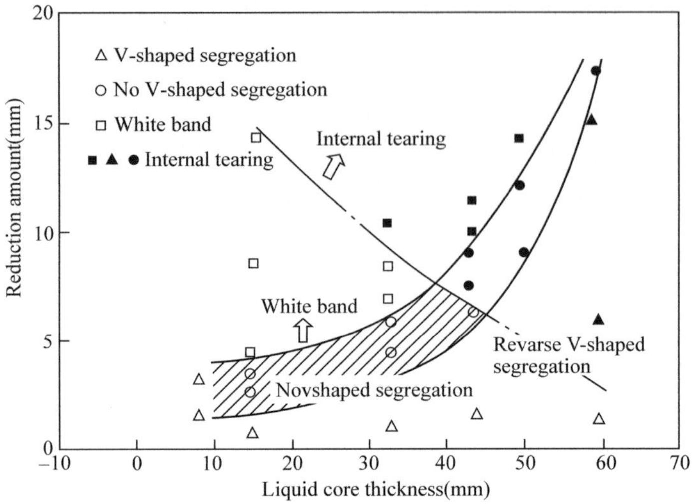

Fig. 1.9 Influence of reduction amount and liquid core thickness on V-shaped segregation (liquid core thickness: 32 mm) [55]

compensate for solidification shrinkage. Simultaneously, an increase in the reduction rate leads to an increase in strain rate, reducing the critical strain, thereby reducing the critical reduction amount. Isobe et al. [56] proposed the use of convex rolls (Crown Roll) for soft reduction to improve central segregation in the continuous casting of large blooms. They established mathematical models for the relationship between reduction force and reduction amount, reduction rate, and reduction efficiency. Subsequently, they analyzed the impact of the arrangement and quantity of reduction rolls on the relationship between reduction force and reduction amount, as well as the load transfer behavior during soft reduction in the fan-shaped segment.

From the perspective of actual production, Finland's Rautaruukki Iron and Steel Company [50] has determined that the optimal reduction amount for casting low-alloy steel in the range of 210–1825 mm is 1.5 mm. Moreover, they have concluded that when the reduction rate is less than 1.00 mm/min, the reduction does not harm the surface quality of the slab. Pohang Iron and Steel Company (POSCO) in South Korea [57] observed that as the reduction amount increases, center segregation continually decreases. However, beyond a reduction amount of 6 mm, there is no further improvement in center segregation, and internal cracks become more prominent. Therefore, a reasonable total reduction amount is determined to be 6 mm. Based on theoretical research and practical exploration conducted by our team, the typical reduction rate for the slab is in the range of 0.55 mm/min to 1.1 mm/min, while for bloom, it is in the range of 1 mm/min to 1.2 mm/min. Generally, the soft reduction total amount for the slab does not exceed 10 mm, and for bloom, it does not exceed 15 mm. However, the reduction amount (reduction rate) is closely related to the steel grade, the cross-section of the cast slab, and the production conditions. For instance, in the production of a 210 mm thick common slab, a soft reduction total amount of

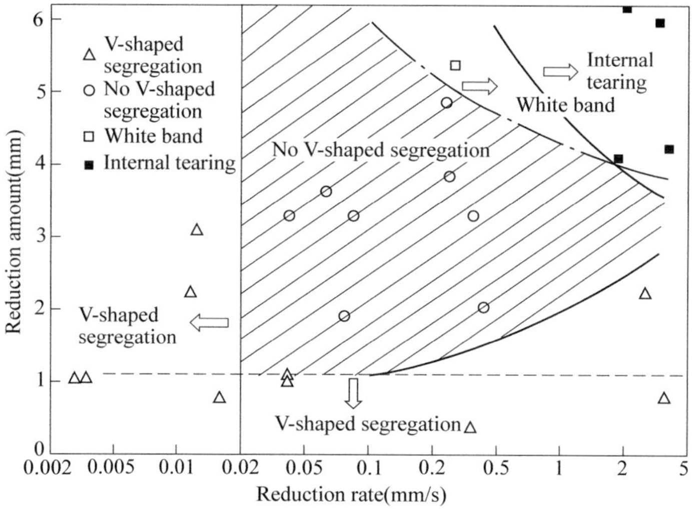

Fig. 1.10 Influence of reduction rate and reduction amount on V-shaped segregation (reduction rate: 0.35 mm/s) [55]

approximately 3 mm is sufficient for achieving optimal process results, especially when producing steel grades like Q234. However, when producing high-carbon and high-alloy steel grades of the same specifications, such as precision die steel for automotive components, an effective resolution of segregation and looseness defects requires a reduction total amount exceeding 7 mm.

Our team [58, 59], based on the compensation principle for solidification shrinkage of the strand (mass non-conservation in the withdrawal direction), has derived the integral models for the reduction amount of large blooms and the reduction rate of the slab. Equation (1.1) provides the theoretical calculation method for the reduction amount of large blooms in the continuous casting straightening machine.

$$R_i = \frac{\Delta A_i}{\eta_i X_i} = \frac{\left(\int_0^{Y_i}\int_0^{X_i}\rho(x, y, z_i)\mathrm{d}x\mathrm{d}y - \int_0^{Y_{i-1}}\int_0^{X_{i-1}}\rho(x, y, z_{i-1})\mathrm{d}x\mathrm{d}y\right)}{\rho_1\eta_i X_i} \tag{1.1}$$

R_i—Minimum theoretical reduction amount of the ith straightening machine;

A_i—Solidification shrinkage area;

η_i—Reduction efficiency;

X—Width of the strand under the ith straightening machine;

$\rho(x, y, z)$—Density of the strand at the (x, y, z) coordinate;

ρ_1—Density of the molten steel.

Using this model, in conjunction with the three-dimensional solidification heat transfer calculation of the strand, the temperature distribution and mass shrinkage along the withdrawal direction were obtained. Theoretical calculations were performed for the liquid core shrinkage of molten steel, liquid-core reduction amount, and surface reduction amount. The model was applied to the continuous casting production practices of bearing steel, cord steel, spring steel, and other large blooms, achieving favorable metallurgical results [59]. In recent years, our team, based on a multiphase solidification model, simulated the formation mechanism of "V"-shaped segregation and center segregation, along with the required deformation to inhibit them (see Chap. 5). The model was further used to investigate the closure effect of shrinkage cavities in the central region of the strand and the corresponding reduction amount demand (see Chap. 4), thereby advancing and refining the theoretical framework of the reduction amount process.

1.2.3 Reduction Efficiency

During the solidification end reduction process, only the deformation applied to the center of the strand can achieve the technological effects of squeezing the liquid core, uniform segregation of solutes, and closing shrinkage cavities. Therefore, the impact of the strand deformation on the reduction rate can be expressed by the reduction efficiency, which also serves as a measure of the consumption level of the solidified shell during the surface reduction of the strand.

In the 1980s, Japanese scholars, including Hayashida et al. [60], proposed that the ratio of the displacement of the solid–liquid interface to the displacement of the strand surface during soft reduction is the reduction efficiency. They used a simplified deformation model to analyze the variation patterns of reduction efficiency at different positions of a 210 mm × 800 mm slab relative to the solidification endpoint under a casting speed of 1.2 m/min. Ito et al. [61] suggested that the ratio of the reduction in the liquid core of the strand to the reduction on the strand surface during soft reduction is the reduction efficiency η (see Fig. 1.11), with the calculation equation as follows:

$$\eta = \frac{h_0 - h_1}{H_0 - H_1} \tag{1.2}$$

Ito et al. employed a three-dimensional finite element method and experimental analysis to investigate the influence of roller diameter and strand dimensions on the reduction efficiency. They derived an empirical equation for reduction efficiency as follows:

$$\eta = \exp(2.36\lambda + 3.73) \times \left(\frac{R}{420}\right)^{0.587} \tag{1.3}$$

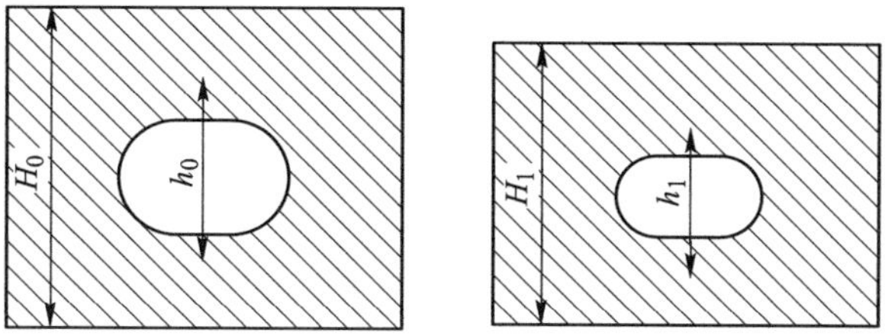

Fig. 1.11 Schematic diagram of reduction efficiency

R—Roller diameter in millimeters (mm);

λ—Index of the strand shape.

Ito et al.'s reduction efficiency model only considers the transfer of reduction in the thickness direction of the strand during the reduction process. Dr. Qiyong Lin from our team [62] proposed using the efficiency of transferring two-dimensional deformation on the surface of the strand to the solidification front during soft reduction to represent reduction efficiency. This more accurately describes the efficiency of transferring the reduction on the strand surface to supplement solidification shrinkage in the presence of the liquid core, as shown in Fig. 1.12. The calculation equation is as follows:

$$\eta = \frac{\Delta A_{\rm i}}{\Delta A_{\rm H}} \tag{1.4}$$

$\Delta A_{\rm i}$—Area of deformation of the liquid core at the solidification front during soft reduction, measured in square millimeters, mm^2;

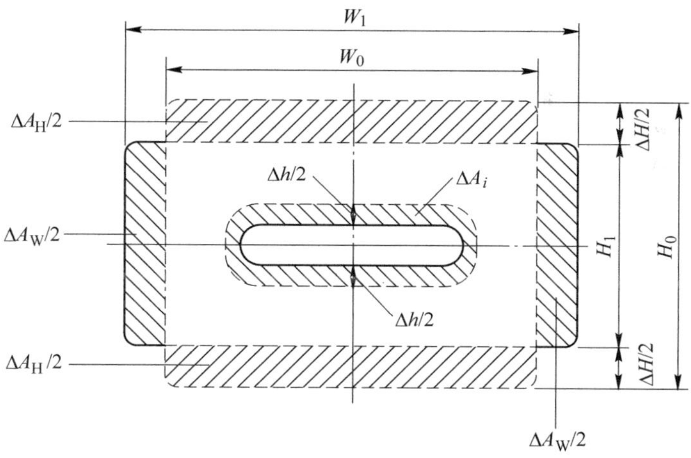

Fig. 1.12 Schematic diagram of transverse deformation of the shell under soft reduction

ΔA_H—Area of deformation on the surface of the strand during soft reduction, measured in square millimeters, mm^2.

However, for wide and thick slabs with non-uniform solidification, the morphology of the liquid core at the solidification end is complex. Moreover, the region under heavy reduction at the solidification end still contains areas with and without a liquid core. Therefore, our team is based on accurately simulating the deformation behavior during the slab reduction process, further evaluating the impact of roll-type structure, force distribution, and reduction distribution on the radial transmission rate of reduction deformation through analyzing equivalent stress distribution, hydrostatic pressure integration, and shrinkage closure on the cross-section [59, 63–66]. Refer to Chap. 4 of this book for more details.

1.3 Developmental Stages and Representative Technologies of Solidification End Reduction

The development of solidification end reduction technology can be roughly divided into three stages. Since the 1960s, Japanese researchers gradually realized that applying deformation at the solidification end of continuous casting products could effectively enhance the internal quality of the strand. They initiated a series of exploratory research activities. As these technologies involved applying reduction at a fixed position at the end of the continuous casting machine, they were termed static soft reduction technologies. From the 1990s onwards, with the development of technologies such as remote adjustment of roll gaps and solidification end prediction, metallurgical equipment companies in Europe, such as the former Voestalpine (now Primetals Technologies), introduced dynamic soft reduction technology. This approach involves dynamically adjusting the reduction zone and amount based on steel grade, casting speed, section, and other factors to accommodate changes in process conditions. In the past decade, driven by the increased demand for high-end, thick-section steel products, continuous casting product sections have gradually widened and thickened. Additionally, the carbon and alloy composition in the steel has significantly increased. Conventional soft reduction deformation is no longer sufficient to meet the requirements for high homogeneity and high-density production of large, high-end steel products. Researchers worldwide have therefore undertaken exploration and industrial application studies on solidification end heavy reduction technology.

1.3.1 Static Soft Reduction

In 1974, Nippon Kokan Kaisha (NKK) Corporation of Japan introduced the concept of mechanical stress soft reduction. They proposed that at the two-phase zone's end of slabs, at least two pairs of rolls should be used, with each pair reducing the strand by 2%. This approach was deemed effective in enhancing the central density of the strand and was applied in production processes [39, 67]. This method represents a typical static soft reduction approach, wherein the roll gaps of the continuous casting machine are set at predetermined fixed values. However, this method can only achieve a stable process effect under fixed casting speeds and pouring temperatures, imposing significant limitations on practical applications. Nevertheless, it laid the foundation for the development of using soft reduction technology to enhance the internal quality of strands. Subsequently, comprehensive experimental and theoretical studies on soft reduction were conducted by researchers worldwide [68, 69].

In the initial stages, roller-type reduction devices employed integral rolls with large diameters and wide roll gaps, leading to bulging of the shell between the front and rear pairs of rolls and insufficient relief of center segregation. Additionally, due to the extended length of the roll body, it was prone to bending deformation during the reduction process. To improve the effectiveness of soft reduction, in 1976, NKK installed a soft reduction fan-shaped section composed of segmented rolls in the soft reduction zone of slab continuous casting machines, achieving favorable results. Since then, soft reduction technology rapidly gained popularity worldwide [70–73]. Figure 1.13 illustrates the schematic diagram of the soft reduction fan-shaped section with small-diameter segmented rolls.

In the late 1980s, NKK introduced the Intentional Bulging and Soft Reduction (IBSR) technology [71]. This involves intentionally enlarging the roll gaps, creating bulging at the solidification end of the continuous casting slab, altering the shape of the solidification endpoint, and subsequently applying soft reduction to achieve desirable strand quality, as shown in Fig. 1.14. This technology helps improve the irregular shell morphology at the end of slab solidification, promotes molten steel flow, and further enhances the effectiveness of soft reduction. However, during the bulging deformation stage, if the shell is thin, significant bulging deformation may cause crack defects in the areas on both sides of the strand.

In the early 1990s, NKK proposed the Disk Roll Soft Reduction (DRSR), also known as the convex roll soft reduction method [72]. This technique stands out by

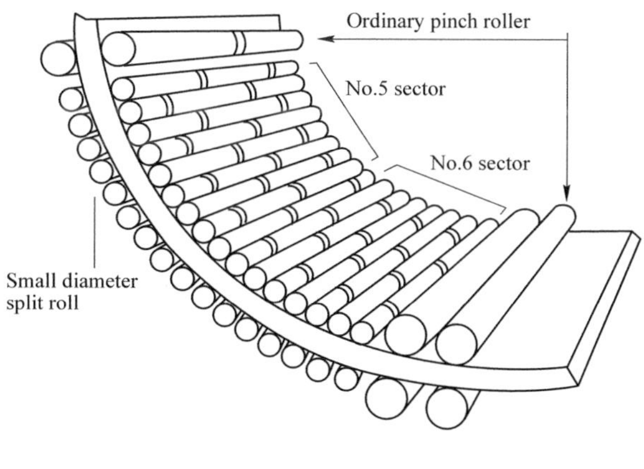

Fig. 1.13 Schematic diagram of soft reduction fan-shaped section with small-diameter segmented rolls [70]

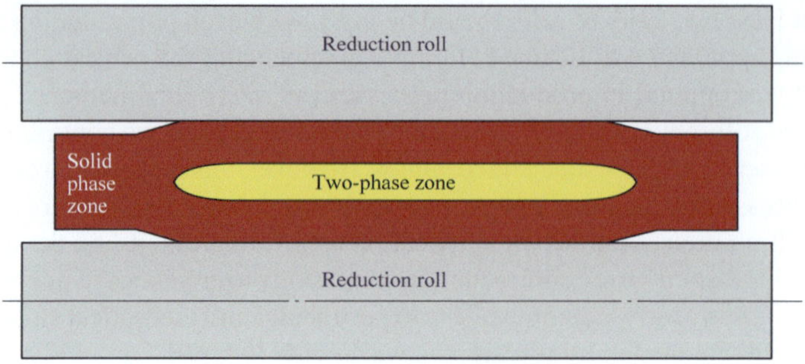

Fig. 1.14 Schematic diagram of intentional bulging reduction

using rolls with convex platforms instead of flat rolls for reduction. As the convex platform area of the casting roll only applies reduction deformation to the central segregation and porous areas of the strand, it avoids the support of the solidified areas on both sides of the strand, enhancing the reduction capacity of the casting roll. POSCO in South Korea applied this technology in the production process of large-section blooms with a cross-section of 400 mm × 500 mm and numerically simulated the stress–strain law of the strand during the convex roll soft reduction process. The industrial test results indicated a significant reduction in central segregation after applying a convex roll reduction of 14 mm [73]. This book also proposes the technology of variable curvature convex rolls, which has been widely applied in large-section bloom continuous casting production lines of enterprises such as Panzhihua Iron and Steel, achieving a more than 2.5 times increase in deformation amount under the same pressure. Moreover, by designing the variable curvature arc, the stress–strain concentration at the raised part during the reduction process was reduced, effectively improving the service life of the casting roll. Detailed design and application are discussed in Chap. 7. Figure 1.15 depicts the schematic diagram of disk roll reduction.

1.3.2 Dynamic Soft Reduction

Early developments in soft reduction processes belonged to the category of static soft reduction, where predetermined static shrinkage roll gaps were used during the casting process, and the roll gaps remained non-adjustable during production. Due to the fixed nature of static soft reduction roll gaps, there were stringent requirements for the stability of the casting process (such as casting speed and steel superheat), necessitating the avoidance of significant changes in the solidification endpoint caused by fluctuations in casting conditions to ensure the implementation of reduction in a reasonable casting flow area. However, as production rhythms and process parameters such as steel grade, casting speed, and steel superheat varied during casting, the

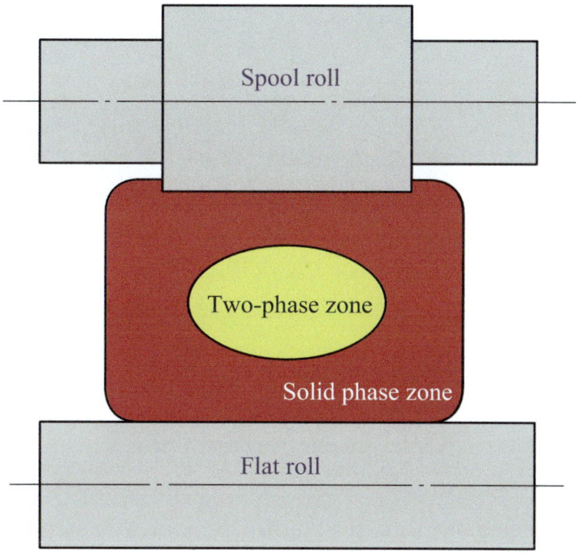

Fig. 1.15 Schematic diagram of disk roll reduction

solidification endpoint was inevitably altered, resulting in the instability of the static soft reduction process.

In the late 1980s, with the development of thin slab near-final continuous casting and rolling short process technologies, it became necessary to reduce the thickness of the thin slab by adjusting the thickness of the liquid-core continuous casting slab within the fan-shaped section in the secondary cooling zone through remotely adjustable rolls. For example, Demag in Italy applied liquid-core reduction functionality in the ISP (Inline Strip Process) production line in Arvedi, allowing a reduction of approximately 15 mm in slab thickness after reduction. The emergence of this technology created the equipment conditions for remote dynamic roll gap adjustment in conventional slab casting.

In 1997, VAI pioneered the application of dynamic soft reduction technology on the No. 6 slab continuous casting machine at Rautaruukki in Finland [74, 75]. The casting machine's straightening device consisted of 15 SMART® (Single Minute Adjustment and Restraining Time) fan-shaped segments with remotely adjustable roll gaps, as depicted in Fig. 1.16. It included an outer frame, inner frame, support frame, drive roll hydraulic adjustment device, inner frame hydraulic adjustment device, drive roll, clamping device, water supply pipelines, etc. The fan-shaped segment formed an angle between the inner and outer frames along the direction of pulling the slab through the hydraulic cylinders at the four corner angles; i.e., the entry roll gap was larger than the exit roll gap. The running slab was reduced in thickness in the thickness direction after passing through the fan-shaped segment, completing the reduction process.

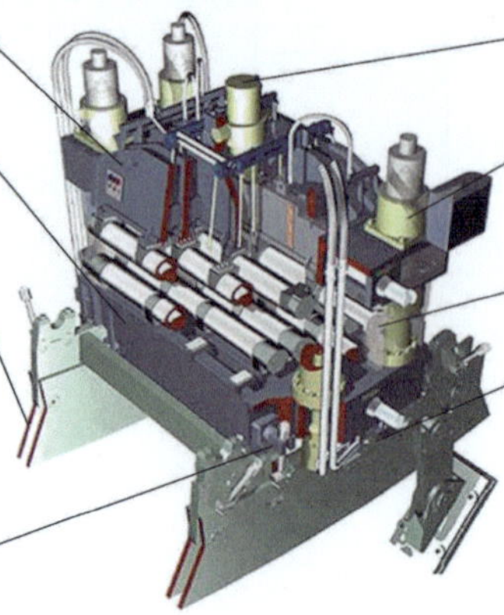

Fig. 1.16 Schematic structure of a caster's fan-shaped segment

Simultaneously, for the first time on this casting machine, dynamic secondary cooling and an online thermal tracking model—DYNOOL® [76]—and an online automatic roll gap setting model ASTC® (Automatic Strand Taper/Thickness Control) [77] were applied. The DYNOOL® model is the predecessor of the DYNACS® (Dynamic Strand Cooling Management System) model.

1.3.2.1 Remote Roll Gap Adjustable Fan-Shaped Segment

Dynamic soft reduction technology requires the rapid remote adjustment of roll gap values within the fan-shaped segment to achieve the implementation of soft reduction along with the changing solidification endpoint position [78]. To meet this requirement, the caster's roller path adopts a segmented structure, with each fan-shaped segment comprising 5 or 7 pairs of casting rolls. The hydraulic drive system is employed to ensure both speed and precision [79]. Currently, most foreign casting machines' fan-shaped segments can achieve hydraulic clamping and remote roll gap adjustment. Dynamic soft reduction technology has been widely applied in continuous casting process design, and mature developments have been industrially applied and commercially promoted, such as VAI's SMART® fan-shaped segment [80], DDD's OPTIMUM® fan-shaped segment, and SMSD's CYBERLINK® fan-shaped segment.

1.3.2.2 Prediction of Solidification Endpoint Position

Soft reduction involves the reduction of a specific region near the solidification endpoint, and determining the solidification endpoint position is a prerequisite for implementing dynamic soft reduction. Currently, there are two main methods used to determine the solidification endpoint position. One method relies on real-time calculations from online thermal tracking models, while the other involves real-time detection using pressure sensors installed on the fan-shaped segment. Thermal tracking model calculation methods require accurate results and short computation times (in the order of seconds). For instance, Danieli's LPC model can calculate the solidification endpoint position online based on factors such as steel grade, casting temperature, and casting speed [81, 82]. VAI's DYNACS® can also predict the solidification endpoint position under different casting conditions [83].

Different companies have proposed sensor detection methods with some variations, but the basic principle involves processing feedback signals from pressure detection of fully solidified and liquid-core strands during the soft reduction process to determine the solidification endpoint position. For example, a German company has developed sensors integrated into CasterCrown, which can predict the solidification endpoint position [84]. The CYBERLINK® fan-shaped segment, using periodic low-amplitude (approximately 2 mm) and low-frequency (approximately 2 Hz) vibrations in the upper frame, can detect the solidification endpoint position online [85]. In 2004, the CYBERLINK® fan-shaped segment was officially put into operation on the new No. 3 slab continuous casting machine at Salzgitter in Germany. DDD also developed the Actual Liquid Core End Monitoring (ALCEM) technology for OPTIMUM® fan-shaped segments, determining the accuracy of the reduction position through pressure feedback signals.

Online detection of strand pressure feedback values through the fan-shaped segment for solidification endpoint positioning has become one of the development trends in dynamic soft reduction technology. However, the use of online "soft measurements" with thermal tracking models remains essential. For instance, DDD uses a technology combining solidification and temperature tracking models with pressure detection. Currently, most domestically introduced continuous casting production lines with dynamic soft reduction functionality do not have fan-shaped segment pressure feedback detection capabilities. From the perspective of system upgrades and new product development, reliance on thermal tracking models for solidification endpoint positioning is necessary. Furthermore, from the perspective of controlling strand quality, real-time calculation of the temperature field is a prerequisite for achieving temperature feedback in the secondary cooling control [86]. Stable secondary cooling control is crucial for the effective implementation of dynamic soft reduction. Therefore, further improvement in the computational accuracy of online thermal tracking models will continue to be a focal point and challenge in research. In recent years, our team has developed a high-precision online detection technology for solidification endpoint morphology and position based on the combination of soft measurements and real detection. This technology achieves accurate prediction of the solidification endpoint under factors such as reduction deformation and solute migration. See Chap. 7 for details.

1.3.2.3 Application of Soft Reduction Technology

China's iron and steel enterprises have also early adopted dynamic soft reduction technology. In 2003, Baosteel Meishan Iron and Steel Company first introduced dynamic soft reduction technology on the No.2 slab continuous casting machine, achieving a center segregation C-grade ratio of 97% in the produced slabs [87]. In the same year, Panzhihua Iron and Steel aimed to improve the internal quality of its heavy rail steel blooms and successfully applied dynamic soft reduction technology on a large bloom continuous casting machine for the first time. The results demonstrated that dynamic soft reduction technology significantly improved the internal quality of heavy rail steel blooms, reducing the carbon segregation index at the center position from 1.17 to 1.05 [88]. Currently, dynamic soft reduction technology has been widely promoted worldwide, becoming a crucial hallmark of modern continuous casting machines. The typical applications and effects of soft reduction technology on slabs and blooms in both domestic and international scenarios are presented in Tables 1.2 and 1.3.

Table 1.2 Typical application effects of slab continuous casting soft reduction technology worldwide

Enterprise	Strand section	Implementation effects
Nippon Steel Fukuyama Works [89]	220 mm × 1950 mm	Uniform reduction in the level of point segregation in slabs
Nippon Steel Corporation [85]	210/240 mm × (960–2200) mm	Reduction in the fluctuation of center segregation in the longitudinal section when $f_S > 0.25$
WISCO Three Refineries [90]	210/230/250 mm × (1350–2150) mm	Elimination of A-grade segregation, with an increase of over 6.1% in the proportion of C-grade ≤ 1.0
Meishan Iron & Steel Co., Ltd. (Baosteel) [91]	210/230 mm × (900–1320) mm	C-grade rate of center segregation reached 100%, center porosity ≤ 0.5 rate reached 98%, and center and triangular area cracks ≤ 0.5 rate reached 96%
Shandong Laigang Group Co., Ltd. [92]	200/250/300 mm × (1600–2500) mm	B-grade 0.5–1.0 level in center segregation for medium carbon steel, center porosity 0.5 level; A-grade 1.0 level in center segregation for high-carbon steel, center porosity 1.0 level; B-grade 0.5 level in center segregation for low carbon alloy and pipeline steel, center porosity 0.5 level
Taishan Iron and Steel Group Co., Ltd. [93]	200 mm × 1600 mm	C and S center segregation indices reduced from 1.30 to 1.05

(continued)

Table 1.2 (continued)

Enterprise	Strand section	Implementation effects
Yangchun New Steel [94]	220 mm × 1500 mm	When the reduction amount is > 1.5 mm and f_S = 0.45–0.75, C-grade 1.0 level in center segregation, and center porosity level less than 0.5
WISCO Echeng Iron & Steel Co., Ltd. [95]	250 mm × 2000 mm	At a casting speed of 0.95 m/min, f_S = 0.4–0.9, and a reduction amount of 6 mm, the average pass rate of C-grade I level in center segregation was 98.0%
Lianyuan Iron and Steel Group Co., Ltd. [96]	230 mm × 1310 mm	C-grade 0.5 level in center porosity and center segregation
Handan Iron and Steel Group [97]	180/220/250 mm × (1400–1900) mm	The qualified rate of slab quality increased from 97.5% to over 99.5%
Xinyu Iron and Steel Group [98]	250 mm × 1870/2070 mm	The proportion of segregation < 0.5 level increased by 4.5%, with an increase of 13%; the proportion of 0.5–1.5 level increased by 15.1%, with an increase of 37.5%; C-grade segregation greater than 1.5 level was only 5.8%
Baosteel Shaoguan Iron & Steel Co., Ltd. [99]	150 mm × (2600–2800) mm	The proportion of 1.0 level porosity and 0.5 level cracks in the center reached over 80%, and the ultrasonic testing pass rate was over 98%
Magang (Group) Holding Co., Ltd. [100]	230 mm × (950–1600) mm	S650MC steel slab center segregation rated C-grade 1.0 level, and the proportion of layering in hot-rolled coil tensile samples decreased to 0.78%
Shougang Group Jingtang United Iron & Steel Co., Ltd. [101]	230 mm × (900–2150) mm	C-grade 0.5 level or below in center segregation of slab, and no internal cracks found in the slabs
Baosteel Bayi Iron & Steel Co., Ltd. [102]	220 mm × 1500 mm	When the reduction amount exceeds 1.5 mm, the center segregation level is C-grade 1.0 level or below, and the center porosity level is less than 0.5
Sanming Iron and Steel Co., Ltd. [103]	180 mm × 1400 mm 220 mm × 1600 mm 250 mm × 1600 mm	Improvement from B-grade in center segregation of slabs without soft reduction
Shougang Group Jingtang United Iron & Steel Co., Ltd. [104]	400 mm × 2400 mm	Fully controlled C-grade 1.5 level or below in center segregation of slabs
Chengde Iron and Steel Group [105]	200 mm × 1500 mm	Approximately 0.5 level in center porosity and center segregation level around C0.5-C1.0, with online defect rate reduced to 0.89%

Table 1.3 Typical application effects of bloom continuous casting soft reduction technology worldwide

Enterprise	Strand section	Implementation effects
Sumitomo Metal Industries, Kokura Works [106]	300 mm × 400 mm	The segregation index of high-carbon steel P70 rods and wires decreased from 0.93 to 0.69; S82 decreased from 1.19 to 0.85; SU2 decreased from 0.9 to 0.2
China Steel Corporation, Taiwan, China [107]	220 mm × 260 mm	The use of soft reduction alleviates and disperses center segregation
Shandong Iron and Steel Group Co., Ltd. [108]	120 mm × 120 mm	After applying soft reduction, the porosity in the center of the billet is reduced to less than 1.5 levels, meeting production standards
OAO OMK-Steel [109]	280 mm × 325 mm 280 mm × 380 mm	The application of dynamic soft reduction reduces the severity of center porosity
Wuhan Iron and Steel Corporation (WISCO) [110]	200 mm × 200 mm	The average center carbon segregation in the billet is reduced to within 1.06, and the peak value of center carbon segregation is lowered to below 1.3
Baotou Iron and Steel Group Special Steel Co., Ltd. [111]	240 mm × 240 mm	The center porosity in the billet is reduced to 1.0–1.5 levels, and V-shaped segregation and center shrinkage are significantly improved. The average carbon segregation index in the billet center is reduced to 1.07–1.13
Panzhihua Iron and Steel Research Institute [112]	360 mm × 450 mm	The proportion of internal defects rated less than one level reaches 99.03%, the average center segregation index is 1.03, and the bloom's surface defect-free rate averages 99.54%
Xingtai Iron and Steel Corp., Ltd. [113]	280 mm × 325 mm	The proportion of center porosity $\leq$ 1.5 levels increases to 90.91%, and porosity in the 0–1 ratio increases to 96.36%; the proportion of center shrinkage $\leq$ 1.5 levels increases to 96.36%
Xingtai Iron and Steel Corp., Ltd. [114]	285 mm × 325 mm	The proportions of center porosity rating $\leq$ 1.5 levels and center shrinkage rating 0–1.5 levels increase to 85.0% and 87.5%, respectively
Panzhihua Iron and Steel Group Co., Ltd. [115]	360 mm × 450 mm	The proportion of center porosity rating $\leq$ 0.5 level in the bloom increases to 100%, the proportion of bloom center without shrinkage increases to 98.74%, and the carbon segregation index on the transverse section of the bloom improves to 0.92–1.06, significantly improving the internal quality of steel No. 45
Anshan Iron and Steel Group Corporation [116]	280 mm × 380 mm	Carbon segregation in blooms decreases to 1.02 in the C center and 1.10 in the S center

(continued)

Table 1.3 (continued)

Enterprise	Strand section	Implementation effects
Qingdao Iron and Steel Group Co., Ltd. [117]	180 mm × 240 mm	Soft reduction significantly improves V-shaped segregation in the billet. The average carbon segregation index in the billet center decreases to 1.07, and the maximum carbon segregation index decreases from 1.27 to 1.15
Benxi Iron and Steel Group Co., Ltd. [118]	350 mm × 470 mm	The center carbon segregation index of high-carbon steels, such as bearing steel, can be reduced to below 1.10; the range of the center carbon segregation index for medium–low carbon steel narrows to 0.95–1.05
Handan Iron and Steel Group [119]	380 mm × 280/325 mm	The proportion of bloom center segregation rating ≤ 1.0 level increases to 94.6%, and the center porosity rating ≤ 1.0 level increases to 89.3%. The proportion of center shrinkage rating ≤ 0.5 level increases to 93.9%
Baosteel Shaoguan Iron & Steel Co., Ltd. [120]	425 mm × 320 mm	After soft reduction, center porosity remains stable at 1.5 levels or below, with an average carbon segregation value of 0.99
Chinh Dai Industrial Ltd., Vietnam [121]	220 mm × 260 mm	The center segregation index on the transverse section of the billet decreases from 1.21 to 1.07, and the overall longitudinal center segregation decreases, averaging 1.10
Jiyuan Iron and Steel Group Co., Ltd. [122]	400 mm × 500 mm	The qualification rate of the center segregation degree in the range of 0.92–1.08 has increased from around 55% initially to over 80% currently

1.3.3 Heavy Reduction

As the dimensions of continuous casting products, such as slabs and blooms, increase in width and thickness, traditional soft reduction processes become less effective in addressing prominent internal quality defects, such as center segregation and shrinkage porosity. In response to this, heavy reduction technology has been further developed based on the foundation of soft reduction processes [123–131]. In comparison with soft reduction processes, the deformation amount in solidification end reduction is generally around 10% of the strand thickness, significantly exceeding the deformation in soft reduction processes. Moreover, the application area of solidification end reduction is extended further backward from the soft reduction base to a certain region after the solidification endpoint of the strand. This extension aims to fully utilize the "external cold, internal hot" advantage near the solidification endpoint, avoiding the formation of cracks in the solidification front. This facilitates the efficient transfer of a large reduction deformation to the core of the strand, enhancing overall strand density and homogeneity. Additionally, significant reduction deformation induces recrystallization of austenite, refining the continuous casting structure and improving the properties associated with rolling.

During the development of solidification end reduction, various modes have emerged. From an equipment implementation perspective, it can be broadly categorized into anvil forging, single-roll reduction, and multiroll reduction structures.

In the late 1990s, Japan's Kawasaki Steel (now Kawasaki Steel Corporation) pioneered the continuous forging technology [123, 124]. While some scholars categorize this technology under soft reduction, the deformation amount in this technology far exceeds that of contemporaneous soft reduction techniques.

As illustrated in Fig. 1.17, continuous forging technology involves continuous forging of the strand by installing a pair of anvils at the solidification endpoint of the casting flow. During the forging process, the inner and outer arc shells are squeezed together, causing the two-phase regions in the casting flow to adhere and promoting the shattering and remelting of dendrites in these regions. The residual molten steel between dendrites flows backward under the squeezing action, remixing with the upstream molten steel. This technology significantly improves macrosegregation, shrinkage, and looseness in continuous casting products, effectively avoiding internal cracks. After the implementation of this technology, the internal quality of 300 mm thick slabs produced is excellent, meeting the requirements for the production of 150 mm thick ultrasonic-tested slabs under a rolling compression ratio of 2:1.

Sumitomo Metal Industries in Japan developed the Porosity Control of Casting Slab (PCCS) technology [125]. As shown in Fig. 1.18, this technology involves applying significant reduction deformation to the strand by installing a pair of rolls at the position of the solidification endpoint. Field trials have shown that applying this technology to the 300 mm thick slabs in continuous casting with about 10 mm large reduction deformation at the end of solidification significantly improves the shrinkage and looseness defects of the strand. This results in the successful production of 600 MPa-grade high-strength 150 mm thick slabs under a low rolling compression

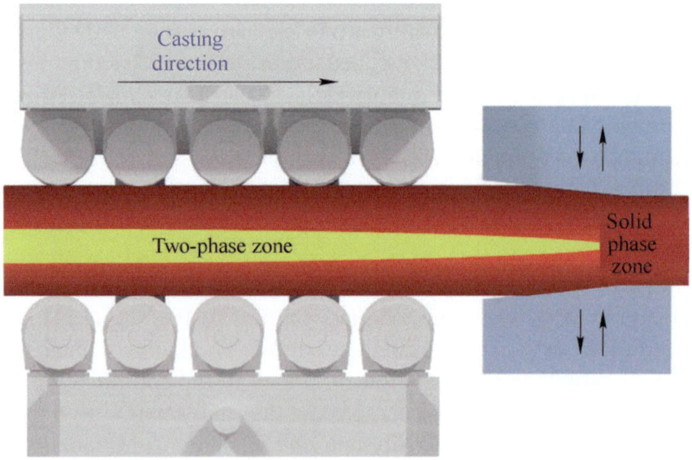

Fig. 1.17 Continuous forging schematic

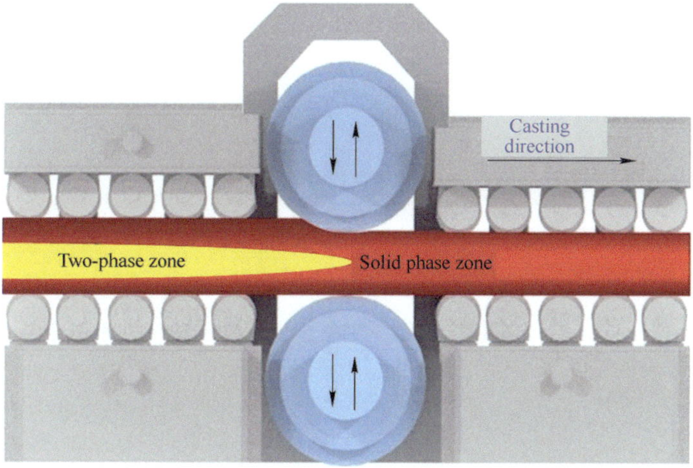

Fig. 1.18 Schematic of Sumitomo metal PCCS technology

ratio of 2:1. China Metallurgical Group Corporation's subsidiary, China Metallurgical Engineering Technology Co., Ltd., proposed a similar liquid core large reduction rolling process to the PCCS implementation. In this process, the slab is not completely solidified at the reduction position, and it continuously extrudes to eliminate solute-segregated molten steel. Concurrently, dendrites are shattered, expanding the equiaxed crystal zone, reducing segregation, and bonding looseness. By applying a 60 mm reduction to a 250 mm thick slab, satisfactory process results were achieved. In the Cao Feidian Phase II project of Shougang Group Co., Ltd., a 20 mm heavy reduction technology was also adopted on 400 mm thick slabs. This not only significantly improved center segregation and porosity in the strand but also refined the internal structure significantly. Taiwan Chung Hung Steel Corp. explored similar technologies, using large-diameter convex rolls for a 10 mm reduction at the tail of the caster. This effectively eliminated center porosity in continuous casting products and significantly improved the ultrasonic testing pass rate and low-temperature impact toughness of thick slabs.

Nippon Steel Corporation in Japan has developed the Nippon Steel Bloom Large Reduction (NS-BLR) technology [126]. Similar to the PCCS technology, this method achieves a heavy reduction using a single-roll process on large blooms. As depicted in Fig. 1.19, the technology involves applying significant reduction deformation to the just-solidified continuous casting bloom by installing a pair of large casting rolls at the solidification endpoint. Due to the convex roll structure used in the reduction roll, it significantly reduces the resistance force from the bloom during the reduction process. Simultaneously, it facilitates the concentration of extrusion deformation in the central region of the bloom, effectively improving macrosegregation, shrinkage, and looseness defects in that area.

POSCO (Pohang Iron and Steel Company) in South Korea has developed a heavy reduction process that facilitates online adjustment of the reduction position, known

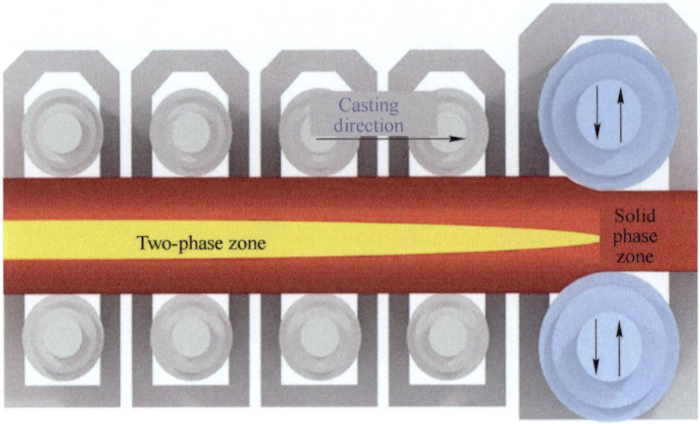

Fig. 1.19 Nippon steel NS-BLR technology

as POSCO Heavy Strand Reduction Process (PosHARP) [127], as shown in Fig. 1.20. Similar to dynamic soft reduction, PosHARP can flexibly adjust its reduction position based on the actual casting conditions. It applies large reduction deformation (5–20 mm/m) to the two-phase areas with low solidification rates (initial solidification rate of 0.05–0.2, final solidification rate of 0.3–0.6). This process forcefully interrupts the solidification process, significantly improving center segregation, shrinkage, and looseness defects in continuous casting products. Upon adopting this technology, POSCO achieved the production of 120 mm thick slabs from 300 mm thick slab blanks. However, PosHARP's reduction position has a lower solidification rate, which may induce mid-cracks and bright bands near the solidification front.

In addition to the heavy reduction technologies already applied in practice, many metallurgical scholars have conducted extensive research and development on heavy reduction technologies from theoretical studies, experimental explorations, and other

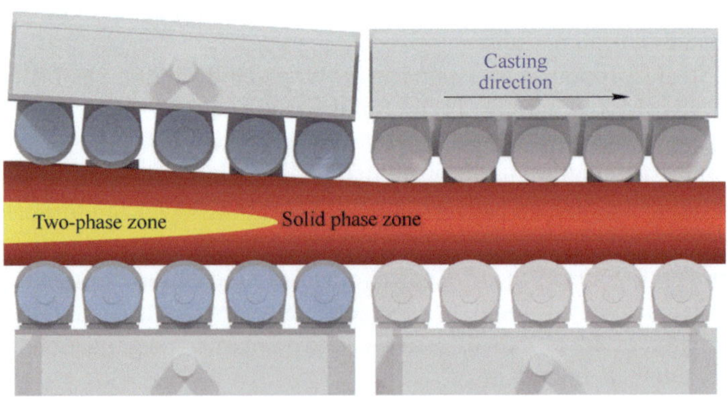

Fig. 1.20 POSCO PosHARP technology schematic

perspectives. Jiongming Zhang's team at the University of Science and Technology Beijing systematically studied the improvement effects of different reduction modes on strand density and structure, proposing the Heavy Reduction Process to Improve Segregation and Porosity (HRPISP) technology [128–130]. Wang Xinhua's team at the University of Science and Technology Beijing developed the Solidification Terminal Advanced Reduction Technique (START) and explored a single-stage 10 mm reduction process on 420 mm thick slab blanks at Xinyu Iron and Steel [131]. These heavy reduction processes aim to achieve effective improvement of center segregation and shrinkage porosity defects by applying significant reduction deformation to the near-solidification endpoint region of the casting flow.

In recent years, our team has collaborated with units such as Pangang and Tanggang, conducting systematic work on the theoretical framework, key technologies and equipment, and engineering applications of heavy reduction processes for large-section continuous casting products. In terms of process mechanism exploration, we have systematically studied the behavior laws of metal flow, heat transfer deformation, shrinkage closure, solute transport, and structural evolution during the heavy reduction process. We have answered key process core questions regarding "where to reduce and how much to reduce". In terms of process control technology, we have developed an integrated technology for heavy reduction at the solidification endpoint, including high-precision solidification endpoint positioning technology based on "soft measurement" + "real detection", a two-stage continuous heavy reduction process that improves center segregation and porosity synchronously, and efficient extrusion control technology that precisely controls driving torque and enhances strand shrinkage closure with "single-point + continuous" heavy reduction. In terms of key equipment, we have developed a reinforced compact fan segment for the heavy reduction of wide and thick slab blanks and a gradually curved convex roll for the heavy reduction of large blooms. Pangang and Tanggang have successfully implemented the engineering application of heavy reduction at the solidification endpoint on large blooms and wide and thick slab blanks, respectively. The process schematics are shown in Figs. 1.21 and 1.22, respectively, and have been promoted and applied in enterprises such as Baowu and Hyundai Steel. After the implementation of this technology, significant improvements have been achieved in strand homogeneity and density, and the grain size of the strand has been refined. It has enabled the production of 150 mm thick high-rise building special thick steel plates from 280 mm thick slab blanks (before reduction) under a rolling compression ratio of only 1.87:1.

1.4 Structure of Each Chapter and Interrelationship

This book consists of seven chapters, systematically elaborating on the technologies for internal quality control of continuous casting product and solidification end reduction (this chapter), the metal rheological characteristics and constitutive models during the reduction process (Chap. 2), the thermal/mechanical behavior laws and risk of cracking during the reduction process (Chap. 3), the closure laws of center

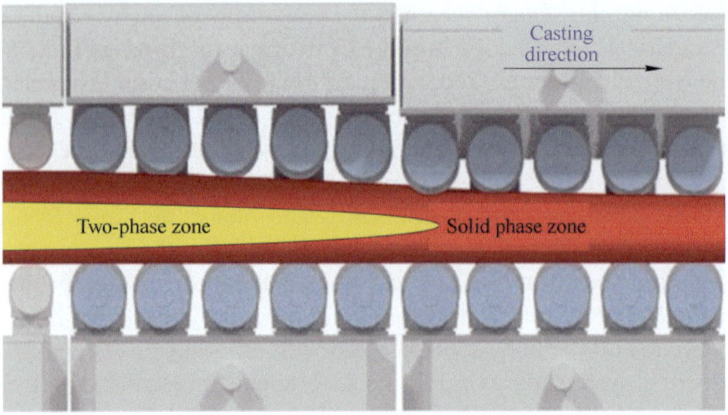

Fig. 1.21 Schematic of continuous dynamic heavy reduction technology at the solidification endpoint for slab blanks

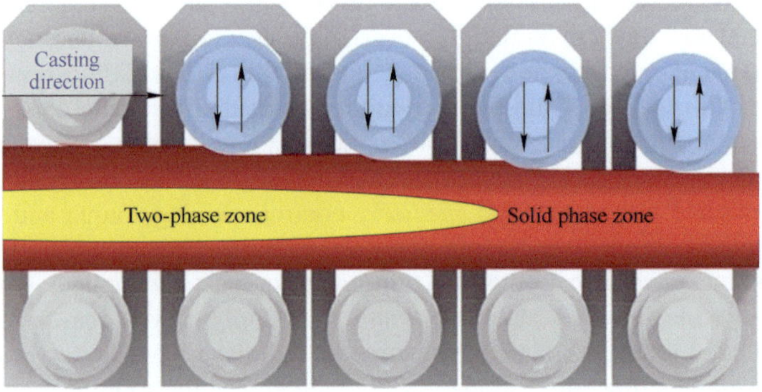

Fig. 1.22 Schematic of continuous dynamic heavy reduction technology at the solidification endpoint for bloom blanks

shrinkage during the reduction process (Chap. 4), the solute transport behavior in continuous casting product under multiple external fields (Chap. 5), the refinement of strand structure during the reduction process and the genetic inheritance of the structure in the transportation and rolling processes (Chap. 6), and the key process equipment and application effects of solidification end reduction (Chap. 7). The interrelationship between the research contents of each chapter is shown in Fig. 1.23.

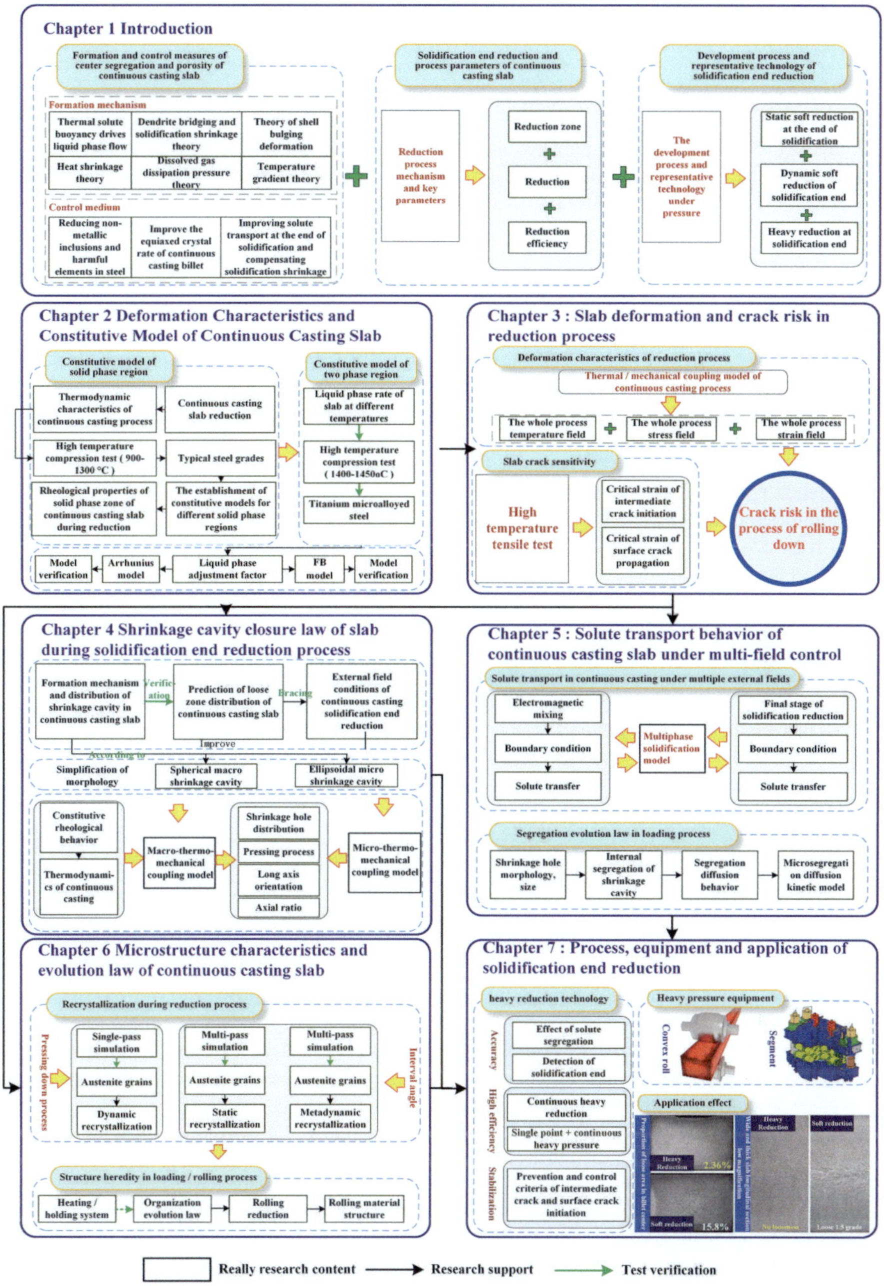

Fig. 1.23 Interrelationship between research contents of each chapter

References

1. Huppert HE, Worster MG (1985) Dynamic solidification of a binary melt. Nature 314(6013):703–707
2. Turner JS (1985) Multicomponent convection. Annu Rev Fluid Mech 17(1):11–44
3. Aboutalebi MR, Hasan M, Guthrie RIL (1995) Coupled turbulent flow, heat, and solute transport in continuous casting processes. Metall Mater Trans B 26(4):731–744
4. Sun HB, Zhang JQ (2014) Study on the macrosegregation behavior for the bloom continuous casting: Model development and validation. Metall Mater Trans B 45(3):1133–1149
5. Suzuki K, Miyamoto T (1978) Study on the formation of "A" segregation in steel ingot. Trans Iron Steel Inst Jpn 18(2):80–89
6. Murao T, Kajitani T, Yamamura H et al (2014) Simulation of the center-line segregation generated by the formation of bridging. ISIJ Int 54(2):359–365
7. Sivesson P, Hallen G, Widell B (1998) Improvement of inner quality of continuously cast billets using electromagnetic stirring and thermal soft reduction. Ironmaking Steelmaking 25(3):239
8. Gabathuler JP, Weinberg F (1983) Fluid flow into a dendritic array under forced convection. Metall Trans B 14(4):733–741
9. Flemings MG, Nereo GE (1967) Macrosegregation. PT. 1, Trans Met Soc AIME 239(9):1449–1461
10. Miyazawa K, Schwerdtfeger K (1981) Macrosegregation in continuously cast steel slabs: preliminary theoretical investigation on the effect of steady state bulging. Arch Eisenhüttenwes 52(11):415–422
11. Kajitani T, Drezet JM, Rappaz M (2001) Numerical simulation of deformation-induced segregation in continuous casting of steel. Metall Mater Trans A 32(6):1479–1491
12. Ludwig A, Kharicha A, Wu MH (2014) Modeling of multiscale and multiphase phenomena in materials processing. Metall Mater Trans B 45(1):36–43
13. Wu MH, Domitner J, Ludwig A (2012) Using a two-phase columnar solidification model to study the principle of mechanical soft reduction in slab casting. Metall Mater Trans A 43(3):945–964
14. Mayer F, Wu MH, Ludwig A (2010) On the formation of centreline segregation in continuous slab casting of steel due to bulging and/or feeding. Steel Res Int 81(8):660–667
15. Domitner J, Wu MH, Kharicha A et al (2014) Modeling the effects of strand surface bulging and mechanical soft reduction on the macrosegregation formation in steel continuous casting. Metall Mater Trans A 45(3):1415–1434
16. Domitner J, Wu MH, Kharicha A et al (2015) Numerical study about the influence of small casting speed variations on the metallurgical length in continuous casting of steel slabs. Steel Res Int 86(3):184–188
17. Guan R, Ji C, Wu CH et al (2019) Numerical modelling of fluid flow and macrosegregation in a continuous casting slab with asymmetrical bulging and mechanical reduction. Int J Heat Mass Transfer 141:503–516
18. Lesoult G, Sella S (1988) Spongy behaviour of alloys during solidification: flow of liquid metal and segregation in the mushy zone. Solid State Phenom 3:167–178
19. Raihle CM, Fredriksson H (1994) On the formation of pipes and centerline segregates in continuously cast billets. Metall Mater Trans B 25(1):123–133
20. El-Bealy M (2000) Modeling of interdendritic strain and macrosegregation for dendritic solidification processes: Part I. Theory and experiments. Metall Mater Trans B 31(2):331–343
21. El-Bealy M (2000) Modeling of interdendritic strain and macrosegregation for dendritic solidification processes: Part II. Computation of interdendritic strain and segregation fields in steel ingots. Metall Mater Trans B 31(2):345–355
22. Pellini WS (1953) Practical heat transfer. AFS Trans 61:603–622
23. Niyama E, Uchida T, Morikawa M (1981) Predicting shrinkage in large steel castings from temperature gradient calculations. Int Cast Met J 6(2):16–22

24. Niyama E (1982) A method of shrinkage prediction and its application to steel casting practice. Int Cast Met J 7(3):52–63
25. Engström G (1983) On the mechanism of macrosegregation formation in continuously cast steels. Metall 12(1):3–12
26. Monroe R (2005) Porosity in castings. AFS Trans 113:519–546
27. Carlson KD, Beckermann C (2009) Prediction of shrinkage pore volume fraction using a dimensionless Niyama criterion. Metall Mater Trans A 40(1):163–175
28. Stefanescu DM (2005) Computer simulation of shrinkage related defects in metal castings—A review. Int J Cast Met Res 18(3):129–143
29. Stefanescu DM, Catalina AV (2011) Physics of microporosity formation in casting alloys–sensitivity analysis for Al–Si alloys. Int J Cast Met Res 24(3–4):144–150
30. Kweon ES, Roh DH, Kim SB et al (2020) Computational modeling of shrinkage porosity formation in spheroidal graphite iron: a proof of concept and experimental validation. Int J Metalcast 14(3):1–9
31. Li D, Chen XQ, Fu P et al (2014) Inclusion flotation-driven channel segregation in solidifying steels. Nat Commun 5(1):1–9
32. Yue Q, Zhang CB, Pei XH (2017) Magnetohydrodynamic flows and heat transfer in a twin-channel induction heating tundish. Ironmaking Steelmaking 44(3):227–236
33. Jiang DB, Zhu MY (2016) Solidification structure and macrosegregation of billet continuous casting process with dual electromagnetic stirrings in mold and final stage of solidification: a numerical study. Metall Mater Trans B 47(6):3446–3458
34. Li WB, Wang F, Qi FS et al (2007) Mathematical model on steel strip-feeding of mold in continuous casting process. Acta Metall Sin 43(11):1191–1194
35. Zhang YH, Zhong HG, Zhai QJ (2017) Research progress of grain refinement and homogenization of solidified metal alloys driven by pulsed electromagnetic fields. J Iron Steel Res 29(4):249–260
36. Jiang DB, Zhu MY (2017) Center segregation with final electromagnetic stirring in billet continuous casting process. Metall Mater Trans B 48(1):444–455
37. Ji C, Zhu MY (2017) Dynamic sequential heavy reduction technology for wide-thick continuous casting slab. In: The 9th Korea–China joint symposium on advanced steel technology, pp 97–99
38. Myyagawa O (1985) Process of the iron and steel technologies in Japan in the past decade. Trans ISIJ 25(7):539–540
39. Miyazama K (2001) Continuous casting of steels in Japan. Sci Technol Adv Mater 2:59–65
40. Takahashi T (1982) Solidification and segregation of steel ingot. Iron Steel 17(3):57–61
41. Takahashi T, Hagiwara I (1965) Study on solidification and segregation of stirred ingot. J Jpn Inst Met 29:1152–1159
42. Ogibayashi S, Yamada M, Mukai T et al. Continuous casting method. U.S. Patent, 4687047, 1987-8-18
43. Bleck W, Wang W, Bülte R (2006) Influence of soft reduction on internal quality of high carbon steel billets. Steel Res Int 77(7):485–491
44. Ito Y, Yamanaka A, Watanabe T (2000) Internal reduction efficiency of continuously cast strand with liquid core. Revue de Métallurgie - Inter J Metall 97(10):1171–1176
45. Yim CH, Park JK, You BD et al (1996) The effect of soft reduction on center segregation in CC slab. ISIJ Int 36(Suppl):S231–S234
46. Qi XX, Zhang G, Jia Q (2011) Studies on technological parameters of optimal soft reduction about superwide slab caster. Adv Mater Res 194:207–212
47. Luo S, Zhu MY, Ji C et al (2010) Characteristics of solute segregation in continuous casting bloom with dynamic soft reduction and determination of soft reduction zone. Ironmaking Steelmaking 37(2):140–146
48. Ji C, Luo S, Zhu M et al (2014) Uneven solidification during wide-thick slab continuous casting process and its influence on soft reduction zone. ISIJ Int 54(1):103–111
49. Zhao PJ, Han HL (2006) Application of soft reduction on new slab caster of Jisteel. Continuous Casting 6:11–12

50. Wu JB (2001) The latest achievement of the slab dynamic soft reduction. World Iron Steel 6:57–61
51. Dong Z (1998) The improvement of center segregation of high carbon bloom, vol 1. Science Technology of Baotou Steel(Group) Corporation, pp 44–48
52. Kelly JE, Michalek KP, O' connor TG et al (1988) Initial development of thermal and stress fields in continuously cast steel billets. Metall Trans A 19A:2589–2602
53. Guan R, Ji C, Zhu M et al (2018) Numerical simulation of V-shaped segregation in continuous casting blooms based on a microsegregation model. Metall Mater Trans B 49(5):2571–2583
54. Sugimaru S, Nakashima J, Miyazawa K et al (1993) Theoretical analysis of the suppression of solidification shrinkage flow in continuously cast steel blooms. Mater Sci Eng A 173(1–2):305–308
55. Zeze M, Misumi H, Nagata S et al (2001) Segregation behavior and deformation behavior during soft reduction of unsolidified steel ingot. Tetsu-to-Hagane 87(2):71–76
56. Isobe K, Maede H, Syukuri K et al (1994) Development of soft reduction technology using crown rolls for improvement of centerline segregation of continuously cast bloom. Tetsu-to-Hagane 80(1):42–47
57. Oh KS, Park JK, Change SH (1995) Development of soft reduction technology for the bloom caster at Pohang Works of POSCO. Steelmaking Conf Proc 178:301–308
58. Lin QY, Zhu MY (2007) Theoretical model and analysis of soft reduction gradient for continuous casting slab. Acta Metall Sin 43(008):847–850
59. Ji C, Luo S, Zhu M (2014) Analysis and application of soft reduction amount for bloom continuous casting process. ISIJ Int 54(3):504–510
60. Hayashida M, Yasuda K, Ogibayashi S et al (1986) Soft reduction efficiency of the strand near the crater end: study on countermeasures for preventing centerline segregation of continuously cast slab VII. Tetsu-to-Hagane 72(12):S1091
61. Ito Y, Yamanaka A, Watanabe T (2000) Internal reduction efficiency of continuously cast strand with liquid core. La Revue de Metallurgie 10:1171–1177
62. Lin QY, Zhu MY (2007) Analysis of reduction efficiency in soft reduction for continuous casting slab. Acta Metall Sin 43(12):1301–1304
63. Ji C, Wu CH, Zhu MY (2016) Thermo-mechanical behavior of the continuous casting bloom in the heavy reduction process. JOM 68(12):3107–3115
64. Li G, Ji C, Zhu M (2021) Prediction of internal crack initiation in continuously cast blooms. Metall Mater Trans B 52(2):1164–1178
65. Wu C, Ji C, Zhu M (2019) Influence of differential roll rotation speed on evolution of internal porosity in continuous casting bloom during heavy reduction. J Mater Process Technol 271:651–659
66. Wu C, Ji C, Zhu M (2019) Closure of internal porosity in continuous casting bloom during heavy reduction process. Metall Mater Trans B 50(6):2867–2883
67. Miyagawa O (1985) Progress of iron and steel technologies in Japan in the past decade. Trans ISIJ 25:539
68. Tsuchida Y (1984) Behavior of semi-macroscopic segregation in continuously cast slabs and technique for reduction the segregation. Trans ISIJ 24(11):899–906
69. Ogibayashi S, Kobayashi M (1991) Influence of reduction with one-piece rolls on center segregation in continuously cast salbs. ISIJ Int 31(12):1400–1407
70. Zheng PR, Zhu X, Gong G, Ji Y, Bei S (1991) Metallurgical Industry Press
71. Xiao YL, Yao LD (1998) Improvement of thick slab quality using No. 6 continuous caster at NKK Fukuyama Factory. Wide Thick Slab 4(4):31–34
72. Miyazaki M, Isobe K, Murao T (2013) Formation mechanism and modeling of centerline segregation. In: Nippon steel technical report, p 104
73. Moon CH, Oh KS, Lee JD et al (2012) Effect of the roll surface profile on centerline segregation in soft reduction process. ISIJ Int 52(7):1266–1272
74. Mörwald K, Pirner K, Jauhola M et al (1998) Basic design features and start-up of rautaruukki's CC No.6. Madrid, Spain. In: 3rd European continuous casting conference (ECCC), Oct.1998.

75. Yahra M, Kang TE, He DH et al (1999) Design features and start-up of the Rautaruukki No.6 slab caster. Iron Steel 34(10):16–19
76. Dittenberger K, Morwald K, Hohenbichler G et al (1998) DYNACS® cooling model-features and operational results. Ironmaking Steelmaking 25(4):323–327
77. Thalhammer M, Federspiel C, Morwald K et al (2002) Operational and economic benefits of SMART®/ASTC technology in continuous casting. In: AISE annual convention, Nashville, USA
78. Hodl H, Thalhammer M, Stiftinger M et al (2003) Advanced equipment for high-performance casters. Metallur Plant Technol Int 3:74–80
79. Zheng Q (2003) Development and research of new technologies in slab continuous casting machines. Hebei Metallur 1:5–10
80. Morwald K, Thalhammer M, Federspiel C et al (2003) Benefits of SMART segment technology and ASTC strand taper control in continuous casting. Steel Times Int 101(4):17–19
81. Morsut L (2003) Technological packages for the effective control of slab casting. Metallur Plant Technol Int 2:44–51
82. Song DF (2005) Application of LPC Model in Dynamic Soft Reduction Control. Metallur Autom 3:57–59
83. Danilo G, Gustavo M, Rubens F et al (2004) Design features and start-up of the high-productivity 2-strand slab caster at Cosipa. Metall Plant Technol Int 4:46–48
84. Watanabe T, Yamashita M (1993) Influence of liquid flow at the final solidification stage on centerline segregation in continuously cast slab. Sumiyomo Metals 45(3):26–39
85. Feng K (2006) Design features of the fan-shaped segment in slab continuous casting machine under soft reduction. Steelmaking 22(2):53–56
86. Lotov AV, Kamenev GK, Berezkin VE, Miettinen K (2005) Optimal control of cooling process in continuous casting of steel using a visualization-based multi-criteria approach. Appl Math Model 29(7):653–672
87. Cheng NL, Chen ZP (2005) Practice of improving internal quality of slab through dynamic soft reduction. Steelmaking 5:29–32
88. Chen Y (2004) Quality control of continuous casting of heavy rail steel. Iron Steel 39(3):23–26
89. Masaoka T, Wang SH (1991) Application of soft reduction technology to improve center segregation in continuous casting slabs. Wuhan Iron Steel Technol 02:27–32
90. Yu ZX, Zheng W, Yang YC (2004) Commissioning of the new wide slab continuous caster at WISCO and its trial production in the first half year. In: Proceedings of the 2004 AISTech continuous casting and hot rolling conference, pp 357–361
91. Ji C (2008) The development and application of dynamic soft reduction technology in continuous casting machines. Benxi Iron and Steel Group Company, Chinese Society for Metals Continuous Casting Sub-Committee. In: Proceedings of the seminar on quality control technology of variety steel continuous casting billets. Benxi Iron and Steel Group Company, Chinese Society for Metals Continuous Casting Sub-Committee: Chinese Society for Metals, pp 87–92
92. Zhang P (2009) Application of dynamic soft reduction technology in continuous casting. Chinese Society for Metals Special Steel Metallurgy Committee, National Electric Furnace Coordination Group, Laigang Special Steel Plant. In: Proceedings of the 2009 Special Steel Metallurgy Committee of the Chinese Society for Metals. Chinese Society for Metals Special Steel Metallurgy Committee, National Electric Furnace Coordination Group, Laigang Special Steel Plant: The Chinese Society for Metals, pp 271–275
93. Xie CC, Wang XH, Zhang JM et al (2009) Force analysis and application of the fan-shaped segment under dynamic soft reduction in slab continuous casting. Spec Steel 30(05):22–24
94. Tian L (2012) Application of dynamic soft reduction technology. Chinese Society for Metals Continuous Casting Steel Branch. In: Proceedings of the 2012 microalloyed steel continuous casting crack control technology seminar. Chinese Society for Metals Continuous Casting Steel Branch: The Chinese Society for Metals, pp 145–149
95. Cheng RJ, Wang ZH, Wang HF et al (2013) Solidification pattern and process optimization of Q345 steel 250 mm × 2000 mm slab continuous casting. Spec Steel 34(2):41–44

96. Cao JX, Yang XZ, Xiao YQ et al (2013) Research and application of soft reduction technology in continuous casting slab at Lianyuan steel. Continuous Cast 1:14–18
97. Li JB (2013) Dynamic soft reduction roll gap adjustment scheme and its application effect at hansteel. The Chinese Society for Metals. In: Proceedings of the ninth China steel annual conference. The Chinese Society for Metals: The Chinese Society for Metals, pp 1267–1273
98. Liu SG, Zou SH, Liao SS et al (2014) Process optimization of dynamic soft reduction technology in slab continuous casting. Jiangxi Metallur 34(3):23–25
99. Yang F, Huang HL, Zeng LY (2014) Research on dynamic soft pressure down technology for wide slab continuous caster at shaoguan steel. Southern Metals 5:20–23
100. Liu QL, Liu GP, Si XM et al (2017) Process optimization of soft reduction for S650MC high strength steel continuous casting. Steelmaking 33(4):52–57
101. Zeng Z, Ji CX, Zhang HY et al (2018) Experimental research on process parameter optimization of soft reduction in slab continuous casting. Continuous Cast 43(4):23–26
102. Chen YJ (2018) Experiment on the impact of soft reduction on the internal quality of Q345 steel slab. Xinjiang Steel 2:16–19
103. Zhou J (2019) Application and practice of dynamic soft reduction system in continuous casting slab at sansteel. Fujian Metallur 48(5):16–19
104. Wang ZM, Zhao J, Wang YL, Hu XT (2019) Research on center segregation control of 400 mm thick slab continuous casting. Continuous Cast 44(6):47–50
105. Lin DS, Li XF, Zhang WT et al (2019) Improvement of internal quality of C610L continuous casting slab by soft reduction. Steelmaking 35(5):50–53
106. Cen YQ (1997) Liquid core reduction process in continuous casting billet. Shanghai Metals 19(5):42–48
107. Dong Z (trans) Improvement of center segregation in high carbon large square billets. Metallur Trans 1:44–48
108. Zhao PJ, Han HL (2002) Application of soft reduction technology on the new slab continuous casting machine at Jigang. Continuous Cast 6:11–12
109. Zhang DD, Li ZY, Li JQ et al (2005) Application of AISTech DYNAGAP soft reduction technology in the six-strand large square billet continuous casting machine. Continuous Cast 2:6–9
110. Wang GJ, Liu ZY, Hong J (2006) Application of static soft reduction technology in high carbon continuous casting square billet production. Steel Res 4:36–39
111. Liu W, Wu W, Liu L et al (2009) Application of static soft reduction technology in GCr15 bearing steel continuous casting production. Spec Steel 30(1):44–45
112. Chen Y, Chen JP, Wu GR et al (2009) Development of key technologies for 360 mm × 450 mm large square billet continuous casting at Pangang. Steel Iron Vanadium Titanium 30(2):61–67
113. Tian XZ, Zhu R, Ji C et al (2010) Application of dynamic soft reduction technology on SWRH82B bloom in continuous casting. Comput Appl Chem 27(12):1707–1710
114. Tian XZ, Zhu R, Ji C et al (2010) Application of dynamic soft reduction technology in process of concasting bloom of bearing steel GCr15. Spec Steel 31(06):26–27
115. Zeng W (2010) Design and application of dynamic soft reduction process parameters for 45# steel in Pangang No. 2 square billet continuous casting. Continuous Cast 2010(2):21–24
116. Chang GH (2010) Improvement of center segregation in curtain line steel billet using soft reduction technology. The Chinese Society for Metals. In: Proceedings of the 2010 national steelmaking-continuous casting production technology conference. The Chinese Society for Metals: The Chinese Society for Metals, pp 377–379
117. Zeng J, Chen WQ, Cao CF et al (2015) Impact of soft reduction on internal quality of 82B steel rectangular billet. Steelmaking 31(4):63–67
118. Zhong XD, Liu J, Zou ZS (2016) Research and practice on improving macro-segregation in 350 mm × 470 mm rectangular billet. Continuous Casting 41(2):62–65
119. Duan YQ, Wang JF, Guo CJ (2016) Analysis and process improvement of center segregation in heavy rail steel continuous casting billet. Special Steel 37(1):25–28
120. Li J, Ye DX, Cheng XW et al (2017) Research on soft reduction technology to improve the internal quality of large bearing steel billets. Sci Technol Econ Market 2017(29):73,119

121. Wang XH, Zhang JB, Tu XK et al (2018) Improvement of internal quality of 220 mm × 260 mm bearing steel large square billet by soft reduction. Steelmaking 34(6):34–39
122. Zhu ZG, Zhao JL, Yan YN (2019) Research and application of soft reduction technology in large square billet bearing steel. Southern Metals 6:22–24
123. Nabeshima S, Nakato H, Fujii T et al (1995) Control of centerline segregation in continuously cast blooms by continuous forging process. ISIJ Int 35(6):673–679
124. Kojima S, Mizota H (1994) Cotinuous forging apparatus for cast strand: USA, US5282374
125. Hiraki S, Yamanaka A, Shirai Y et al (2009) Development of new continuous casting technology (PCCS) for very thick plate. Materia Jpn 48(1):20–22
126. Matsuoka Y, Miura Y, Higashi H et al (2015) NSENGI's new developed bloom continuous casting technology for improving internal quality of special bar quality (NS Bloom Large Reduction). In: The METEC and 2nd ESTAD, pp 307–318
127. Yim CH, Won YM, Park JK et al (2012) Continuous cast slab and method for manufacturing the same: USA, US8245760B2. 2012.
128. Zhao XK, Zhang JM, Lei SW et al (2013) Dynamic recrystallization (DRX) analysis of heavy reduction process with extra-thickness slabs. Steel Res Int 85(5):811–823
129. Zhao XK, Zhang JM, Lei SW et al (2014) The position study of heavy reduction process for improving centerline segregation or porosity with extra-thickness slabs. Steel Res Int 85(4):645–658
130. Zhao XK, Zhang JM, Lei SW et al (2014) Finite-Element analysis of porosity closure by heavy reduction process combined with ultra-heavy plates rolling. Steel Res Int 85(11):1533–1543
131. Xu ZG, Wang XH, Jiang M (2017) Investigation on improvement of center porosity with heavy reduction in continuously cast thick slabs. Steel Res Int 88(2):1–12

Chapter 2
The Flow Properties and Constitutive Models of Continuous Casting Strand for Reduction Process

2.1 The Deformation Properties and Constitutive Models of Solid Zone

Most studies on constitutive relations, which requires a wide range of applicable conditions and easy convergence, are based on a large number of experimental data. Due to the different lattice type and crystal structure of the material itself, the deformation process presents different phenomena. Therefore, the constitutive models applicable to different materials and different processing conditions are different.

In the study of solid mechanics, the constitutive relation is mainly composed of stress–strain relationship, yield criterion, and hardening law. The determination of yield criterion and hardening law depends on the stress–strain relationship of materials. Therefore, the research on the stress–strain relationship in the deformation process of materials is the focus of constitutive relationship research.

Its general expression can be expressed as:

$$\sigma = f(\varepsilon, \dot{\varepsilon}, T) \tag{2.1}$$

where ε is the true strain; $\dot{\varepsilon}$ is strain; T is the temperature.

2.1.1 Flow Properties of Solid Zone During Reduction Process

The ranges of key variables such as strain rate and temperature are prerequisites for establishing the constitutive relationship and constructing the constitutive model.

In this section, the thermomechanical properties of bearing steel GCr15 have been studied by mathematical modeling. Selecting 25 nodes in a 1/4 section of the bloom,

C. Ji et al., *Solidification End Reduction of Continuous Casting Process*, Key Steel Manufacturing Technologies by Collaborative Innovation,
https://doi.org/10.1007/978-981-96-8544-8_2

the strain rate and strain were calculated at each node by a 3-D thermomechanical-coupled FE model. The locations of the 25 nodes are shown in Fig. 2.1a. The variation ranges of the strain and strain rates under the HR process are shown in Fig. 2.1b, c, respectively.

The strain rate range for GCr15 steel was approximately 10^{-3}–10^{-1} s^{-1}, and the maximum strain did not exceed 0.2 at 900–1300 °C during the HR deformation process by above the figure.

The Gleeble thermal/mechanical simulation machine was used to measure the thermal deformation characteristics of continuous casting strand during the compression and tensile process, which is the main physical simulation to investigate the evolution of microstructure and mechanical properties of metal materials. Combine the reduction characteristics of continuous casting, the compressive tests were performed to determine the flow curve of GCr15 bearing steel.

As shown in Fig. 2.2, the specimen orientation was perpendicular to the casting direction, avoiding the central area with serious segregation and porosity. The sample was processed into a cylinder with a radius of 8 and a length of 12 mm, respectively. The chemical composition of the GCr15 steel in this study is shown in Table 2.1.

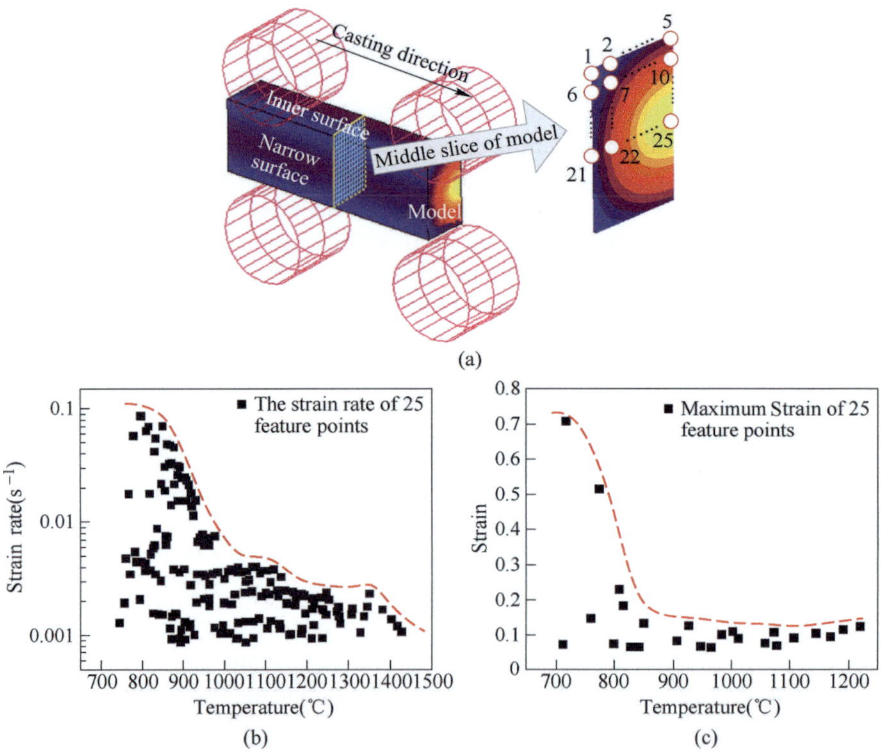

Fig. 2.1 Deformation characteristics during compression process. **a** The schematic diagram of feature points; **b** the strain rate of 25 feature points; **c** maximum strain of 25 feature points

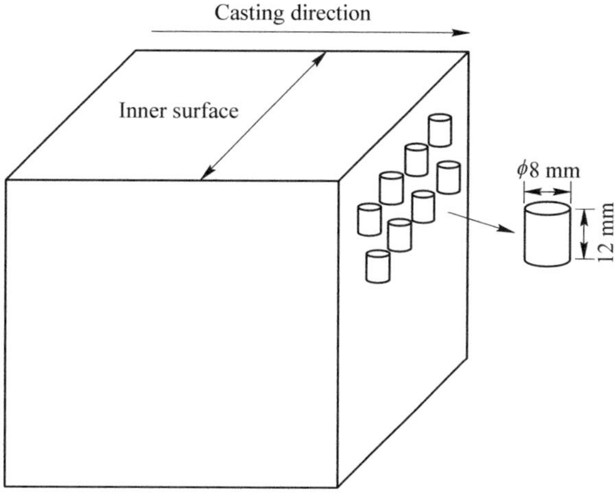

Fig. 2.2 Sampling location of the GCr15 bloom

Table 2.1 Chemical composition of the GCr15 bearing steel (wt%)

C	S	P	Mn	Cr	Si	Ni	Cu	Mo	Al
1.03	0.002	0.012	0.263	1.36	0.194	0.01	< 0.01	< 0.01	0.012

The solidus temperature of GCr15 bearing steel is approximately 1330.5 °C by empirical equation calculation [1, 2]. That would provide the reference for the compression experiment. The tests were performed at temperatures from 750 to 1300 °C in 50 °C increments.

Single-pass isothermal hot compression tests were shown in Fig. 2.3. In the isothermal hot compression tests, the specimen was heated to the experiment temperature at a heating rate of 10 °C/s. The temperature was maintained for 360 s, and the specimen was subsequently compressed at a constant temperature and a constant strain rate to a true strain of 0.7. After the experiment is finished, the sample is quenched to room temperature by spray quenching.

Figure 2.4 shows the true stress–strain curves under different temperatures with strain rates of (a) 0.001 s^{-1}, (b) 0.01 s^{-1}, and (c) 0.1 s^{-1}. The influence of temperature and strain rate on flow stress is substantial. The flow stress increases with decreasing deformation temperature and increasing strain rate, and the peak strain increases with decreasing deformation temperature and increasing strain rate. All curves show peak and flow softening, especially at lower temperatures. Initially, the flow stress increased with increasing strain; the flow stress then reached a peak. Finally, the flow stress decreased with increasing strain, and some curves reached a steady state. This phenomenon indicates that dynamic recrystallization (DRX) occurred during the hot deformation process under HR deformation conditions.

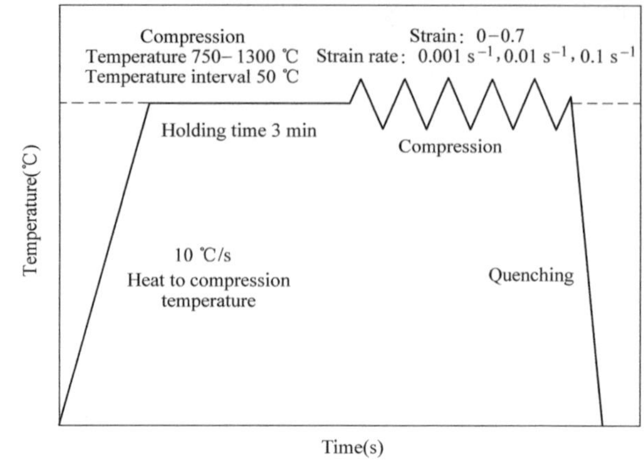

Fig. 2.3 Hot compression test procedure

2.1.2 *Constitutive Models and Applicability*

The constitutive model is mainly divided into empirical model, physical model, and intelligent prediction model. Among these models, empirical constitutive model is the most widely used. In this section, taking GCr15 bearing steel as an example, several common constitutive models are selected to predict the rheological behavior of metal.

2.1.2.1 Johnson-Cook Model

Johnson-Cook model (J-C model) is an empirical constitutive model proposed by G. R. Johnson and W. H. Cook in 1983 for the common problems such as high strain rate, large strain, and high temperature of ductile materials in engineering field. The model has the advantages of a simple structure, few parameters, and being easy to solve. Many software programs have the parameters used by the model, and the basic form of its expression is also very suitable for numerical simulation. The effect of the three variables of strain, strain rate, and temperature on the yield stress and failure of the material is expressed by the model adopting the functional relationship of continuous multiplication. The model is characterized by simple structure and easy fitting and has strong universality in engineering field.

The J-C model equation is expressed as follows:

$$\sigma = \left[A + B\varepsilon^{n}\right]\left[1 + C\ln\left(\frac{\dot{\varepsilon}}{\dot{\varepsilon}_{0}}\right)\right]\left[1 - \left(\frac{T - T_{\mathrm{r}}}{T_{\mathrm{m}} - T_{\mathrm{r}}}\right)^{m}\right] \tag{2.2}$$

The austenitizing complete temperature of GCr15 bearing steel is 900 °C. The steel is in the cementite and austenite two-phase zone at the temperature from 750 to 900 °C, and it is in the austenite phase at the temperature from 900 to 1300 °C.

Fig. 2.4 True stress–strain curves for GCr15 steel under different temperatures and strain rates. **a** Strain rate of 0.001 s^{-1}; **b** strain rate of 0.01 s^{-1}; **c** strain rate of 0.1 s^{-1}

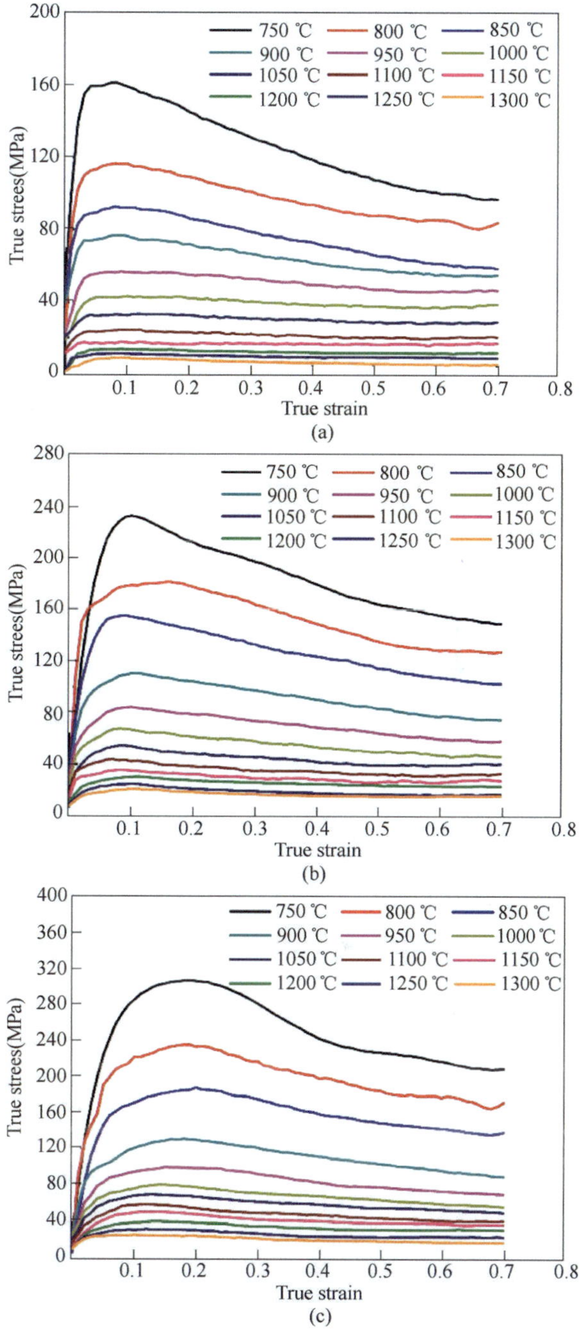

This leads to the different properties of materials in different temperature ranges, so the temperatures are divided into the two intervals for discussion. The reference temperature is 750 °C at the temperature from 750 to 900 °C, and the reference temperature is 900 °C at the temperature from 750 to 900 °C. The reference strain rate is 0.001 s^{-1}.

The parameter solution at temperature from 750 to 900 °C is taken as an example. The solving process is as follows:

When the strain rate is equal to the reference strain rate and the temperature is the reference temperature. Equation (2.2) then can be simplified as:

$$\sigma = A + B\varepsilon^{n} \tag{2.3}$$

The parameter A is the initial yield stress of material in the equation when the effective plastic is zero. According to the method of Mirzadeh, the parameter A could be ignored at this time. Equation (2.2) would be further simplified as:

$$\text{In}\,\sigma = \text{In}\,B + n\,\text{In}\,\varepsilon \tag{2.4}$$

This method is widely used to solve the constitutive model parameters, and the equation is transformed into a linear relationship.

In Eq. (2.4), the parameter n is the slope of $\ln\sigma - n\ln\varepsilon$ relation, that is, $n = \partial\ln\sigma/\partial\ln\varepsilon$ and ln$B$ is the intercept. The stress–strain curves data at 750 °C and 0.001 s^{-1} were substituted into Eq. (2.4) to calculate the parameters n and B.

When the temperature is the reference temperature, the third term of Eq. (2.2) is 1, and Eq. (2.2) can be transformed into:

$$\sigma = (B\varepsilon^{n})\left(1 + C\,\text{In}\frac{\dot{\varepsilon}}{0.01}\right) \tag{2.5}$$

Taking the logarithm of both sides of the equation:

$$\frac{\sigma}{B\varepsilon^{n}} - 1 = C\ln\frac{\dot{\varepsilon}}{0.001} \tag{2.6}$$

In Eq. (2.6), C is the slopes of the $\left(\frac{\sigma}{B\varepsilon^{n}} - 1\right) - \ln\frac{\dot{\varepsilon}}{0.001}$. In this example, the selection of strain value is 0.05, 0.1, 0.15, 0.2, 0.25, 0.3, 0.35, 0.4, 0.45, 0.5, 0.55, 0.6, 0.65, and 0.7.

When the strain rate is equal to the reference strain rate, that is, the strain rate is 0.001 s^{-1}, the second term in Eq. (2.2) is 1, which can be simplified to:

$$\sigma = (B\varepsilon^{n})\left[1 - \left(\frac{T - 1023}{580}\right)\right] \tag{2.7}$$

Taking the logarithm of both sides of the equation:

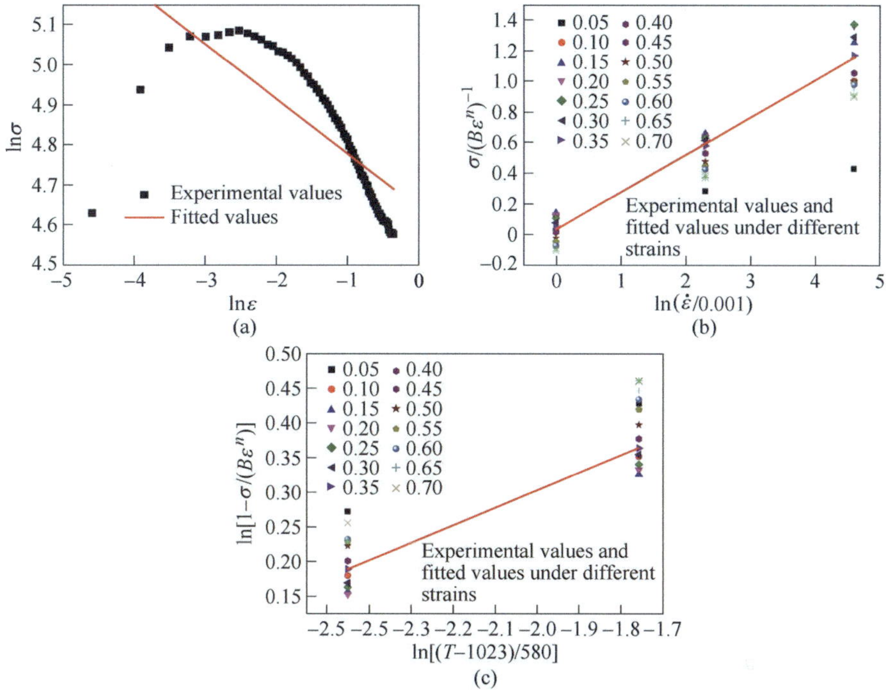

Fig. 2.5 Solution parameters by least squares. **a** Parameter n, **b** parameter C, **c** parameter m

$$\sigma = \left(1 - \frac{\sigma}{B\varepsilon^n}\right) m \ \mathrm{In}\left(\frac{T - 1023}{580}\right) \tag{2.8}$$

The value of m could be obtained by calculating the slopes of the $\ln\left[1 - \frac{\sigma}{B\varepsilon^n}\right] - m\ln\left(\frac{T-1023}{580}\right)$. Substitute the values of n and B into Eq. (2.8), and the result is shown in Fig. 2.5.

All model parameters have been obtained through the above calculations, and the model parameters of the two-temperature zone are substituted into Eq. (2.9) and Eq. (2.10).

The J-C model at the temperature from 750 to 850 °C is expressed as follows:

$$\sigma = \left[103.71\varepsilon^{-0.1365}\right]\left[1 + 0.2448\ln\left(\frac{\dot{\varepsilon}}{0.001}\right)\right]\left[1 - \left(\frac{T - 1023}{1603 - 1023}\right)^{0.2526}\right] \tag{2.9}$$

The J-C model at the temperature from 900 to 1300 °C is expressed as follows:

$$\sigma = \left[56.69\varepsilon^{-0.09153}\right]\left[1 + 0.1716\ln\left(\frac{\dot{\varepsilon}}{0.001}\right)\right]\left[1 - \left(\frac{T - 1173}{1603 - 1173}\right)^{0.7703}\right] \quad (2.10)$$

Substitute different temperature and strain values into the constitutive model, and then compare the predicted stress value obtained with the experimental value. Figure 2.6 shows that the overall deviation of the constitutive model at the temperature from 750 to 850 °C is large, and the data are in good agreement only at the reference temperature of 750 °C. It can be seen from the comparison diagram that the predicted value of the constitutive model shows a monotonically decreasing trend, and the decay trend of stress gradually slows down. The true stress–strain curve trend is characterized by the "concave function", which cannot describe the stress rise stage at low strain levels.

As shown in Fig. 2.6, the consistency between the predicted value of the model and the experimental value decreases significantly with the increase of temperature, which is related to the selection of reference temperature T_r in the calibration process of model parameters. The J-C model must take the lowest temperature of the research system as the reference temperature during parameter calibration; otherwise, the

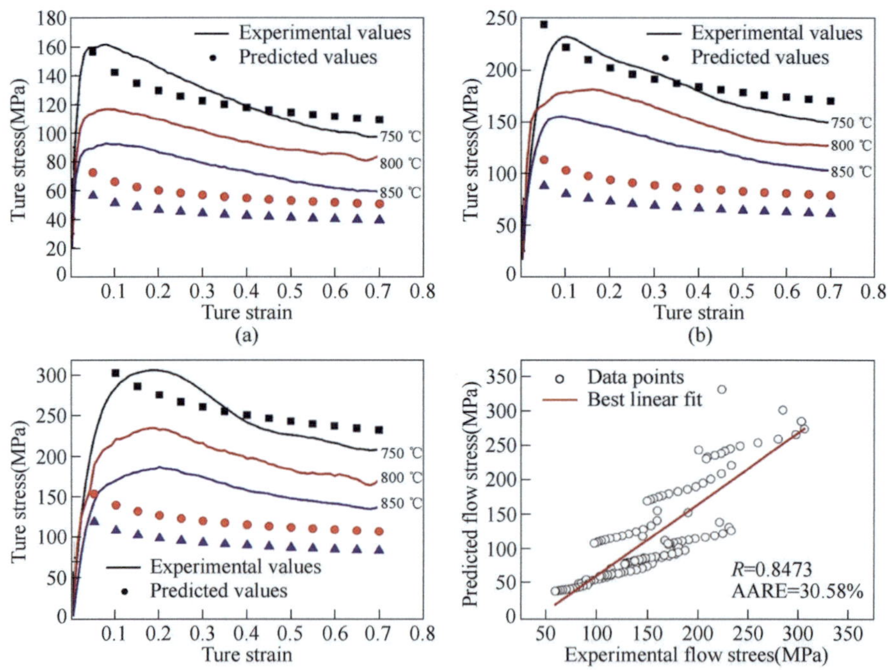

Fig. 2.6 Comparison between the predicted and experimental values of the J-C model at the temperature from 750 to 850 °C. **a** Strain rate 0.001 s^{-1}, **b** strain rate 0.01 s^{-1}, **c** strain rate 0.1 s^{-1}, **d** error analysis

calculation process cannot be carried out smoothly, and this is also a defect of the model. Since GCr15 bearing steel is not completely austenitized in the range from 750 to 850 °C, and this temperature region is a two-phase zone of cementite + austenite. This also shows that the J-C model cannot describe the flow stress change in the cementite + austenite two-phase zone.

Standard statistical parameters such as the correlation coefficient (R) and the average absolute relative error (AARE) were used to evaluate the accuracy of the constitutive equation:

$$R = \frac{\sum_{i=1}^{N} (E_i - \overline{E})(P_i - \overline{P})}{\sqrt{\sum_{i=1}^{N} (E_i - \overline{E})^2 \sum_{i=1}^{N} (P_i - \overline{P})^2}} \tag{2.11}$$

$$\text{AARE} = \frac{1}{N} \sum_{i=1}^{N} \left| \frac{E_i - P_i}{E_i} \right| \times 100\% \tag{2.12}$$

where E_i is the experimental value and P_i is the predicted value of the constitutive equations; and are the mean values of E and P, respectively; and N is the total number of data sets employed in this study.

R is used to represent the strength of the linear relationship between the experimental and predicted values, but it does not necessarily represent the degree of fit. This is because the predicted values may be all high or all low. AARE value is an average relative error between the predicted value and the experimental value, which can better characterize the agreement between the overall predicted value and the experimental value.

In this paper, R and AARE are combined to evaluate the accuracy of the predicted value of the flow stress of the constitutive model. In the temperature range from 750 to 850 °C, the correlation coefficient between the predicted value of J-C model and the experimental value is $R = 0.8473$, and AARE $= 30.58\%$. It can be seen from Fig. 2.8 that when the strain rate is 0.001 s^{-1}, the predicted value of the model at a temperature of 900–1000 °C is in good agreement with the experimental value, and the agreement is poor at a temperature of 1050–1300 °C.

When the strain rate is 0.01 s^{-1} at temperature from 950 to 1200 °C, the predicted value of the model is in good agreement with the experimental value at low stress, and the description of the stress–strain curve with an obvious downward trend is poor. When the strain rate is 0.1 s^{-1} at that temperature range, the predicted value of the model deviates greatly. According to the comparison between the model prediction value and the experimental value in the two temperature ranges of 750–850 °C and 900–1000 °C, the average relative error is large. Therefore, the J-C model is not suitable for GCr15 to predict the rheological curve of continuous casting bloom during reduction process (Fig. 2.7).

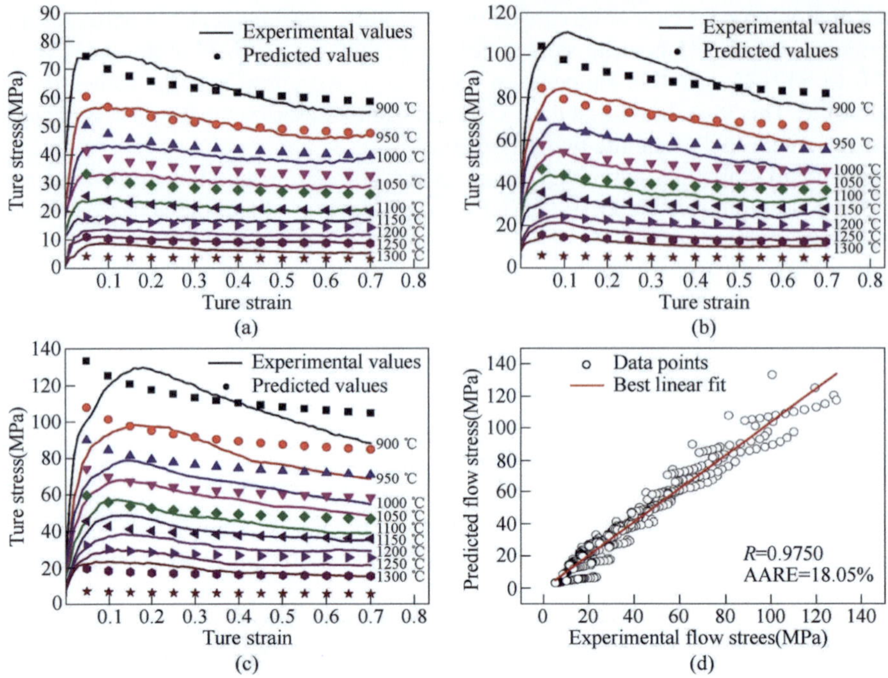

Fig. 2.7 Comparison between the predicted and experimental values of the J-C model at the temperature from 900 to 1300 °C. **a** Strain rate 0.001 s^{-1}, **b** strain rate 0.01 s^{-1}, **c** strain rate 0.1 s^{-1}, **d** error analysis

2.1.2.2 Modified J-C Model

In the J-C model, the influences of temperature, strain and strain rate on stress are independent of each other, so the influences on stress are expressed in the form of the product of the three. However, in fact, the effects of temperature, strain, and strain rate on stress are interrelated, so many researchers modify the J-C model in various ways to improve the prediction accuracy of the model. Li et al. [3] proposed a quadratic polynomial to fit the nonlinear relationship between stress and strain under the conditions of reference temperature and reference strain rate. At the same time, the coupling effect of temperature and strain rate on stress is also considered, and a modified J-C model is proposed, whose basic expression is as follows:

$$\sigma = (A_1 + B_1 + B_2\varepsilon^2)\,(1 + C_1\,\mathrm{In}\dot{\varepsilon}^*)\,\exp[-\lambda_1 + \lambda_2\mathrm{In}\dot{\varepsilon}^*](T - T_r) \qquad (2.13)$$

where A_1, B_1, B_2, C_1, λ_1 and λ_2 are material constants; σ is the true stress, MPa; ε is the strain; $\dot{\varepsilon}^*=\dot{\varepsilon}/\dot{\varepsilon}_0$, $\dot{\varepsilon}$ is the strain rate, s^{-1}; $\dot{\varepsilon}_0$ is the reference strain rate, s^{-1}; T is the temperature, K; T_r is the reference temperature, K.

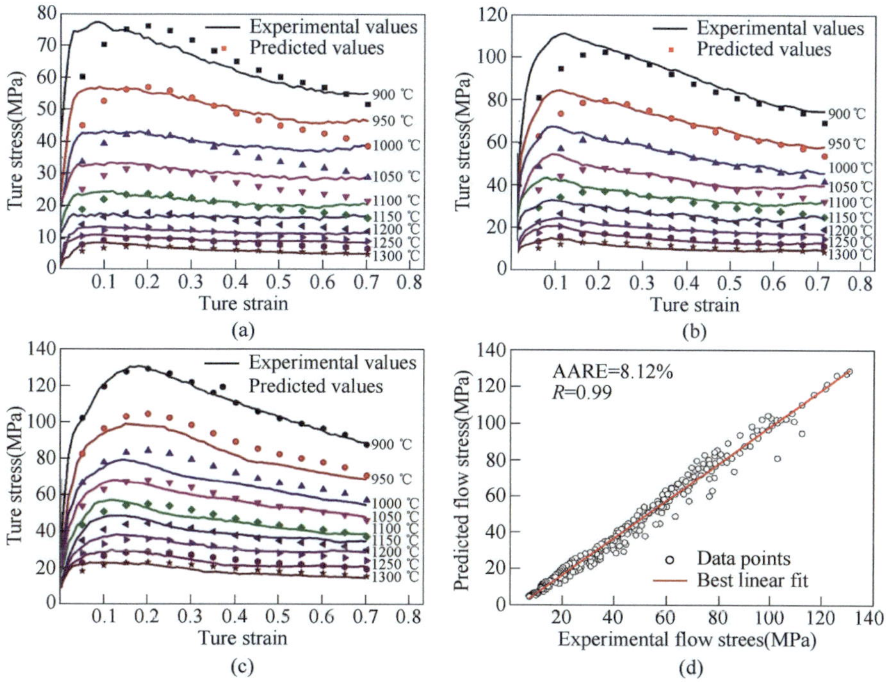

Fig. 2.8 Comparison between the predicted and experimental values of the modified J-C model at different strain rate. **a** Strain rate 0.001 s^{-1}, **b** strain rate 0.01 s^{-1}, **c** strain rate 0.1 s^{-1}, **d** error analysis

The modified J-C model has three components, where $f(\varepsilon) = \left[A_1 + B_1 + B_2\varepsilon^2\right]$ represents the effect of strain strengthening. The strain rate strengthening term is represented $f(\dot{\varepsilon}) = (1 + C_1 \cdot \ln \dot{\varepsilon}^*)$, which indicates that the stress increases as the strain rate increases. The thermal softening effect of temperature on stress is represented $f(T, \varepsilon) = \exp[-(\lambda_1 + \lambda_2 \ln \dot{\varepsilon}^*)(T - T_r)]$, and the coupling effect of temperature and strain rate on stress is considered. The solution process of each parameter is the same as that of J-C model, so it is not repeated here.

By substituting different temperatures, strains, and strain rates into the modified J-C model, the stress values under different deformation conditions were calculated. The accuracy of the model was verified by comparing and analyzing the predicted values of the model and the experimental values. As shown in Fig. 2.8, the overall difference between the predicted value of the modified J-C model and the experimental value is not large, which can basically describe the metal rheological law of bearing steel under deformation conditions. With the increase of strain, the predicted value reaches the peak stress and then decreases and finally reaches a stable state. The predicted value scatter shows an obvious dynamic recrystallization phenomenon, which is consistent with the curve trend of the experimental value. As shown in Fig. 2.8d, the predictive correlation coefficient and average relative error of the revised J-C model are 0.99 and 8.12%, respectively.

2.1.2.3 Zerilli-Armstrong Model

The Zerilli-Armstrong model [4] (Z-A model) is a new constitutive relation proposed by Zerilli and Armstrong in 1987. The model considers strain hardening, strain rate hardening, and heat softening, which combines them into a constitutive relation with a certain degree of accuracy. The corresponding constitutive relation expressions are shown in Eq. (2.14) and Eq. (2.15), respectively:

$$\sigma = C_0 + C_2 \varepsilon^{1/2} \exp(-C_3 + C_4 T \cdot \ln \dot{\varepsilon}) \tag{2.14}$$

$$\sigma = C_0 + C_1 \exp(-C_3 + C_4 T \cdot \ln \dot{\varepsilon}) + C_5 \varepsilon^n \tag{2.15}$$

where C_0, C_1, C_2, C_3, C_4, C_5, and n are material constants; σ is the true stress, MPa; ε is the strain; $\dot{\varepsilon}$ is the strain rate, s^{-1}; T is the temperature, °C.

Although the Z-A model is proposed for the crystal structure, it has several defects. First, the crystal structure of the material needs to be defined in the construction of the model, and it changes dynamically over a large temperature deformation range. Second, Z-A model is not suitable for temperatures higher than $0.6T_m$ (T_m is the melting point), and − 273 °C is not available when the final Z-A model parameters are solved. Moreover, the Z-A model does not consider the absolute effect of strain rate on stress at − 273 °C, that is, no matter how the strain rate changes at − 273 °C, the stress value is a constant, which is inconsistent with common sense. Therefore, Z-A model is not suitable for describing the deformation behavior of continuous casting.

2.1.2.4 Modified Zerilli-Armstrong Model

In view of some defects of Z-A model, Samantaray et al. [5] proposed a modified Z-A model, whose constitutive model is shown in Eq. (2.16).

The modified Z-A model expression is as follows:

$$\sigma = \left[C_1 + C_2 \varepsilon^n\right] \exp\left[-(C_3 + C_4 \varepsilon) \cdot T^* + (C_5 + C_6 T^*) \cdot \ln \dot{\varepsilon}^*\right] \tag{2.16}$$

where C_0, C_1, C_2, C_3, C_4, C_5, and n are material constants; σ is the true stress, MPa; ε is the strain; $T^* = T - T_r$. Other parameters are defined in the preceding section.

In Eq. (2.16), the first square bracket on the right of the equation indicates the effect of strain hardening on stress. C_1 is the yield stress at reference temperature and reference strain rate (unit: MPa). The flow stress corresponding to 0.2% strain of metal material without obvious yield platform is the yield stress. Parameter C_3

Table 2.2 Parameters for the modified Z-A model for GCr15

Temperature (°C)	900–1300
C_1	80
C_2	14.03542124
C_3	0.004068028
C_4	0.000285058
C_5	0.085469958
C_6	0.000341295
n	− 0.410341985

represents the influence of temperature on flow stress. C_4 represents the effect of temperature and strain coupling on flow stress. C_5 represents the effect of strain rate strengthening on flow stress. C_6 represents the effect of temperature and strain rate coupling on flow stress. Taking bearing steel as an example, the intercept method is used to solve each parameter. The results of parameter solving of the modified Z-A model are shown in Table 2.2, and the solving process is shown in Fig. 2.9.

The predicted value of Z-A model was modified under different strain, strain rate, and temperature. It can be seen in Fig. 2.10 that the predicted value of the modified

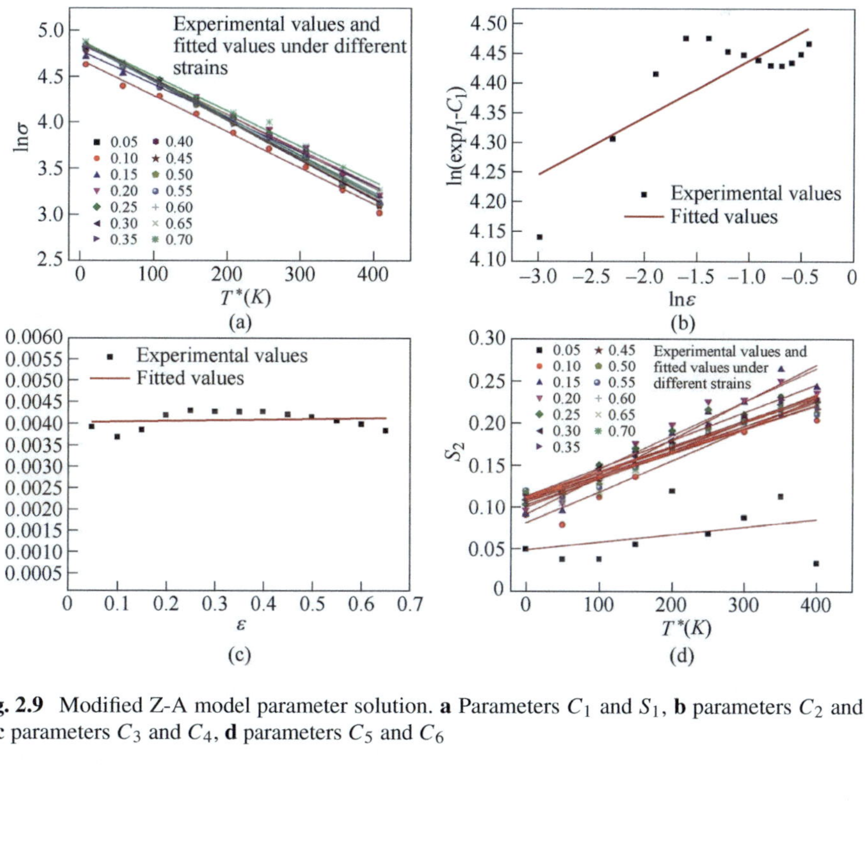

Fig. 2.9 Modified Z-A model parameter solution. **a** Parameters C_1 and S_1, **b** parameters C_2 and n, **c** parameters C_3 and C_4, **d** parameters C_5 and C_6

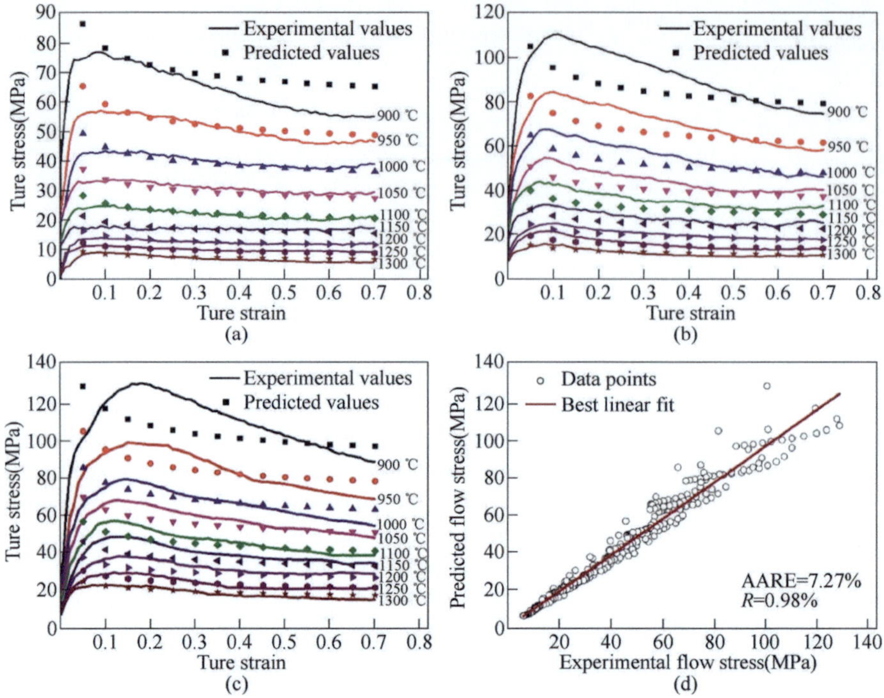

Fig. 2.10 Comparison between the predicted and experimental values of the modified Z-A model for bearing steel at different strain rate. **a** Strain rate 0.001 s^{-1}, **b** strain rate 0.01 s^{-1}, **c** strain rate 0.1 s^{-1}, **d** error analysis

Z-A model generally presents a downward trend, especially at 900–1300 °C. At low strain, the predicted value of the model deviates greatly from the experimental value. Compared with the experimental value, the scatter plot of the predicted value of the model shows a "leading" state of the peak stress; that is, the peak strain of the predicted value of the model should change less than that of the experimental value. When the temperature is greater than 1000 °C, the predicted value of the model is in good agreement with the experimental value. Combined with error analysis, the mean relative error (7.27%) is small, but the correlation coefficient R is slightly lower, only 0.98. Although the average relative error of the modified Z-A model for bearing steel is small, the scatterplot of the predicted values of the model cannot describe the metal rheology of bearing steel under this deformation condition.

2.1.2.5 Arrhenius Model

The temperature and strain rate of metal during thermal deformation determine the atomic diffusion rate and the dislocation density accumulation rate, respectively. In order to more truly reflect the constitutive relationship in the thermal deformation process of metal materials, Sellars and Tegart et al. [6] proposed the Arrhenius equation. This equation describes the relationship between strain rate, temperature, and flow stress [7, 8]. The model considers deformation temperature and activation energy of deformation, and its general expression is as follows:

$$\dot{\varepsilon} = A \exp\left(-\frac{Q}{RT}\right) F(\sigma) \tag{2.17}$$

where $D = D_0 \exp(-Q_{sd}/RT)$ is the function of stress.

$$F(\sigma) = \begin{cases} \sigma^{n_1} & \alpha\sigma < 0.8 \\ \exp(\beta\sigma) & \alpha\sigma > 1.2 \\ [\sinh(\alpha\sigma)]^n & \text{For all } \sigma \end{cases} \tag{2.18}$$

where $G = G_0[1 + \eta(T - 300)/T_M]$ is the strain rate, s^{-1}; σ is the true stress, MPa; R is the ideal gas constant, $R = 8.314$ J/(mol K); T is the absolute temperature, K; Q is the activation energy of hot deformation, J/mol; n is the material's stress index; and A, α, β, and n_1 are material constants, where $\alpha = \beta/n_1$.

For low stress levels ($\alpha\sigma < 0.8$), the stress function $F(\sigma)$ can choose the form of a power function. For high stress levels ($\alpha\sigma > 1.2$), the stress function $F(\sigma)$ can be selected as an exponential function. The third hyperbolic sine function is applicable to all stress states, and the equations of low stress level and high stress level can be regarded as the equations obtained by Taylor expansion of hyperbolic sine function according to the stress state. The deformation activation energy Q in the equation is a physical quantity that represents the difficulty of rearrangement and combination of microscopic atoms in the thermal deformation process of the material, and its value is affected by many factors such as chemical composition, microstructure, deformation rate, and deformation temperature. In order to describe the influence of strain rate and temperature on deformation behavior, Zenner-Hollomon factor Z can be used. The Z factor is a temperature compensated strain rate factor with the following form [9]:

$$Z = \dot{\varepsilon} \exp\left(\frac{Q}{RT}\right) \tag{2.19}$$

Equations (2.17)–(2.19) give Eq. (2.20):

$$Z = A \cdot [\sinh(\alpha\sigma)]^n \tag{2.20}$$

The relationship between Z parameter and flow stress can be obtained from Eqs. (2.19) and (2.20):

$$\sigma = \frac{1}{\alpha} \ln\left\{\left(\frac{Z}{A}\right)^{1/n} + \left[\left(\frac{Z}{A}\right)^{2/n} + 1\right]^{1/2}\right\} \tag{2.21}$$

Many calculation methods in the process of constitutive model calculation are based on the idea of "average calculation". The temperature range of the constitutive model studied by predecessors is small, yet the accuracy of the constitutive model can be guaranteed. However, the experimental temperature span is large, and the accuracy of conventional constitutive model is difficult to guarantee. In order to improve the accuracy of the model, the characteristics of large temperature span and obvious difference in solidification structure of continuous casting were taken into account. Similar to the solving process of the J-C model, two temperature intervals are divided to solve the Arrhenius constitutive model, namely 750–850 °C and 900–1300 °C, which are discussed in detail in Sect. 2.1.2.1. The parameter solving process is similar to the J-C model. The results of parameter solving are shown in Tables 2.3 and 2.4, and the solving process is shown in Fig. 2.11.

The parameters under different stress conditions are fitted linearly to calculate the specific values of each material parameter, which are substituted into the constitutive model. The stress values predicted by the constitutive model at different temperatures and strain rates are calculated. As shown in Fig. 2.12, the predicted values of the model are in good agreement with the experimental values under the temperature range of 750–850 °C, and the values of R and AARE are 0.9921 and 3.74%, respectively.

As can be seen from Fig. 2.12, there are obvious deviations in the predicted values of the 900 °C model under the strain rate of 0.01 and 0.1 s^{-1}. This is due to the large deformation temperature span in the 900–1300 °C range, which far exceeds the traditional thermal deformation research temperature span range (200 °C). The calculation process of each parameter has an average calculation process, so there is a large deviation at a lower temperature (900 °C). In addition, due to the segregation phenomenon in the continuous casting process of GCr15 bloom, the composition of

Table 2.3 Polynomial fitting coefficients of the material parameters (750–850 °C)

Strain	Q (J/mol)	ln$A$	n	α
0.05	452,519.3	44.28	8.001	0.006702
0.1	366,060.7	35.15	5.609	0.006185
0.15	315,821.2	29.76	4.931	0.006094
0.2	279,503.3	25.82	4.522	0.006131
0.25	277,725.7	25.63	4.455	0.006384
0.3	271,844.9	24.97	4.461	0.006646
0.35	266,460.3	24.35	4.532	0.006965
0.4	261,337.6	23.74	4.655	0.007307
0.45	245,867.3	22.02	4.567	0.007602
0.5	252,322.9	22.78	4.524	0.007882
0.55	251,890.8	22.75	4.464	0.008082
0.6	250,224.9	22.57	4.465	0.008268
0.65	253,398.9	22.89	4.455	0.008510
0.7	248,703.2	22.39	4.466	0.008537

Table 2.4 Polynomial fitting coefficients of the material parameters (900–1300 °C)

Strain	Q (J/mol)	ln$A$	n	α
0.05	568,182.5	43.66	5.398	0.02777
0.1	461,409.1	34.63	4.240	0.02696
0.15	441,658.1	33.13	3.881	0.02775
0.2	446,308.2	33.60	3.854	0.02845
0.25	471,488.3	35.77	3.997	0.02977
0.3	488,100.9	37.21	4.091	0.03116
0.35	498,701.1	38.09	4.139	0.03265
0.4	502,608.7	38.42	4.207	0.03361
0.45	513,794.8	39.38	4.360	0.03418
0.5	512,181.4	39.26	4.408	0.03488
0.55	509,431.6	38.99	4.504	0.03496
0.6	519,207.9	39.81	4.698	0.03501
0.65	531,141.7	40.75	4.914	0.03492
0.7	558,243.2	43.01	5.302	0.03426

the sample is slightly different. In the process of the experiment, the sample cannot be complete austenitizing at 900 °C, which will also affect the overall law of the data. The predicted value of the constitutive model has a certain deviation with the strain of 0.001 s^{-1} and the true strain of 0.05, 0.1, and 0.15. At the strain rate of 0.01 s^{-1}, the predicted values were in good agreement with the experimental values except that the predicted values at 900 °C were poor. The predicted values of 950 and 1000 °C at the strain rate of 0.1 s^{-1} are biased when the strain is lower than 0.15. In summary, in the temperature range of 900–1300 °C, the overall predicted value of the model has high accuracy, the correlation coefficient $R = 0.9921$ in the error analysis, and the average relative error AARE = 5.76%, which can fully meet the needs of the finite element simulation calculation process (Fig. 2.13).

2.1.2.6 Simplified Arrhenius Model

As can be seen from the conventional solution of the Arrhenius model, the solution process of the model parameters corresponding to each stress value is very complicated, requiring a lot of fitting and regression calculation, consuming a lot of time. In the actual application process, for different materials and different steel grades, it is often necessary to use the above method to establish the corresponding constitutive model, and its calculation is extremely large. In order to simplify the material parameter solving process, shorten the calculation time and improve the calculation efficiency, the conventional solving process of the Arrhenius model is simplified in this section.

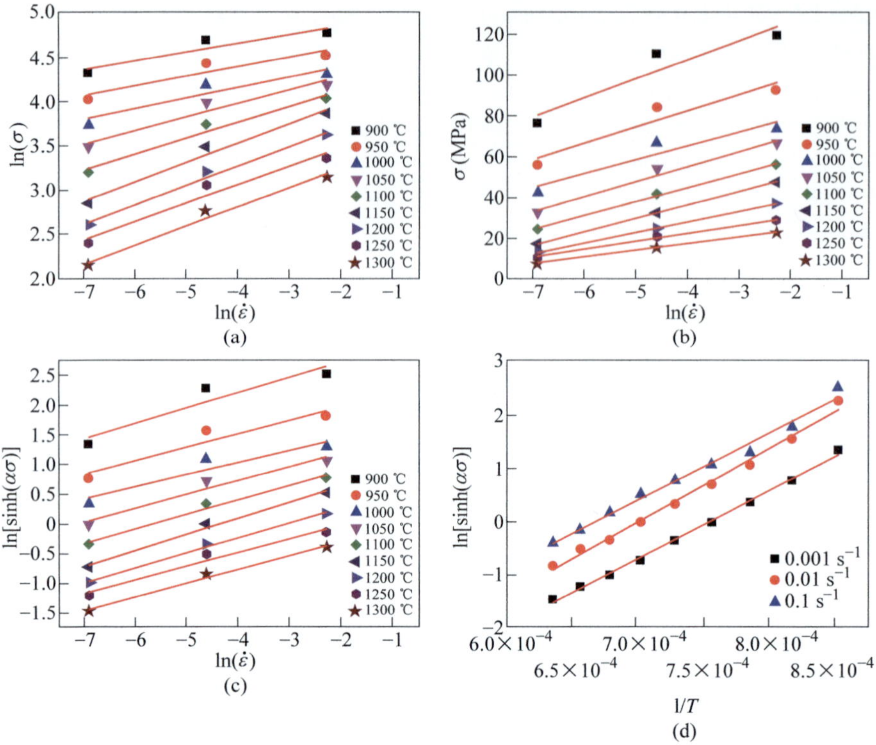

Fig. 2.11 Detailed least squares fitting process for a strain of 0.1 in the temperature range from 1173 K (900 °C) to 1573 K (1300 °C). **a** Fitting process of material parameter n_1, **b** fitting process of material parameter β, **c** fitting process of material parameter n, and **d** fitting process of material parameter Q

The specific method of simplifying the Arrhenius model calculation process [10] is shown in Eq. (2.22). Firstly, the activation energy Q_p and parameter α_p in hyperbolic sine function under peak stress condition are calculated. These two parameters are taken as constants that do not change with strain, and then the material parameters n and ln$A$ are calculated for different strain conditions. This will greatly shorten the calculation process and calculation time, but also reduce the probability of errors in the calculation process.

$$Z_\mathrm{p} = \dot{\varepsilon}\exp\left(\frac{Q_\mathrm{p}}{RT}\right) = A_\mathrm{p}\left[\sinh\left(\alpha_\mathrm{p}\sigma_\mathrm{p}\right)\right]^n \tag{2.22}$$

The calculation process of activation energy Q_p and parameter α_p under the condition of peak stress is the same as that in the previous section. It is only necessary to sort out the curve peak stress of different strain rates and different temperatures and put it into the above calculation process. The detailed process will not be repeated.

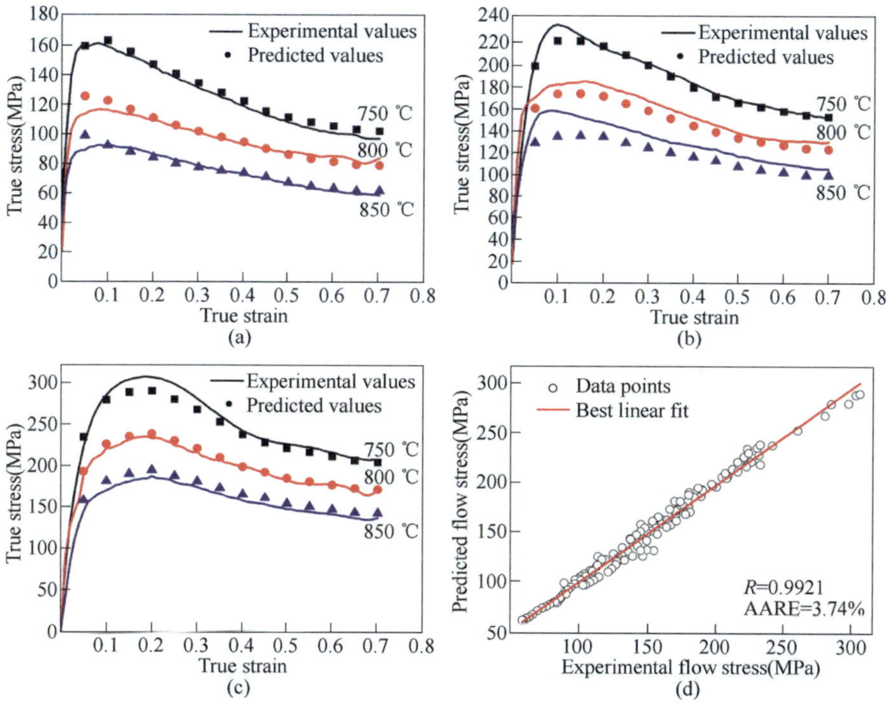

Fig. 2.12 Comparison between the experimental and predicted values at 750–850 °C with a **a** strain rate of 0.001 s^{-1}, **b** strain rate of 0.01 s^{-1}, **c** strain rate of 0.1 s^{-1}, and the **d** error analysis

750–850 °C:

$$Z_p = \dot{\varepsilon} \exp\left(\frac{323{,}676.1}{RT}\right) = 1.987 \times 10^{13} \times \left[\sinh(0.005885\sigma_p)\right]^{5.0472} \tag{2.23}$$

900–1300 °C:

$$Z_p = \dot{\varepsilon} \exp\left(\frac{446{,}719.4}{RT}\right) = 3.437 \times 10^{14} \times \left[\sinh(0.02708\sigma_p)\right]^{3.987} \tag{2.24}$$

By substituting the material parameters corresponding to different stress values into the constitutive model, the predicted flow stress values at different temperatures and strain rates can be calculated. The comparison between the experimental value of the Arrhenius model with simplified algorithm in the temperature ranging from 750 to 850 °C is shown in Fig. 2.14. From the comparison of data in the figure, it can be seen that the predicted value of the model is in good agreement with experimental value with the strain rate of 0.001 s^{-1}.

The deviation is large under the low stress state of 800 and 850 °C. At 750 °C, the overall softening trend is obvious, but there is a slight deviation. As can be seen from

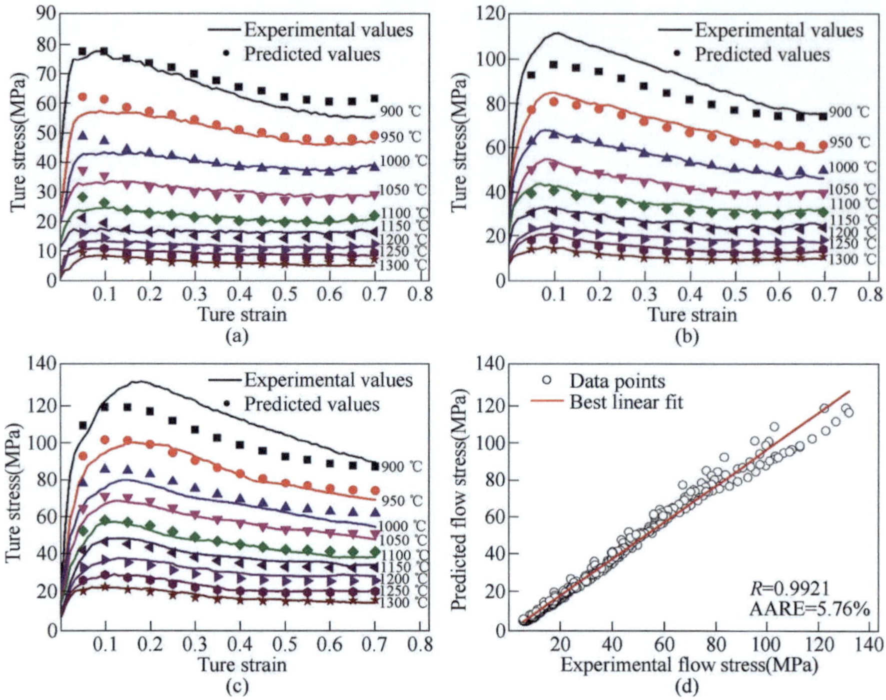

Fig. 2.13 Comparison between the experimental and predicted values at 900–1300 °C with a **a** strain rate of 0.001 s^{-1}, **b** strain rate of 0.01 s^{-1}, **c** strain rate of 0.1 s^{-1}, and the **d** error analysis

Fig. 2.14, the peak stress at 750 °C is prominent, the predicted value of the model has a large deviation from the experimental value, and the stress level at 850 °C is generally low. When the strain rate is 0.001 s^{-1}, the predicted values with temperature of 750–850 °C are in good agreement with the experimental values, and the accuracy required for finite element simulation can be fully achieved. All predicted deviations under the temperature of 750–850 °C were evaluated, and the correlation coefficient *R* and mean relative error AARE were 0.9872 and 5.17%, respectively. The overall error in the temperature range is small, the simplified Arrhenius model meets the requirements of use.

As shown in Fig. 2.15, the predicted stress value of model has a good agreement with that measure in the experiment with the strain rate of 0.001 s^{-1}. When the strain is 0.05, the deviation of the predicted stress mainly shows that the stress value is larger at 1050–1300 °C, and the stress is smaller at 900 °C. When the curve strain of 1250 and 1300 °C exceeds 0.4, the predicted stress value of the constitutive model is generally high. At the strain rate of 0.01 s^{-1}, the flow stress curve at 900 °C is generally low, and the predicted stress at both ends of the curve at 1200–1300 °C is also high. At the strain rate of 0.1 s^{-1}, the predicted stress error trend is similar to that at the strain rate of 0.01 s^{-1}. From the overall error analysis in the temperature range of 900–1300 °C, the correlation coefficient $R = 0.9920$, the average relative

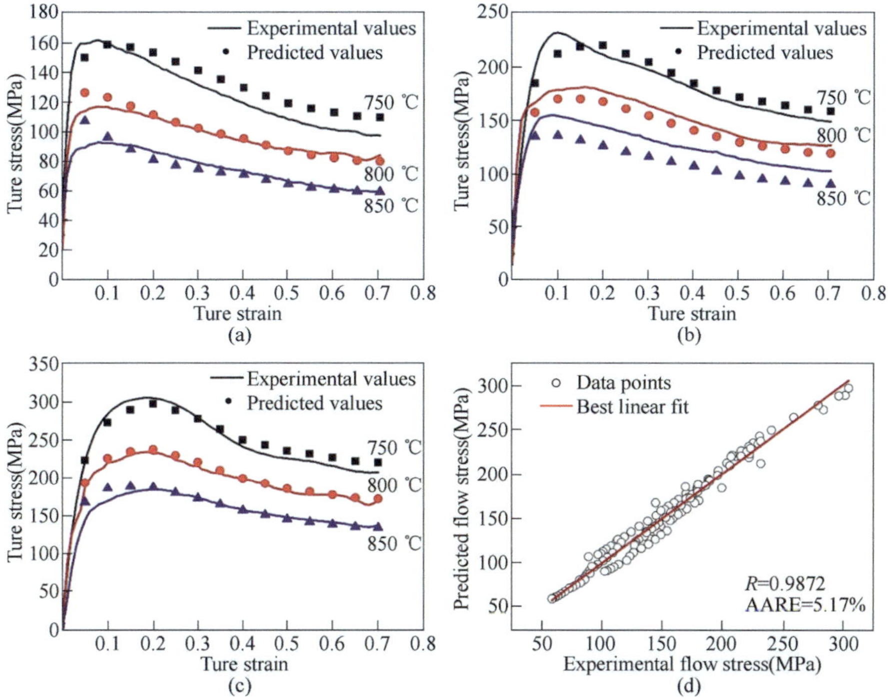

Fig. 2.14 Comparison between the experimental and predicted values at 750–850 °C with a **a** strain rate of 0.001 s^{-1}, **b** strain rate of 0.01 s^{-1}, **c** strain rate of 0.1 s^{-1}, and the **d** error analysis

error AARE = 7.97%, the error level is slightly higher than the AARE = 5.76% of the conventional solution of the Arrhenius model. However, the calculation process greatly reduces the complexity of the calculation, shortens the calculation time, and meets the needs of the thermomechanical simulation.

2.1.2.7 Modified Arrhenius Model

The conventional Arrhenius constitutive model is phenomenological constitutive model. The material parameters Q, A, n, and α of the constitutive model are calculated based on the stress–strain curve. The model assumes that the structure of metal materials is constant during deformation and does not consider the change of microstructure during high-temperature deformation. In this chapter, the crystallographic parameters related to dislocation slip and climb are introduced for the defects which do not contain physical significance in Arrhenius equation. The Arrhenius constitutive model was modified based on the physical meaning, and the parameters of the constitutive model were calibrated.

In recent years, some scholars [9–13] have found that in the process of deformation at high temperature, when the deformation mechanism is controlled by slip and climb

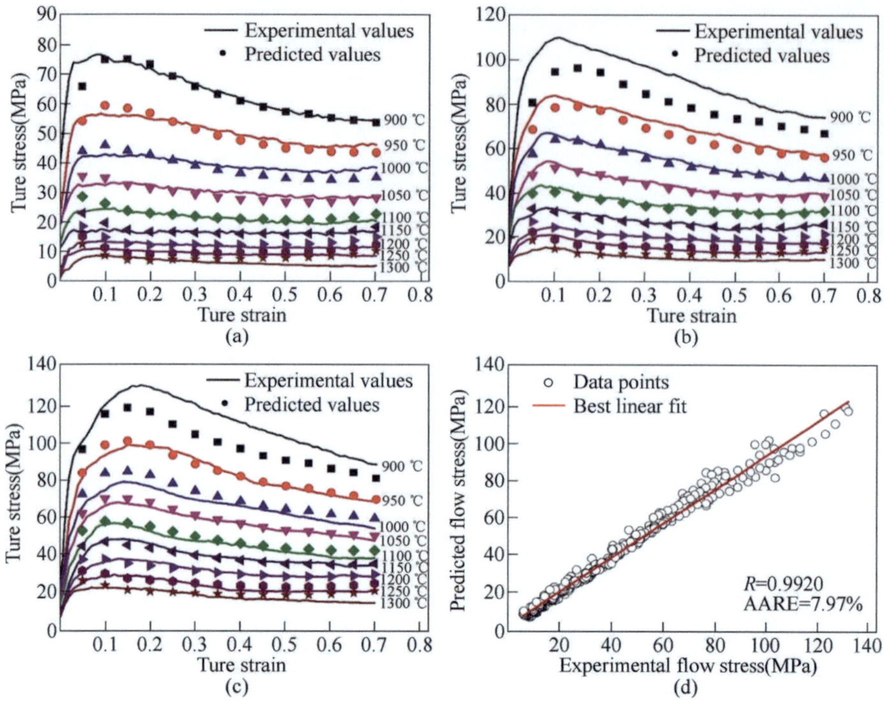

Fig. 2.15 Comparison between the experimental and predicted values at 900–1300 °C with a **a** strain rate of 0.001 s^{-1}, **b** strain rate of 0.01 s^{-1}, **c** strain rate of 0.1 s^{-1}, and the **d** error analysis

of dislocation hyperbolic sine index n and self-diffusion coefficient D can be used to describe the change of flow stress. Its expression is shown in Eq. (2.25). Some calculation results and key parameters are shown in Table 2.5:

$$D = D_0 \exp(-Q_{sd}/RT) \tag{2.25}$$

Table 2.5 Relevant lattice parameters such as self-diffusion coefficient and shear modulus are required [13]

Materials	D_0 (m^2/s)	Q_{sd} (KJ/mol)	η	G_0 (MPa)	T_M (K)
AISI 304/316 steel	3.7×10^{-5}	280	− 0.85	81,000	1680
1%CrMoV steel	1.9×10^{-4}	239	− 1.09	81,000	1573
γ-Fe	1.8×10^{-5}	270	− 0.91	81,000	1184–1665
Al	1.7×10^{-4}	142	− 0.50	25,400	933
Mg	1.0×10^{-4}	135	− 0.49	16,600	924
Cu	2.0×10^{-5}	197	− 0.54	42,100	1356
Ni	1.9×10^{-4}	284	− 0.64	78,900	1726

where D_0 refers to the pre-factor; Q_{sd} is the self-diffusion activation energy.

After the elastic modulus E and self-diffusion coefficient D are introduced and modified based on physical significance, the Arrhenius constitutive model can be expressed as:

$$\frac{\dot{\varepsilon}}{D} = \begin{cases} B'\left(\dfrac{\sigma}{E}\right)^{n'} \\ B''\exp\left(\dfrac{\beta'\sigma}{E}\right) \\ B\left[\sinh\left(\dfrac{\alpha'\sigma}{E}\right)\right]^{n} \end{cases} \tag{2.26}$$

$$E = 2G(1+\nu) \tag{2.27}$$

$$G = G_0[1+\eta(T-300)/T_M] \tag{2.28}$$

where E is the elastic modulus, MPa; υ is Poisson's ratio, with the value 0.3. G is the shear modulus, MPa; G_0 is the shear modulus at a temperature of 300 K, MPa; B', B'', B, n', n, β', α' are all material parameters.

Since GCr15 parameters cannot be obtained, this paper adopts the parameter of 1%CrMoV with similar physical properties to GCr15 (as shown in Table 2.5) and replaces the parameters in Table 2.5 into Eqs. (2.25) and (2.28) to obtain:

$$D = 1.9\times10^4\times\exp(-239{,}000/RT) \tag{2.29}$$

$$G = 81{,}000\times[1-1.09\times(T-300)/1603] \tag{2.30}$$

Take logarithms of both sides of Eq. (2.26):

$$\ln\left(\frac{\dot{\varepsilon}}{D}\right) = \begin{cases} n'\ln\left(\dfrac{\sigma}{E}\right)+\ln B' \\ \beta'\left(\dfrac{\sigma}{E}\right)+\ln \mathrm{B}'' \\ n\ln\left[\sinh\left(\dfrac{\alpha'\sigma}{E}\right)\right]+\ln B \end{cases} \tag{2.31}$$

The least square method is used to establish the functional relationship of $\ln(\dot{\varepsilon}/D)-\ln(\sigma/E)$. $\ln(\dot{\varepsilon}/D)$ is the vertical coordinate, $\ln(\sigma/E)$ is the horizontal coordinate. The slope of the line is calculated as $n' = 9.031$. A similar method is used to establish the functional relationship of $\ln(\dot{\varepsilon}/\mathrm{D})-\sigma/\mathrm{E}$. $\ln(\dot{\varepsilon}/D)$ is the vertical coordinate, σ/E is the horizontal coordinate. The slope of the line is calculated as $\beta' = 7992.7$. At this point, we can calculate $\alpha = \beta'/n' = 885.03$.

The stress values at different temperatures and strain rates can be calculated by substituting material parameters into the model. As shown in Fig. 2.16, it can be found by comparing the predicted value of the model with the experimental value. When the strain rate is 0.001 s^{-1} in the temperature range of 750–850 °C, the predicted value of the model is in good agreement with the experimental value, but there is a large deviation when the strain is lower than 0.1. At the strain rates of 0.01 and 0.1 s^{-1}, the predicted value of the model is low. The main reason for the deviation is the introduction of material parameters related to dislocation slip and climb, such as self-diffusion coefficient *D*. Due to the presence of cementite and austenite two-phase zone of GCr15 bearing steel under 750–850 °C, the deformation mechanism is more complicated, so the predicted value of the model has a large deviation. As can be seen from Fig. 2.16d, in the error analysis of this model, the correlation coefficient $R = 0.9628$, and the average relative error AARE = 15.44%. The overall error is large, which cannot meet the requirements of heat-force simulation calculation.

As can be seen from Fig. 2.17, the predicted value of the model is in good agreement with the experimental value, especially when the strain rates are 0.001 and 0.01 s^{-1}, the predicted value of the model is accurate. At strain rate of 0.001 s^{-1} and temperature of 900 °C with strain of 0.05, there is a large deviation in the predicted value. The predicted value of the model is low on the whole under the strain rate is

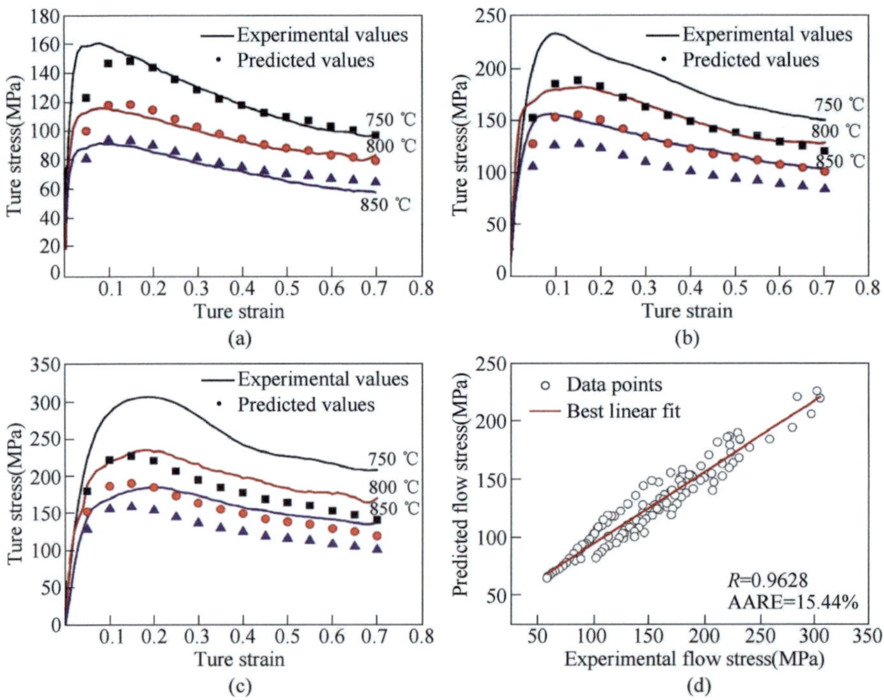

Fig. 2.16 Comparison between the experimental and predicted values at 750–850 °C with a **a** strain rate of 0.001 s^{-1}, **b** strain rate of 0.01 s^{-1}, **c** strain rate of 0.1 s^{-1}, and the **d** error analysis

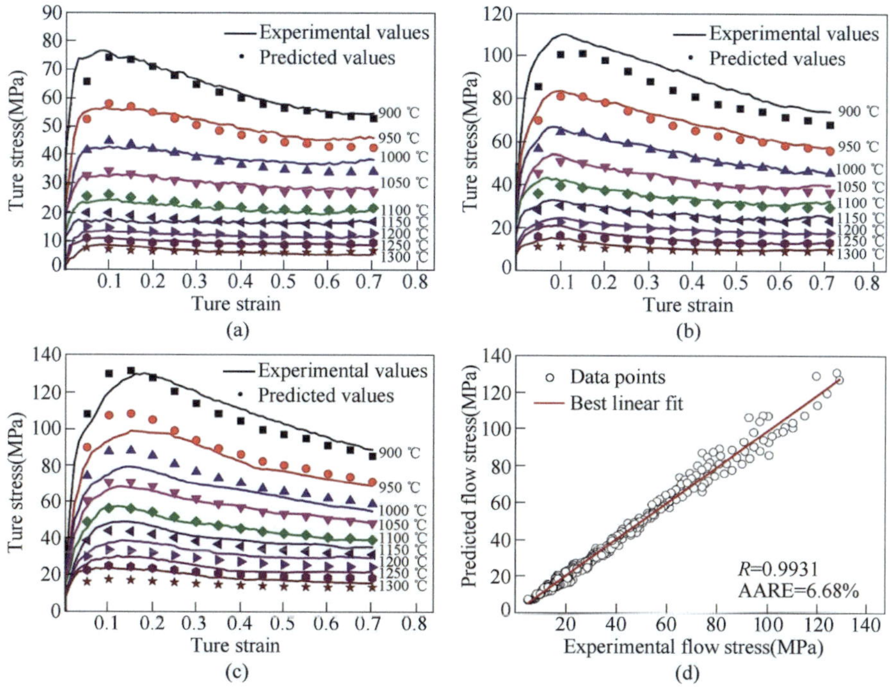

Fig. 2.17 Comparison between the experimental and predicted values at 750–850 °C with a **a** strain rate of 0.001 s^{-1}, **b** strain rate of 0.01 s^{-1}, **c** strain rate of 0.1 s^{-1}, and the **d** error analysis

0.01 s^{-1} and temperature of 900 °C, and the predicted value of the model is small deviation when the strain is lower than 0.25. There is a high agreement at other temperatures and strains. There is a deviation in the low strain region at the strain rate of 0.1 s^{-1} and temperature of 900–1000 °C. The predicted value of the model at 1150–1300 °C is generally low, and the predicted value at 0.1 s^{-1} is generally large. According to the model error analysis under all strain rates in the temperature segment, the correlation coefficient $R = 0.9931$, the average relative error AARE = 6.68%, and the overall error are small in the acceptable error range, which can meet the application requirements of finite element simulation calculation.

2.1.2.8 Comprehensive and Comparative Analysis of the Error of Each Model

Table 2.6 shows the error comparison of GCr15 bearing steel under different models in the 900–1300 °C solid-phase region.

As can be seen from Table 2.6, the Arrhenius model has the highest prediction accuracy, and its average relative error AARE is 5.76% and correlation coefficient R is 0.99. This model can well reflect the stress–strain change of GGr15 bearing

Table 2.6 Error comparison of GCr15 bearing steel under different models in solid-phase region at 900–1300 °C

Model	AARE (%)	*R*
J-C	18.05	0.975
Modified J-C	8.12	0.99
Modified Z-A	7.27	0.98
Arrhenius	5.76	0.99
Simplified Arrhenius	7.97	0.992
Modified Arrhenius	6.68	0.9628

steel during the compression process, while the J-C model has the worst prediction accuracy, with the average relative error AARE is 18.05% and correlation coefficient *R* is 0.975. The model does not reflect the change of stress and strain well.

2.1.3 Constitutive Model of Typical Steel Grades

The difference in the lattice type of metal materials makes its physical properties also different. The macroscopic phenomena of metal materials with different crystal structures in the loading process are different, which leads to the different constitutive models adapted to them. In this section, titanium microalloyed steel, structural steel, and pipeline steel are taken as examples, and the deformation characteristics of casting billets at 900–1300 °C are determined by the method of Sect. 2.1.2, and the applicable constitutive model is given.

2.1.3.1 Titanium Microalloyed Steel

The samples in this experiment were taken from the continuous casting billet of titanium microalloyed steel from a domestic steel plant, and the steel composition is shown in Table 2.7.

As can be seen from Fig. 2.18, during the deformation process, the bad stress of continuous casting of titanium microalloyed steel always increases with the increase of strain. With the increase of strain, the stress increases first at a larger rate and then at a smaller rate, and there is no obvious softening process in the whole process. This indicates that dynamic recovery and dynamic recrystallization are not very significant, but only slow down the rate of rapid stress increase caused by work hardening, and the softening effect caused by dynamic recovery and dynamic recrystallization does not exceed the hardening effect caused by work hardening.

Table 2.7 Chemical composition of titanium microalloyed steel (wt%)

C	Mo	Ti	N
0.16	1.3	0.37	0.2

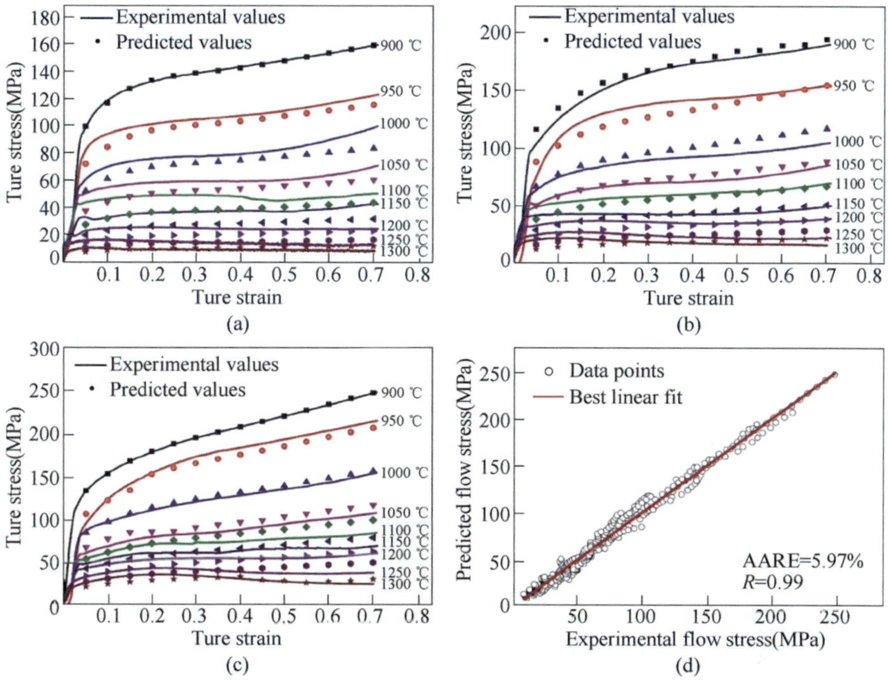

Fig. 2.18 Comparison between the experimental and predicted values of modified J-C model at different strain rate. **a** Strain rate of 0.001 s^{-1}, **b** strain rate of 0.01 s^{-1}, **c** strain rate of 0.1 s^{-1}, and the **d** error analysis

The constitutive model was established by the method in Sect. 2.1.2. The errors of each model are shown in Table 2.8. As can be seen from the table, the revised J-C model has the highest prediction accuracy, with the average relative error AARE is 5.97% and correlation coefficient R is 0.95. This is because compared with other models, the modified J-C model only considers the influence of strain hardening and strain rate strengthening on the flow stress, ignoring the influence of heat softening, that is, temperature on the stress. Due to the high iron content of this steel, a large amount of carbide is precipitated by deformation during high-temperature compression, which enhances the work hardening effect and cancels out the softening effect of the original hot deformation. If the softening effect is considered at this time, the overall accuracy of the model will be reduced.

2.1.3.2 Microalloy Structural Steel Q345E

The sample of this experiment was taken from the microalloyed steel Q345E. Table 2.9 shows the steel components.

Table 2.8 Error comparison of Ti microalloyed steel under different models in the solid phase of 900–1300 °C

Model	AARE (%)	*R*
J-C	20.15	0.98
Modified J-C	5.97	0.95
Modified Z-A	23	0.99
Arrhenius	10.02	0.99
Simplified Arrhenius	9.97	0.99
Modified Arrhenius	7.68	0.97

Table 2.9 Chemical composition of microalloyed steel (wt.%)

C	Si	Mn	P
0.17	0.31	1.5	0.014

As shown in Fig. 2.19, structural steel Q345E has a relatively small amount of microalloy precipitates due to the small content of microalloying elements. Under the influence of recrystallization at the later stage of deformation, the overall stress–strain curve presents a dynamic recovery phenomenon; that is, the stress increases rapidly and then gently and finally maintains a stable state. At this time, the hardening effect caused by work hardening and the softening effect caused by dynamic recovery reach a dynamic balance.

The prediction error pairs of each constitutive model for the rheological properties of Q345E metal are shown in Table 2.10. The deviation between the predicted value and the experimental value of the Arrhenius model is small, with the average relative error being only 4.72% and the correlation coefficient *R* being 0.99. The Arrhenius model has high prediction accuracy. Compared to other models, the relationship between strain and model parameters is considered in the process of solving the Arrhenius model parameters, so it can fully reflect the dynamic stress–strain recovery of metal materials in the later stage.

2.1.3.3 Pipeline Steel X65MS

The test sample was taken from the pipeline steel of a domestic steel plant, and the steel composition is shown in Table 2.11.

The strain curve of pipeline steel flow was measured by compression method, as shown in Fig. 2.20. The alloy content of pipeline steel is between Q345E and titanium microalloyed steel, the stress increases rapidly in the early stage of compression, and the work hardening is remarkable. Then it gradually increases and finally maintains a stable state, but at this time, the dynamic recovery strain begins to lag behind that of Q345E.

The constitutive model of pipeline steel was established using the method in Sect. 2.1.2. The errors of each model are shown in Table 2.12. As can be seen from the table, the Arrhenius model has the smallest error, with an average relative error of

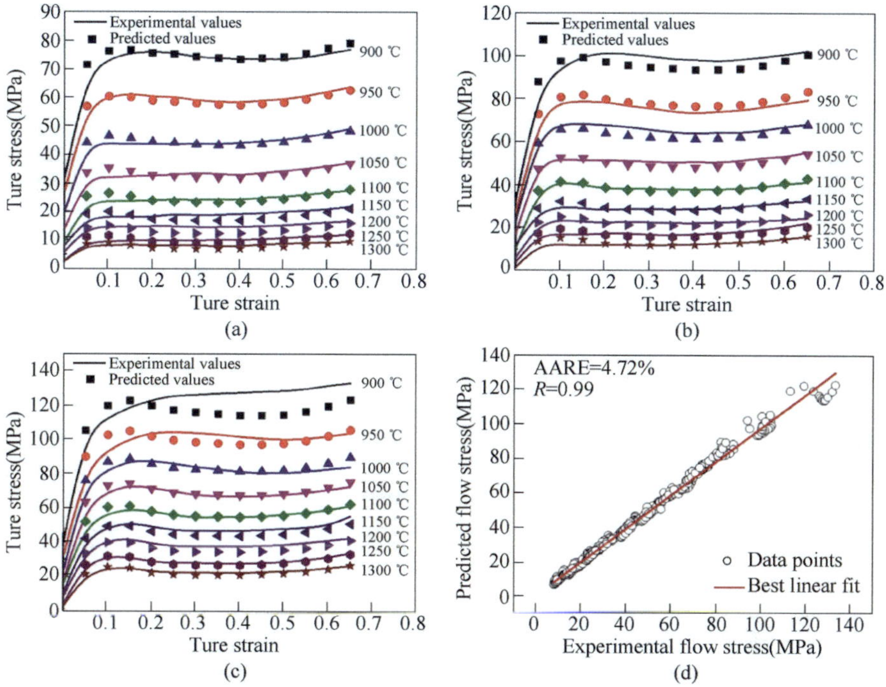

Fig. 2.19 Comparison between the experimental and predicted values of Arrhenius model at different strain rate. **a** Strain rate of 0.001 s^{-1}, **b** strain rate of 0.01 s^{-1}, **c** strain rate of 0.1 s^{-1}, and the **d** error analysis

Table 2.10 Comparison of errors in the solid phase of Q345E at temperature of 900–1300 °C

Model	AARE (%)	*R*
J-C	10.63	0.99
Modified J-C	7.8	0.99
Modified J-C	7.16	0.99
Arrhenius	4.72	0.99
Simplified Arrhenius	8.97	0.99
Modified Arrhenius	6.93	0.98

Table 2.11 Chemical composition of X65MS pipeline (wt%)

Chemical composition													
C	Si	Mn	P	S	Al	Cu	Nb	Ni	Cr	Mo	Ti	V	Ca
0.05	0.25	1.3	0.01	0.001	0.03	0.25	0.045	0.2	0.2	0.05	0.013	0.03	0.002

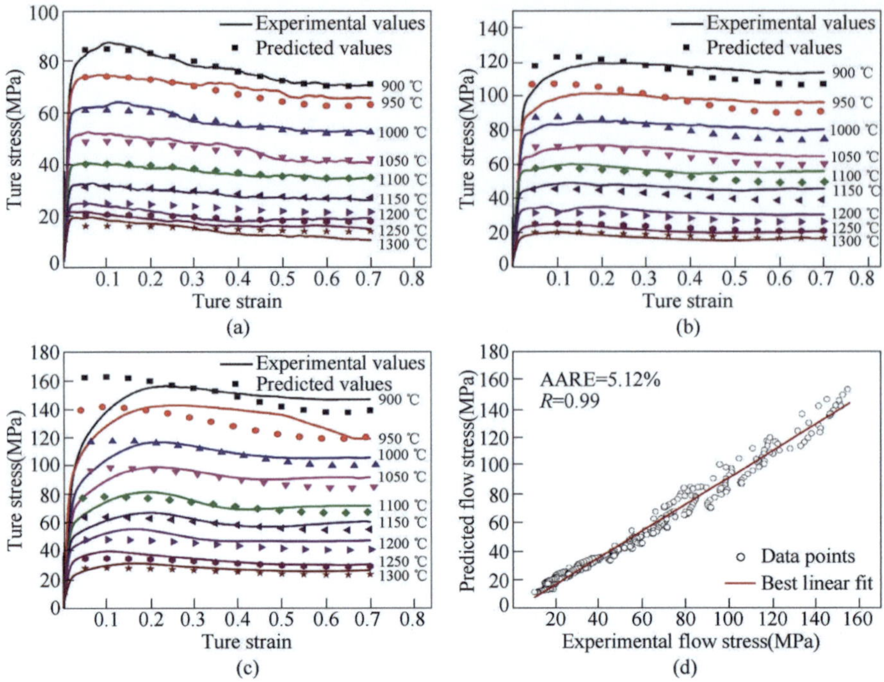

Fig. 2.20 Comparison between the experimental and predicted values of Arrhenius model at different strain rate. **a** Strain rate of 0.001 s^{-1}, **b** strain rate of 0.01 s^{-1}, **c** strain rate of 0.1 s^{-1}, and the **d** error analysis

5.12% and a correlation coefficient of 0.99. Its prediction error and error analysis are shown in Fig. 2.21d. Due to the high temperature of the sample during deformation, the softening effect of the sample is caused. Therefore, the Arrhenius model considering the coupling effect of temperature and strain rate on the flow stress can accurately predict the rheological characteristics of pipeline steel deformation process. Compared with Q345E, there are more species number of precipitated species, the dynamic recovery softening effect lags should the model prediction error be slightly larger.

Table 2.12 Comparison of errors in the solid phase of pipeline steel at temperature of 900–1300 °C

Model	AARE (%)	*R*
J-C	7.63	0.99
Modified J-C	6.02	0.99
Modified Z-A	8.58	0.99
Arrhenius	5.12	0.99
Simplified Arrhenius	6.78	0.99
Modified Arrhenius	7.85	0.98

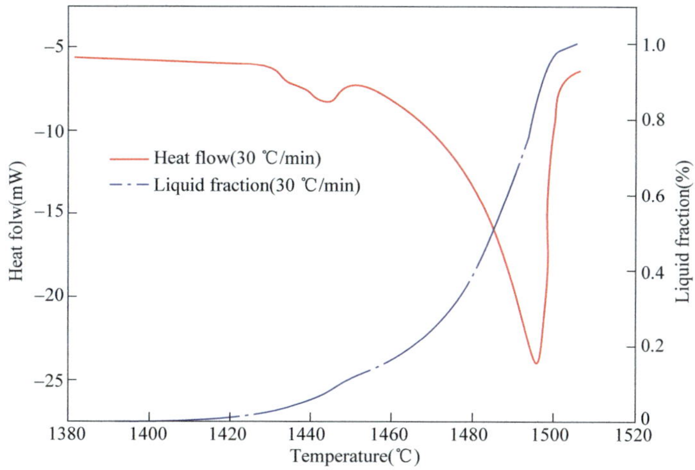

Fig. 2.21 Variation in the liquid fraction of Ti microalloyed steel with temperature

2.2 Mushy Zone Constitutive Model

The coexistence of solid and liquid phases is a prominent feature of the end solidification reduction. The presence of a liquid phase in the core of continuous casting affects the deformation of the continuous casting. In particular, with the thickening of the section of continuous casting, the thickness of the mushy zone and the required deformation amount increase, and the influence of the mushy zone deformation on the whole deformation of the billet cannot be ignored. The influence of liquid fraction on deformation must be considered during the establishment of the mushy zone constitutive model. Therefore, this section will take titanium microalloying steel as an example to analyze the microstructure particularity caused by the presence of liquid phase in the two-phase region deformation process and adopt the "phenomenological" method to construct the two-phase region constitutive model.

2.2.1 Rheological Characteristics of Metals in the Mushy Zone

Compared with the rheological characteristics of metals in the solid-phase zone, there is a certain liquid phase in the mushy region. Therefore, the specimen will have a significant "softening" effect during the deformation process, resulting in flow stress fluctuation or even instability during the compression process, and the stress value cannot be detected. Therefore, the liquid-phase fraction of the specimen must be strictly controlled during the experiment.

Table 2.7 lists the components of titanium microalloyed steel. Firstly, the liquid-phase curve and the liquid-phase fraction corresponding to different temperatures were determined by the heat flow curve detected by DSC. According to the DSC measured heat flow curve in Fig. 2.21, combined with the calculation equation:

$$f_L = \frac{\int_{T_S}^{T} W(T)\mathrm{d}T}{\int_{T_S}^{T_L} W(T)\mathrm{d}T} \tag{2.32}$$

From the calculation results, the corresponding liquid-phase fraction at 1425 °C, 1430 °C, 1435 °C, 1440 °C, and 1450 °C were 2.16%, 2.85%, 4.25%, 6.15%, and 12.06%, respectively.

The experimental scheme is designed according to the target temperature, as shown in Fig. 2.22. Before the compression experiment, all samples were heated to 1300 °C at a rate of 10 °C/s and held for 20 s to reduce the temperature gradient in the samples. Then it was heated to the target temperature at 1 °C/s, held for 5 s, compressed and deformed at 0.1, 1, and 5 s^{-1} until the strain reached 0.2. After the compression and deformation were completed, it was quickly cooled to room temperature by air injection, and the structure morphology was retained for subsequent observation.

As shown in Fig. 2.23, when the strain rate is 0.1 s^{-1}, the stress increases slowly with the increase of strain, and when the strain is greater than 0.18, the stress increases at a larger growth rate. When the strain rate is 1 and 5 s^{-1}, the stress increases rapidly with the increase of strain at first. When the strain reaches 0.04, the stress changes little with the increase of strain. Due to the presence of the liquid phase in the high-temperature region, the stress value is greatly reduced compared with the solid-phase region, even if the strain rate is as high as 5 s^{-1}, the maximum stress value is not more than 20 MPa. Therefore, the presence of the liquid phase greatly reduces the deformation resistance required for compression experiments and significantly "softens" the pattern. Moreover, it can be obviously seen that with the increase of

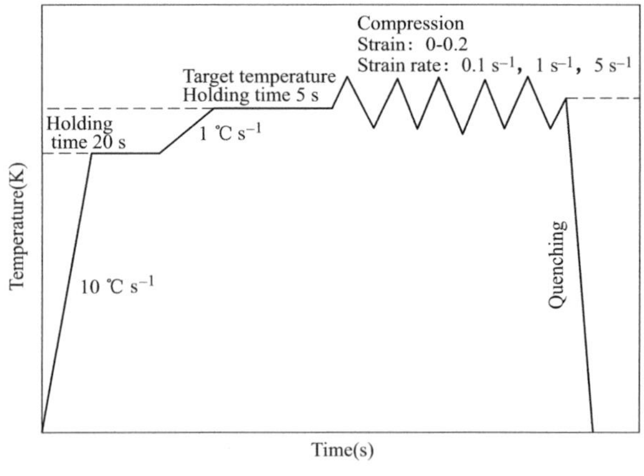

Fig. 2.22 Diagram showing the hot working compression process in the mushy zone

temperature, the content of the liquid phase in the sample increases, and the softening effect of the sample is enhanced.

Figure 2.24 shows the port morphology of titanium microalloyed steel specimen containing part of the liquid phase after compression deformation. The liquid phase in the sample is extruded during deformation and becomes radial after cooling.

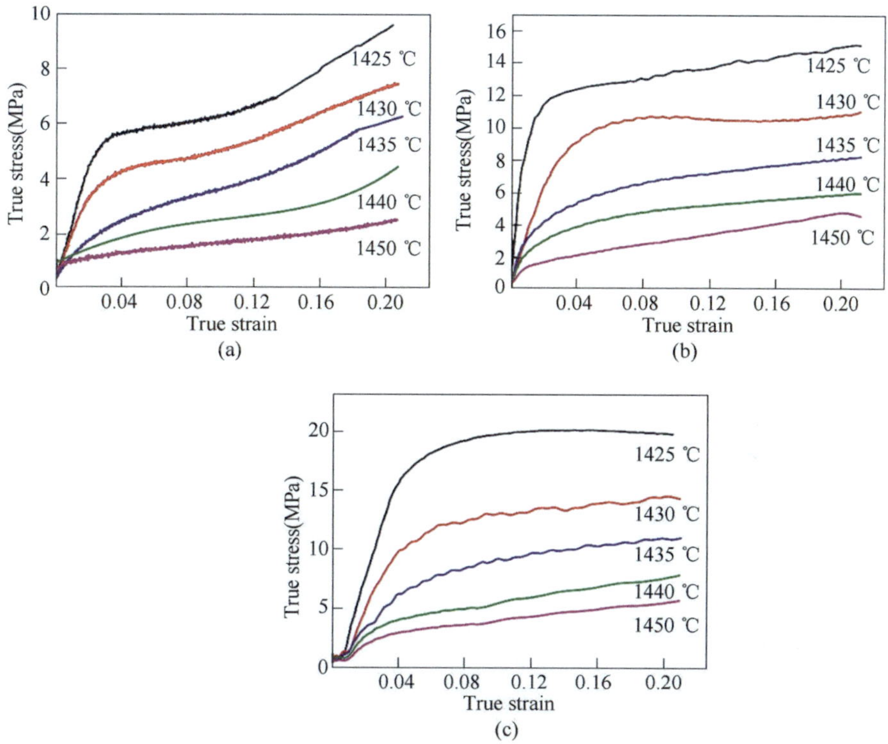

Fig. 2.23 True stress–strain curves for Ti microalloyed steel at 1698–1723 K and different strain rates of **a** 0.1 s^{-1}, **b** 1 s^{-1}, and **c** 5 s^{-1}

Fig. 2.24 Compressed specimens of Ti-Mo steel under different deformation conditions

2.2.2 Microstructure Evolution of Mushy Zone

Figure 2.25 shows a metallographic picture of the whole cross-section of the specimen perpendicular to the compression direction after compression at a temperature of 1430 °C and strain rate of 1 s^{-1}. Martensite appeared in the red dashed box. Many black areas were located at the edges (e.g., the green dashed zone). The middle part of the sample was composed of fine deformed grains, and equiaxial recrystallized grains were located all around these fine deformed grains. There were many black areas around the small grains at the boundary of the sample. Based on the above, this paper divided the microstructure of the compressed specimen in the mushy region into two parts: the center and the boundary.

As illustrated in Fig. 2.26, there were many small dendrites in the black area. The elemental distribution in the black area was observed at the same time. Ti was evenly distributed in the middle portions of the sample, and the grain boundary was slightly C-rich. However, in the black area of the boundary, Ti and C were seriously enriched. The contents of C (36.5%) and Ti (2%) in the black area were much higher than those of C (13.8%) and Ti (0.2%) at the grain boundary. It can be inferred that the black regions were the first to melt during the heating process, and these regions were converted from the solid phase to the liquid phase. In the process of deformation, the liquid phase was squeezed from the middle to the boundary to form a black region.

Figure 2.27 shows the solute distribution during the solidification process in the black region of the sample, in which the solidification process is non-equilibrium solidification. During solidification, the solid–liquid interface pushed forward, and

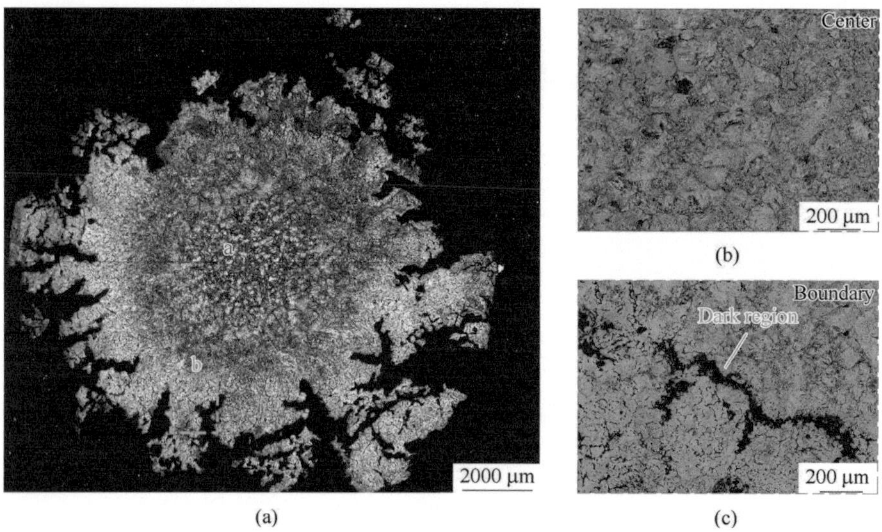

Fig. 2.25 Microstructure of the deformed specimen of Ti-Mo steel at 1430 °C and 0.1 s^{-1}. **a** Overall structure morphology, **b** microstructure of specimen center, **c** microstructure of specimen boundary

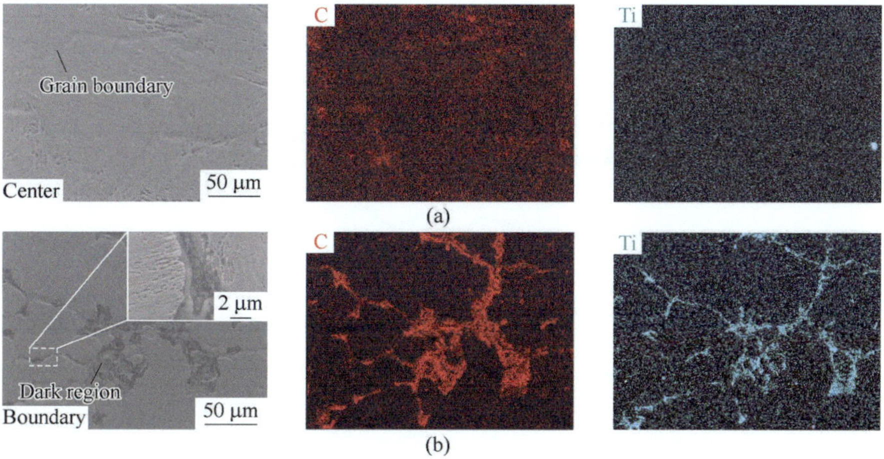

Fig. 2.26 SEM images and EDS mapping of the deformed specimen of Ti microalloyed steel at 1430 °C and 0.1 s^{-1}. **a** Center and **b** boundary

the solute element diffused into the liquid phase and the solid phase; however, more solute diffused into the liquid phase than into the solid phase. The sample was extruded and deformed by the external force, and the external force caused the liquid phase to flow. When the liquid phase flowed, the dendrite grew toward the direction of heat flow. The solute flowed in the same direction as the liquid, so the final solidified part had a high concentration of elements. At the same temperature, the solubility of the solute was much higher in the liquid phase than in the solid phase. However, in the non-equilibrium solidification process, the diffusion coefficient of solute elements in the solid phase was generally several orders of magnitude smaller than that in the liquid phase. Therefore, the elements in the liquid phase surrounded by the solid phase were still relatively more concentrated after solidification.

Figure 2.28 depicts the microstructure at different strain rates and at 1430 °C. When the strain rate was 0.1 s^{-1}, unlike at the boundary part, obvious lath martensite appeared in the middle of the specimen. This occurred because more liquid phase was squeezed from the middle part to the boundary part during the deformation process. There was a large amount of liquid phase in the outer boundary, and almost no liquid phase was present in the inner middle part. The solubility of the liquid relative to carbon was much greater than that of the solid phase. The liquid-phase outflow caused the imbalance in the carbon distribution in the middle part, resulting in the carbon-poor region in the middle of the specimen, which led to the formation of martensite.

There were many small grains in the middle of the specimen at strain rates of 1 and 5 s^{-1} and many recrystallized grains in the middle of the small grains. The applied stress was used not only to overcome the sliding and rotation of solid particles between liquid phases but also to deform solid particles and form finer recrystallized grains. As the strain rate increased, many fine grains at the boundary were separated

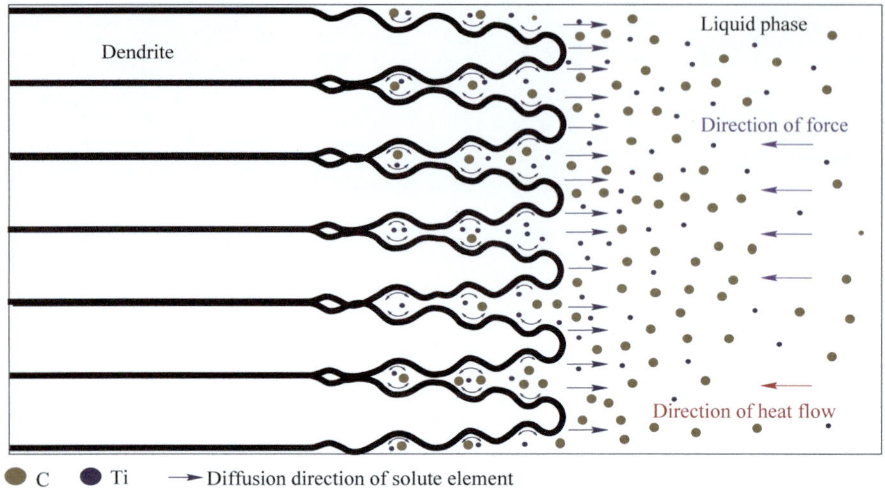

Fig. 2.27 Distribution of solutes in the black region during unidirectional solidification

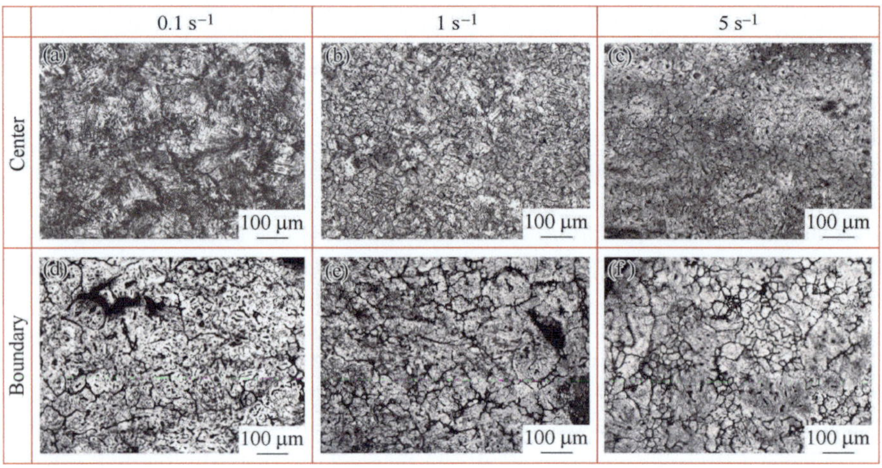

Fig. 2.28 Microstructure of the deformation specimen of Ti microalloyed steel at 1430 °C and different strain rates. **a** 0.1 s^{-1}, **b** 1 s^{-1}, **c** 5 s^{-1}, **d** 0.1 s^{-1}, **e** 1 s^{-1}, and **f** 5 s^{-1}

by the liquid phase. The higher the strain rate was, the faster the deformation and the shorter the time needed to reach the same deformation amount. Small grains could be aggregated into large grains, and the liquid phase did not participate in the deformation process. At this time, the main deformation mechanism was the plastic deformation of solid particles and the grain boundary slip of the solid phase, which led to a greater deformation resistance at a high strain rate, similar to the deformation mechanism in the thixotropic process of aluminum alloy.

At the same time, as the strain rate increased, the microstructure of the specimen showed more recrystallized grains. According to the stress–strain curve, the stress increased slowly with strain at 0.1 s^{-1}. In the whole process, the stress–strain curve showed the law of work hardening, and the softening effect caused by dynamic recrystallization and dynamic recovery was not obvious. within contrast to the stress–strain curves obtained under deformation conditions of 1430 °C and 0.1 s^{-1}, the stress obtained at 1430 °C and a strain rate of and 1 s^{-1} reached its peak and then decreased with an increase in strain, and the curve showed an obvious dynamic recrystallization rule.

The main reason for this phenomenon was that the strain rate was high, the time required to reach the same deformation amount was short, and the liquid phase had no time to participate in the deformation process. Compared with that obtained at a low strain rate, the plastic deformation of solid particles was larger, and the accumulated dislocation density and energy promoted the dynamic recrystallization of solid particles. At the same time, the liquid phase around the solid particles also had a return temperature, which was conducive to atomic diffusion.

2.2.3 Constitutive Equation of Continuous Casting in the Mushy Zone

2.2.3.1 Fields-Backofen Model with a Liquid-Phase Adjustment Factor

Fields and Bachofen proposed the Fields-Backofen (F-B) model to describe the effect of strain hardening and strain rate strengthening on flow stress. However, the softening effect of temperature was not considered in the F-B model. Zhang et al. [14] modified the F-B model by using an exponential function to represent the softening effect of temperature and strain on flow stress. The exponential form was used to represent the softening effect of temperature to modify the F-B model in this article.

To accurately understand the law of the rheological properties of metals during the deformation process in the mushy zone, this study modified the F-B model based on the exponential function of temperature and strain on the softening effect of rheological stress. The liquid phase had a significant effect on the rheological stress because the deformation of the sample occurred between the solid and liquid phases. Therefore, an empirical model describing the deformation of the mushy region was established by introducing a liquid-phase adjustment factor. The model form is as follows:

$$\sigma = \alpha_1 \exp(\alpha_2/T)\dot{\varepsilon}^m \varepsilon^n (1 - \beta f_{\mathrm{L}})^K \tag{2.33}$$

where σ is the true stress, MPa; ε is the strain; $\dot{\varepsilon}$ is the strain rate, s^{-1}; T is the absolute temperature, K; m, n, α_1 and α_2 are material constants; K is the liquid fraction factor; f_{L} is the liquid fraction; β is a geometric parameter; and $(1-\beta f_{\mathrm{L}})^K$ is the liquid-phase adjustment factor. The value of β can be adjusted according to the model prediction accuracy. The liquid-phase ratio is less than 50% that of the deformed specimen, and the value of β is generally set as 2.

Taking the logarithm of Eq. (2.33),

$$\ln\sigma = \ln\alpha_1 + \alpha_2/T + m\ln\dot{\varepsilon} + n\ln\varepsilon + L\ln(1 - \beta f_L) \tag{2.34}$$

The slopes of the functional relationships of $\ln\sigma - \ln\dot{\varepsilon}$, $\ln\sigma - \ln\varepsilon$, $\ln\sigma - \ln(1 - \beta f_L)$ and $\ln\sigma - 1/T$ are the parameters *m, n, L,* and α_2, which are calculated by the intercept of the $\ln\sigma - \ln\varepsilon$ functional relationship. The calculated values were all averaged. The parameters *m, L,* and α_2 were calculated with a strain of 0.04 as an example, and the calculation results are shown in Fig. 2.29. Table 2.13 shows the calculation results of model parameters.

Firstly, a liquid adjustment factor ($\beta = 2.0$) was selected, and the model parameters were obtained through the above calculation, and the strain, strain rate, and

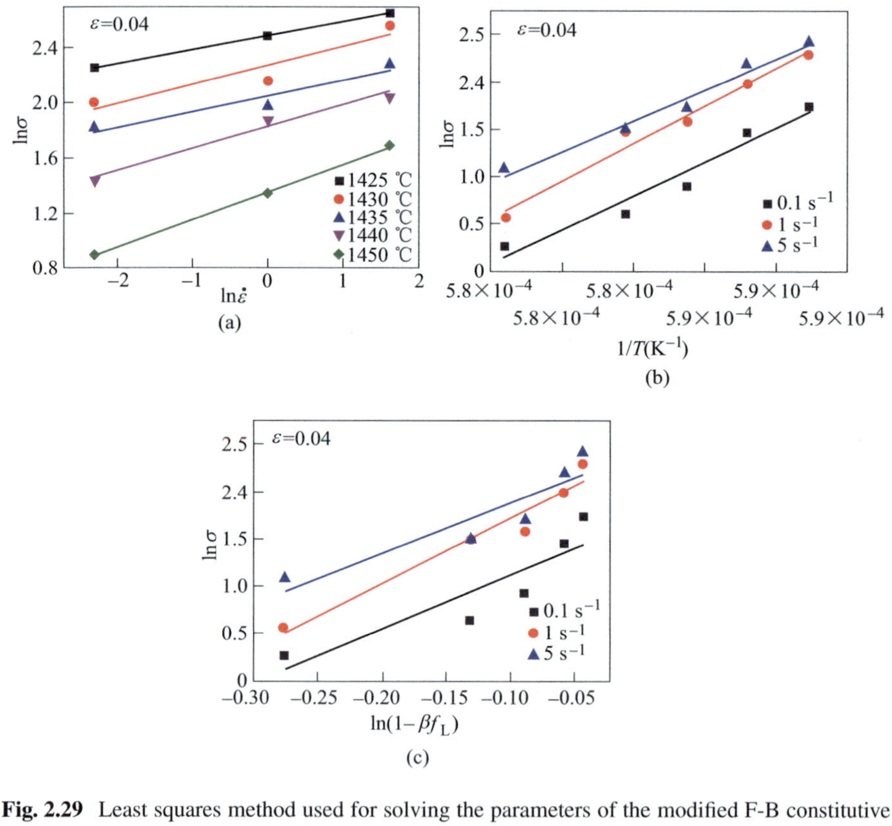

Fig. 2.29 Least squares method used for solving the parameters of the modified F-B constitutive model with the liquid adjustment factor. **a** Solution process of parameter *m*, **b** solution process of parameter α_2, **c** solution process of parameter *L*

Table 2.13 Parameters of F-B model were modified by adding a liquid adjustment factor

Parameter	425–1450 °C
a_1	7.4411×10^{-39}
a_2	155,351.7956
n	0.30816
m	0.19529
l	5.25857
β	0.8 and 2.0

temperature values were substituted to calculate the predicted values of the model, and the predicted values of the model were compared and analyzed with the experimental values. As shown in Fig. 2.30, the predicted value of the model is greater than the experimental value under the strain rate of 0.1 s^{-1} and temperature of 1425–1440 °C. The predicted value of the model is smaller than the experimental value at 1450 °C. The predicted values are much higher than the experimental values with strain rates 1 and 5 s^{-1}, the temperature of 1425 °C and 1430 °C, and the other temperature predicted values are lower than the experimental values. It can be seen that the prediction accuracy of the modified F-B model is low with a liquid adjustment factor ($\beta = 2.0$), and it is not suitable for the characterization of metal rheology in the temperature range of the two-phase region of the steel.

The prediction accuracy of the model can be improved by adjusting the β value of the liquid adjustment factor. The prediction accuracy of the model was studied when the β value of [15] was 0.9, 1.2, and 1.5, respectively, and the conclusion was drawn that the prediction accuracy of the model increased with the decrease of the β value. Therefore, the liquid-phase regulatory factor $\beta = 0.8$ was further selected for analysis, and the predicted results were shown in Fig. 2.31.

Compared with Figs. 2.30 and 2.31, it can be seen that when the temperature is low (less than 1435 °C), the prediction error decreases significantly after the β value drops from 2.0 to 0.8. However, when the temperature is higher (greater than or equal to 1435 °C), the prediction error increases and the model accuracy decreases after the β value drops to 0.8. The β-value is reduced from 2.0 to 0.8, the model error is reduced from 39.87 to 35.45%, but it still cannot meet the needs of use. Therefore, the modified F-B model is not suitable for the characterization of metal rheology in the two-phase region of 0.37% microalloyed steel when the liquid-phase regulating factor $\beta = 0.8$.

2.2.3.2 Modified Arrhenius Model with a Liquid Adjustment Factor

The Arrhenius model is a widely used empirical constitutive model in the solid-phase region. Researchers [16, 17] have studied constitutive behaviors in the high-temperature mushy region of magnesium alloy by adding a liquid adjustment factor to the Arrhenius model. Considering an Arrhenius model with a liquid-phase adjustment

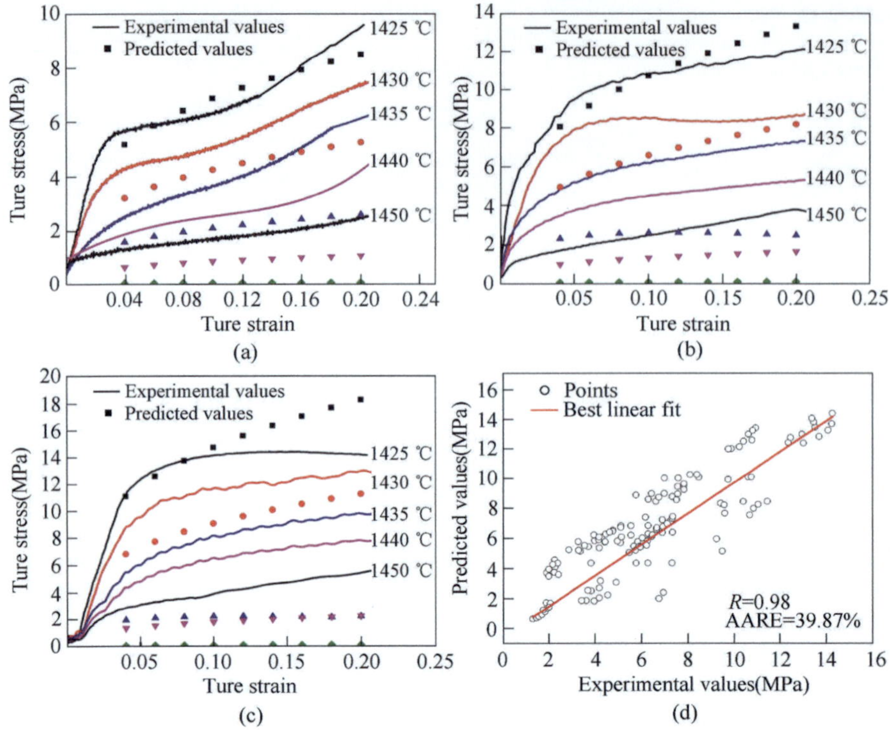

Fig. 2.30 Comparison between experimental and predicted flow stress data by the modified F-B model with a liquid adjustment factor ($\beta = 2$) at **a** strain rates of 0.1 s^{-1}, **b** 1 s^{-1}, and **c** 5 s^{-1}, **d** error analysis

factor, the constitutive behavior in the high-temperature mushy zone of magnesium alloy can be accurately predicted.

The Arrhenius model was coupled with a liquid adjustment factor in all stress states:

$$(1 - \beta f_L)^K \dot{\varepsilon} = A \cdot [\sin h(\alpha\sigma)]^n \cdot \exp\left(-\frac{Q}{RT}\right) \quad (2.35)$$

where σ is the true stress, MPa; $\dot{\varepsilon}$ is the strain rate, s^{-1}; R is the ideal gas constant, $R = 8.314$ J/(mol K); T is the absolute temperature, K; Q is the activation energy of hot deformation, J/mol; n is the material's stress index; and A, α, β, and n_1 are material constants, where $\alpha = \beta/n_1$. When the liquid-phase ratio is lower than 0.5, the value of β is 2 for magnesium alloy [17] and 1.5 for aluminum alloy [18], which can be adjusted according to the prediction accuracy of the model. f_L is the liquid-phase ratio, K is the material constant related to the liquid phase, and $(1 - \beta f_L)^K$ is the liquid adjustment factor. In this paper, β was 0.8.

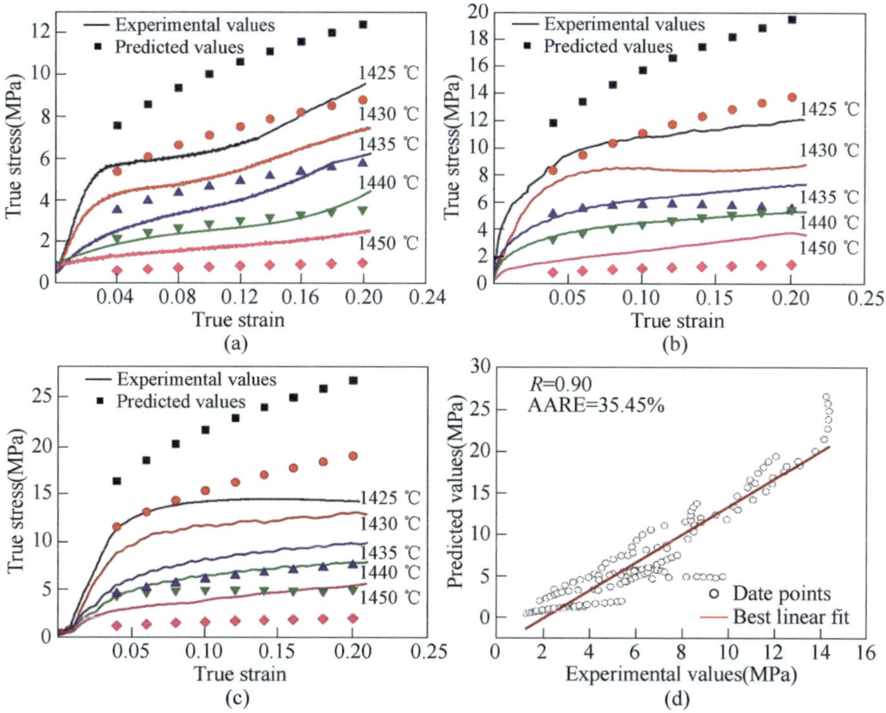

Fig. 2.31 Comparison between experimental and predicted flow stress data by the modified F-B model with a liquid adjustment factor ($\beta = 0.8$) at **a** strain rates of 0.1 s^{-1}, **b** 1 s^{-1}, and **c** 5 s^{-1}, **d** error analysis

The procedure for obtaining parameters α and n is similar to Sect. 2.1.2.5 and will not be repeated here. The calculated predicted values of stress, strain, and strain rate are shown in Fig. 2.32. It can be seen that when the temperature is equal to 1425 and 1430 °C, the predicted values at high strain rate (1, 5 s^{-1}) are lower than the measured values. This indicates that the model does not fully consider the work hardening effect of high strain rates. In the high-temperature range (no less than 1435 °C), the prediction accuracy of the model is higher. The average relative error of the model is 10.81%.

The original Arrhenius model does not show a sufficient strain rate strengthening effect. To improve the prediction accuracy of the model, the original Arrhenius model was revised. As the strain rate increased, the deviation between the predicted value and the experimental value increased. The strengthening effect caused by the increase in the strain rate as not well displayed in the low-power number strain rate strengthening term. Because plastic deformation mainly occurred in the solid phase with increasing strain rate, the rapid increase and accumulation of dislocations in the solid phase caused the flow stress to increase. Nevertheless, as the low-power number strain rate strengthening cannot meet the ideal requirements due to the existence of

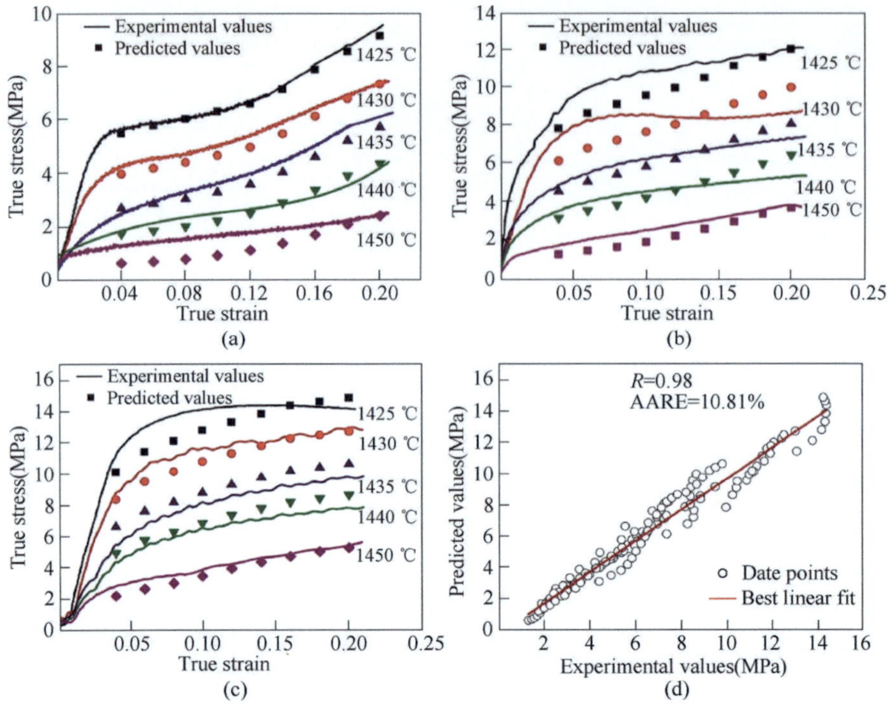

Fig. 2.32 Comparison between experimental flow stress data and flow stress data predicted by using an Arrhenius model with a liquid factor for Ti microalloyed steel at strain rates of **a** 0.1 s^{-1}, **b** 1 s^{-1}, **c** and 5 s^{-1}, and **d** error analysis

the liquid phase, the power of the strain rate strengthening term was increased. Based on previous research [19], by multiplying $\dot{\varepsilon}^{1/3}$ in the original model.

$$(1-\beta f_{\mathrm{L}})^{K}\dot{\varepsilon}^{4/3}=A\cdot[\sin h(\alpha\sigma)]\cdot\exp\left(-\frac{Q}{RT}\right) \tag{2.36}$$

The strain rate strengthening effect failed to achieve the expected goal, and different strain rate strengthening effects were thus adopted for different strain rates. The equation used for the Arrhenius model with the addition of a liquid adjustment factor is as follows (Fig. 2.33):

$$\begin{aligned}(1-\beta f_{\mathrm{L}})^{\kappa}\dot{\varepsilon}^{4/3}&=A[\sinh(\alpha\sigma)]^{n}\exp\left(-\tfrac{Q}{RT}\right) \quad \dot{\varepsilon}>1\,\mathrm{s}^{-1}\\(1-\beta f_{\mathrm{L}})^{\kappa}\dot{\varepsilon}&=A[\sinh(\alpha\sigma)]^{n}\exp\left(-\tfrac{Q}{RT}\right) \quad \dot{\varepsilon}\leq 1\,\mathrm{s}^{-1}\end{aligned} \tag{2.37}$$

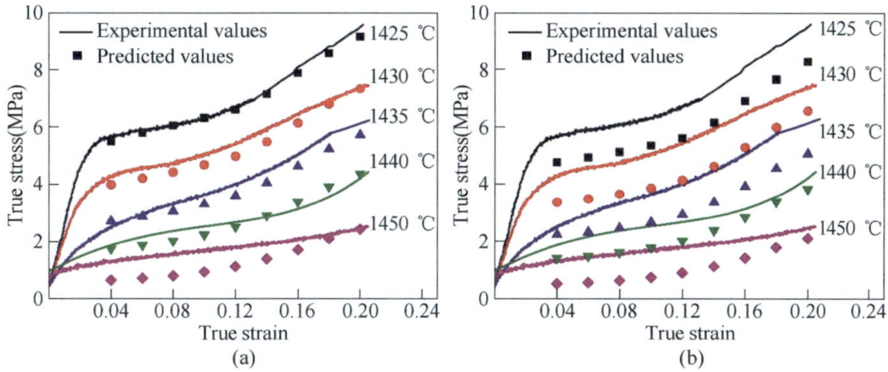

Fig. 2.33 Comparison between the experimental flow stress data and flow stress data predicted by the Arrhenius model with a liquid adjustment factor for high-titanium steel at a strain rate of 0.1 s^{-1}: **a** 4/3-times function and **b** 1-time function

As shown in Fig. 2.34, the deviation between the predicted value and the experimental value decreases at the strain rate of 5 s^{-1} with the temperature of 1425–1430 °C and 1450 °C. However, at the strain rate of 1 s^{-1} and the temperature of 1430 °C, the variation trend of the scatter plot of the predicted value of the model does not show the softening behavior presented by the test curve. It is shown that the Arrhenius model cannot reflect the dynamic recrystallization softening behavior even if the power index of ε is adjusted. Comprehensive analysis shows that the average relative difference decreases from 10.81 to 8.34% after adjusting ε power index. The prediction accuracy of the model meets the demand of accurately predicting metal rheological behavior in the two-phase region of titanium microalloyed steel.

2.2.3.3 Compression Deformation Mechanism of Ti–Mo Steel in the Mushy Zone

Figure 2.35 illustrates the rheological properties of metals in the mushy region. An external load was applied to deform the material when it contained a certain liquid phase. The external load force caused the solid particles to rotate and slide in the liquid phase, and the liquid phase acted as a lubricant at this time. Second, the external load force acted on the solid particles, causing the solid particles to slip and deform. The deformation energy led to the recrystallization of the solid particles, and finally, the external load crushed the dendrites. A small amount of liquid existed between the solid particles at a low temperature, which formed thin and discontinuous liquid films.

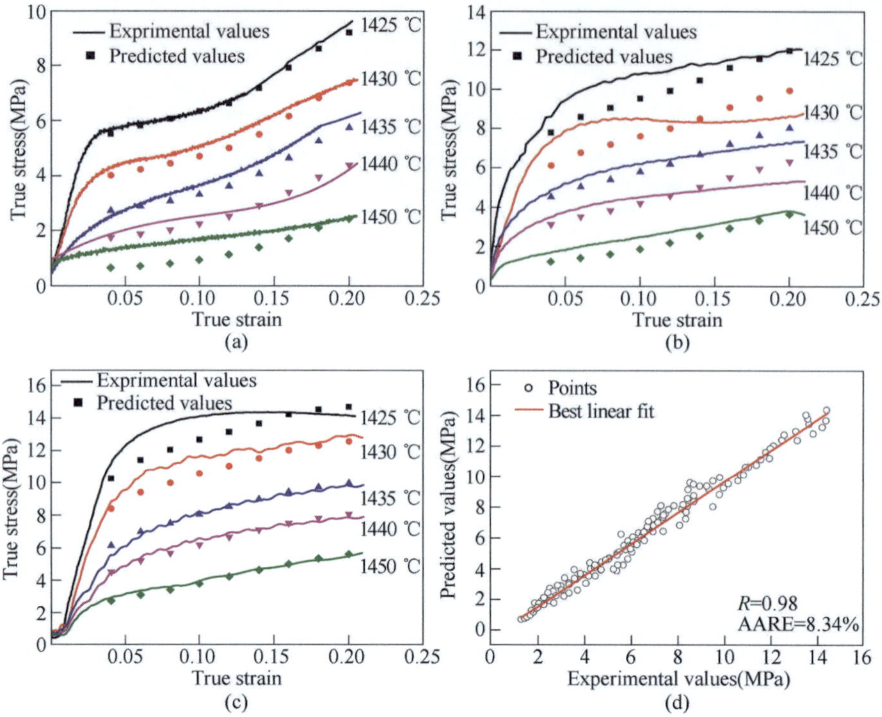

Fig. 2.34 Comparison between experimental flow stress data and flow stress data predicted by using an Arrhenius model with a liquid factor for Ti microalloyed steel at strain rates of **a** 0.1 s^{-1}, **b** 1 s^{-1}, **c** and 5 s^{-1}, and **d** error analysis

As shown in Fig. 2.35a, the thin and discontinuous liquid film was extruded, and solid particles gathered due to external forces during the process of deformation. The liquid film had a limited lubrication effect on the deformation of the solid particles, and the metal strength was improved. Figure 2.35a indicates that the liquid-phase content was higher at high temperatures and was more evenly distributed around solid particles than at low temperature. Although the liquid layer between the solid particles was thinner during the deformation process, the liquid phase still had a good lubrication effect on the movement between the solid particles, so the metal strength was low. The stress–strain curve shows that the stress decreased with an increase in temperature.

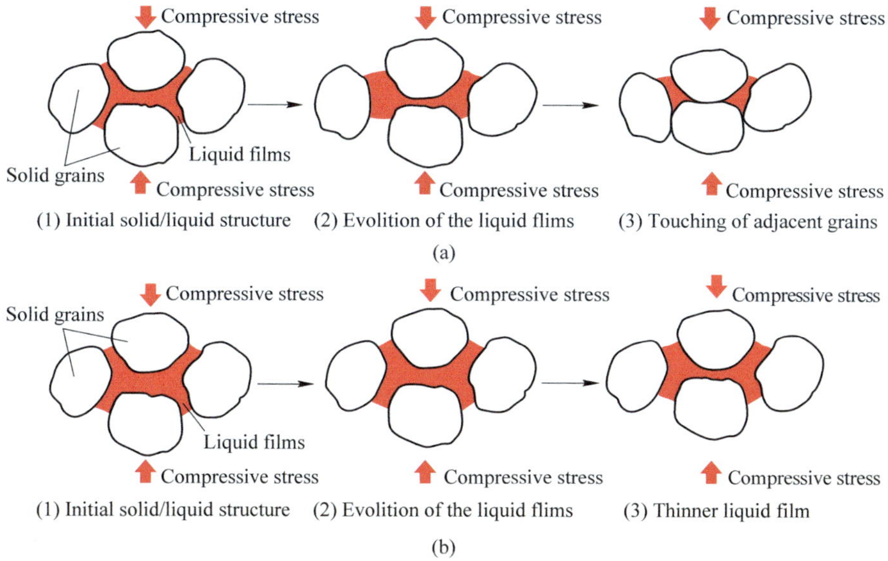

Fig. 2.35 Transformation of the solid/liquid structure in the dense deformation zone during hot compression. **a** Changes of more liquid and liquid films during deformation. **b** Changes in more liquid and liquid films during deformation

References

1. Chen JX (1991) Manual for continuous casting steel. Metallurgical Industry Press, Beijing
2. Bhadeshia HKDH (2012) Steels for bearings. Prog Mater Sci 57(2):268–435
3. Li HY, Xiao XF, Duan JY et al (2013) A modified Johnson-Cook model for elevated temperature flow behavior of T24 steel. J Mater Sci Eng A 577:138–146
4. Zerilli FJ, Armstrong RW (1987) Dislocation-mechanics-based constitutive relations for material dynamics calculations. Jpn J Appl Phys 61(5):1816–1825
5. Xiao X, Liu GQ, Hu BF et al (2012) A comparative study on Arrhenius-type constitutive equations and artificial neural network model to predict high-temperature deformation behavior in 12Cr3WV steel. Comput Mater Sci 62(11):227–234
6. Sellars CM, Tegart WJM (1972) Hot workability. Int. Metall. Rev. 17(1):1–24
7. Zhang C, Zhang L, Shen W et al (2016) Study on constitutive modeling and processing maps for hot deformation of medium carbon Cr–Ni–Mo alloyed steel. Eng Mater Des 90:804–814
8. Zuo Q, Liu F, Wang L et al (2015) Prediction of hot deformation behavior in Ni-based alloy considering the effect of initial microstructure. Prog Nat Sci: Mater Int 25:66–77
9. Rakesh SV, Sivaprasad PV et al (2009) Constitutive equations to predict high temperature flow stress in a Ti-modified austenitic stainless steel. Mater Sci Eng A 500(1):114–121
10. Mirzadeh H (2015) A simplified approach for developing constitutive equations for modeling and prediction of hot deformation flow stress. Metall Mater Trans A 46(9):1–11
11. Cai J, Li F, Liu T et al (2011) Constitutive equations for elevated temperature flow stress of Ti–6Al–4V alloy considering the effect of strain. Eng Mater Des 32(3):1144–1151
12. Mirzadeh H, Cabrera JM, Najafizadeh A (2011) Constitutive relationships for hot deformation of austenite. Acta Mater 59(16):6441–6448
13. Frost HJ, Ashby MF (1982) Deformation-mechanism maps: the plasticity and creep of metals and ceramics. Pergamon Press 9(2):224–225

14. Zhang XH (2003) Experimental and numerical study of magnesium alloy during hot working process. Ph.D. thesis. Shanghai Jiaotong University
15. Wang JJ, Phillion AB, Lu GM (2014) Development of a visco-plastic constitutive modeling for thixoforming of AA6061 in semi-solid state. J Alloys Compd 609:290–295
16. Chen G, Lin YF, Yao SJ et al (2016) Constitutive behavior of aluminum alloy in a wide temperature range from warm to semi-solid regions. J Alloys Compd 674:203–205
17. Tang Q, Zhou MY, Fan LL et al (2018) Constitutive behavior of AZ80 M magnesium alloy compressed at elevated temperature and containing a small fraction of liquid. J Vac 155:476–489
18. Xu Y, Chen C, Jia JB et al (2018) Constitutive behavior of a SIMA processed magnesium alloy by employing repetitive upsetting-extrusion (RUE). J Alloys Compd 748:694–705
19. Lin YC, Chen MS, Zhong J (2008) Constitutive modeling for elevated temperature flow behavior of 42CrMo steel. Comput Mater Sci 42(3):470–477

Chapter 3
Deformation Behavior and Crack Risk During Heavy Reduction Process

The accurate description of the deformation behavior for casting strand during heavy reduction process, such as deformation transfers efficiency at center of the strand or stress and strain distribution at different positions of section, which has a significant effect on both law of solute squeezing out of mushy zone at the solidification end of casting and prediction of improvement law of center porosity and shrinkage. At present, the thermal–mechanical coupled FE simulation method has been applied to study law of the deformation behavior during the reduction process by most scholars. Moreover, as an important qualification for reduction process technological design, crack risk during reduction process is also an important problem which is concerned by scholars. The initiation of cracks is due to the comprehensive factors from inside and outside, the main inside factor is poor ductility, Suzuki et al. [1] proposed the theory that there are three brittle zones during steel solidification. The outside factor is stress concentration, such as mechanical stress of bending and straightening process or thermal stress during cooling process. Crack initiation occurs when the tensile deformation of steel exceeds certain limits in its material properties during reduction process. According to the characteristics of reduction deformation and solidification, the cracks during reduction process can be divided into two parts, one is the initiation of cracks in the brittle zone I at the front of the solidification interface, another is the surface crack propagation, especially the transverse crack propagation at the corner in the brittle zone III.

In this chapter, the establishment method of 3-D thermal–mechanical coupled FE model during reduction process was introduced in detail, stress and strain distribution of blooms and wide-thick slabs during reduction process at the solidification end of casting has been analyzed systematically, determination of critical strain of initiation and propagation of intermediate cracks and surface cracks has been proposed. Take bearing steel GCr15 bloom and micro-alloyed steel Q345E wide-thick slab for example, risk and control criterion of crack initiation and propagation under different reduction process were analyzed through a crack risk prediction model which was developed based on the analysis of the thermos-mechanical behavior.

C. Ji et al., *Solidification End Reduction of Continuous Casting Process*, Key Steel Manufacturing Technologies by Collaborative Innovation,
https://doi.org/10.1007/978-981-96-8544-8_3

3.1 Thermo-Mechanical Behavior of Entire Continuous Casting Process

3.1.1 Thermal/Mechanical Model and Reduction Deformation Rule of Continuous Casting Bloom

3.1.1.1 Modeling Parameters

The 3D thermal/mechanical coupling simulation model of entire continuous casting process was established based on the bloom caster of 320 mm × 425 mm section dimensions, the total length of each strand is approximately 26.5 m, the cooling zones includes the initial cooling zone of the mold, 4 secondary cooling zones and 1 air-cooling zone. The detailed cooling zone parameters of the casting machine are shown in Table 3.1. A total of 9 tension levelers are arranged within the range of 17.7–26.5 m from the meniscus in the air-cooling zone of the casting machine. Each tension leveler is equipped with a pair of 500 mm diameter casting rolls. The opening degree of the roll gap and rotation speed of the roller can be flexibly adjusted in a certain range by the corresponding hydraulic cylinder and the driving motor to complete the solidification end reduction. The distance between tension levelers is 1.1 m. The specific casting flow positions of each tension leveler are shown in Table 3.2. Bearing steel GCr15 was selected as the steel for the specific study. The casting speed was 0.58 m/min, the specific water of secondary cooling was 0.22 L/kg, the casting temperature was 1497 °C, and the reduction amount of the No.1–No.9 tension levelers was 0\1\1\1\2\3\4\4\2 mm, respectively.

Table 3.1 Cooling zone parameters of continuous casting machine

Cooling zone		Distance from meniscus (m)		Length (m)
		Initial position	End position	
Primary cooling	Mold height	0.00	0.78	0.78
Secondary cooling	Zone 1	0.78	1.13	0.35
	Zone 2	1.13	2.63	1.50
	Zone 3	2.63	4.73	2.10
	Zone 4	4.73	7.93	3.2
Air cooling	Air-cooling zone	7.93	26.5	18.57

Table 3.2 Distance of each tension leveler from meniscus

No.	1	2	3	4	5	6	7	8	9
Distance (m)	17.7	18.8	19.1	21.0	22.1	23.2	24.3	25.4	26.5

3.1.1.2 3D Thermal–Mechanical Coupling Simulation Model

Finite Element Method (FEM) is the most commonly used numerical simulation method, and its application fields have covered the analysis and coupling analysis of various physical fields such as temperature field, electromagnetic field, flow field and stress field. With the development of commercial finite element software such as ANSYS, ABAQUS and MSC.Marc, the workload and computational efficiency of researchers have been greatly reduced. In this section, the commercial finite element software MSC.Marc was adopted to establish a thermal/mechanical model for continuous casting strand, and the interaction between solidification heat transfer and deformation during continuous casting was systematically studied, as well as the cumulative effect of deformation on the stress and strain variation of the strand at different positions.

A. **Geometric model**

The roll row geometry model was established according to the casting machine drawings and imported into the casting strand mesh, as shown in Fig. 3.1. Due to the symmetry of the casting strand along its width direction, the 3D thermal–mechanical coupling simulation model of the entire process corresponded to half of the actual casting strand. The characterized casting strand is width 212.5 mm, thickness 320 mm and length 3000 mm, divided by tetrahedral mesh. The average mesh size is 15 mm, and the total number of elements is 249,284.

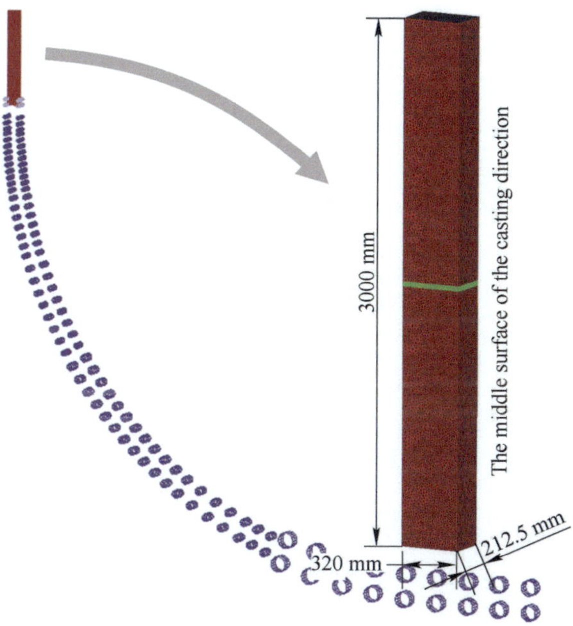

Fig. 3.1 3D thermal–mechanical coupling simulation model of the entire process

In the actual continuous casting process, the strand is continuously pulled, but in the simulation process, due to the limited length of the model of the strand and the bite restriction of the strand head, the position of the head and the end of the strand cannot truly reflect the deformation of the strand when it is in contact with the casting roller. Therefore, the intermediate surface is set in the middle of the drawing direction of the casting strand model (the middle slice position of the finite element model of the casting billet in Fig. 3.1). The intermediate surface is far enough away from the head and tail positions of the casting strand, and the biting effect has little influence on the position, which can better reflect the real deformation law of the casting strand.

B. **Governing equations**

The differential equation of heat transfer and solidification is usually used in the thermal–mechanical coupled model. As is shown in Fig. 3.2, the microelement is taken at the meniscus of the mold along the center of the strand. Its width, thickness, and height are dx, dy, and dz, respectively, and it moves together with the strand along the casting direction.

Heat input and output of microelement is shown as follow:

(1) The heat that molten steel brings into the microelement from the top (surface dx, dz):

$$\rho vcT\mathrm{d}x\mathrm{d}z \tag{3.1}$$

(2) Heat transfer from center of wide surface of the strand to the microelement (surface dy, dz):

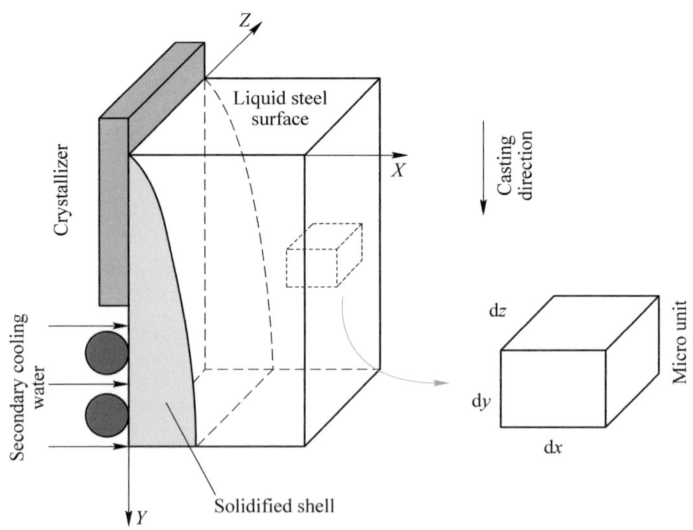

Fig. 3.2 Schematic diagram of solidification for continuous casting strand

$$-k_{\text{eff}}\frac{\partial T}{\partial x}\text{d}y\text{d}z \tag{3.2}$$

(3) Heat transfer from center of narrow surface of the strand to the microelement (surface dx, dy):

$$-k_{\text{eff}}\frac{\partial T}{\partial z}\text{d}x\text{d}y \tag{3.3}$$

(4) The heat stored inside the cell:

$$-\rho c\frac{\partial T}{\partial t}\text{d}x\text{d}y\text{d}z \tag{3.4}$$

(5) Heat transferred downward by the microelement (surface dx, dz):

$$-\rho vc\left(T+\frac{\partial T}{\partial t}\text{d}y\right)\text{d}x\text{d}y \tag{3.5}$$

(6) Heat transfer from the wide surface of the microelement (surface dy, dz):

$$-\left[k_{\text{eff}}\frac{\partial T}{\partial x}+\frac{\partial}{\partial x}\left(k_{\text{eff}}\frac{\partial T}{\partial x}\right)\text{d}x\right]\text{d}y\text{d}z \tag{3.6}$$

(7) Heat transfer from the narrow surface of the microelement (surface dy, dz):

$$-\left[k_{\text{eff}}\frac{\partial T}{\partial z}+\frac{\partial}{\partial z}\left(k_{\text{eff}}\frac{\partial T}{\partial z}\right)\text{d}z\right]\text{d}x\text{d}y \tag{3.7}$$

(8) The latent heat of solidification, S_0: put all terms blow into the heat balance equation, a 3-D transient heat conduction equation was employed to describe the heat transfer behavior as follows:

$$\rho c\frac{\partial T}{\partial t}=-\rho vc\frac{\partial T}{\partial y}+\frac{\partial}{\partial z}\left(k_{\text{eff}}\frac{\partial T}{\partial z}\right)+\frac{\partial}{\partial x}\left(k_{\text{eff}}\frac{\partial T}{\partial x}\right)+S_0 \tag{3.8}$$

where T and t are the temperature in °C and calculation time in s, respectively; ρ, c, and k_{eff} are the density in kg/m^3, specific heat in J/(kg °C), and heat conductivity in W/(m °C), respectively; x, y and z are the width, thickness, and length of the strand, respectively.

C. **Boundary conditions**

Molten steel enters the mold through the nozzle from the tundish, the casting temperature and tundish temperature can be equal to the initial temperature of molten steel in mold when ignoring calories burned during this period, therefore, temperature of molten steel before entering the mold satisfies the following equation:

$$T_0 = T_{\text{tundish}} = T_\text{c} \tag{3.9}$$

where T_0, T_{tundish}, and T_c are the initial temperature in °C, tundish temperature in °C and pouring temperature in °C, respectively.

The symmetry plane in mold is regarded as an adiabatic condition according to the symmetry of inner face:

$$\left.\frac{\mathrm{d}U}{\mathrm{d}Z}\right|_{\text{symm}} = 0 \tag{3.10}$$

where U is heat and Z is direction of inner face.

Mold is also called primary cooling zone, the contact position is cooled to form the shell due to the rapid heat exchange in contact area between steel and mold, the cooling process in this zone adopts the second boundary condition. In the proposed model, the heat transfer boundary conditions in mold are applied by the heat flux density equation proposed by Savage et al. [2], as shown in Eq. (3.11):

$$q = 2.688 - B\sqrt{t} \tag{3.11}$$

where q is the heat flux on the surface of the strand with a unit of MW/m^2; B is the coefficient related to the cooling condition of mold; t is cooling time in mold in s.

The surface of the strand is cooled by spraying cooling water or mist after entering the secondary cooling zone from the mold, based on the characteristics of each cooling mode, the corresponding cooling heat transfer coefficient of solidification heat transfer model is calculated as the third boundary condition.

Heat transfer between the surface of strand and cooling water or mist is carried out by means of convection and evaporation of cooling water, the heat transfer coefficient was calculated by the equation proposed by Nozaki et al. [3] shown as follow:

$$h^i_{\text{spray}} = \alpha_i W_i^{0.55}(1 - 0.075T_\text{w}) \tag{3.12}$$

where i is the ith secondary cooling zone; α_i is the correction coefficient of the ith secondary cooling zone; h^i_{spray}, W_i and T_w is the heat transfer coefficient between strand surface and cooling water in W/(m^2 °C), flow density of cooling zone in L/(m^2 min), and temperature of cooling water or mist in °C, respectively.

The heat transfer mode is mainly the mean of radiation between the strand and air when the strand enters the air-cooling zone from the secondary cooling zone, the heat transfer coefficient is shown as Eq. 3.13:

$$h_{\text{rad}} = \varepsilon\sigma\left(T^2_{\text{surf}} + T^2_{\text{env}}\right)(T_{\text{surf}} + T_{\text{env}}) \tag{3.13}$$

where h_{rad} is the i_{th} radiation heat transfer coefficient in W/(m^2 K); ε is radiation coefficient of 0.8; σ is Boltzmann's constant of 5.67×10^{-8} W/(m^2 K); T_{surf} and T_{env} is strand surface temperature in K and ambient temperature in K, respectively.

D. Thermal material properties

Physical parameters mainly include thermal material properties and mechanical parameters. Among them, mechanical parameters mainly include the elastic modulus, Poisson's ratio, and the constitutive model that describes the relationship between stress and strain, all parameters are directly related to temperature. Therefore, accurate calculation of temperature field is the most important prerequisite to ensure accurate prediction of deformation characteristics strand.

During the continuous casting process, the molten steel gradually becomes solid from the mold to the outlet of the casting machine through the mold cooling, secondary cold water cooling, and air cooling. The cooling conditions of the whole solidification process are complex and varied, and the temperature span is large. Therefore, the boundary conditions of solidification heat transfer and thermal physical parameters directly determine the accuracy of the model. In fact, since Mizikar first realized the numerical solution of two-dimensional solidification heat transfer model of casting billet by using finite difference method in the late 1960s [4], the solidification heat transfer model algorithm has become very mature in the following decades. The results of finite difference, finite volume and finite element (including commercial finite element software) are almost identical under the same thermal physical parameters and boundary conditions. Therefore, the research focus of numerical simulation of heat transfer in continuous casting solidification mainly focuses on boundary condition treatment, accurate determination of physical property parameters (including the influence of liquid convection on heat transfer and solidification latent heat treatment, etc.), and how to achieve high-precision and real-time calculation of multi-dimensional temperature field in production process.

In order to obtain the physical property parameters of bearing steel GCr15 in the range of solid-phase and liquidus temperatures, a one-dimensional finite difference model was established which can accurately predict the δ/γ phase transition process and solute redistribution. Further, the thermal conductivity, enthalpy and density in the solid–liquid two phases can be obtained according to the calculation method of key physical property parameters of phase separation [5].

Based on the work of Ueshima et al. [6], it is assumed that the cross-section of solidified dendrites is a regular hexagon, and 1/6 of its cross-section is used as the calculation domain, as shown in Fig. 3.3a. The length of the dendrite stem radius is $\lambda/2$, and this region is divided into 100 nodes, as shown in Fig. 3.3b, c.

According to Fick's second law, the governing equation for solute diffusion is expressed as follows:

$$\frac{\partial C_i^{\mathrm{s}}}{\partial t} = \frac{\partial}{\partial x}\left(D_i^{\mathrm{s}}(T)\frac{\partial C_i^{\mathrm{s}}}{\partial x_i}\right) \tag{3.14}$$

where, $D_i^{\mathrm{s}}(T)$ is the diffusion coefficient of solute element i in the solid phase, m^2/s; C_i^{s} is the solute concentration of solute element i in the solid phase, wt%.

In the process of dendrite growth, the melt flow between dendrites is ignored and the total mass of solute element i is assumed to be constant. Where, the primary

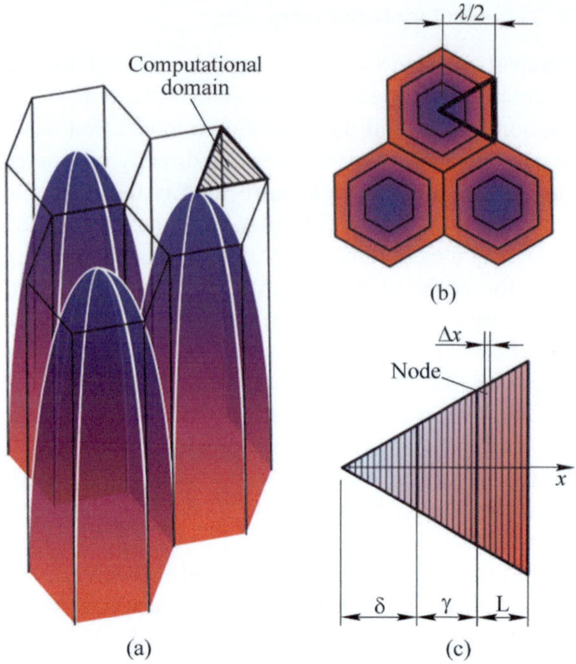

Fig. 3.3 Dendrite morphology and calculation domain of regular hexagonal section. **a** Overview of dendrite arrays; **b** dendrite cross-section; **c** modeling domain structure

dendrite arm distance λ is calculated according to Eq. (3.15) [7]:

$$\lambda = 278.748 \cdot C_{\mathrm{R}}^{-0.206278} \cdot w_{\mathrm{C}}^{n} \tag{3.15}$$

where, C_{R} represents the equivalent cooling rate; w_{C} stands for the nominal carbon content of steel (wt%); n is the Calculation coefficient, which can be obtained from Eq. (3.16).

$$n = \begin{cases} -0.316225 + 2.0325C_{\mathrm{R}} & 0 \le C_{\mathrm{R}} \le 0.15 \\ -0.0819 - 0.491666C_{\mathrm{R}} & 0.15 < C_{\mathrm{R}} \le 1.0 \end{cases} \tag{3.16}$$

In the equation, both liquidus temperature T_{l} and δ/γ phase transition temperature point T_{Ar4} are determined by the concentration of solute (mass) in the interface node, and the specific calculation equation is as follows [6, 8]:

$$T_{\mathrm{l}} = 1536 - \sum_{i}^{N} m_i w(C)$$

$$T_{\mathrm{Ar4}} = 1392 - \sum_{i}^{N} n_i w(C) \tag{3.17}$$

where m_i is the liquidus slope; n_i is the δ/γ interface in Fe-*i* binary phase diagram; N is the number of solute elements calculated in the model; C is the mass concentration of solute, %.

According to the preliminary work of our team [9], the average cooling rate of the billet in the continuous casting process is about 1.78 °C/s, but the cooling rate in the two-phase zone at the solidification end is lower than this value. In order to accurately reflect the actual solidification end, the cooling rate of 0.25 °C/s was selected for calculation. The evolution law of phase fraction and interdendritic solute segregation ratio in the solidification process of bearing steel GCr15 is shown in Fig. 3.4a. It can be seen that during the solidification process of bearing steel GCr15, the molten steel directly forms the γ phase and does not form the δ phase. Compared with the initial stage of solidification, the element segregation rate increases rapidly at the end of solidification. Figure 3.4b–d show the density, enthalpy and thermal conductivity of bearing steel GCr15 calculated by weighted phase fraction equation, respectively [5].

As for mechanical parameters, the constitutive model of bearing steel GCr15 has been discussed in detail in Chap. 2. In addition, elastic modulus and Poisson's ratio need to be determined. The equation for calculating the elastic modulus E of the model [10, 11] is as follows:

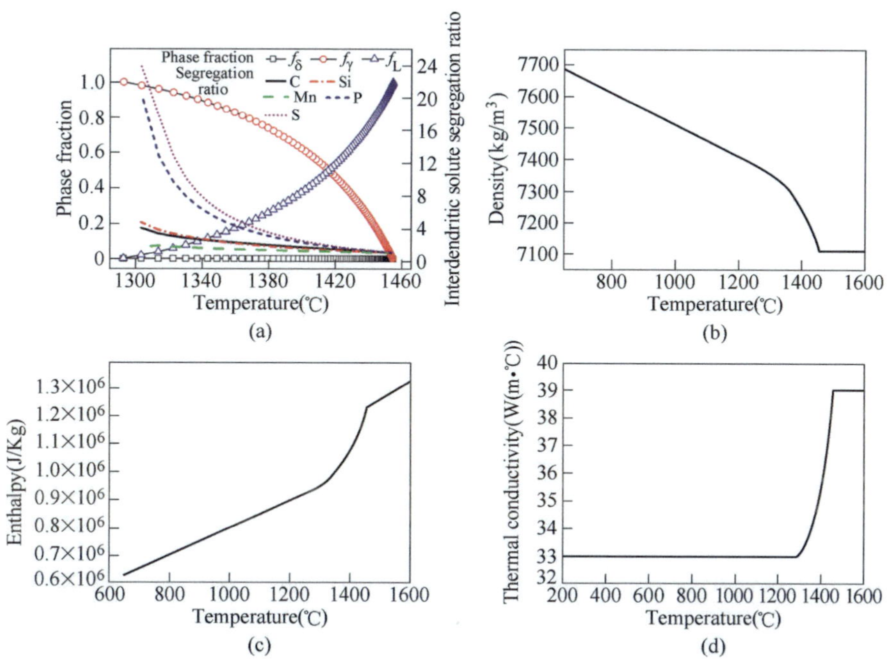

Fig. 3.4 Phase fraction, microsolute segregation and material thermal property parameters of bearing steel GCr15. **a** Phase fraction and microsolute segregation; **b** density; **c** enthalpy; and **d** thermal conductivity

When $T \leq 900$ °C,

$$E = 347.6525 - 0.350305T \quad (3.18)$$

When $900°C < T \leq T_s$,

$$E = 968 - 2.33T + 1.9 \times 10^{-3}T^2 - 5.18 \times 10^{-7}T^3 \quad (3.19)$$

There is Zero Strength Temperature (ZST) in the temperature interval of the two-phase zone, and the strength of the billet is zero above ZST, and if the elastic modulus in the model is set to zero, the calculation will be terminated. Therefore, it is necessary to modify the elastic modulus above ZST, which is generally set to a minimum value greater than zero, and 0.0013 MPa is taken in this model.

Poisson's ratio v in the model is calculated using the equation determined by Uehara et al. [12] based on experimental data:

$$v = 0.278 + 8.23 \times 10^{-5}T \quad (3.20)$$

Poisson's ratio is also involved in the treatment problem in the two-phase zone, and generally, this equation is used to calculate at ZST temperature. Poisson's ratio can be located 0.5 at the liquid phase solidification temperature, and it is linearly interpolated between ZST and liquid phase solidification temperature.

3.1.1.3 Model Validation

In order to ensure the accuracy of the analysis, the calculation results need to be verified. At present, surface temperature measurement (infrared temperature measurement or subcutaneous embedded thermocouple) and strand shell thickness measurement (nail shooting method or tracer method) are mainly used. For the thermal–mechanical coupled FE model, the accuracy of deformation behavior can also be verified by the measured results of pressure reduction.

Figure 3.5 shows the measured values of strand surface temperature (center position of side of generous strand) and shell thickness obtained by infrared temperature measurement and nail shooting under this working condition (320 mm × 425 mm section bearing steel GCr15, casting speed 0.58 m/min, and superheat of 30°C), and the comparison results with the calculated values of the model. It can be seen that in terms of solidification heat transfer, the surface temperature and shell thickness calculated by the model are in good agreement with the measured values at corresponding positions, and the relative errors are less than 2.9% and 3.1%, respectively.

Further, the actual output pressure of the hydraulic cylinder of each tension leveler in the production process is compared with the cast roller reaction force calculated by the model to verify the accuracy of deformation behavior prediction. Figure 3.6 shows the comparison between the calculated value of compact force and the measured value under the condition that the reduction of No.5 roll and No.7 roll is 2 mm and 4 mm

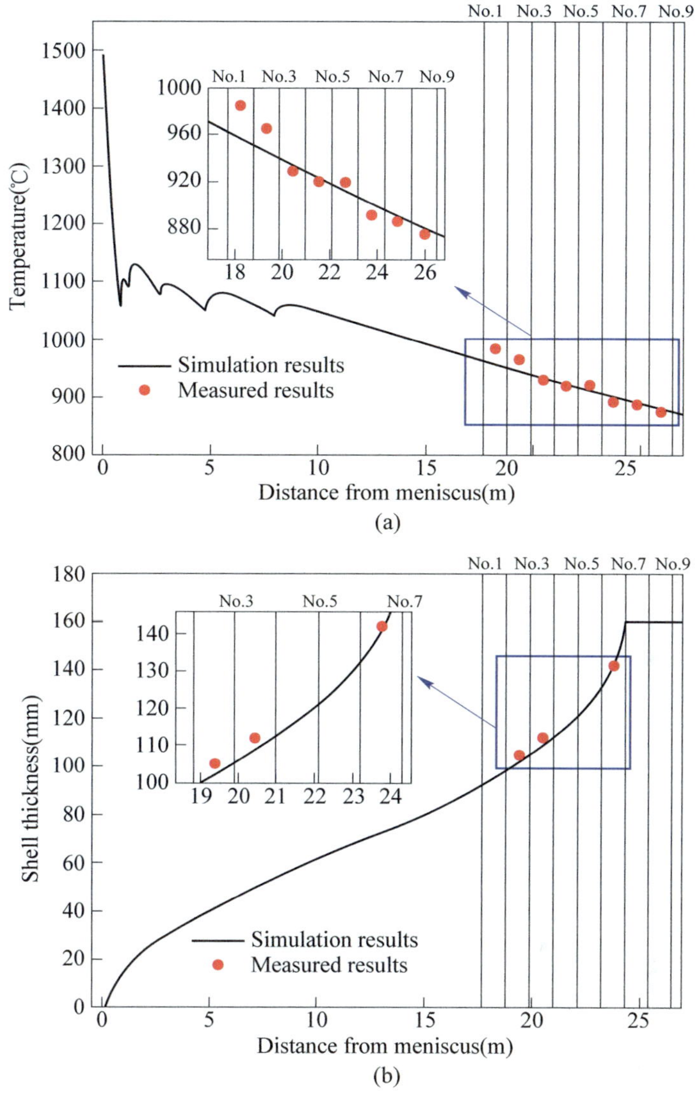

Fig. 3.5 Comparison of side center temperature (**a**) and shell thickness (**b**) between measured and calculated values

respectively. Among them, Fig. 3.6a, b, respectively, show the contact reaction force between No.5 roll and No.7 roll in the model, and Fig. 3.6c shows the actual output pressure of hydraulic cylinder of tension leveler No.5 roll and No.7 roll during the production process. It can be seen that the calculated contact reaction of the No.5 roll is 582.3 kN, while the actual output pressure of the No.5 tension leveler is 597.8 kN, and its relative error is 2.59%. The calculated contact reaction force of the No.7

roll is 986.4 kN, the actual output pressure of the No.7 tension leveler is 1012.8 kN, and the relative error is 2.61%.

3.1.1.4 Thermal/Mechanical Behavior of the Entire Process of Continuous Casting Bloom

During the continuous casting process, the strand is changed by mechanical force mainly in the bending, straightening and pressing processes, the equivalent plastic strain and Mises equivalent stress in the thermal–mechanical model reflects the plastic deformation of strand. The equivalent plastic strain clouds were extracted from the bending section, straightening section and the end of the caster, as shown in Fig. 3.7.

As shown in Fig. 3.7a, the equivalent plastic strain of the strand is larger in the inner and outer arcs of the bending section and smaller in the center of the thickness. As shown in Fig. 3.7b, the stress increased significantly after the contact between the strand and the roller, the strain variation is most obvious at the corner and the euivalent strain on the outer arc side is larger than that in the inner side. As is shown in Fig. 3.7c, at the end of the caster, the equivalent plastic strain of the strand increased under the pressure of the casting roller, and the strain in the center of the inner and outer arcs was larger due to the direct contact with the casting roller, and the strain in the center of the narrow surface was obviously smaller than that in the corner. The strain decreases first and then increases along the thickness direction at the position of the narrow plane.

Figure 3.8 shows the Mises equivalent stress scores extracted from the bending section, straightening section and the end of caster. As is shown in Fig. 3.8a, in the bending section, the stress is large at the head of the outer arc side and the middle of the length direction of inner arc side of the strand under the similar three-point top bending force of the inner and outer arcs of strand. This is because when the strand enters the bending section from the mold, the outer arc strand head puts bending stress directly, oppositely, the roller of the middle of the length direction of inner arc side block the deformation of the strand. As is shown in Fig. 3.8b, during the straightening process, the strand changes from bending shape to flat shape gradually under the action of the straightening roller. The surface stress of the strand increases first and then decreases when it contacts the straightening roller, the stress is obviously larger than that of the bending process, especially, the maximum stress concentrated occurs in the inner and outer arc corners at the contact position of the straightening roller. As is shown in Fig. 3.8c, at the end of the caster, the deformation of the strand under pressure produced by rolling down is significantly larger than that of the bending, the stress concentration at the corner of strand is more obvious during the straightening process, which means that the crack risk is highest.

Temperature cloud map and equivalent plastic strain cloud map during reduction process are shown in Fig. 3.9a, b, respectively. As is shown in Fig. 3.9b, the strain of the side surface of the strand was small in the inner and outer arc and large at the center of the casting strand, which is opposite to the distribution of the equivalent plastic strain on the narrow surface shown in Fig. 3.7. As is shown in Fig. 3.9a, The

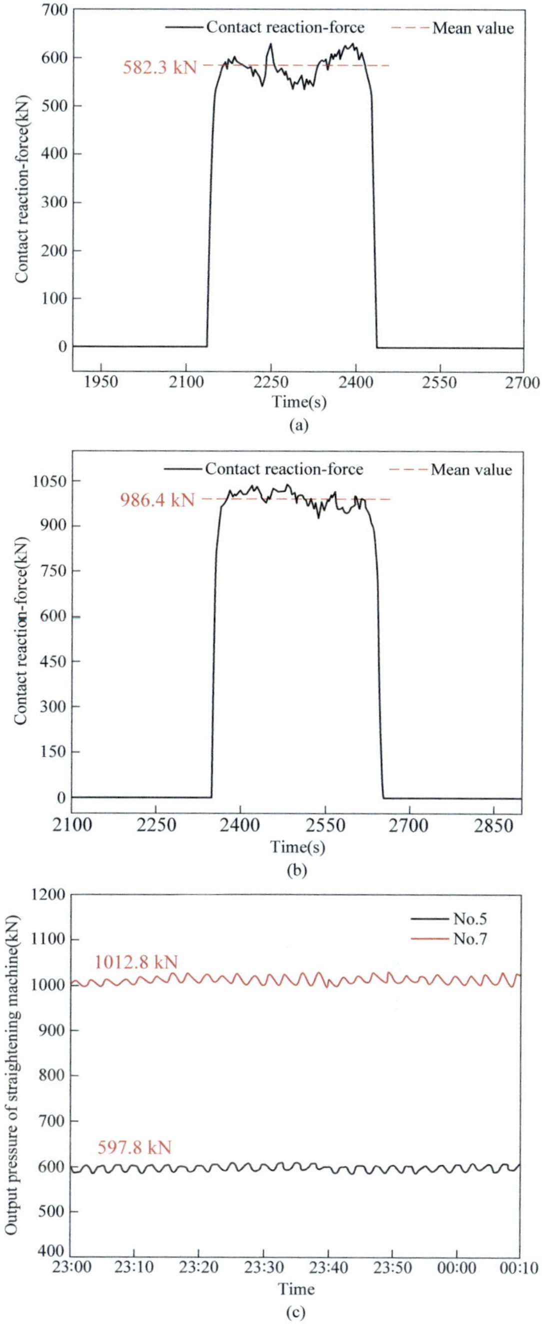

Fig. 3.6 Calculate the contact reaction and the actual output pressure. **a** The No.5 roller calculated reaction, **b** the No.7 roller calculated reaction, **c** the output pressure of the tension leveler

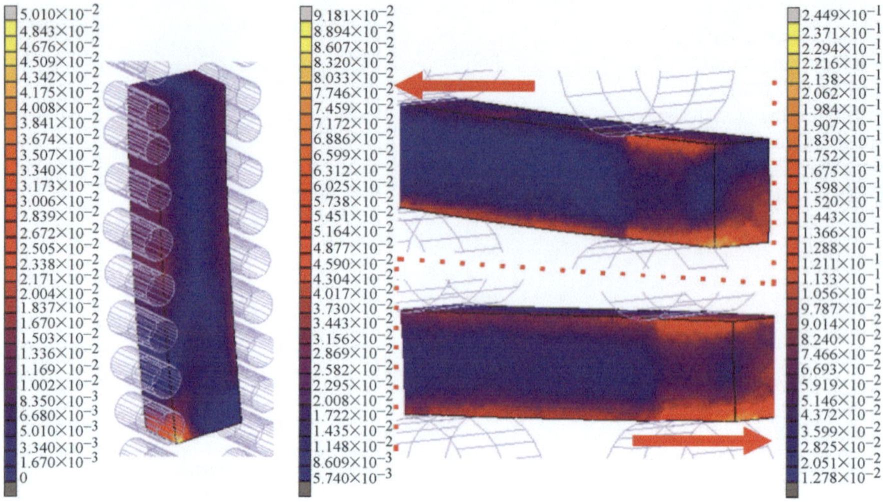

(a) (b) (c)

Fig. 3.7 Equivalent plastic strain cloud diagram of bearing steel bloom at each strand positions. **a** Bending section; **b** straightening section; **c** end of the caster

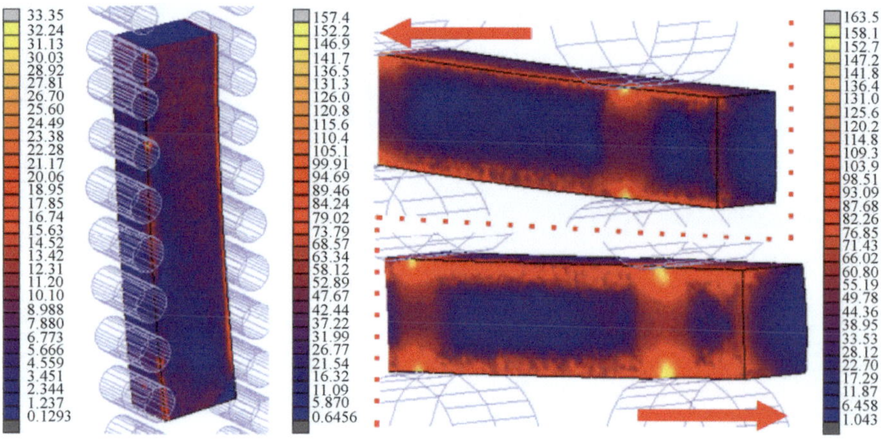

(a) (b) (c)

Fig. 3.8 Equivalent von mises stress cloud diagram of bearing steel bloom at each strand positions. **a** Bending section; **b** straightening section; **c** end of the caster

outer arc temperature is about 950 °C and that at the center is about 1350 °C, there is a temperature gradient about 400 °C between surface and center of the strand, which is significantly unconducive to the efficient transfer of deformation to the center of the strand. The temperature of the narrow surface is lower and temperature gradient is small, so the strain at inner and outer arc sides is large while the strain at the center is small.

As shown in Fig. 3.10, in order to describe the change of strand deformation with casting flow position more clearly, the center of the strand on the middle surface of the model (point A), the inner arc 1/4 (point B, the thickness of the strand

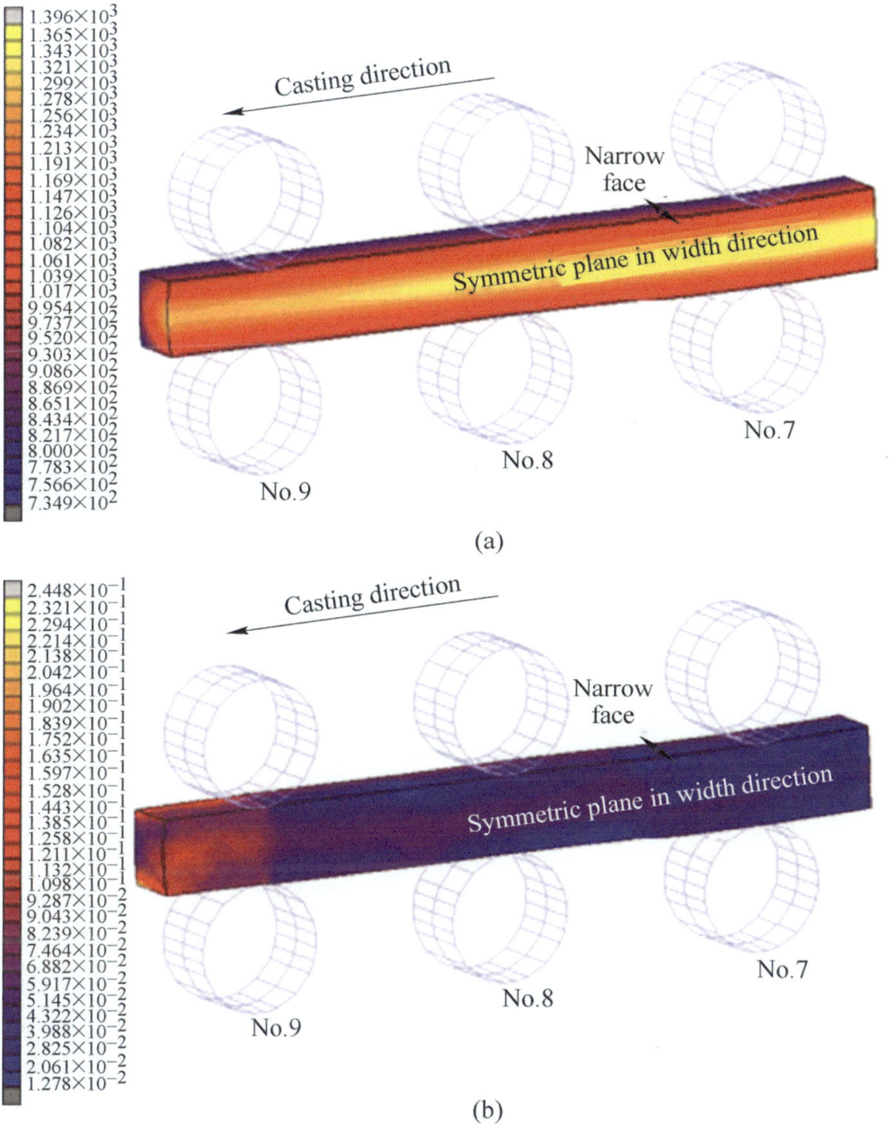

Fig. 3.9 The bearing steel bloom width symmetry plane. **a** Temperature cloud map; **b** equivalent plastic strain cloud map

1/4), the center of the inner arc surface (point *C*), inner arc corner (point *D*) and the narrow surface center (point *E*) were studied, the variation of strain and stress at each position during continuous casting was analyzed. As is shown in Fig. 3.10a, with the increase of distance from meniscus, the equivalent plastic strain at each position of strand increases gradually. During the process of top bending, the equivalent plastic strain of the strand increases obviously, the strain at the corner of the inner arc and center of the strand are larger than those at the center of the narrow face and inner arc. Considering the temperature distribution of the strand, the relatively high and uniform overall temperature of the strand is unfavorable for the transfer of deformation to the center. Therefore, the equivalent strain at the center of the narrow surface and strand is small, and the strain on the inner and outer arcs is large. The strain decreases from the inner and outer arcs to the thickness center. After the strand enters the arc zone (2–18 m from the meniscus), the strand moves in the arc-shaped roller and the strain basically keeps stable. When the strand entered the straightening zone, the equivalent plastic strain at the center of the billet, the center of the inner arc and the center of the narrow surface showed a trend of slow increase. The equivalent of 1/4 position of the inner arc was basically unchanged, while that of the corner of the inner arc decreased.

As shown in Fig. 3.10b, Mises equivalent stress showed a rising trend at all locations and fluctuated greatly at the position of direct contact between strand surface and roller. Along the direction of casting flow, the stress variation trend of the strand is basically consistent with the strain variation trend, the largest stress value appears at the corner of the strand and fluctuates wildly with a peak value of 170–180 MPa. This is mainly due to the stress concentration at the corner during the deformation process, especially the stress increases obviously when the strand is in contact with the roller during the pressing process and the stress decreases rapidly after the strand passes through the roller.

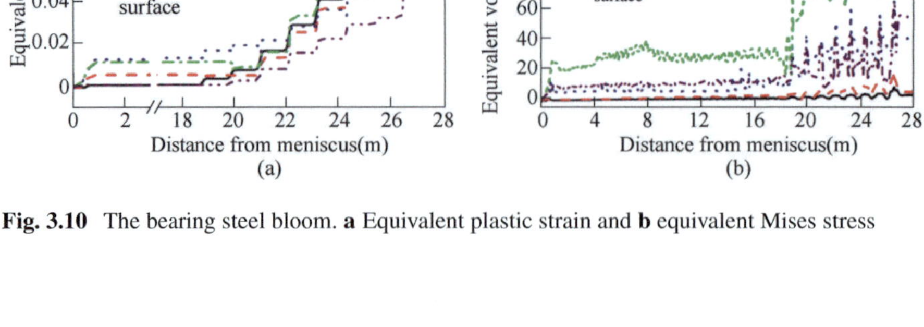

Fig. 3.10 The bearing steel bloom. **a** Equivalent plastic strain and **b** equivalent Mises stress

3.1.2 *Thermal/Mechanical Model and Reduction Deformation Rule of Continuous Casting Bloom*

3.1.2.1 Modeling Parameters

The wide and thick slab casting machine with 300 mm × 2200 mm section (cold slab section) in a steel plant was taken as the research object. The casting machine was a vertical curved continuous casting machine, consisting of 14 sector sections, including 1 curved section (No.1 section), 5 curved sections (No.2–No.6 sections), 2 straightening sections (No.7 section, No.8 section) and 6 horizontal sections (No.9–No.14 sections). The length of the casting is about 35.29 m. Table 3.3 lists the cooling zones. Table 3.4 lists the inlet and outlet positions of the horizontal lower pressure section. According to the above conditions, the thermal/mechanical model of the entire process of slab continuous casting is established, and the deformation behaviors during the slab continuous casting process as well as those of the micro - alloy are computed and analyzed.

Table 3.3 Cooling zone parameters of wide and thick slab caster

Cooling zone		Distance from meniscus (m)		Length (m)
		Initial position	End position	
Primary cooling	Mold height	0.00	0.88	0.88
Secondary cooling	Zone 1	0.88	1.52	0.64
	Zone 2	1.52	2.71	1.19
	Zone 3	2.71	4.00	1.29
	Zone 4	4.00	6.02	2.02
	Zone 5	6.02	10.01	3.99
	Zone 6	10.01	13.91	3.90
	Zone 7	13.91	20.83	6.92
	Zone 8	20.83	27.97	7.14
Air cooling	Air-cooling zone	27.97	35.29	7.32
Foot roll section	–	0.88	1.41	0.53

Table 3.4 Position of each horizontal section casting flow (m)

Segment	9	10	11	12	13	14
Entrance	21.2	23.5	25.9	28.3	30.7	33.1
Exit	23.2	25.5	27.9	30.3	32.7	35.1

3.1.2.2 3D Thermal–Mechanical Coupling Simulation Model

The thermal/mechanical model of the entire process of wide and thick slab continuous casting is similar to that of the generous slab model established in Sect. 3.1.1. Firstly, the position of each casting roll is determined through the CAD drawing of the casting machine, then the roll row geometric model is established through secondary development, and the model is introduced into the casting slab grid to complete the model establishment, as shown in Fig. 3.11.

Micro-alloyed steel Q345E was selected as the specific steel type, the drawing speed was 0.80 m/min, and the reduction amount of No.9–No.14 sector sections, are 1 mm/2 mm/3 mm/4 mm/5 mm, respectively. In order to save calculation time, storage space and improve calculation efficiency, considering the symmetry of casting billet, this model is a width direction 1/2 model. In the model, the width of the casting blank is 1100 mm, the thickness is 300 mm, and the length is 1000 mm. The tetrahedral element is used to divide the casting slab, the size of the element is 15 mm, and the total number of the casting elements is 370,604. In the same way, the deformation of the center section in the direction of the length of the slab during the pressing process is analyzed to avoid the effect of biting.

Similarly, the physical property parameters of micro-alloyed steel Q345E were determined using the method in Sect. 3.1.1, as shown in Fig. 3.12.

The constitutive model of mechanical parameters is shown in Chap. 2. The calculation methods of elastic modulus and Poisson's ratio are the same as those in Sect. 3.1.1. In addition, the governing equations and solving conditions of the model are similar to the treatment methods of thick and wide slab continuous casting billets in Sect. 3.1.1.

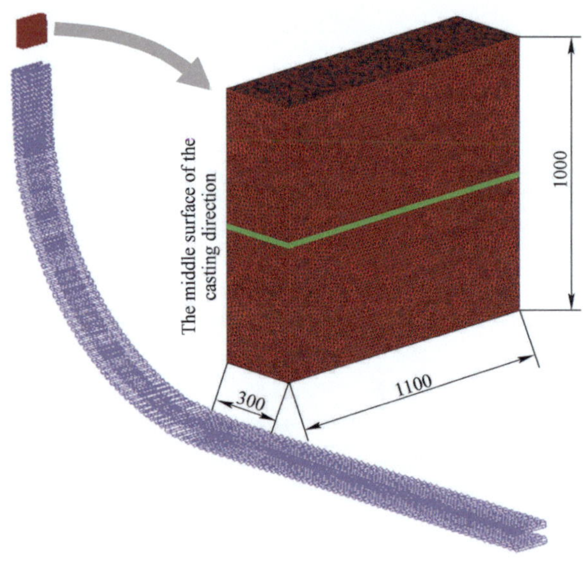

Fig. 3.11 The micro-alloyed steel wide-thick slab full continuous casting process model (mm)

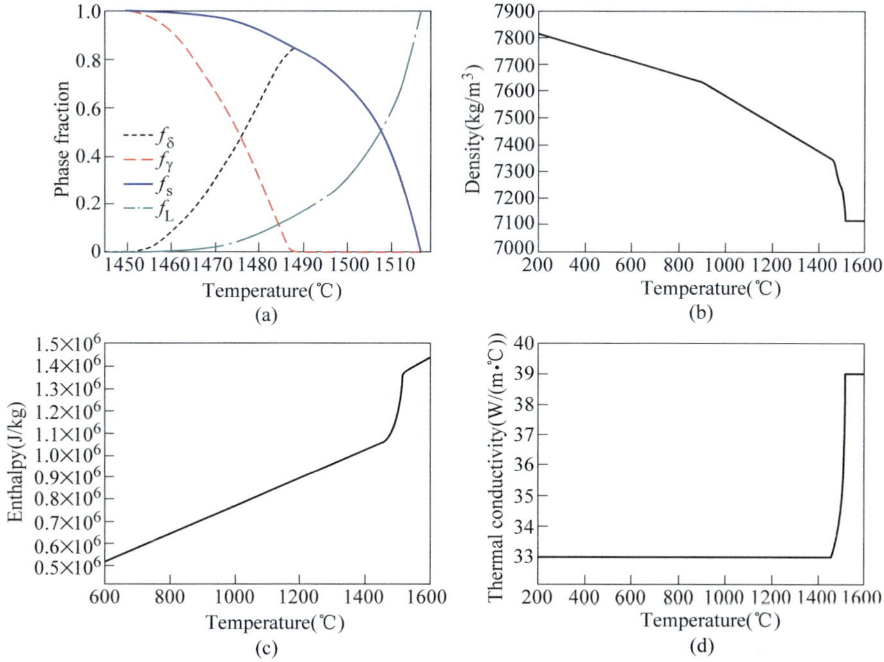

Fig. 3.12 Thermal conductivity, density, and enthalpy of micro-alloyed steel as a function of temperature

3.1.2.3 Model Validation

To verify the accuracy of the thermal/mechanical calculation model for wide and thick slabs, Fig. 3.13 compares the actual measured values of surface temperature and shell thickness with the calculated values of the heat transfer model. It can be seen that the wide surface 1/8 ($P_{1/8}$) temperature, wide surface center ($P_{1/2}$) temperature and shell thickness calculated by the solidification heat transfer model are in good agreement with the actual measured values. Further calculation shows that the relative errors between the calculated values of the model and the measured results are 1.79% and 1.19%, respectively. It shows that the heat transfer model can accurately calculate the cooling solidification process.

In order to verify the accuracy of the model's prediction of deformation behavior, the reduction force of the hydraulic cylinder in a single segment during the reduction process at the end of solidification was compared with the calculated value, as shown in Fig. 3.14, where the reduction force is the total slab force of the four clamping cylinders in the segment. It can be seen that the reduction force required in the sector section of the process of reduction increases with the increase of the reduction amount, and the relative error between the calculated value of the model and the measured value is less than 3.2%, so the calculation accuracy of the model is good.

Fig. 3.13 Comparison of the measured and calculated values of the 1/8 temperature (**a**) and shell thickness (**b**) in the center and wide surface of the slab

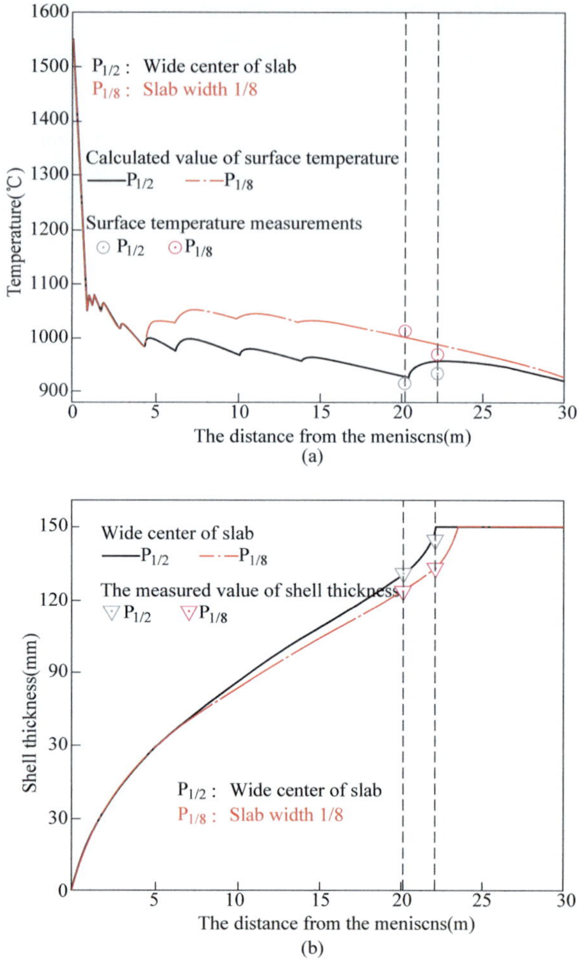

3.1.2.4 Thermal/Mechanical Behavior of the Entire Process of Continuous Casting Slab

Similarly to the analytical methods of continuous casting bloom, the deformation behavior of the strand at position of bending and straightening section was analyzed by extracting clouds of the equivalent plastic strain and Mises equivalent stress. Figure 3.15 shows the equivalent plastic strain cloud diagram at each position of the wide-thick slab casting process, inner arc rollers were hidden to avoid blocking the clouds. As is shown in Fig. 3.15a, in the bending section, the strand is bent because of the top bending effect in the vertical state. The strain at the strand head is larger due to direct bending of the roller. After a certain distance of the strand head, the equivalent plastic strain distribution region is stable, the strain at the inner and outer arcs is

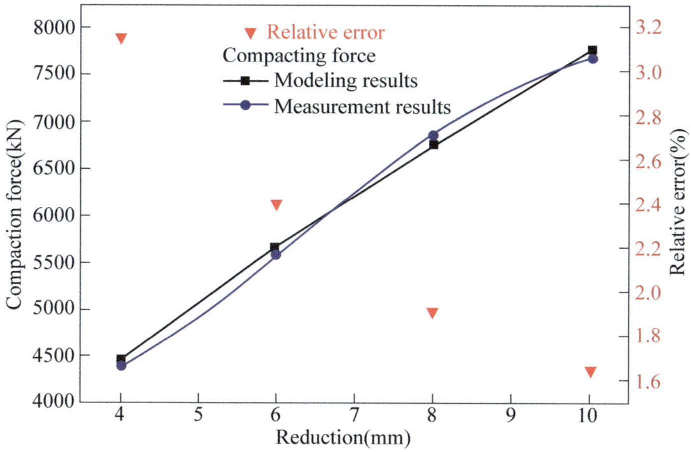

Fig. 3.14 Comparison of measured and calculated reduction force of wide and thick slabs

large and the strain at the center of the thickness is small. As is shown in Fig. 3.15b, in the straightening section, the stress increases obviously after the contact between the slab and the roller, and the equivalent plastic strain on the outer arc side is larger than that on the inner arc side. As is shown in Fig. 3.15c, at the end of the caster, the equivalent plastic strain of the strand increases under the pressure of the casting roller; Due to the direct contact with the casting roller, the inner and outer arc side has large strain, the strain at the center of the narrow plane is obviously smaller than that at the corner.

Figure 3.16 shows the Mises equivalent stress values extracted from the bending section, straightening section and the end of caster. Similarly to the continuous casting bloom, as is shown in Fig. 3.16a, under the similar three-point top bending force of the inner and outer arcs of slab, the stress is large at the head of the outer arc side and the middle of the length direction of inner arc side of the slab. As is shown in Fig. 3.16b, during the straightening process, the stress at the corners of the inner and outer arcs of the slab is on the high level, the stress at the contact position with the roller fluctuates slightly, and the stress distribution at other positions of the slab is small and uniform. As is shown in Fig. 3.16c, at the end of the caster, the slab is deformed by the rolling down under pressure, the stress concentration at the corner of the slab is obvious during the contact process of the slab.

As shown in Fig. 3.17, in order to describe the change of slab deformation with casting flow position more clearly, the center of the slab on the middle surface of the model (point *A*), the inner arc 1/4 (point *B*, the thickness of the slab 1/4), the center of the inner arc surface (point *C*), inner arc corner (point *D*) and the narrow surface center (point *E*) were studied, the curves of equivalent plastic strain and Mises equivalent stress at each point were shown in Fig. 3.17a, b, respectively.

As is shown in Fig. 3.17a, the equivalent plastic strain increases gradually with the casting flow position, and the strain begins to increase in the top bending process; In

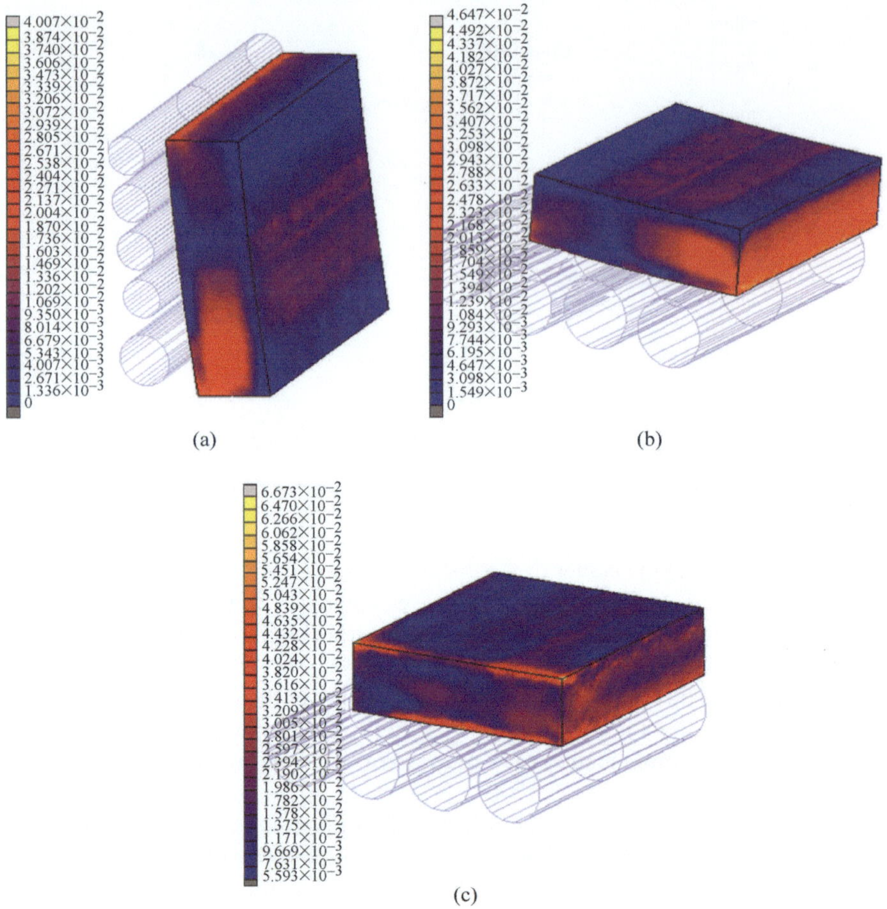

Fig. 3.15 Equivalent plastic strain cloud diagram of micro-alloyed steel wide-thick slab at each strand positions. **a** Bending section; **b** straightening section; c end of the caster

the arc section, the strain remains unchanged. In the straightening process, the strain at each point begins to change, but the strain at each point increases step by step after entering the reduction process. As is shown in Fig. 3.17b, the trend of Mises equivalent stress was similar to that of the equivalent plastic strain, however, the stress fluctuates greatly under the pressure of each roller, similar to the continuous casting bloom, the strain fluctuation at the corner of the wide-thick slab is most significant. The stress fluctuation is about 20 MPa in the arc section, 50 MPa in the straightening process and 150 MPa in the reduction process.

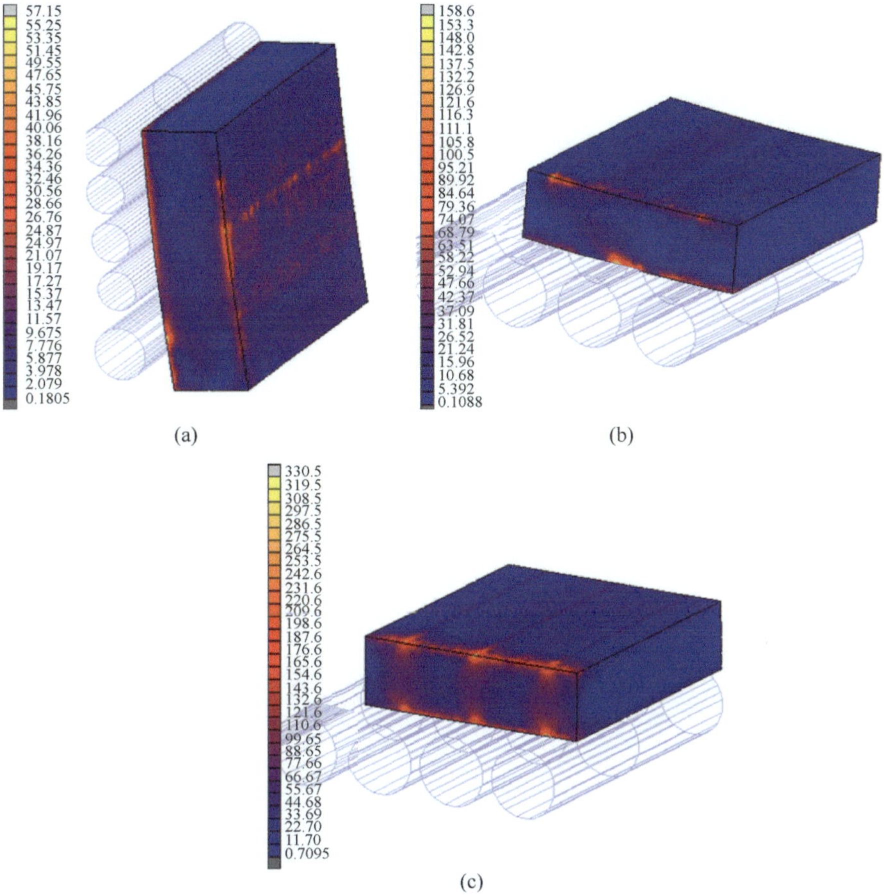

Fig. 3.16 Equivalent plastic stress cloud diagram of micro-alloyed steel wide-thick slab at each strand positions. **a** Bending section; **b** straightening section; **c** end of the caster

3.2 Critical Criterion of Crack Initiation and Propagation

It can be seen from the analysis in the previous section that the compression deformation will greatly increase the stress and strain of the strand, and the crack defect will be formed if strain exceeds the critical value of the critical criterion of crack initiation and propagation. Crack is one of the three quality defects of the continuous casting billet, as well as center segregation and porosity. Most of them will be further expanded in the subsequent heating and rolling process, resulting in a decline in the completion rate. In fact, the risk of cracking caused by reduction has always been a major concern for producers and has limited the application of the reduction process.

In order to avoid the occurrence of the reduction crack, the deformation of the reduction process must be confined within the critical strain leading to the casting

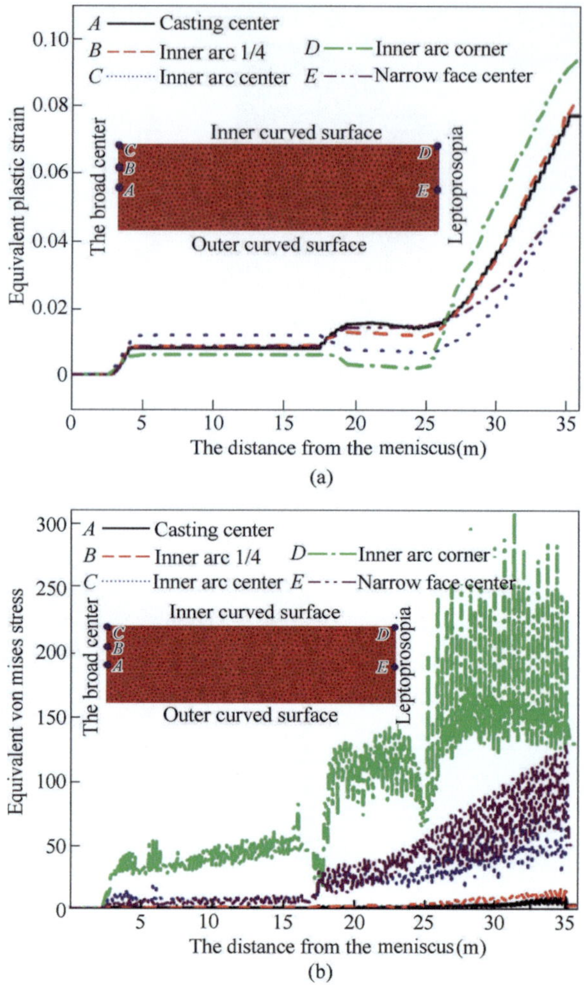

Fig. 3.17 The micro-alloyed steel wide-thick slab. **a** Equivalent plastic strain, **b** equivalent Mises stress curve

crack. On the one hand, it is necessary to accurately characterize the effects of different reduction process conditions on the stress and strain of the strand (as described in Sect. 3.1). On the other hand, it is necessary to accurately determine the critical criterion for crack initiation and propagation. As early as the 1980s, a lot of pioneering work has been done by Japanese researchers. Miyazaki et al. [13] studied the critical strain of intermediate crack formation through the top bending experiment of steel ingot with liquid core. Yasunaka et al. [14] studied the critical strain of surface crack formation through the top bending experiment of steel ingot with liquid core. Matsumiya et al. [15] believed that the top bending test required a large sample and the measurement of critical strain is tedious, so the melting bending test was adopted to measure the critical strain value of intermediate crack initiation. Bernhard et al. [16] and Rowan et al. [17] determined the critical criterion crack formation by the Submerged Split Chill Contraction test. In this method, the steel connected with sensors was placed in molten steel pool, and the critical criterion for crack formation was determined by measuring the strain of blank shell formed by cooling. In addition, Won [18], Kato [19], Yamanaka [20], Yu [21], Punnose [22] and Enos [23] all adopted a relatively simple and accurate tensile method to determine the critical value of crack generation. These critical crack measurement methods provide important research support for avoiding the crack defect of continuous casting strand. However, most of the cracks occur in the initial solidification process of mold, so most of the above researchers focus on the surface cracks in the initial solidification stage.

3.2.1 Critical Criterion of Intermediate Crack Initiation

The intermediate cracks of strand are mostly located in the thickness direction of 1/4 strand. When the billet still has a large number of liquid cores, the solid–liquid interface front crack is formed due to the large amount of deformation under reduction. Clyne et al. [24] believe that the temperature range between zero plastic temperature (ZDT $= f_s$, 0.99) and zero strength temperature (ZST $= f_s$, 0.75) can be divided into mass/liquid feeding zone (LIT-ZST) and cracking zone (ZDT-LIT) by liquid impenetrable temperature (LIT). In the mass/liquid feeding zone, cracks will be eliminated by filling of liquid steel even if they are formed. However, there are a lot of solute segregation elements in the molten steel, which appear as "black lines". In cracking zone, there is no liquid supplement or incomplete penetration after the formation of intermediate cracks, and the "black lines" occurs along with cracks.

3.2.1.1 Determining Scheme of Critical Strain of Intermediate Crack Initiation

According to the formation mechanism of intermediate crack initiation, the critical strain of intermediate crack initiation was measured by tensile test performed on a Thercmemastor-Z/100 thermodynamic simulation system. The tensile specimens were obtained from GCr15 bearing steel and Q345E micro-alloyed steel. As shown in Fig. 3.18, the specimen orientation was parallel to the casting direction, and the porous area was avoided. The specimen structure is shown in Fig. 3.19.

According to the formation mechanism of intermediate crack, firstly, the solidus temperature of different steel grades was calculated theoretically. Then several groups of temperatures were taken near the solidus temperature for high temperature tensile tests. The sample was broken and the ZDT were determined according to the fracture morphology of the samples. Finally, tensile tests of different strain variables were carried out under ZDT, and the crack generation of samples with different strain variables was detected by metallographic microscope and scanning microscope, the critical strain of intermediate crack initiation was finally determined. The theoretical

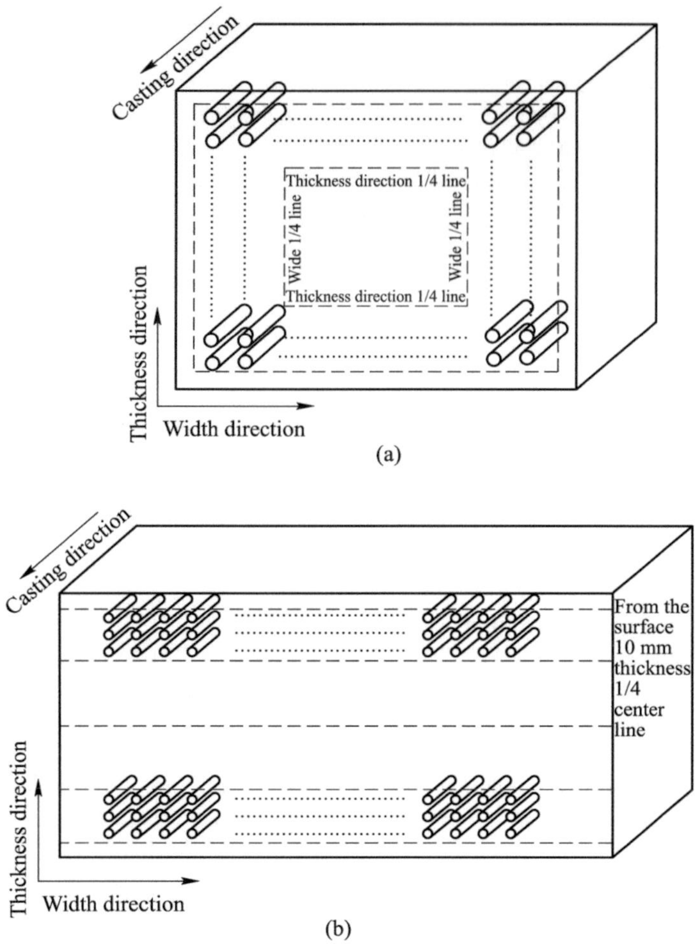

Fig. 3.18 Schematic diagram of sampling position of tensile specimen for intermediate crack initiation test. **a** Bearing steel bloom and **b** micro-alloyed steel wide-thick slab

temperature range of intermediate crack initiation of bearing steel and micro-alloy steel is shown in Fig. 3.20.

The experimental procedure of bearing steel is shown in Fig. 3.21a, and the specimen was heated to 1200 °C at a heating rate of 10 °C/s and maintained isothermally for 20 s to reach a uniform temperature distribution; then, the specimen was heated to 10 °C below the test temperature. Subsequently, the specimen was heated to the test temperature at a heating rate of 0.5 °C/s and maintained isothermally for 10 s, after which tension was applied on the specimen. The strain rate of the continuously cast bloom was approximately 10^{-3} to 10^{-1} s^{-1} and the tensile strain rate used in the experiment was 6×10^{-3} s^{-1}.

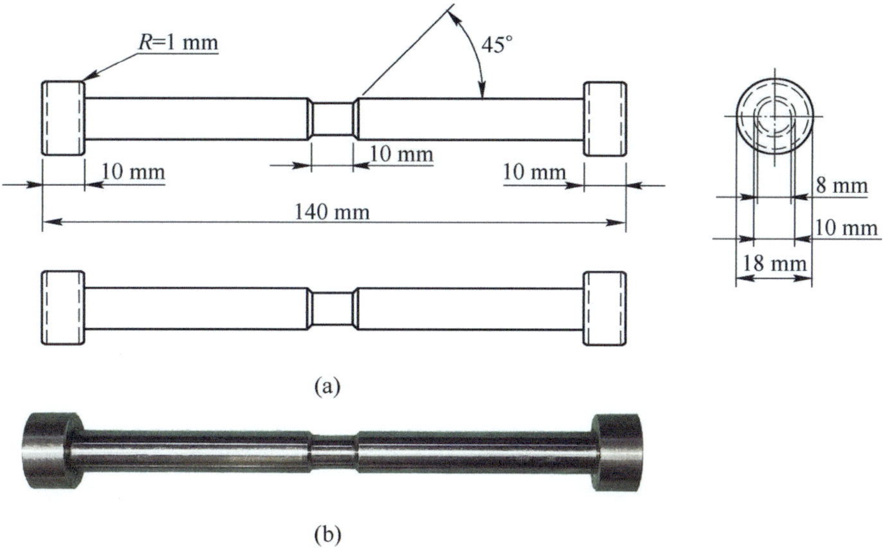

Fig. 3.19 Intermediate crack initiation test specimen. **a** Schematic and **b** photo

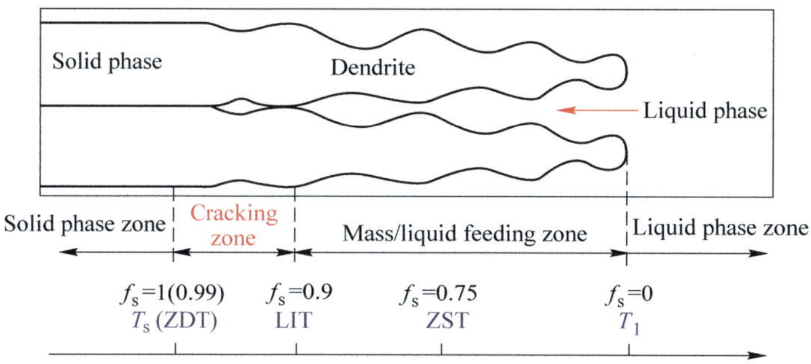

Fig. 3.20 Schematic diagram of theoretical temperature interval for crack formation

Because the solidus temperature of micro-alloyed steel is higher than that of the bearing steel, the experimental heating system is slightly different, as shown in Fig. 3.21b, the specimen was heated to 1300 °C at a heating rate of 10 °C/s and maintained isothermally for 20 s to reach a uniform temperature distribution; then, the specimen was heated to 10 °C below the test temperature. Subsequently, the specimen was heated to the test temperature at a heating rate of 0.5 °C/s and maintained isothermally for 10 s, and then the tension was applied with a strain rate of $6 \times 10^{-3}\ s^{-1}$.

During the experiment, the ZDT of the sample was determined by the fracture morphology and necking phenomenon. After the sample temperature was determined, the critical strain range of crack initiation was roughly determined according

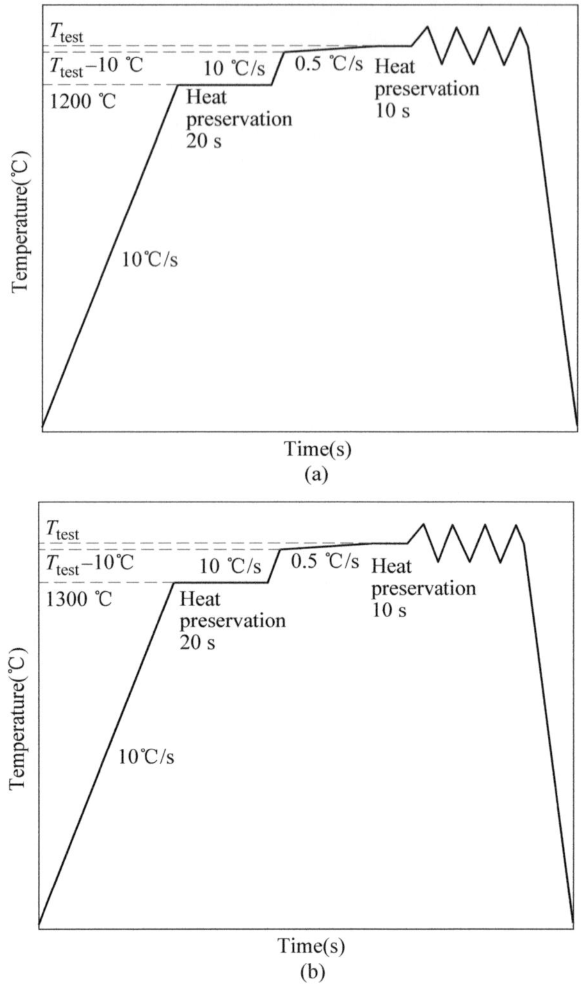

Fig. 3.21 Heating system for intermediate crack initiation test. **a** Bearing steel and **b** micro-alloyed steel

to the stress–strain curve; The experimental schemes were designed for different strains at ZDT, and the crack generation of samples was detected by metallographic microscope and scanning electron microscope to determine the critical strain of intermediate crack initiation.

3.2.1.2 Determination of Critical Strain of Intermediate Crack Initiation in Bearing Steel

Element segregation at the end of solidification leads to a decrease in the solidus temperature, and leads to ZDT being lower than the solidus temperature [25]. According to experience, the experimental temperature is usually 30 °C lower than the theoretical temperature. Therefore, three groups of experiments were carried out between 1310 and 1320 °C, specimens are shown in Fig. 3.22. Figure 3.23 shows the tensile specimens and scanning electron microscopy (SEM) images of the fractures, at 1310 °C, there is no obvious necking phenomenon in the experiment, which indicates that the temperature is near the solid-phase line. At 1315 °C, the sample has no obvious necking phenomenon, and there is no large melting area at the fracture, so it should be about ZDT. When the specimen is stretched at 1320 °C, the melting phenomenon at the fracture is obvious, the liquid phase rate is high, and the temperature should be higher than LIT.

Figure 3.24 shows the enlarged details of the tensile test fracture at different temperatures. When the tensile test was stretched at 1310 °C, the fracture was fine-grained and the grain boundaries were obvious. When stretched at 1315 °C, the small particles at the fracture are relatively round, and there are fine dendrites in some places, indicating that the sample is in the two-phase region at this temperature. When stretched at 1320 °C, the dendrite area at the fracture of the sample is relatively large, indicating that the liquid phase has been melted and the liquid phase rate is high, and the temperature should be higher than LIT. According to the above analysis,

Fig. 3.22 Tensile specimens and fractures of bearing steel at different temperatures

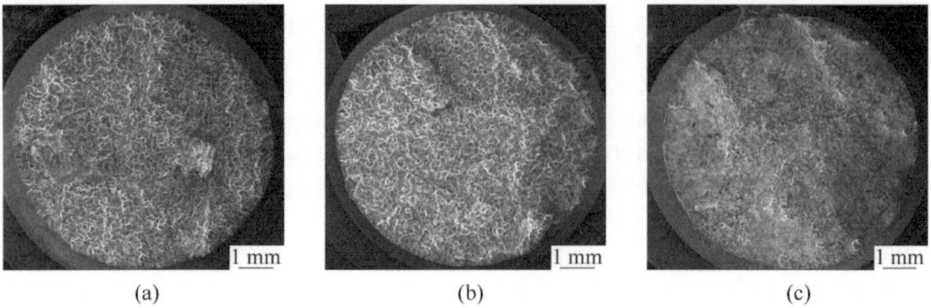

Fig. 3.23 Tensile specimens and fracture SEM images at **a** 1310 °C, **b** 1315 °C, and **c** 1320 °C

it can be determined that the bearing steel GCr15 sample is in the ZDT-LIT range at 1315 °C, and the critical strain of intermediate crack initiation can be measured at 1315 °C.

Since the stress during the tensile test at 1315 °C is small, it is difficult to directly obtain the stress–strain curve, so the strain peak value cannot be determined directly according to the stress–strain curve. Therefore, tensile tests of strain 0.02, 0.03, 0.04, 0.05 and 0.06 are conducted directly at 1315 °C. The critical strain of intermediate crack initiation was determined by detecting the micro-morphology of each sample after incision.

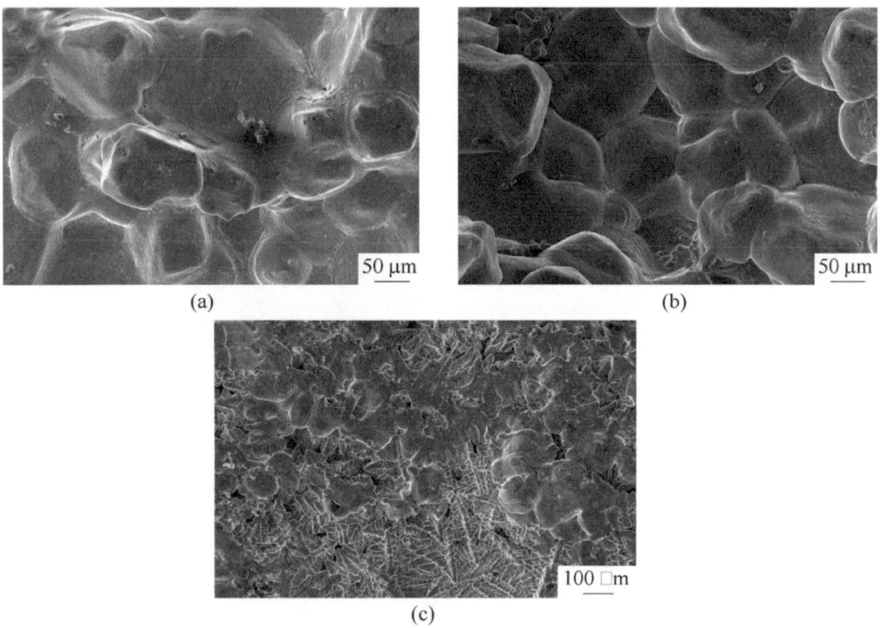

Fig. 3.24 The enlarged details of the fractures at **a** 1310 °C, **b** 1315 °C, and **c** 1320 °C

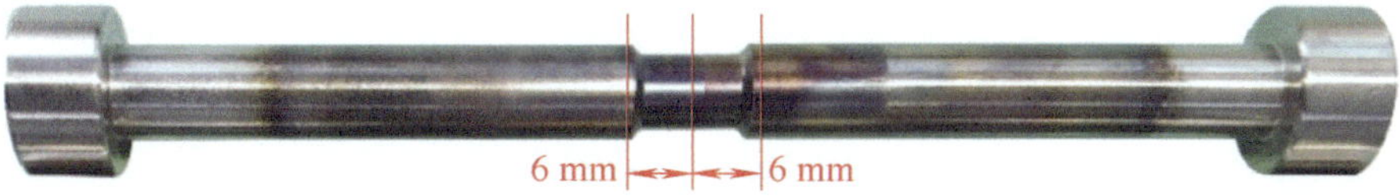

Fig. 3.25 Tensile specimen and processing diagram of bearing steel at 1315 °C

The specimens were cut in the manner shown in Fig. 3.25, and crack formation was detected after grinding and polishing. Metallographs of the specimens with different strains applied at 1315 °C are shown in Fig. 3.26. When the strain variables were 0.02 and 0.03, no obvious defects were formed in the samples. When the strain variable rises to 0.04, a small group of shrinkage holes are formed in the sample. When the strain variable further reaches 0.05, fine cracks have formed in the sample. When the strain variable reaches 0.06, the cracks in the sample are very clear. This is exactly consistent with the conclusion of Enos et al. [24] that cracks are formed by the growth and connection of small shrinkage holes.

To further characterize the formation and propagation of cracks, as shown in Fig. 3.27, specimens with strains of 0.05 and 0.06 were characterized by SEM. When the strain was 0.05, cracks began to form at the grain boundaries, but these cracks were small and discontinuous and did not completely cover the grain boundaries. When the strain increased to 0.06, the cracks further propagated and completely covered the grain boundaries. At 1315 °C, with a tensile strain rate of 6×10^{-3} s^{-1}, the cracks initiated at a tensile strain of 0.05 and subsequently propagated when the tensile strain exceeded 0.05. Therefore, at 1315 °C and a tensile strain rate of 6×10^{-3} s^{-1}, the critical strain of internal crack initiation for GCr15 steel was 0.05.

According to the theory of intermediate crack formation in Fig. 3.20, the possibility of crack formation is small below ZDT. This is mainly because when the temperature is lower than ZDT ($f_s = 0.99$) temperature, the material is a solid phase, has a certain strength, plasticity and ductility, and is not easy to form cracks when the stress variable is small. In order to further determine the risk and critical strain of intermediate crack formation at temperatures lower than ZDT, high temperature tensile experiments were carried out at 1310 °C. Figure 3.28 shows the stress–strain curve obtained by pulling the sample at 1310 °C. It can be seen that the stress peak occurs when the strain variable reaches 0.073. As the strain increases further, the stress begins to decrease rapidly. It can be seen that at 1310 °C, cracks are most likely to form when the strain variable reaches 0.073, and expand with the continuous increase of strain. In other words, when cracks expand to a certain limit, the material will be damaged, and the curve shows a sharp drop in stress, leading to the fracture of the sample.

In order to further investigate the critical strain of crack formation at 1310 °C, high temperature tensile experiments with variables 0.07, 0.08 and 0.10 were carried out at this temperature. The samples after tensile were shown in Fig. 3.29. The three samples were processed into metallographic samples according to the method shown in Fig. 3.25, and the metallographic test results are shown in Fig. 3.30.

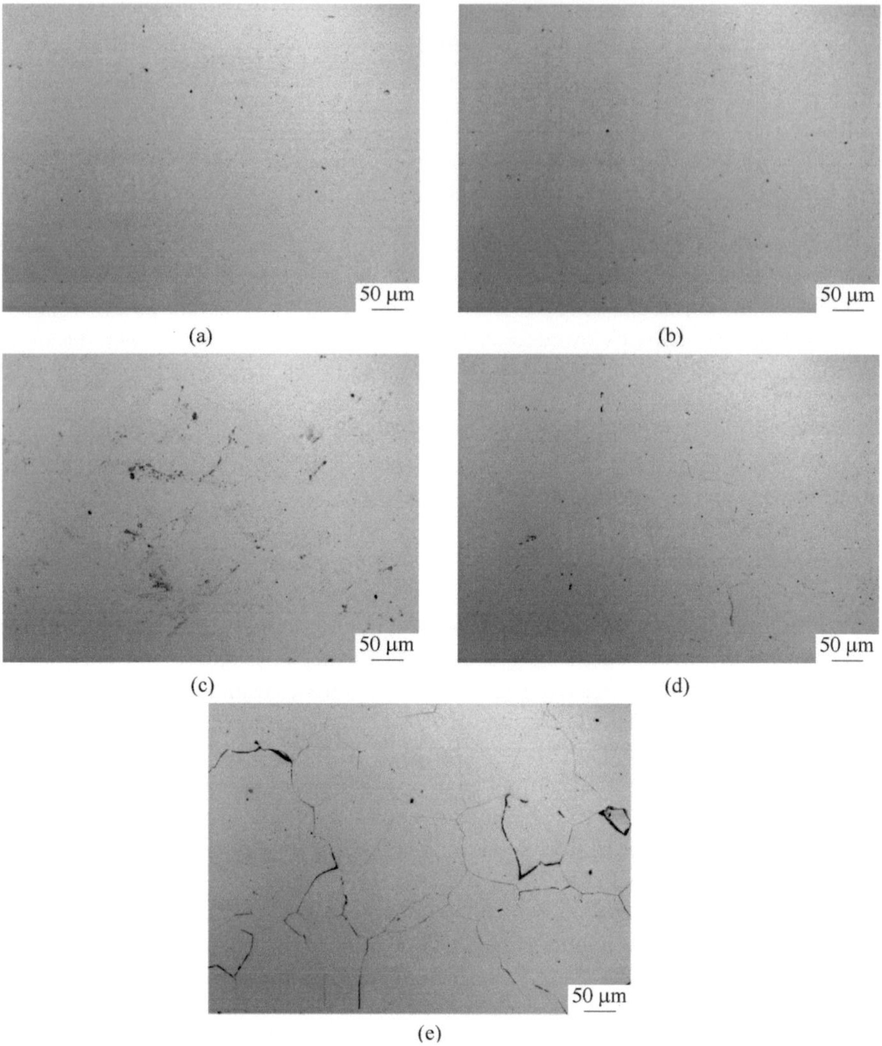

Fig. 3.26 Metallographs of tensile specimens of GCr15 steel with different strains applied at 1315 °C. **a** $\varepsilon = 0.02$; **b** $\varepsilon = 0.03$; **c** $\varepsilon = 0.04$; **d** $\varepsilon = 0.05$; **e** $\varepsilon = 0.06$

As seen from Fig. 3.30, when the tensile strain variable of 1310 °C is 0.07, no crack defects appear in the sample. When the strain variable is 0.08, small cracks appear in part of the sample. When the strain variable reaches 0.10, the cracks along the grain boundary are formed by the connection of small cracks. Similarly, in order to further characterize the cracks formed at the stress variables 0.08 and 0.10 at 1310 °C, SEM characterization was performed, as shown in Fig. 3.31. It can be clearly observed that the crack initiation begins when the tensile strain variable of 1310 °C reaches 0.08, and the crack is smaller than that when the tensile strain variable of 1315 °C is 0.05.

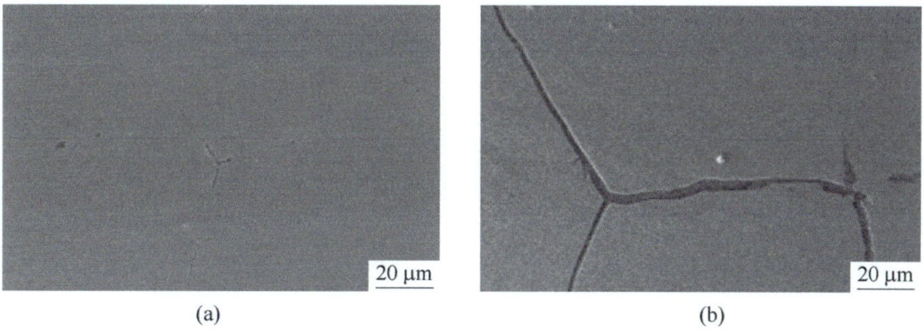

Fig. 3.27 SEM images of tensile specimens of GCr15 steel with different strains applied at 1315 °C. **a** $\varepsilon = 0.05$; **b** $\varepsilon = 0.06$

Fig. 3.28 Tensile stress–strain curve of GCr15 steel at 1310 °C

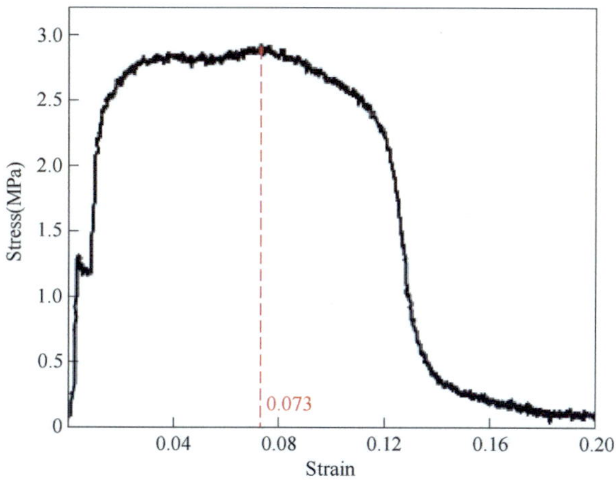

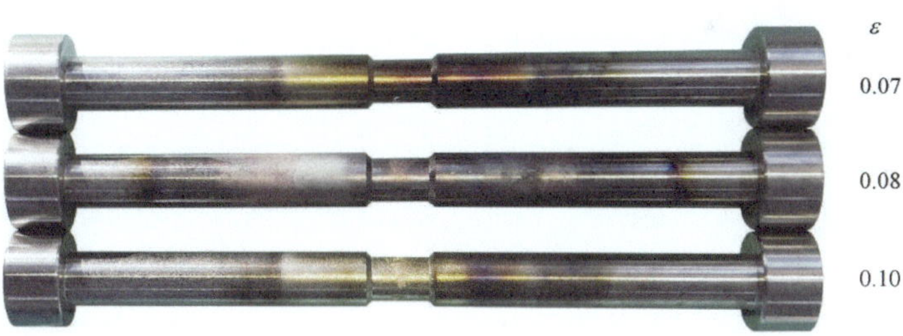

Fig. 3.29 Tensile specimens of GCr15 steel with different strains applied at 1310 °C

At the 1310 °C stress variable of 0.10, the crack expanded slightly compared with that at the stress variable of 0.08, but the overall shape remained relatively small.

By comparing Fig. 3.26 with Fig. 3.27 and Fig. 3.30 with Fig. 3.31., it can be seen that cracks begin to form when the strain variable reaches 0.05 at 1315 °C, which is

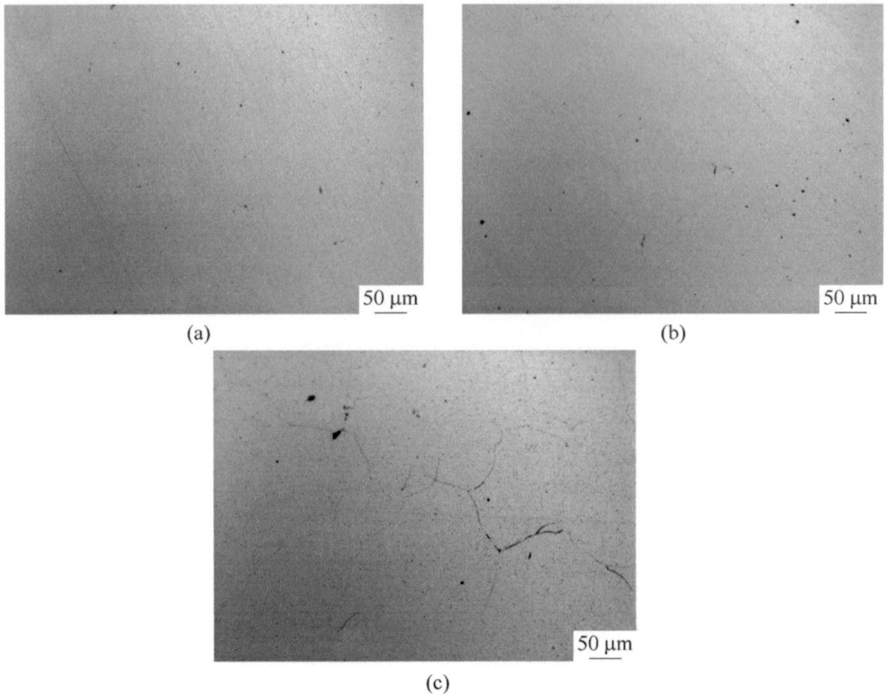

Fig. 3.30 Metallographs of tensile specimens of GCr15 steel with different strains applied at 1310 °C. **a** $\varepsilon = 0.07$; **b** $\varepsilon = 0.08$; **c** $\varepsilon = 0.10$

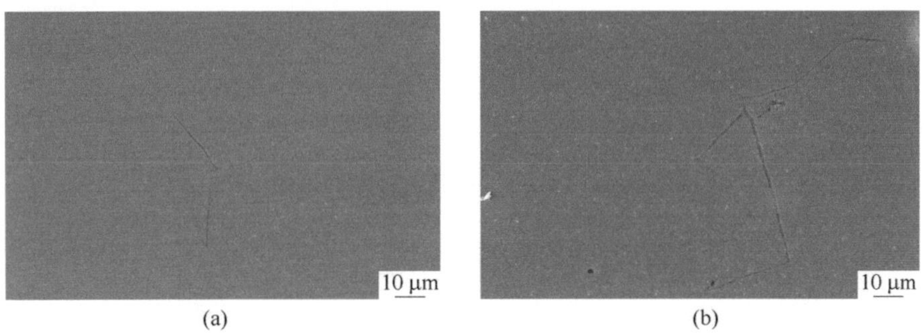

Fig. 3.31 SEM images of tensile specimens of GCr15 steel with different strains applied at 1310 °C. **a** $\varepsilon = 0.08$; **b** $\varepsilon = 0.10$

the critical strain at this temperature. When the strain variable reaches 0.06, cracks are more obvious. When the strain variable reaches 0.08 at 1310 °C, the crack begins to form, which is the critical strain at this temperature. When the strain variable reaches 0.10, the crack is more obvious. Moreover, the cracks formed at the 1310 °C strain variable of 0.10 are finer than those formed at the 1315 °C strain variable of

0.06, that is, the intermediate cracks are easier to form above ZDT, while the cracks are more difficult to form when the temperature is lower than ZDT.

3.2.1.3 Determination of Critical Strain of Intermediate Crack Initiation in Micro-alloyed Steel

Similar to bearing steel, the ZDT of micro-alloyed steel Q345E may also be tens of degrees Celsius lower than the theoretical temperature. Therefore, five groups of experiments were performed between 1420 and 1440 °C, and tensile samples were shown in Fig. 3.32. From the tensile test samples and the appearance of the fracture, the tensile sample shows a necking phenomenon when the temperature is 1420, 1425 and 1430 °C. At 1435 °C, the sample has no obvious necking and should be around ZDT. At 1440 °C, the melting phenomenon at the fracture is obvious and the liquid rate is higher and the temperature should be greater than the ZDT.

In order to observe the fracture morphology more clearly, SEM was used to observe the experimental fracture, as shown in Fig. 3.33. It is clearly seen that the fracture of the tensile sample is fine granular, and the edges and corners of the small particles are slightly melted but the liquid phase can hardly be seen at 1435 °C, At this time, the solidus rate should be close to 1. At 1440 °C, the fracture melting phenomenon is obvious and the solid-phase ratio is low, which should be higher than ZDT. According to the above analysis, the micro-alloyed steel should be in the actual ZDT-LIT range at 1435 °C, the critical strain of crack initiation in micro-alloyed steel Q345E should be measured at 1435 °C.

Fig. 3.32 Tensile specimen and its fracture surface at different temperatures of micro-alloyed steel

Fig. 3.33 SEM morphology of fractures surface of micro-alloyed steel at different temperatures. **a** 1435 °C, **b** 1440 °C

Figure 3.34 shows the stress–strain curve of tensile test at 1435 °C, it can be seen that the stress peak occurs when the strain variable reaches 0.043, but with the increase of strain, the stress value decreases rapidly, which is caused by crack initiation and propagation until fracture. Therefore, it can be determined that the critical strain of crack initiation should be less than 0.043. Correspondingly, tensile tests with strain variables of 0.01, 0.02, 0.03, and 0.04 were carried out at 1435 °C, respectively, and the critical strain for crack initiation was determined by detecting the presence of cracks in each sample.

As is shown in Fig. 3.35, metallographic observation was performed on the samples with different strain variables.

When the strains are 0.01 and 0.02, no obvious defects are formed. When the strain is 0.3, the sample forms a small shrinkage group, when the strain reaches 0.4, short cracks are formed in the sample. In order to further characterize the short cracks at strain of 0.04, SEM observation was performed on the sample as shown in Fig. 3.36.

Fig. 3.34 Tensile stress–strain curve of micro-alloyed steel at 1435 °C

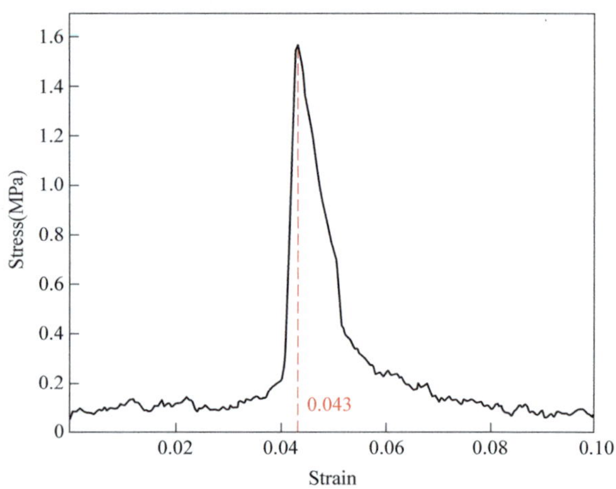

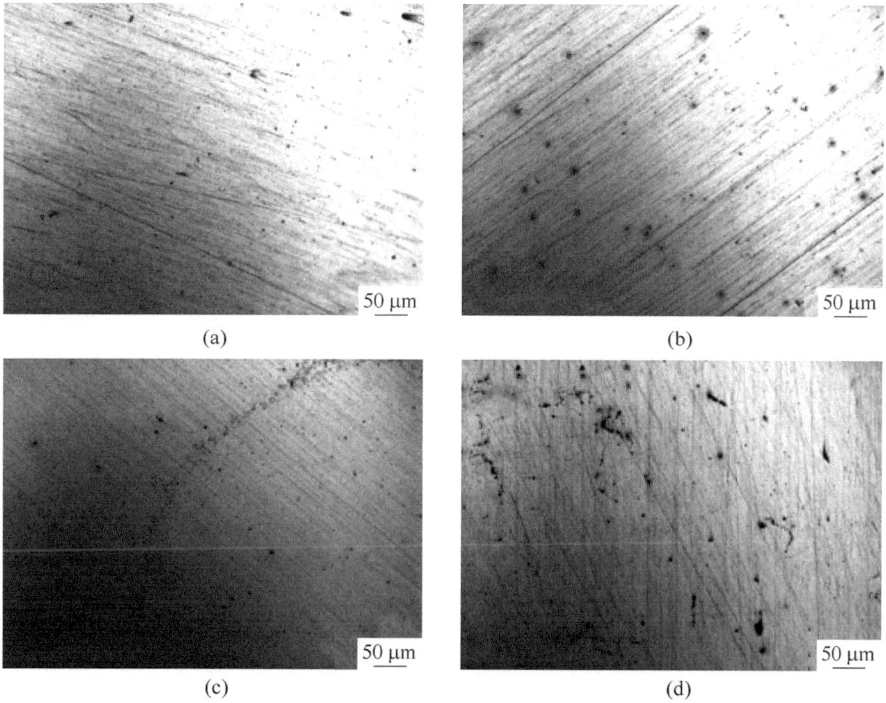

Fig. 3.35 Metallograph of tensile specimens of micro-alloyed steel under different strains at 1435 °C

Figure 3.36a shows the crack morphology magnified 2000 times. Small shrinkage holes can be observed near the crack. Figure 3.36b shows the crack morphology magnified 4000 times and small shrinkage holes can be observed more clearly.

In summary, the ZDT of micro-alloyed steel Q345E is 1435 °C and the critical strain of intermediate crack initiation is 0.04.

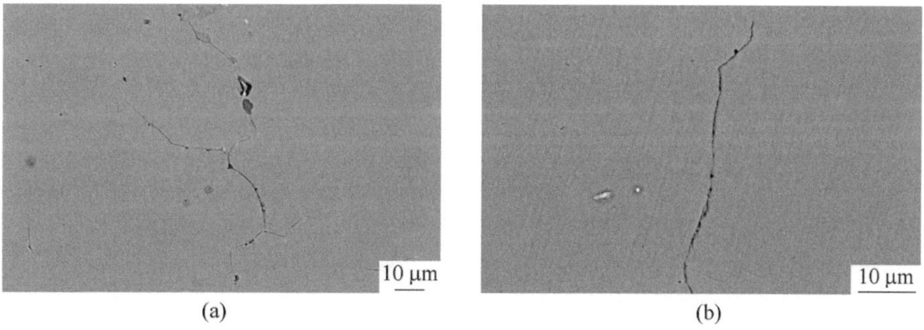

Fig. 3.36 SEM photo of crack of micro-alloyed steel at 1435 °C. **a** 2000×; **b** 4000×

3.2.2 Critical Criterion of Surface Crack Propagation

In general, most of the surface cracks occur in the initial solidification process of the mold. As for the cracks in the corner mold, the large shrinkage of the strand in the mold decreases the heat transfer rate of the corner shell, which leads to the coarse austenite grains in the corner. At the same time, the vibration mark is formed under the action of static pressure of molten steel and slag pressure, short cracks can form easily under the stress concentration at the bottom of the vibration mark during the bending process [26]. In the subsequent secondary cooling process, the grain boundary strength is greatly weakened by the precipitation of microalloying carbonitrides on the austenite grain boundary and the proeutectoid ferrite film in the phase transformation process [27, 28]. If further pressure is applied at this time, the surface short crack will propagate inevitably when the deformation exceeds the critical Criterion.

3.2.2.1 Determining Scheme of Critical Strain of Corner Crack Propagation

According to the mechanism of corner crack propagation, the critical strain of corner crack propagation was measured by pre-cracked specimens tensile test performed on a Thercmemastor-Z/100 thermodynamic simulation system. The steel used in the experiment is titanium micro-alloyed steel, which has a higher crack sensitivity than Q345E, the major chemical composition of titanium micro-alloyed steel can be found in Chap. 2. Pre-cracked tensile specimens were shown in Fig. 3.37.

Different from the intermediate crack initiation, the temperature of the corner crack propagation is lower than 900 °C during the reduction process, and the microstructure of the billet directly determines its thermoplasticity. Based on the history of the strand, the temperature of the corner decreases (at mold and foot roller area) and raises up again (at foot roller narrow surface where cooling water disappears), which is significantly different from the microstructure obtained from strand cooled directly. In order to fully consider the effect of the heat history during slab cooling on the microstructure of corner, the experimental heat history should be designed according to the changing law of temperature field in the continuous casting process of titanium micro-alloyed steel.

A 3D thermal–mechanical coupling simulation model of titanium micro-alloyed steel continuous casting slab was established (see Sect. 3.1 for details). Temperature curve of the slab and the local strain and temperature cloud diagram of reduction process was shown in Fig. 3.38. Based on the thermal history of the corner, the heating system for corner crack propagation test was designed, as shown in Fig. 3.39. The specimen was heated to 1200 °C at a heating rate of 10 °C/s and maintained isothermally for 60 s in order to make the austenite grains grow up fully, which close to that of the actual surface of strand at 1200 °C. Subsequently, the temperature decreased from 770 °C at a cooling rate of 5.5 °C/s which was equal to cooling

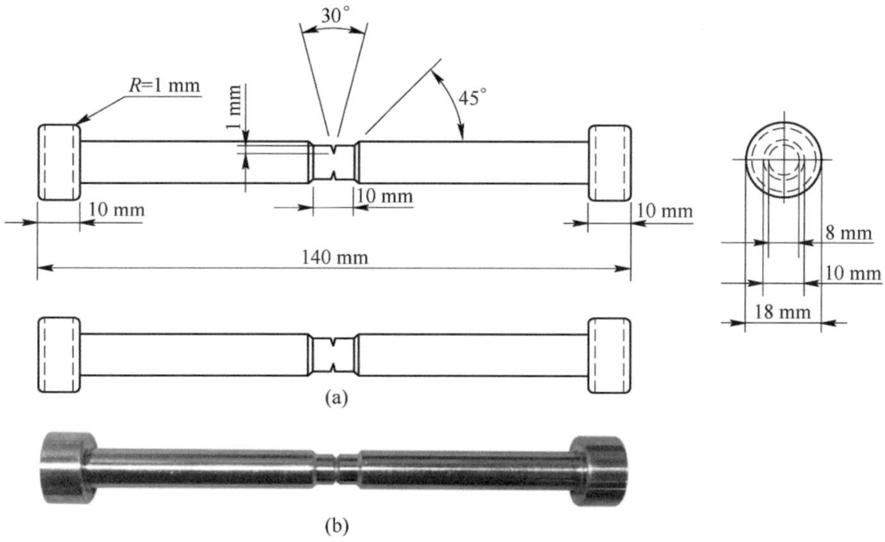

Fig. 3.37 Surface crack propagation test tensile specimen. **a** Schematic and **b** photo

rate at the corner of the casting slab between the old and the foot roller area; The temperature return to 870 °C at a heating rate of 0.65 °C based on the temperature filed of the foot roller area; The temperature of specimens then decrease to the test temperature at a cooling rate of 0.1 °C/s, after which tension was applied on the specimen with a strain rate of 10^{-2} s^{-1}.

As is shown in Fig. 3.39, strain concentration occurred at the corner of the slab in the straightening section, and the corresponding temperature was 750–850 °C. Therefore, corner crack propagation experiments were carried out at 750, 800 and 850 °C to determine the critical criteria for corner crack propagation.

3.2.2.2 Determination of Critical Strain of Corner Crack Propagation in 0.37% Ti Micro-alloyed Steel

Since the critical strain of crack propagation during the process of compression was tested, the tensile sample with prefabricated crack was designed and machined according to Fig. 3.40. According to the scheme in the above section, firstly, the continuous casting billet of titanium micro-alloyed steel is subjected to tensile test at 750 °C, and the stress–strain curve is shown in Fig. 3.41. It can be seen that the stress firstly increases and then decreases with the increase of strain, and the stress peak occurs when the strain reaches 0.143, the stress begins to decrease when the strain continues to increase, and the decreasing rate increases obviously when the strain reaches about 0.2, which indicates that the crack is very likely to expand near the strain 0.143. In order to determine the critical strain of crack growth, tensile

Fig. 3.38 **a** Temperature curve of titanium steel wide-thick slab and **b** strain and temperature cloud diagram of titanium steel wide-thick slab

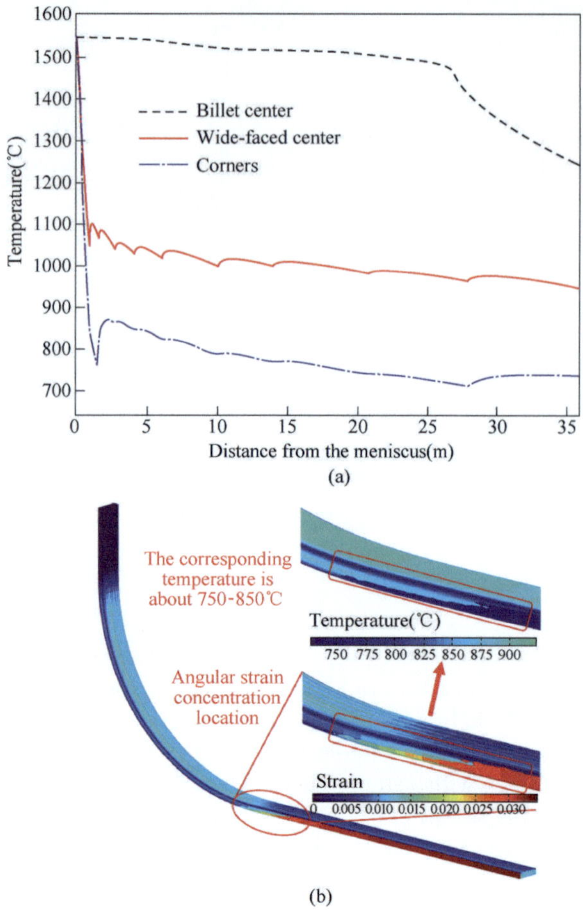

Fig. 3.39 Heating system for corner crack propagation test

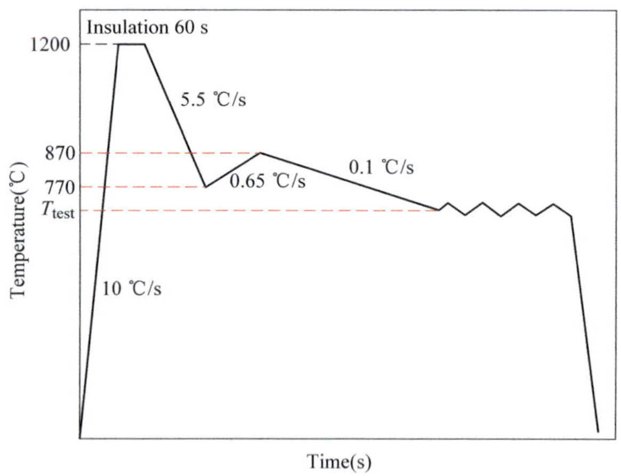

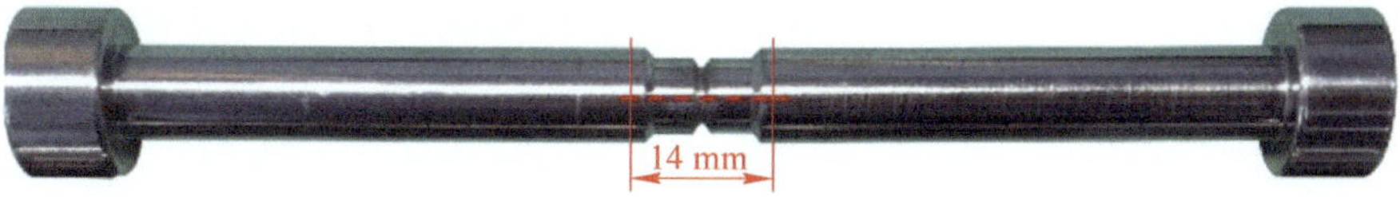

Fig. 3.40 Tensile sample of 0.37% titanium micro-alloyed steel

Fig. 3.41 Tensile stress–strain curve of high titanium steel at 750 °C

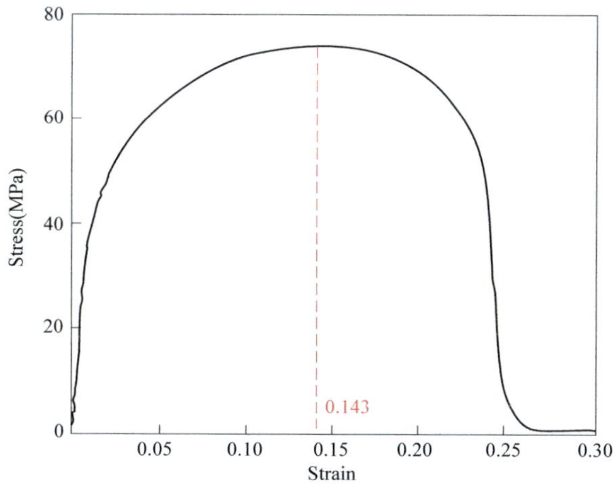

experiments with the corresponding variables 0.05, 0.10, 0.12, 0.143 and 0.18 were carried out at 750 °C, respectively. The crack growth of each sample observed by metallographic microscope is shown in Fig. 3.42.

As is shown in Fig. 3.42, the pre-crack does not propagate with the strain less than 0.143. When the strain reaches 0.143, the crack propagation phenomenon occurs in the pre-cracked sample, but the crack propagation length is small, only about 40 μm. When the strain increases to 0.18, the crack propagation is obvious, and the crack propagation length is about 100 μm. Combined with the stress–strain curve in Fig. 3.41, the crack begins to propagate at the strain corresponding to the stress peak, and the crack propagation leads to the degradation of the material properties, and then the stress decreases. Therefore, the critical strain of crack propagation should be the strain value when the stress–strain curve begins to decline, that is, the strain value corresponding to the stress peak.

Tensile tests were also carried out at 800 and 850 °C, and the strain values corresponding to the stress peak values were determined after the stress–strain curves were obtained, and the tensile tests and metallographic microstructure of the corresponding strain variables were observed. The test results were shown in Fig. 3.43b, c, respectively. The corresponding strain variables of the stress peak are 0.149 and 0.158, respectively. In conclusion, it can be determined that the critical strains of corner crack propagation of titanium micro-alloyed steel at 750 °C, 800 °C and 850 °C are 0.143, 0.149, and 0.158, respectively.

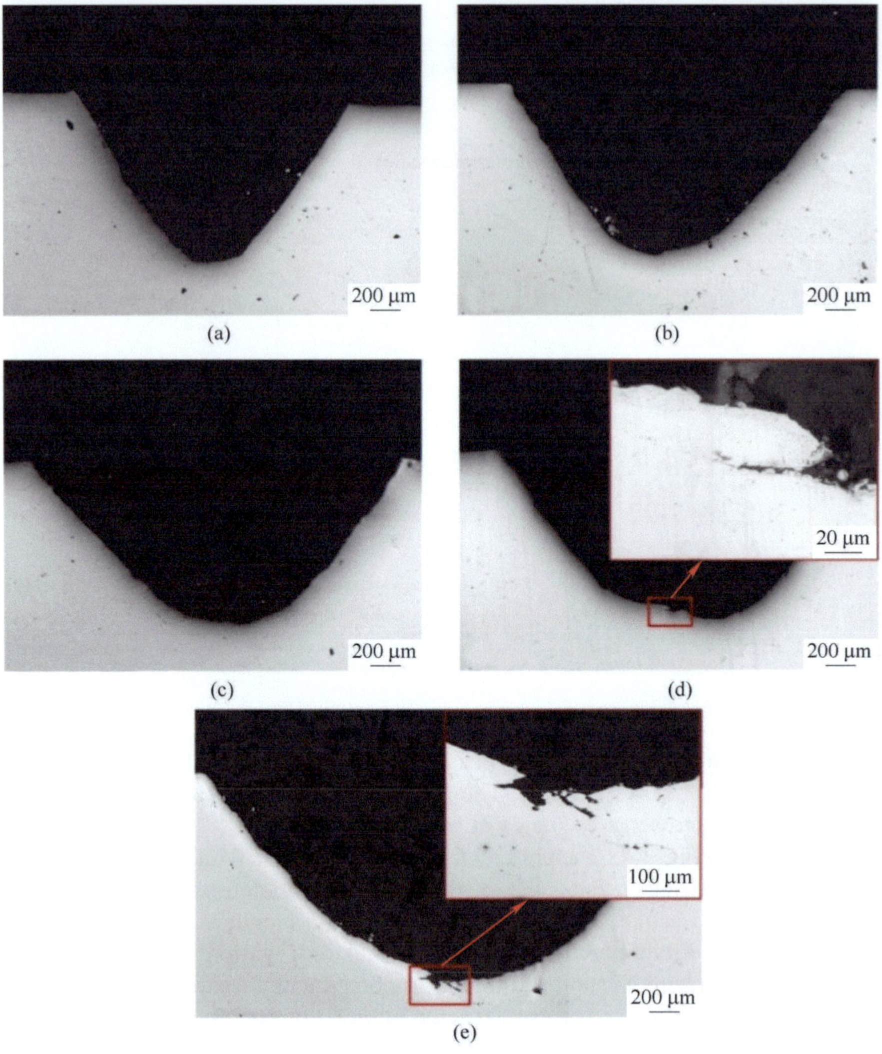

Fig. 3.42 Metallographic pictures of tensile samples of high titanium steel at 750 °C. **a** $\varepsilon = 0.05$; **b** $\varepsilon = 0.10$; **c** $\varepsilon = 0.12$; **d** $\varepsilon = 0.143$; **e** $\varepsilon = 0.18$

3.3 Prediction of Crack Risk During Heavy Reduction Process

Combined with the simulation results of 3D thermal–mechanical coupling simulation model in Sect. 3.1 and the experimental results of critical criterion in Sect. 3.2, a comprehensive analysis of the risk of crack initiation and propagation during the entire continuous casting process was conducted., which provided quantitative data

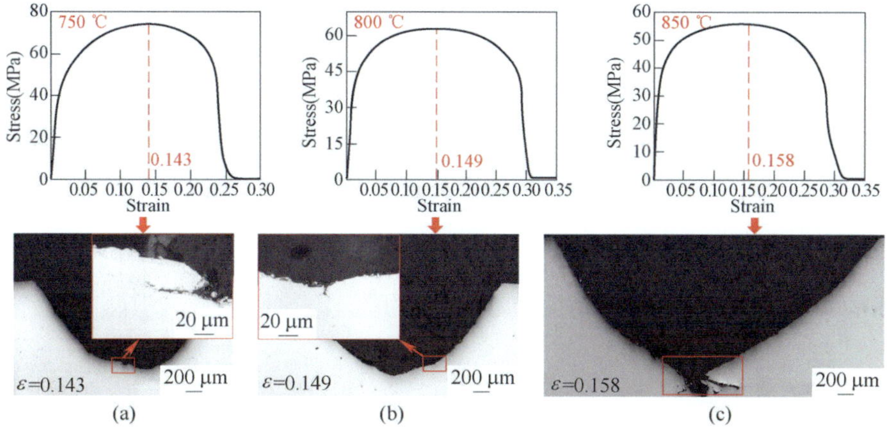

Fig. 3.43 Stress–strain curves and crack growth metallographic pictures of high titanium steel at different temperatures. **a** 750 °C; **b** 800 °C; **c** 850 °C

reference for the prevention and control of reduction crack. In particular, based on the 3D thermal–mechanical coupling simulation model developed in Sect. 3.1, secondary development interface of MSC. Marc was used to extract the required data. Subsequently, the tensile strain of key nodes was compared with the crack critical criterion, as a result, the crack risk coefficient of each node was determined and the crack risk contour was plotted. The calculation process is shown in Fig. 3.44.

3.3.1 Risk Prediction of Intermediate Crack Initiation of Bearing Steel GCr15 Bloom

Based on critical criterion of crack initiation in bearing steel GCr15 bloom measured in Sect. 3.2.1.2, the simulation results of temperature and deformation in Sect. 3.1.1 and the crack risk prediction model in Fig. 3.44, the risk of crack initiation of GCr15 bearing steel with 320 mm × 425 mm section was analyzed during the process of casting with a drawing speed of 0.58 m/min and a superheat of 30 °C.

The condition of the intermediate crack initiation condition is the strand temperature lies within the range of ZDT-LIT, and the tensile strain of the strand exceeds the critical strain. Therefore, in the process of thermal–mechanical coupled FE simulation, whether each node satisfies the conditions of intermediate crack initiation is judged by incremental steps one by one. Subsequently, the total incremental steps of each node meeting the conditions of intermediate crack initiation are counted to characterize the risk of intermediate crack initiation. The simulation scheme of different pressure reduction simulated by different rollers is shown in Fig. 3.45.

As depicted in Fig. 3.45, the crack risk prediction model was to calculate the crack risk when single roll was pressed separately for 5mm, and the crack risk contour map

Fig. 3.44 Crack risk model calculation flowchart

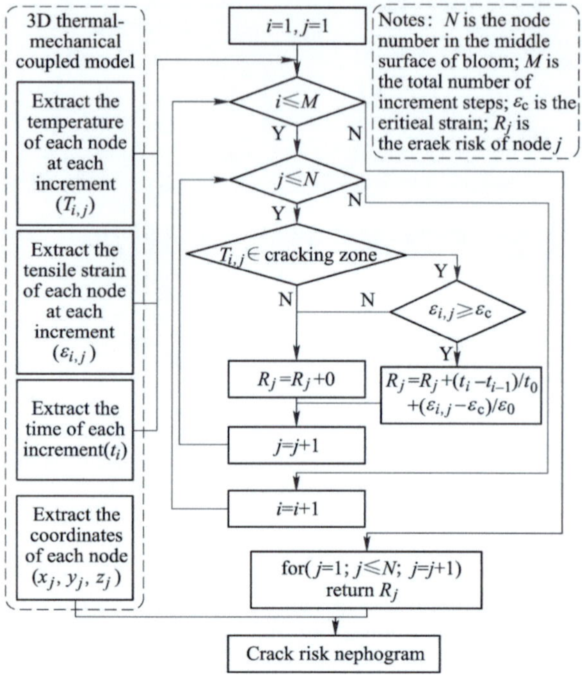

Roll number	Depressed amount(mm)											
	1	3	5	7	9	11	13	15	17	19	21	23
1	▲		▲									
2		▲	▲									
3			▲									
4			▲	▲	▲	▲						
5			▲				▲	▲				
6			▲						▲			
7			▲							▲	▲	
8			▲									▲
9			▲									▲

Fig. ▲ The scenario has been simulated ■ The scenario is not simulated

Fig. 3.45 Single roll reduction experimental scheme

was drawn. Cracks emerged when No.1 roller and No.2 roller were pressed down 5 mm, there is no crack risk when No.3–No.9 rollers were pressed down 5 mm. In order to further investigate the influence of single roll pressure on intermediate crack, the other conditions marked in Fig. 3.45 were simulated and the crack risk contour map was generated.

After simulating each condition in Fig. 3.45, the crack risk prediction model was utilized to calculate the risk coefficient of middle crack initiation. The intermediate crack initiation risk occurred when No.1 roller was pressed down 5 mm, No.3 roller was pressed down 5 mm, No.7 roller was pressed down 17 mm and 21 mm, and no crack risk occurred in the other calculation examples. The intermediate crack initiation risk contour is shown in Fig. 3.46. It can be observed that when No.1 roller and No.2 roller were pressed down 5 mm, the crack risk is higher and the crack risk area is larger. The allowable reduction amount increases with the roller moves backward. When No.6 roller was pressed down 17 mm, there is a small crack risk about 50 mm away from the thickness center. When No.7 roller was pressed down 19 mm and 21 mm, there is a risk of crack in a small area at the center of strand. The comparison shows that when the roller moves backward, the maximum allowable reduction amount increases and the crack risk area decreases. The crack risk of each condition was statistically analyzed, and the relationship between the reduction amount and crack risk was obtained, as shown in Fig. 3.47.

Figure 3.48 presents the curve of central solid fraction of bearing steel at 0.58 m/min with the position of casting flow. The central solid fraction of strand at the position of No.1-No.9 rollers is shown in Table 3.5. Combining Figs. 3.46 and 3.47, it can be noticed that when No.1 roller and No.2 roller were pressed down, the central solid fraction of the casting blank is relatively low ($f_s = 0.31$–0.37), and the maximum allowable reduction amount is small. With the increase of solid fraction ($f_s = 0.43$–0.51), the allowable reduction amount of No.4 roller and No.5 roller increases obviously. A small crack risk emerged at some areas when No.6 roller ($f_s = 0.69$) was pressed down 17 mm. At the position of No.7 roller ($f_s = 0.82$), the solid fraction in the center of the billet is relatively high, and the crack initiation risk exists when 19 mm and 21 mm are pressed. At the position of No.8 roller and No.9 roller, the strand has completely solidified, and there is no risk of intermediate crack initiation when the amount of reduction continues to increase.

Combined with the risk analysis of intermediate crack initiation of bearing steel, it can be seen that when the bearing steel bloom is produced on site, the reduction amount of No.1 roller and No.2 roller should be kept within a narrow range; The reduction amount of No.3 roller and No.4 roller can be increased gradually and that of No.6 roller and No.7 should not be increased too much; No.8 roller and No.9 can perform a large amount of reduction to reduce porosity, shrinkage and other defects.

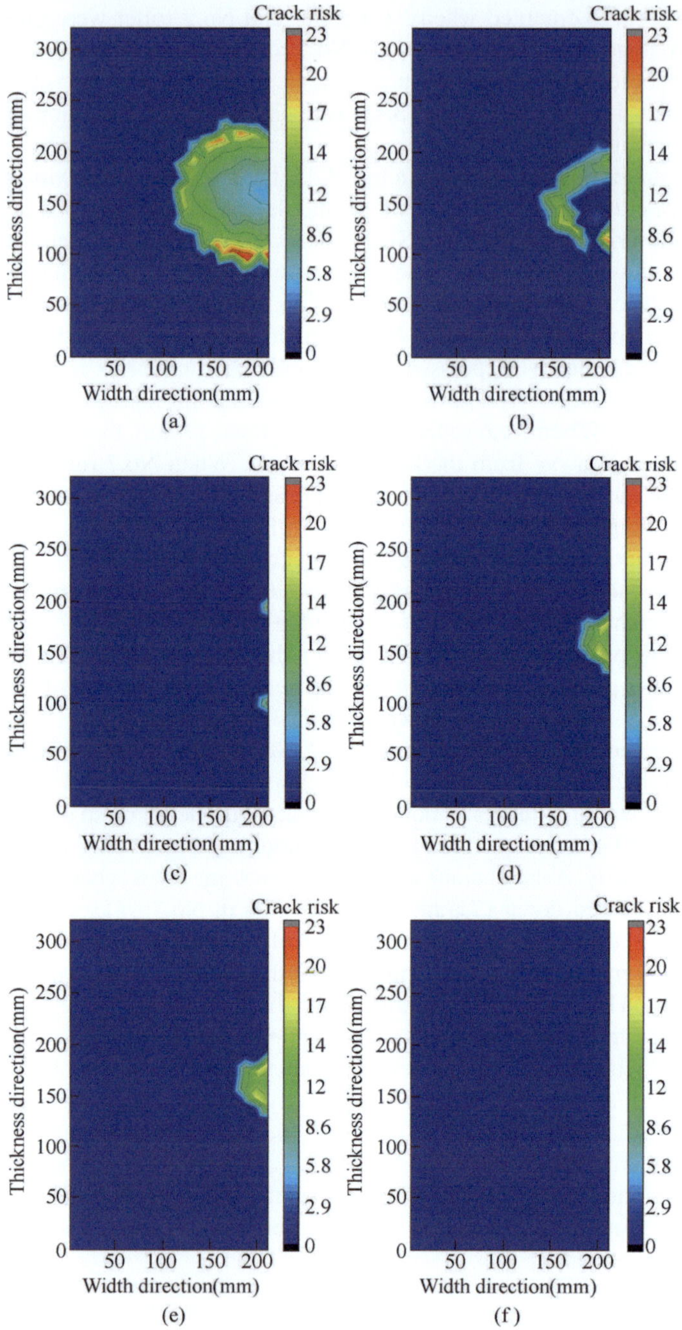

Fig. 3.46 Risk map of intermediate crack initiation in bearing steel bloom under single roll reduction. **a** No.1 roll down 5 mm; **b** No.2 roll down 5 mm; **c** No.6 roll down 17 mm; **d** No.7 roll down 19 mm; **e** No.7 roll down 21 mm; **f** the rest of the studies

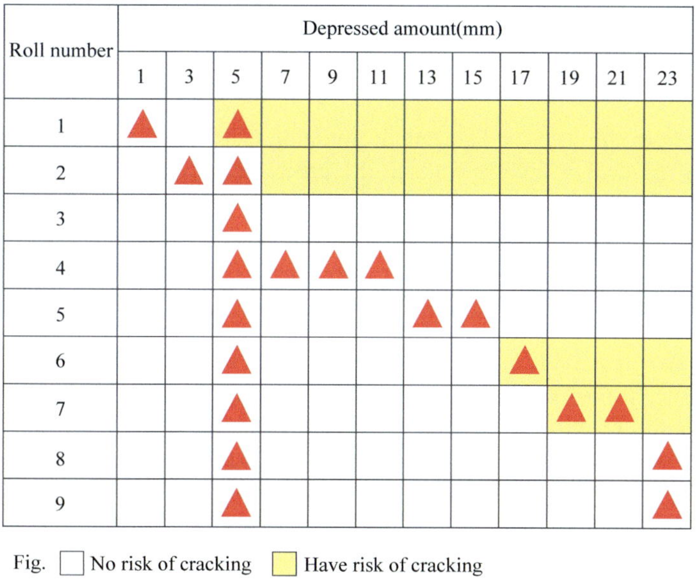

Fig. 3.47 Relation diagram between reduction and crack risk of each tensioner

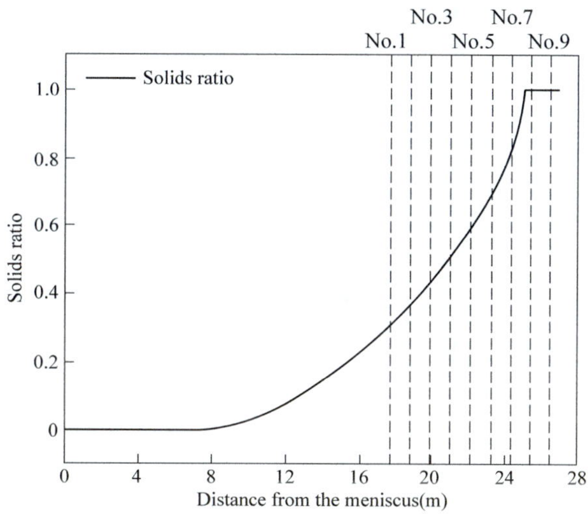

Fig. 3.48 The bearing steel solid-phase rate change curve at casting speed of 0.58 m/min

Table 3.5 Solid-phase ratio at the bloom center at each withdrawal machine

Roller number	No.1	No.2	No.3	No.4	No.5	No.6	No.7	No.8	No.9
Solid fraction	0.31	0.37	0.43	0.51	0.59	0.69	0.82	1.0	1.0

3.3.2 Risk Prediction of Intermediate Crack Initiation of Wide-Thick Steel Q345E Slab

In this section, the intermediate crack initiation risk for wide and thick slabs of microalloying steel is analyzed using the critical criteria for intermediate cracks measured in Sect. 3.2.1.3, the simulation results of temperature and deformation in Sect. 3.1.2, and the crack risk prediction model described in Sect. 3.3. Firstly, the deformation behavior of micro-alloyed steel slabs under 20, 30, 40 and 50 mm was simulated by thermal/mechanical coupling model. On this basis, the risks of intermediate crack initiation under 20, 30, 40 and 50 mm were analyzed. The downsizing process is shown in Table 3.6. Figure 3.49 illustrates the change curve of solid-phase ratio with the position of casting flow. Figure 3.50 shows the cloud diagram of crack risk at the exit of each sector segment for each downsizing plan.

When Case 1 was pressed down 20 mm, Case 2 was pressed down 30 mm and Case 3 was pressed down 40 mm, the risk coefficients of cracks initiation of the wide-thick slab of micro-alloyed steel were all zero. The probability of cracks initiation is extremely low under these three processes. When Case 4 was pressed down 50 mm,

Table 3.6 The micro-alloyed steel wide-thick slab reduction process

Reduction	Reduction amount (mm)						Total amount (mm)
Case	Seg.9	Seg.10	Seg.11	Seg.12	Seg.13	Seg.14	
Case 1	1	2	3	4	5	5	20
Case 2	2	3	4	5	7	9	30
Case 3	2	4	7	8	9	10	40
Case 4	2	6	9	10	11	12	50

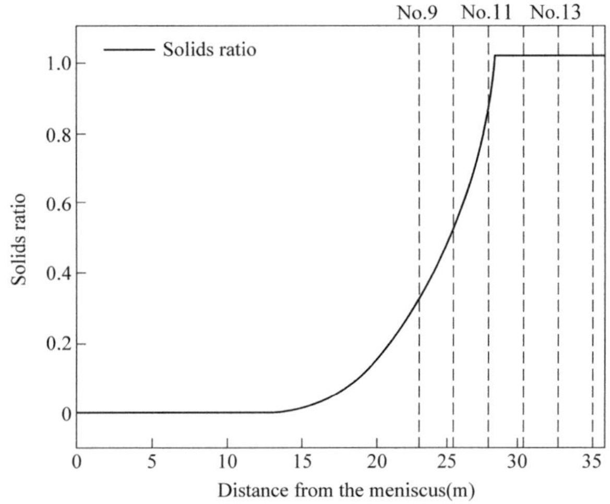

Fig. 3.49 The micro-alloyed steel solid-phase rate change curve at casting speed of 0.80 m/min

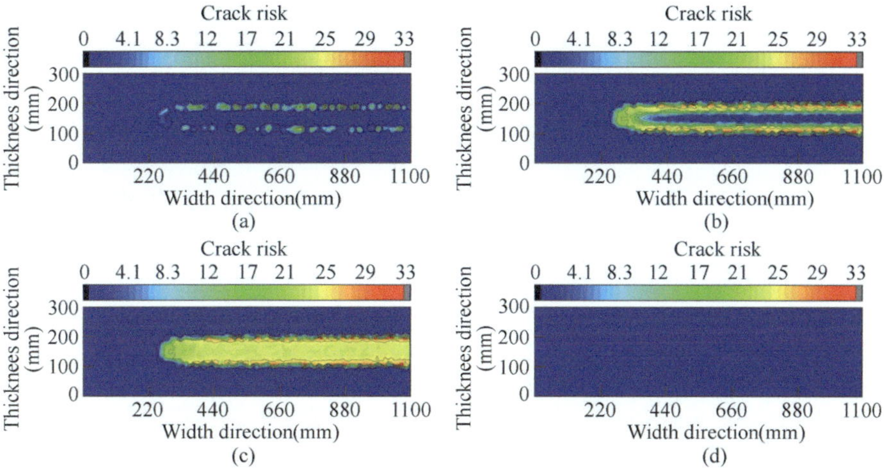

Fig. 3.50 Risk contour of intermediate crack initiation in micro-alloyed steel wide-thick slab. **a** Case 4-Seg.10; **b** Case 4-Seg.11; **c** Case 4-Seg.12; **d** the rest of the studies

the crack risk at the exit of Seg.10 was shown in Fig. 3.50a; At the exit of Seg.11, the crack risk increased further as depicted in Fig. 3.50b, c showed the contour of crack risk at the exit of Seg.12; The crack risk at the exit of Seg.13 and Seg.14 is the same as that at the outlet of Seg.12; The risk coefficient of intermediate crack initiation in the other cases was zero, and the cloud map was shown in Fig. 3.50d.

The comparison of risk contour shows that with the increase of the total reduction amount, the risk area of crack initiation gets wider. This is primarily because the increase of the total reduction amount leads to the deformation exceeding the allowable limit value when the solid fraction is low, resulting in the formation of cracks near the solid–liquid interface. When Case 4 was pressed at the Seg.10, the intermediate crack distributed along the solid–liquid interface due to the low solid fraction at this time, although the amount of reduction is only 6 mm. The crack propagates to the center of the strand with the increasing of reduction amount at the Seg.11. On the one hand, this is because the solid–liquid interface shrinks to the center of the strand during the cooling process. Excessive reduction amount leads to the risk of crack occurrence near the center of the strand; On the other hand, the cracks formed by the Seg.10 may continue to propagate toward the center of the strand in the subsequent reduction process, and ultimately lead to a larger crack risk area. The end of solidification was located at Seg.12, and the large amount of reduction at Seg.12 led to the further propagation of the crack risk area, as shown in Fig. 3.50c. The reduction amount further increases at Seg.13 and Seg.14, but the crack risk contour diagram remains the same as that at the 12th section, indicating that the intermediate crack initiation risk is very low after the solidification end point, and the intermediate crack initiation risk does not increase after the solidification end point. Similar to the production of bearing steel GCr15 continuous casting bloom, when the solid fraction is low, the reduction should be controlled at a low level. When the solid fraction is higher, the reduction amount can be increased appropriately. The porosity and shrinkage of strand can be improved by applying more pressure after the end of solidification.

3.3.3 Risk Prediction of Surface Crack Propagation of Ti Micro-alloyed Steel

The crack occurrence rate at the corner of the wide-thick slab of micro-alloyed steel is relatively high, and it is easy to propagate during the reduction process at the end of solidification, which affects the quality of the reduction process. In this section, titanium micro-alloyed steel is chosen as the research object and the risk of corner crack propagation is analyzed based on the critical criterion of corner crack propagation measured in Sect. 3.2.2.2.

The transverse crack propagation of corner was mainly affected by the tensile strain, the crack propagation occurs while the tensile strain exceeds the critical strain. The reduction position of the wide-thick slab caster is in the horizontal section and the drawing direction of the horizontal is the x direction in the model. Therefore, the risk of crack propagation can be analyzed according to the strain in the x direction of corners. The strain in the x direction of titanium micro-alloyed steel with the reduction Case 1- Case 4 in Table 3.5 was obtained, as shown in Fig. 3.51.

It can be observed that the strain in the x direction of the inner and outer corners of the arc increases step by step with the increase of the distance from the meniscus, the reduction amount increases gradually. The strain in the x direction of the corner increases with the increasing of the reduction amount from Case 1- Case 4. The total reduction amounts of patterns listed in Table 3.4 are 20 mm, 30 mm, 40 mm and 50 mm, the corresponding maximum strains in x direction of corners are 0.068, 0.094, 0.115 and 0.139, respectively. The critical strains of corner crack propagation in titanium micro-alloyed steel measured in Sect. 3.2.2 at 750 °C, 800 °C and 850 °C are 0.143, 0.149 and 0.158, respectively. Compared with Fig. 3.51, it can be seen that the strain in the x direction of the inner and outer arc corners of Case 1- Case 4 after reduction is less than 0.14, so 0.37% titanium micro-alloyed steel has a low risk of crack propagation in corners of Case 1- Case 4. At for Case 4, the strain at the corner position in the x direction is close to 0.14. If the amount of reduction is further increased, the crack propagation at the corner may be triggered.

It can be concluded from the above analysis that the corner crack propagation risk of titanium micro-alloy steel is low in Case 1- Case 4. Nevertheless, the corner crack propagation may be initiated when the reduction amount is continuously increased. Therefore, in the production of wide-thick slabs of this kind of steel, the amount of reduction should be controlled to avert corner crack caused by an excessive reduction amount.

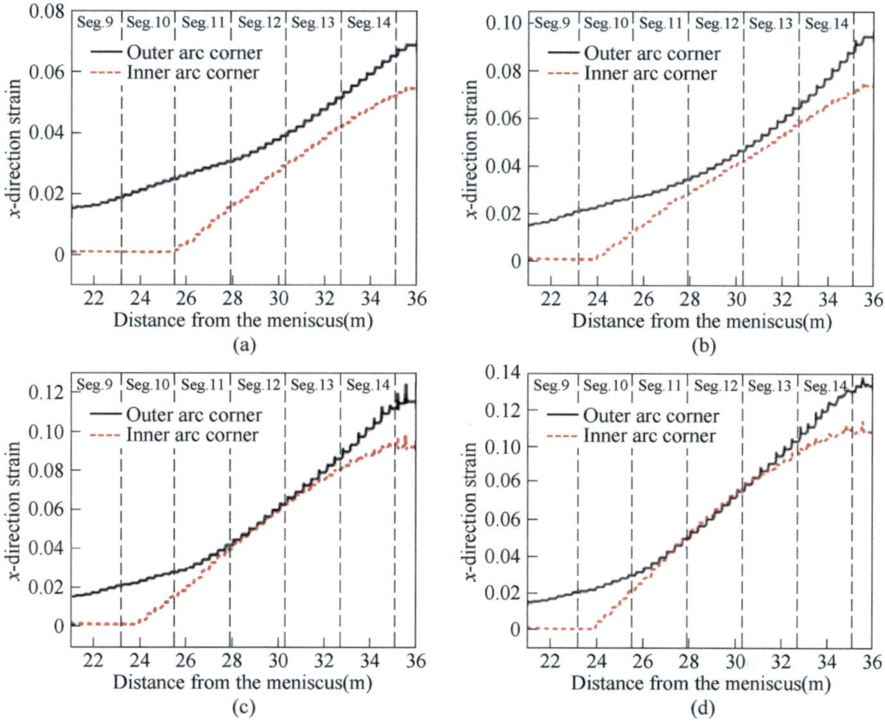

Fig. 3.51 Strain in the x direction at the corner position of high titanium steel wide-thick slab. **a** Pressed down 20 mm; **b** pressed down 30 mm; **c** pressed down 40 mm; **d** pressed down 50 mm

References

1. Suzuki HG, Nishimura S, Yamaguchi S (1982) Characteristics of hot ductility in steels subjected to the melting and solidification. Trans Iron Steel Inst Jpn 22(1):48–56
2. Savage J, Pritchard W (1954) The problem of rupture of the billet in the continuous casting of steel. J. Iron Steel Inst 178(3):269–277
3. Nozaki T, Matsuno J, Murata K et al (1978) A secondary cooling pattern for preventing surface cracks of continuous casting slab. Trans Iron Steel Inst Jpn 18(6):330–338
4. Mizikar EA (1967) Mathematical heat transfer model for solidification of continuously cast steel slabs. Trans Metallur Soc AIME 239:1747–1753
5. Li CS, Thomas BG (2004) Thermomechanical finite-element model of shell behavior in continuous casting of steel. Metall Mater Trans B 35(6):1151–1172
6. Ueshima Y, Mizoguchi S, Matsumiya T et al (1986) Analysis of solute distribution in dendrites of carbon steel with δ/γ transformation during solidification. Metall Trans B 17(4):845–859
7. ElBealy M, Thomas BG (1996) Prediction of dendrite arm spacing for low alloy steel casting processes. Metall Mater Trans B 27(4):689–693
8. Won YM, Thomas BG (2001) Simple microsegregation model for steel solidification. Metall Mater Trans A 32(7):1755–1767
9. Wang WL, Zhu MY, Cai ZZ et al (2012) Micro-segregation behavior of solute elements in the mushy zone of continuous casting wide-thick slab. Steel Res Int 83(12):1152–1162
10. Mixwkami H, Murakami K (1977) Mechanical properties of continuously cast steel at high temperature. Nihan Kohan Corporation Testu-to-Hagan6 (Iron and Steel) 63:146–265

11. Wu CH, Ji C, Zhu MY (2018) Numerical simulation of bulging deformation for wide-thick slab under uneven cooling conditions. Metall Mater Trans B 49(3):1346–1359
12. Uehara M (1983) Mathematical modelling of the unbending of continuously cast steel slabs. University of British Columbia, Vancouver
13. Miyazaki J, Narita K, Nozaki T (1981) On the internal cracks caused by the bending test of small ingot. Trans Iron Steel Inst Jpn. 21:B210
14. Hiroyuki Y, Kiichi N, Takasuke M et al (1981) On the surface cracks caused by the bending test of small ingot. Trans Iron Steel Inst Jpn 21:B470
15. Matsumiya T, Ito M, Kajioka H et al (1986) An evaluation of critical strain for internal crack formation in continuously cast slabs. Trans Iron Steel Inst Jpn 26(6):540–546
16. Bernhard C, Xia G (2006) Influence of alloying elements on the thermal contraction of peritectic steels during initial solidification. Ironmaking Steelmaking 33(1):52–56
17. Rowan M, Thomas BG, Pierer R et al (2011) Measuring mechanical behavior of steel during solidification; modeling the SSCC test. Metall Mater Trans B 42(4):837–851
18. Won YM, Yeo TJ, Seol DJ et al (2000) A new criterion for internal crack formation in continuously cast steels. Metall Mater Trans B 31(4):779–794
19. Kato T, Ito Y, Kawamoto M et al (2003) Prevention of slab surface transverse cracking by microstructure control. ISIJ Int 43(11):1742–1750
20. Yamanaka A, Nakajima K, Okamura K (1995) Critical strain for internal crack formation in continuous casting. Ironmaking Steelmaking 22(6):508–512
21. Yu CH, Suzuki M, Shibata H et al (1996) Simulation of crack formation on solidifying steel shell in continuous casting mold. ISIJ Int 36(SI):159–162
22. Punnose S, Mukhopadhyay A, Sarkar R et al (2013) Determination of critical strain for rapid crack growth during tensile deformation in aluminide coated near-a titanium alloy using infrared thermography. Mater Sci Eng Proc Conf 576:217–221
23. Enos D, Scully J (2002) A critical-strain criterion for hydrogen embrittlement of cold-drawn, ultrafine pearlitic steel. Metall Mater Trans A 33(4):1151–1166
24. Clyne TW, Wolf M, Kurz W (1982) The effect of melt composition on solidification cracking of steel, with particular reference to continuous casting. Metall Trans 13(2):259–266
25. Kim K, Han H, Yeo T et al (1997) Analysis of surface and internal cracks in continuously cast beam blank. Ironmaking Steelmaking 24(3):249–256
26. Meng XN, Zhu MY (2009) Effect of continuous casting mold oscillation parameters on slab surface cracks formation under high casting speed. Iron & Steel 44(8):34–38
27. Ji C, Cai Z Z, Luo S, et al (2012) Control technology for wide-thick slab corner crack during continuous casting process of micro-alloy steel. In: Micro-alloy steel continuous casting crack control technology seminar. Jiu Jiang, pp 1–10
28. Ji C, Cai ZZ, Wang WL et al (2014) Effect of transverse distribution of secondary cooling water on corner cracks in wide thick slab continuous casting process. Ironmaking Steelmaking 41(5):360–368

Chapter 4
Closure of Internal Porosity in Continuous Casting Strand During Heavy Reduction Process

Shrinkage cavity is a common internal defect in continuous casting billets [1], which can significantly reduce the mechanical properties and service life of metallic materials. As is known, defects such as shrinkage cavities are the sources of crack initiation [2]. The implementation of unreasonable reduction in the bending and straightening section of the continuous caster is extremely likely to cause crack propagation at the shrinkage cavity. After the continuous casting billet is heated by the heating furnace, the shrinkage cavity is prone to oxidation. Once the shrinkage cavity defect cannot be welded during rolling, it will expand into crack, seriously affecting the yield of the product. According to the size of the shrinkage cavity, it can be divided into micro-shrinkage cavity and macro-shrinkage cavity. Shrinkage cavity can't be removed through the subsequent heat-treatment processes. Therefore, the best strategy to reduce the porosity of the casting billet is to better understand the formation of shrinkage cavities and thus determine appropriate strategies to suppress its impact on the quality of the casting billet.

Predecessors mainly focused on studying the evolution behavior of spherical macro-shrinkage cavities under conditions such as forging and rolling [3, 4]. In recent years, with the emergence and promotion of the heavy reduction technology at the end of continuous casting solidification, scholars have successively started to study the evolution effect of heavy reduction implementation on shrinkage cavities [5, 6]. Based on the metal rheological characteristics obtained in Chap. 2 and the continuous casting billet deformation model established in Chap. 3, our team has established a finite element model for the mechanical reduction deformation of the continuous casting billet at the end of solidification with preset micro- and macro-shrinkage cavities [7–10], which is used to reveal the law of increasing the density of the casting billet during the reduction process and provide quantitative data support for reasonably determining the reduction amount and reduction interval.

In view of this, based on the distribution law of the heavy reduction implementation equipment and the porosity zone and the morphology of the shrinkage cavity for wide and thick slabs and large blooms, this chapter uses numerical simulation

C. Ji et al., *Solidification End Reduction of Continuous Casting Process*, Key Steel Manufacturing Technologies by Collaborative Innovation,
https://doi.org/10.1007/978-981-96-8544-8_4

methods to study the evolution law of shrinkage cavities during the heavy reduction process at the end of solidification.

4.1 Morphology, Distribution, and Prediction of Porosity in Central Region of Continuous Casting Billet

4.1.1 Formation and Prediction Method of Porosity in Solidification Process of Continuous Casting Billet

During the solidification process, molten steel continuously flows to the solidification area to compensate for solidification shrinkage. If the molten steel cannot effectively fill the relevant area, shrinkage cavities will be generated [9]. However, the flowability of molten steel continuously decreases as the metal gradually solidifies. When the dendrites in the solidification area form a fixed network structure, the replenishment ability of molten steel will be further reduced until it cannot be replenished [10, 11].

Predicting the distribution area and size of shrinkage cavities is a challenging task, and people attempt to analyze it through complex three-dimensional numerical models. However, due to the mathematical complexity of these models and the lack of reliable databases, many researchers have developed simpler "criterion functions" to predict the defect formation probability at various locations in the casting billet. The criterion function is a simple rule related to cooling rate, solidification rate, temperature gradient, etc. The shrinkage cavity prediction by different methods can be summarized as follows:

(1) Thermal model [12]: Solve the energy transport equation to determine the last solidification area or the area where molten steel replenishment is restricted.
(2) Thermal/volume calculation model [13]: Predict the location of the free surface and the last solidification area by solving the energy transport equation and mass conservation.
(3) Thermal/fluid flow model [14]: Solve the mass and energy transport equations to predict the location of the free surface and the last solidification area.
(4) Gas pore nucleation and growth model [15]: When dendrites form a continuous solid-phase grid structure, calculate the nucleation and growth of pores; liquid flow is described as flow through a porous medium.

This paper uses the Niyama correction model based on the thermal/fluid flow model. This model not only considers local thermal conditions but also the solidification characteristics of metals [16, 17].

$$Ny^* = C_\lambda \dot{T}^{\frac{1}{3}} Ny \sqrt{\frac{\Delta p_{\mathrm{cr}}}{\mu_1 \beta \Delta T_{\mathrm{f}}}} \tag{4.1}$$

where Δp_{cr} is the critical pressure drop; C_λ is the material property parameter; $\dot{T}$ is cooling rate; Ny is Niyama criteria for porosity formation probability, $Ny = G/\dot{T}$; μ_l is the dynamic viscosity; β is the solidification shrinkage; ΔT_f is the cooling interval.

4.1.2 Distribution and Morphology of Porosity in Wide-Thick Continuous Casting Slab

4.1.2.1 Distribution and Prediction of Porosity Zones in Wide-Thick Slab

The porosity zone at the end of solidification of wide-thick slabs is unevenly distributed on the cross-section. This is mainly due to the different solidification characteristics and processes. Generally speaking, due to the nozzle impact within the continuous casting mold for wide-thick slabs and the uneven distribution of the cooling water flow density in the width direction, the positions at 1/8 to 1/4 of the width solidify last [18]. To observe the internal defects of the slab, a high-frequency ultrasonic microscope (SAM) was used to detect the cross-section of a micro-alloyed steel Q345E wide-thick slab with a cross-sectional dimension of 2000 mm × 280 mm. The morphology is shown in Fig. 4.1a. In Fig. 4.1a, the dark part represents the matrix, and the light dots represent internal defects (here specifically referring to shrinkage cavities). As can be seen from Fig. 4.1a, most of the shrinkage cavities are distributed within the dotted line box. This area is the porosity zone and is symmetrically distributed along the center line.

In order to observe the distribution of the porosity zones in wide-thick slab more clearly and intuitively, the grid of 10 mm × 10 mm is used to divide Fig. 4.1a. Equation (4.2) is applied to calculate the porosities within each grid, and ultimately, the porosities distribution map is drawn using image processing software, as shown in Fig. 4.1b.

$$
\begin{aligned}
D &= \frac{S_T - S_V}{S_T} \times 100\% \\
&= (1 - \frac{S_V}{S_T}) \times 100\% = 1 - P
\end{aligned}
\tag{4.2}
$$

where S_T is total region detected, mm^2; S_V is total porosity region detected, mm^2; D is density; P is porosity.

In terms of thickness direction, the porosity zone of wide and thick slabs is mainly distributed in the 1/3 thickness region symmetrical along the thickness center line. In terms of width direction, there is a transition area from the narrow face to 1/8 of the width in the porosity zone, and it is discontinuous, which is caused by the non-uniform solidification in the width direction. At the same time, the 1/8 area in the width direction of wide and thick slabs is the area with the most serious porosity.

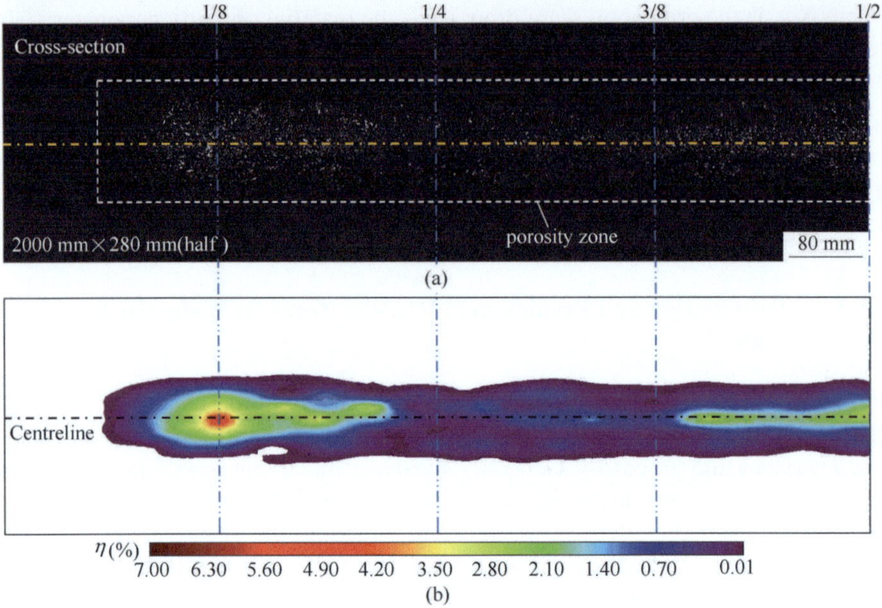

Fig. 4.1 The microstructure of wide-thick slab porosity zone distribution diagrams is presented in this section. **a** The scanning acoustic microscopy (SAM) image; **b** the porosity distribution cloud map

Based on the solidification process of Ni-based alloys, Plancher et al. [19] proposed two formation mechanisms of shrinkage cavities: one is the shrinkage cavity between secondary dendrite arms. In the high-temperature section at the initial stage of solidification, due to the poor flow of high-temperature melt perpendicular to the primary dendrite trunk, micro-shrinkage cavities are formed between the secondary dendrite arms during the growth and healing of dendrites, which are mainly micro-shrinkage cavities. The other is the shrinkage cavity between dendrite gaps. In the low-temperature section at the late stage of solidification, the thermal shrinkage of dendrites leads to the formation of gaps between dendrites, and the molten steel cannot be replenished in time to produce shrinkage cavities, which can produce macro-shrinkage cavities. Takahashi et al. [20] believe that when the solid fraction f_s reaches 0.31, the dendrite grid forms and hinders the flow of molten steel. When the solid fraction reaches 0.67, the dendrite network structure will completely block the flow of molten steel. Based on the research of the above two, it can be considered that when the solid fraction $0.31 < f_s \leq 0.6$ in the central area of the continuous casting billet, it is mainly the micro-shrinkage cavity between secondary dendrite arms; when the solidification end point at the center is $f_s > 0.67$, it is mainly the shrinkage cavity between dendrite gaps (including micro-shrinkage cavities and macro-shrinkage cavities).

The liquidus temperature of the micro-alloyed steel wide and thick slab is 1517.7 °C, the solidus temperature is 1467.5 °C, so the cooling interval ΔT_0 is

50.2 °C. The density of the molten steel of micro-alloyed steel is 7030.14 g/mm^3, and the density after solidification of the molten steel is 7670.69 g/mm^3. Therefore, the solidification shrinkage β is 0.0911. Substituting the above parameters and the solidification heat transfer simulation results into the Niyama correction model of Eq. (4.1), and taking the solid fraction $f_s = 0.67$ of the aforementioned research conclusions as the basis, the distribution area of macro-shrinkage cavities between dendrite gaps can be obtained, as shown in Fig. 4.2a. If according to the aforementioned research conclusion, taking the solid fraction $f_s = 0.31$–0.67 as the basis, the distribution area of micro-shrinkage cavities between secondary dendrite arms can be obtained, which includes the entire porosity zone including macro and micro, as shown in Fig. 4.2b.

Figure 4.2c presents the measured porosity volume fraction within the porosity zones (in agreement with Fig. 4.1b). It can be clearly seen that the distribution of porosity zones predicted by the Niyama correction model shows a high degree of

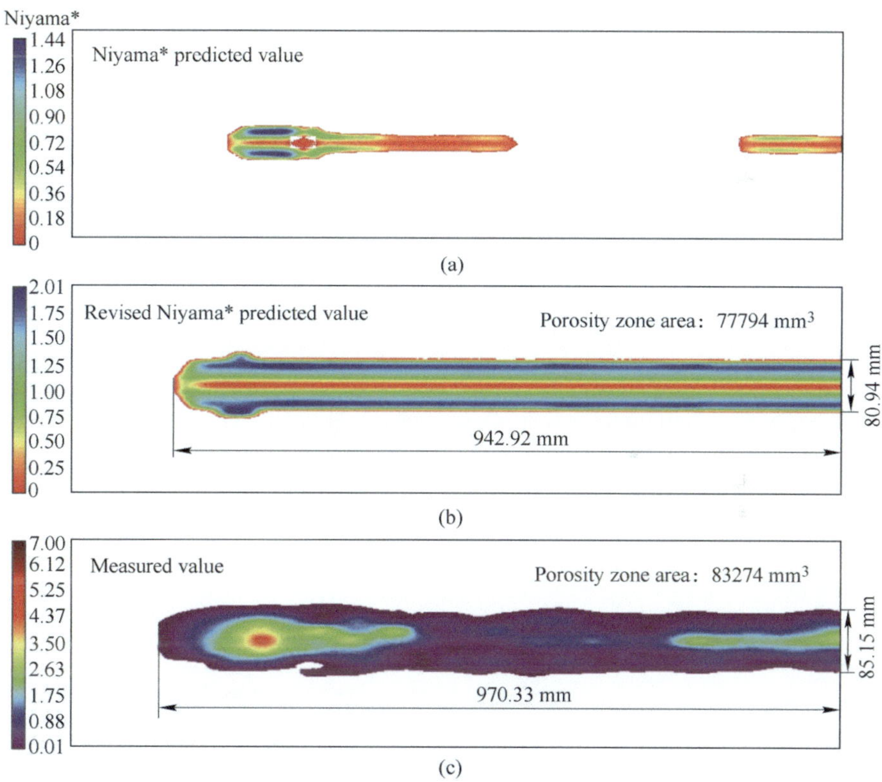

Fig. 4.2 A comparison between the predicted macroscopic and microporosities distributions based on the Niyama correction model and the actual measured porosity zones. **a** Macroporosities distribution in the interdendritic spaces; **b** microporosities distribution between secondary dendrite arms; **c** distribution of the measured porosity zones

consistency with the distribution of porosity zones obtained from the measurements. Moreover, the positions of the large porosities in the macroporosity distribution are found to be basically in line with the positions where a high porosity volume fraction was measured. According to the quantitative comparison, the relative error of the Niyama correction model's predicted porosity zones distribution value compared to the measured value is 2.82% in the width direction of the transverse section, and 4.94% in the thickness direction, indicating higher predictive accuracy in the width direction. Furthermore, the relative error between the predicted values and experimental values based on the porosity zones region ratio is 6.58%, which fully demonstrates that the Niyama correction model can accurately predict the distribution and severity of the porosity zones in the wide-thick slab.

4.1.2.2 Porosity Morphology in Porosity Zones of Wide-Thick Continuous Casting Slab

Based on the distribution of the porosity zones, further investigation of the morphology of porosities within the porosity zones in wide-thick slab was conducted. The ultrasonic scan images of the characteristic positions of the wide-thick slab are processed as shown in Fig. 4.3. The processing workflow of the characteristic positions using Avizo software is illustrated in Fig. 4.4, and the three-dimensional reconstruction of the porosities morphology is presented in Fig. 4.5. As shown in Fig. 4.3, based on the solidification characteristics and porosities distribution of the wide-thick slab, sampling is performed at positions $W_{1/8}$ (A), $W_{1/4}$ (B), and $W_{1/2}$ (C) in the width direction of the transverse section of the specimen. The sample dimensions are 50 mm in width and 93.3 mm in thickness.

Figure 4.4a shows the schematic diagram of the ultrasonic scan after binarization processing. After the preliminary three-dimensional reconstruction of these images, as shown in Fig. 4.4b, many small black dots are observed. These small black dots are caused by the clutter detected in the ultrasonic scan. Therefore, it is necessary to set a threshold within the Avizo software to filter out the clutter and carry out

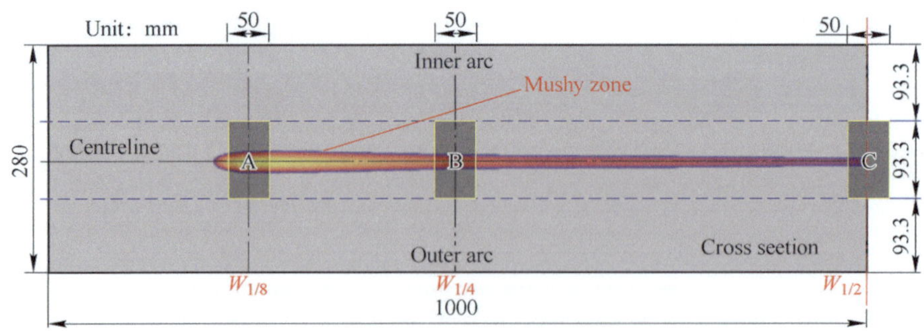

Fig. 4.3 Three-dimensional reconstruction sampling schematic of wide-thick slab billet for microalloy steel

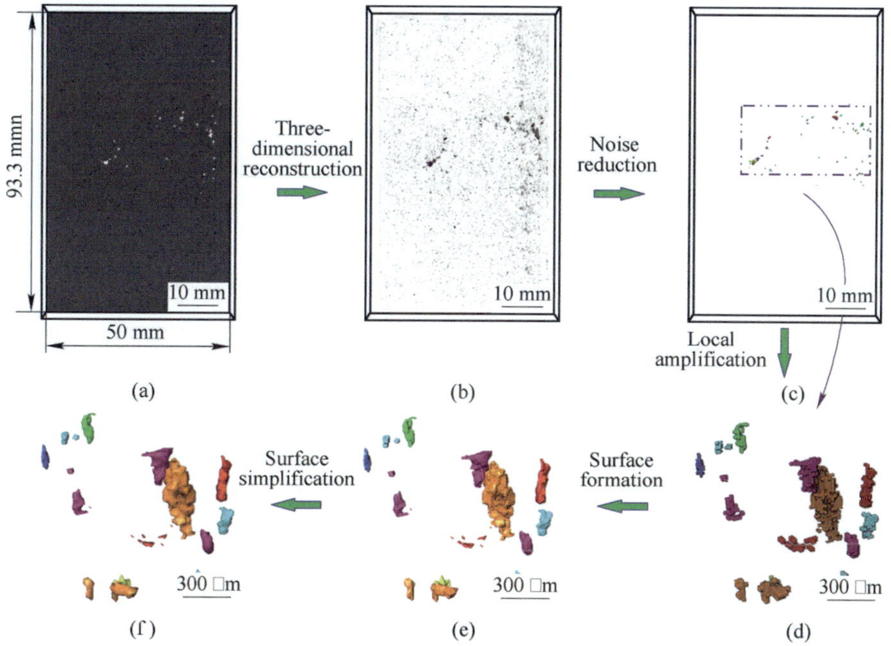

Fig. 4.4 Three-dimensional reconstruction process diagram of porosity

denoising operations, thereby obtaining Fig. 4.4c. Figure 4.4a shows the schematic diagram of ultrasonic scanning after binarization processing. After the preliminary three-dimensional reconstruction processing of these images, as shown in Fig. 4.4b, there are many small black dots in the image. These small black dots are caused by the noise detected in the ultrasonic scanning; thus, a threshold needs to be set in the Avizo software to filter out the noise, that is, denoising processing to obtain Fig. 4.4c. The local position of the porosity zones is enlarged to obtain Fig. 4.4d. The constructed porosities morphology is observed to be composed of a stack of pixels, demonstrating extreme irregularity in the morphology of the porosities. Building upon Fig. 4.4d, a smoother surface is generated using surface generation technology as shown in Fig. 4.4e, covering the porosities morphology with a layer of surface. However, the surface of the porosities still exhibits sharp transitions. To address this, surface simplification technology is utilized to transform the sharp transitions of the porosities surface into smooth ones, as depicted in Fig. 4.4f.

After the image processing steps shown in Fig. 4.4, three-dimensional reconstructions of the porosity morphology at different positions of the wide-thick slab, as depicted in Fig. 4.5, were obtained. Based on the porosity morphology, the porosities can be classified into three types: macroporosities with a size greater than or equal to 1 mm, microporosities with a size smaller than 1mm, and aggregated porosities formed by the aggregation of numerous porosities. To observe the morphology of the porosities more accurately, the porosities at the characteristic positions in

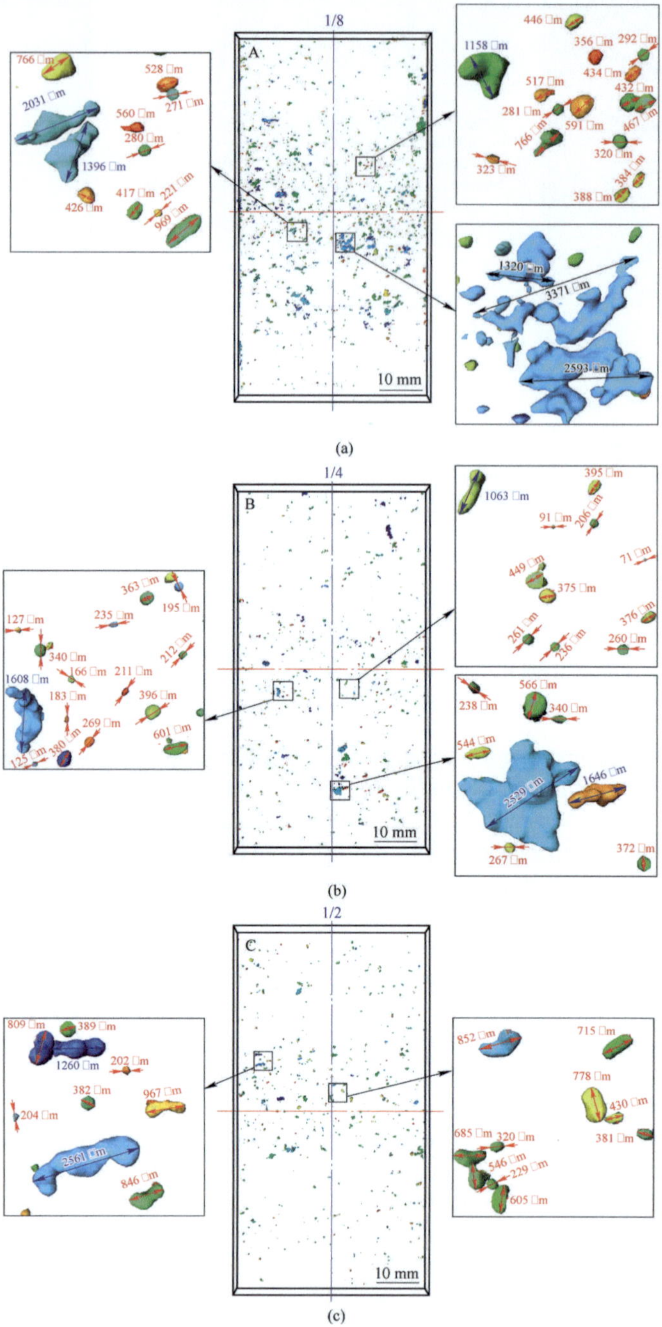

Fig. 4.5 Reconstruction of three-dimensional shrinkage cavity morphology at various locations on wide slab ingots. **a** 1/8 width location (Sample A); **b** 1/4 width location (Sample B); **c** 1/2 width location (Sample C)

Fig. 4.5 were locally magnified. From the magnified images, it was observed that the morphology of microporosities (size less than 1000 μm) is approximately ellipsoidal, but with different axial ratios, and microporosities are distributed in all three samples. The morphology of macroporosities (size greater than 1000 μm) is highly irregular, and a large number of them are located in the 1/8 width direction of Sample A. Aggregated porosities (formed by the aggregation of two or more porosities with a size greater than 1000 μm) are also mainly present in the 1/8 width direction of Sample A, and are formed by the aggregation of macroporosities and microporosities. They are usually of a huge size and extremely irregular in morphology. To quantitatively describe the porosities in the samples, the Avizo software was used to statistically analyze the volume, quantity, and types of porosities, and the statistical results are shown in Fig. 4.6. The Sample A at the 1/8 width direction has the highest number of porosities, while Sample B and Sample C have fewer porosities.

In Fig. 4.6a, it is evident that the number of macroporosities and aggregate porosities inclusions is significantly higher and larger in size in the Sample A located at 1/8 widthwise, compared to other locations. These porosity inclusions are concentrated

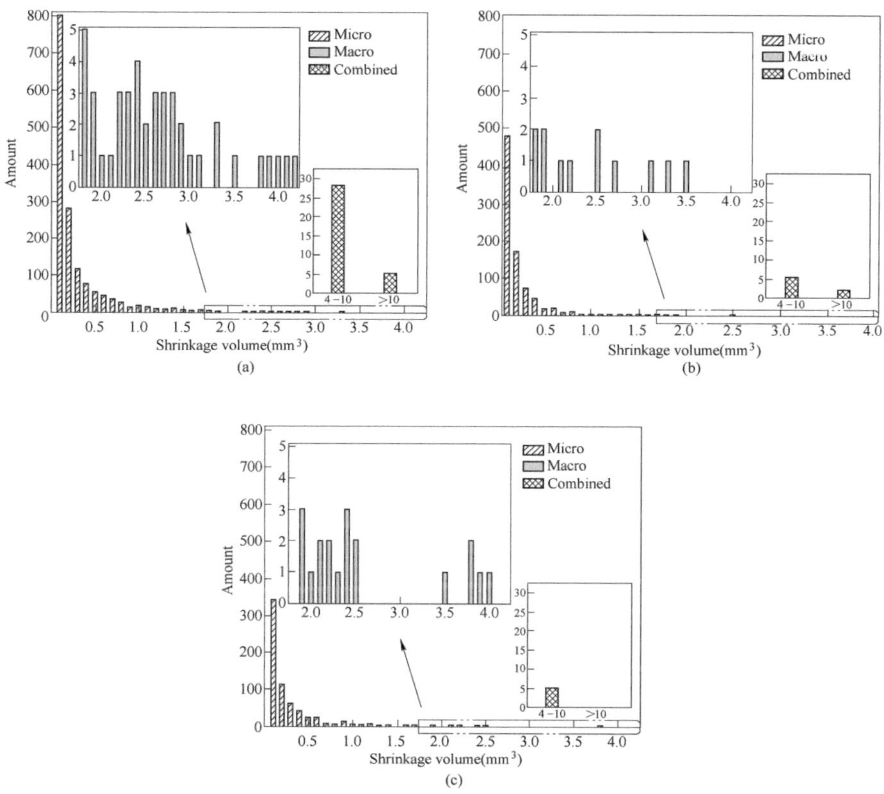

Fig. 4.6 Volume-based quantitative statistics of various types of porosities. **a** 1/8 width (Sample A); **b** 1/4 width (Sample B); **c** 1/2 width (Sample C)

in this sample. Figure 4.6b and c demonstrate that the quantity of porosities in the Sample B and Sample C at 1/4 and 1/3 widthwise is significantly lower than that in the Sample A. Through statistical analysis of the porosity quantity, it is observed that smaller porosities are more numerous and have a wider distribution. Conversely, larger porosities are less numerous and have a more limited distribution.

The 1/8 region in the width direction of the thick slab is the last to solidify. The temperature and cooling rate in this region not only meet the conditions for the formation of microscopic shrinkage, but also are the only conditions that meet the formation of macroscopic shrinkage. Therefore, this region not only has the highest number of microscopic shrinkages, but is also the main region where macroscopic shrinkages exist. Due to the difficulty of supplementing the adjacent secondary dendrite arms in the mushy region with longer steel liquid, a convoluted macroscopic shrinkage morphology is formed. The formation of microscopic shrinkage occurs throughout the entire solidification process. As the fluidity of the steel is good in the early stage of solidification and is easy to fill, there are almost no microscopic shrinkages in the 1/3 region from the skin to the center of the slab. Under the combined action of secondary dendrite arms and interfacial tension, the morphology of the microscopic shrinkage is approximately ellipsoidal with a large axis ratio difference.

4.1.3 Distribution and Morphology of Porosity in Continuous Casting Bloom

4.1.3.1 Distribution of Porosity Zones in Continuous Casting Bloom

The width dimension of the bloom is relatively smaller compared to that of the wide—thick slab, consequently leading to a more uniform flow density distribution and solidification process in the width direction of the bloom. Compared with the non-uniform solidification characteristics of the wide-thick slab, the solidification end of the bloom is located in the central region of the ingot. In order to observe the quality defects inside the bloom, high-frequency ultrasonic microscope (SAM) was used to detect the transverse section of a 425 mm × 320 mm gear steel bloom. The distribution of porosity zones in the bloom is shown in Fig. 4.7a, with the dark regions representing the ingot matrix and the light spots representing porosity. It is evident that most porosities are distributed within the dashed line frame. The porosity distribution cloud map of the continuous casting bloom was obtained using Eq. (4.2). From Fig. 4.7b, it can be seen that the closer the porosity is to the central position, the larger and denser it becomes, similar to the distribution pattern of the wide-thick slab.

It can be clearly observed that, compared with wide-thick slab, the porosity zones of the bloom mainly distribute within the 1/3 region symmetrically along the central line (the central region of the bloom) in both the width and thickness directions. In both width and thickness directions, there exists a transitional region from the bloom

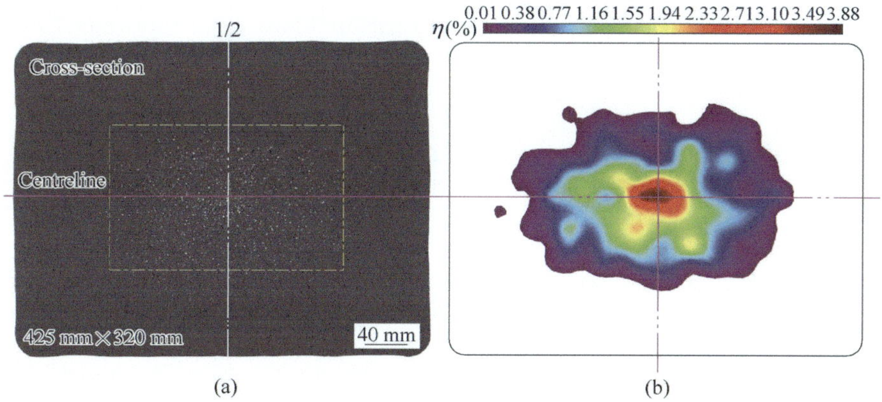

Fig. 4.7 Distribution map of porosity zone in bloom. **a** SAM scanning image; **b** porosity distribution cloud map

shell to the central region where the porosity zones gradually increase. The central region of the bloom is the area that is most severely impacted by porosity zones.

The Niyama* correction model was also applied to predict the distribution of porosity zones on the transverse section of the bloom, distinguishing between the macroporosities distribution region and the microporosities distribution region. Figure 4.8 shows the distribution map of porosity zones in the bloom.

By comparison, the relative error of the predicted porosity zones distribution values in the width direction of the transverse section of the bloom is 0.71%, while the error in the thickness direction is 4.81%. Similar to wide-thick slab, the model demonstrates higher prediction accuracy in the width direction than in the thickness direction. Further comparison based on the region ratio of the porosity zones indicates a relative error of 6.94% between the predicted and experimental values, suggesting

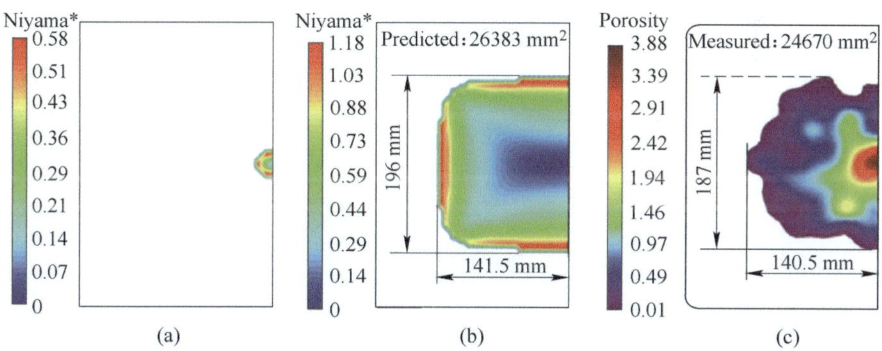

Fig. 4.8 Cloud map of porosity zones distribution in bloom. **a** The macroporosities distribution region; **b** the microporosities distribution region; **c** real measurement distribution of porosity zones

that the model can effectively predict the distribution law of porosity zones in the bloom.

4.1.3.2 Morphology of Porosity in Continuous Casting Bloom

As shown in Fig. 4.9, the bloom is divided into nine small specimens. The width and thickness of the small specimens are 141.6 mm and 106.6 mm, respectively (with respect to the transverse section of the bloom). For this transverse section of the bloom, it is determined through solidification heat transfer calculations that the two-phase region of the bloom is mainly distributed in Sample B2, which is also the primary region for the distribution of the slab's porosity zone. Meanwhile, particular attention ought to be devoted to the surrounding eight specimens near the two-phase region of Sample B2, and three-dimensional reconstruction and color rendering are implemented according to the method shown in Fig. 4.4.

Figure 4.10 presents the morphology of the porosities in the center and adjacent regions of the bloom. It can be observed that the morphology of the porosities is consistent with the pattern observed in the wide-thick slab. Microporosities with a size smaller than 1 mm are approximately ellipsoidal, while macroporosities with a size larger than 1 mm have complex and irregular shapes. Macroporosities mainly exist in Sample B2, that is, the central region, while Microporosities are dispersed throughout the entire porosity zones. To visually display the quantity of porosities of different sizes, the Avizo software is used to statistically analyze the different sizes of porosities in the four specimens, and the results are shown in Fig. 4.11.

Compared with the wide-thick slab, the center region of the bloom represents the solidification end region. It is the primary distribution region for macroporosities, and

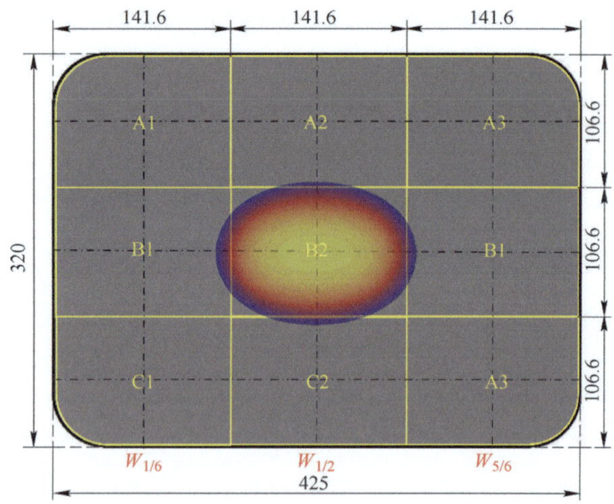

Fig. 4.9 Sampling diagram of three-dimensional reconstruction of gear steel bloom (unit: mm)

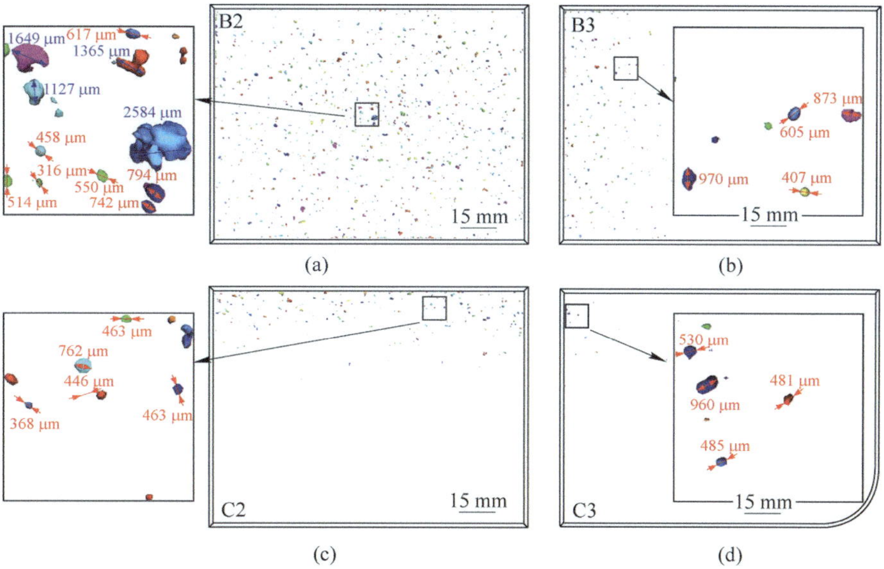

Fig. 4.10 Three-dimensional reconstruction of porosities in bloom. **a** Sample B2, central region; **b** Sample B3, right side of central region; **c** Sample C2, lower side of central region; **d** Sample C3, lower right side of central region

its morphology is no different from that of the wide-thick slab. The quantity of Microporosities gradually decreases from the center to the surface, and its morphology are also consistent with those of the wide-thick slab.

4.2 Evolution Law of Porosities During Heavy Reduction of Wide-Thick Slabs

4.2.1 Study on the Evolution Law of Macroporosities

4.2.1.1 Thermal/Mechanical Coupling Simulation Model of Reduction Process with Macroporosities

Figure 4.12 shows the longitudinal section low magnification of the actual production billet. Combined with Sect. 4.1, the central macroporosities in the wide-thick slab mainly concentrate in the region near the slab's thickness centerline region, and the shrinkage size is usually less than 5 mm. To draw regular conclusions, the macroporosities are simplified to spherical. In the three-dimensional thermal/mechanical coupling model of the wide-thick slab, spherical cavities with a diameter of 3 mm are preset at different positions across the slab's width to quantitatively reveal the

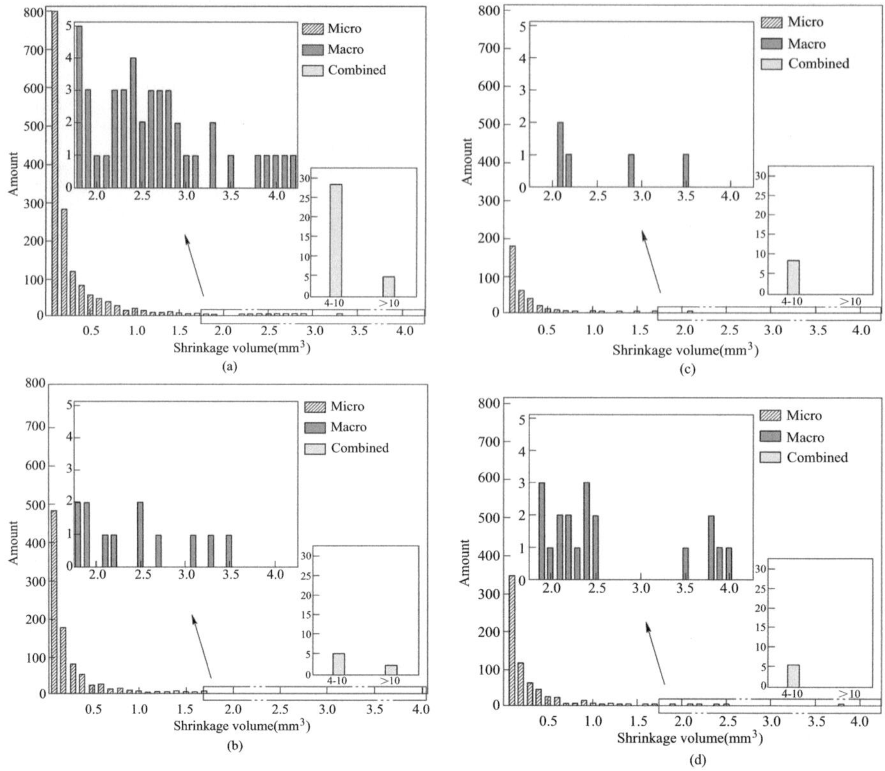

Fig. 4.11 Based on volume quantitative statistics of different types of porosities in continuous casting bloom. **a** Sample B2, central region; **b** Sample B3, right side of central region; **c** Sample C2, lower side of central region; **d** Sample C3, lower right side of central region

evolution law of the macroporosities (size larger than 1 mm) of the solidification end reduction process. The model containing spherical porosity is depicted in Fig. 4.13.

Figure 4.13 presents the morphology of the spherical porosity ($P_{1/2}$) located at the 1/2 position of the billet width. Additionally, due to the uneven cooling in the width direction of the wide-thick slab, when the slab width 1/2 is completely solidified (i.e., solid fraction $f_s = 1.0$ at the center position of the width), there are still obvious liquid regions near the width 1/8, as shown in Fig. 4.14. This means that the temperature field distribution in this region is significantly different from other regions and may affect the evolution law of shrinkage in this region during the reduction process. Therefore, in addition to the spherical porosity ($P_{1/2}$) at the width 1/2 position mentioned above, spherical porosity of the same size is also preset at the width 1/8 position in the three-dimensional model to study the evolution law of porosity in this region during the reduction process. The distribution position and morphology of the two spherical porosities in the transverse section of the three-dimensional thermal/mechanical coupling finite element model are shown in Fig. 4.14.

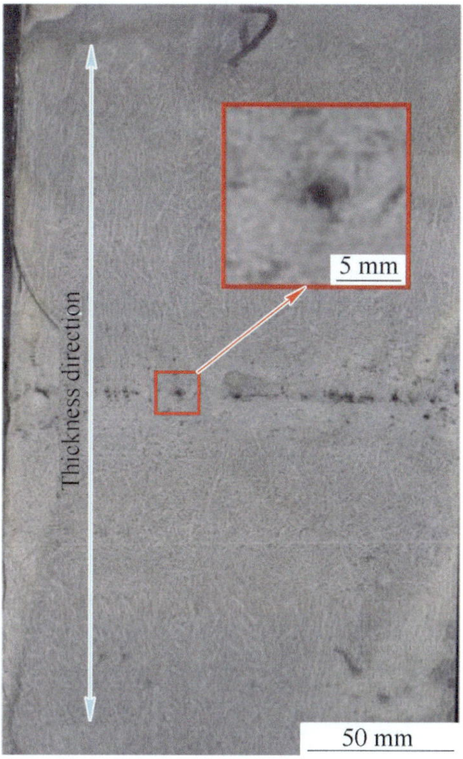

Fig. 4.12 Low power of the longitudinal section of the wide and thick slab

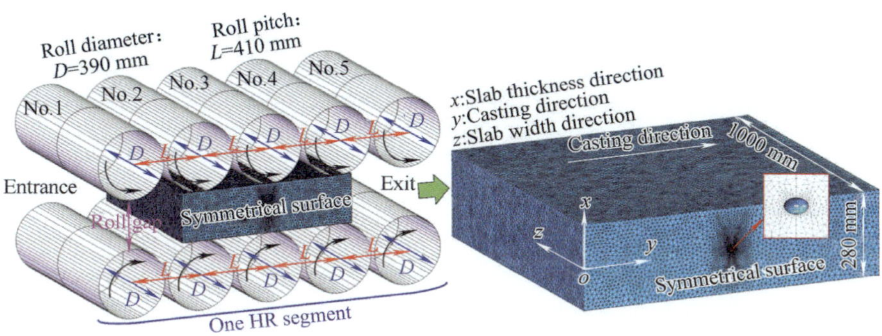

Fig. 4.13 Three-dimensional thermal/mechanical coupling model under heavy reduction of wide-thick slab

As shown in Fig. 4.15, to quantitatively describe the different direction evolution laws of the macroporosities after reduction, the deformation along the three coordinate axes (x, y, z axes representing the thickness direction, casting direction, and width direction of the thick slab, respectively) of the porosity before and after reduction are defined based on the true strain definition:

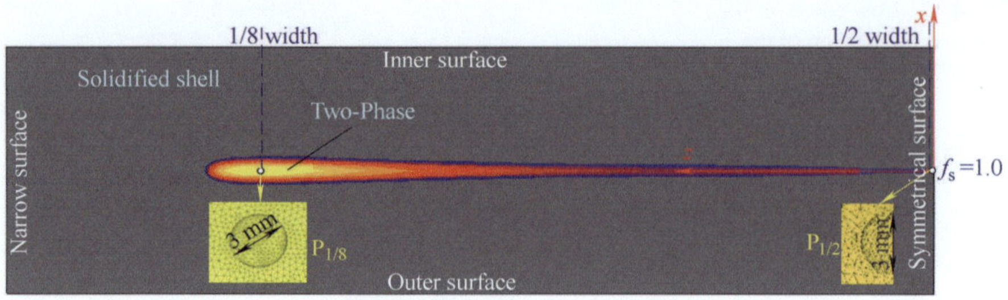

Fig. 4.14 Transverse section solidification morphology when the solid fraction (fs) reaches 1.0 at the center in the width-thick slab, and the distribution and morphology of two porosities in the three-dimensional thermal/mechanical coupling model

$$\begin{aligned}\Delta l_x &= 1n\left(\frac{L'_x}{L_x}\right)\\ \Delta l_y &= 1n\left(\frac{L'_y}{L_y}\right)\\ \Delta l_z &= 1n\left(\frac{L'_z}{L_z}\right)\end{aligned} \tag{4.3}$$

where Δl_x, Δl_y, and Δl_z denotes the deformation degree of porosity along the thickness direction, casting direction and width direction of the slab; where L_x, L_y, and L_z denotes the original porosity dimension along three axes direction, as schematically shown in Fig. 4.15; L'_x, L'_y, and L'_z denote internal porosity dimension along three axes direction after HR.

In order to quantitatively describe the overall deformation law of porosity during the reduction process and to characterize the effectiveness of the reduction process in improving porosity, the porosity closure ratio is defined based on the deviation ratio change of the porosity along the thickness direction of the slab (i.e., the direction in which the reduction is applied) before and after reduction:

$$\eta_s = \frac{2L_x}{L_y + L_z} - \frac{2L'_x}{L'_y + L'_z} \tag{4.4}$$

Fig. 4.15 Porosity size schematic diagram

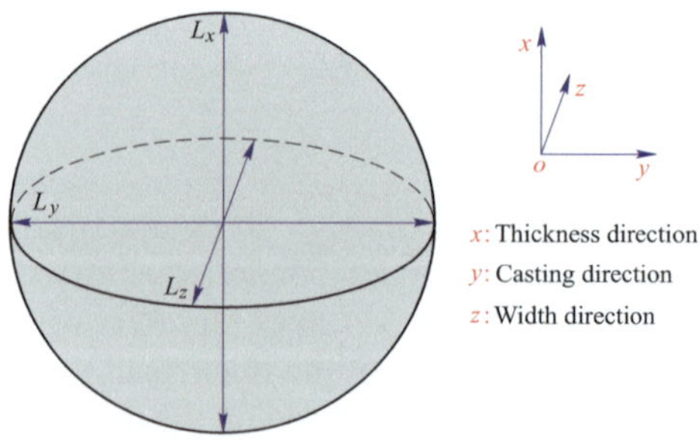

where η_s denote the closure degree of porosity after reduction.

The value range of η_s is from 0 to 1.0, and the larger the value of η_s, the better the reduction effect on porosity.

4.2.1.2 The Influence of Reduction Amounts on the Evolution of Porosity

Using the three-dimensional thermal/mechanical coupling model for wide-thick slab described in the previous section, the axial deformation and closure characteristics of porosity after uniform application of different reduction amounts in single segments at the solidification end ($f_s = 1.0$) were first calculated. The axial deformations and closures at the half width position ($P_{1/2}$) and the eighth width position ($P_{1/8}$) of the specimen as shown in Fig. 4.16 were analyzed.

It can be observed from Fig. 4.16a–c that with the increase of reduction amount at the solidification end, the axial deformations (Δl_x) in the thickness direction, (Δl_y) in the casting direction, and (Δl_z) in the width direction of the porosity continuously increase. Among them, Δl_x and Δl_z are less than 0 while Δl_y is greater than 0, indicating that the size of the porosity decreases in the thickness and width directions, but increases in the casting direction after reduction. Compared with the axial deformations in the casting direction (Δl_y) and the width direction (Δl_z), the axial

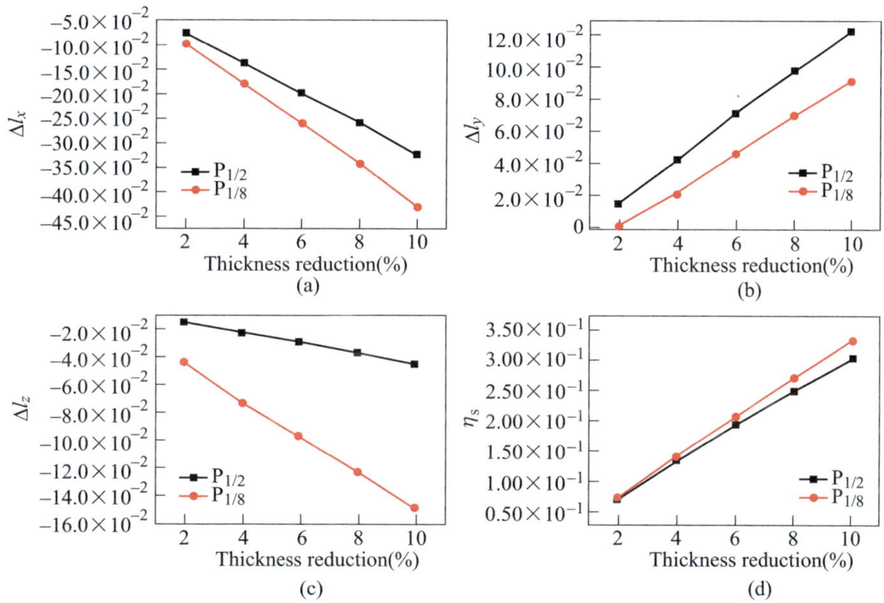

Fig. 4.16 Axial deformation and closure of porosities after different amounts of reduction. **a** Δl_x; **b** Δl_y; **c** Δl_z; **d** η_s

deformation in the thickness direction (Δl_x) is significantly larger, indicating that the thickness direction is the main deformation direction of the porosity after reduction. It can be seen from Fig. 4.16d that with the increase of reduction amount, the closure degree (η_s) of the porosity continuously increases, indicating that the porosity can be continuously improved with the increase of reduction amount during the reduction process.

The comparison of the evolution laws of porosities $P_{1/2}$ and $P_{1/8}$ at different positions in the width direction revealed significant differences between the two. After implementing a 10% reduction deformation, the closure degree of porosities $P_{1/8}$ was 9.7% higher than that of $P_{1/2}$. This indicates that during the reduction process at the solidification end of wide-thick slab, the porosities in the vicinity of the 1/8 width position can be more effectively enhanced compared to those at the 1/2 width position. The possible reasons accounting for the differences in the evolution laws between $P_{1/2}$ and $P_{1/8}$ during the reduction process are: (1) the different width distribution positions of $P_{1/2}$ and $P_{1/8}$; (2) the temperature field distribution differences between the 1/2 and 1/8 width regions of the wide-thick slab.

To reveal the influence of the position of porosities on the evolution laws during the reduction process, under the assumption of uniformly cooled conditions in the width direction of the ingot (i.e., assuming the uniform distribution of secondary cooling water density along the width of the ingot and neglecting the non-uniform distribution under actual operating conditions), the closure laws of porosities at $P_{1/2}$ and $P_{1/8}$ were calculated and compared, as shown in Fig. 4.17. Since the thickness direction of the ingot is the primary deformation direction for the porosities, only the thickness direction deformation (Δl_x) of $P_{1/2}$ and $P_{1/8}$ was compared as shown in Fig. 4.17a. It can be observed that under uniform cooling conditions, the differences in the thickness direction deformation and the closure degree of porosities between $P_{1/2}$ and $P_{1/8}$ are not significant. This indicates that the different distribution positions of $P_{1/2}$ and $P_{1/8}$ are not the main reasons for their different evolution laws.

To reveal the influence of the temperature field distribution at the 1/2 and 1/8 width positions on the evolution laws of $P_{1/2}$ and $P_{1/8}$ during the reduction process, temperature and corresponding position changes in temperature differences between

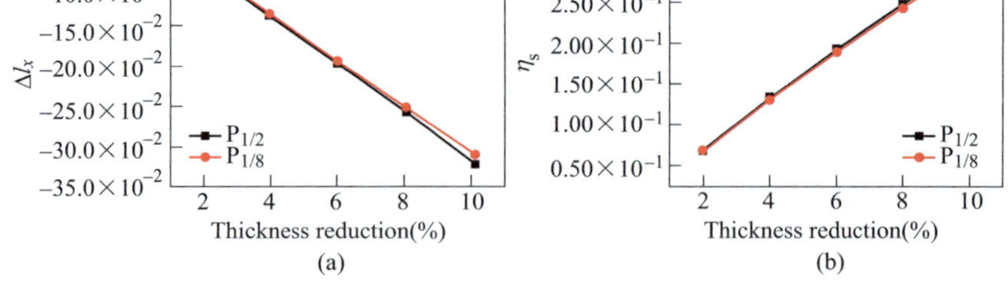

Fig. 4.17 Evolution of porosity after reduction the solidification end of wide-thick slab during uniform cooling. **a** Δl_x; **b** η_s

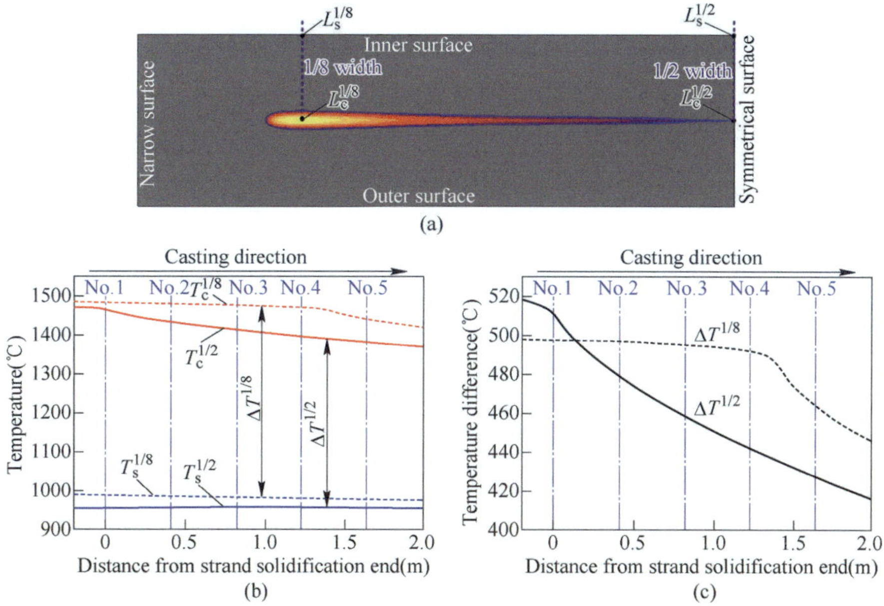

Fig. 4.18 Distribution of feature point positions in slab transverse section (**a**) and temperature of feature point positions during reduction process (**b**) in relation to temperature difference between inside and outside of the ingot (**c**)

the inside and outside of the ingot were extracted as shown in Fig. 4.18a. The results are presented in Fig. 4.18b and c. In Fig. 4.18b, $T_s^{1/8}$, $T_c^{1/8}$, $T_s^{1/2}$, $T_c^{1/2}$ represent the temperature changes at the characteristic points $L_s^{1/8}$, $L_c^{1/8}$, $L_s^{1/2}$, and $L_c^{1/2}$ during the reduction process. In Fig. 4.18c, $\Delta T^{1/8}$ ($\Delta T^{1/8} = T_c^{1/8} - T_s^{1/8}$) and $\Delta T^{1/2}$ ($\Delta T^{1/2} = T_c^{1/2} - T_s^{1/2}$) represent the temperature differences between the width center position and the corresponding thickness center position at the 1/8 and 1/2 width positions of the ingot.

From Fig. 4.18b, it can be seen that during the reduction process at the solidification end, the temperature of various characteristic points in the ingot continuously decreases from the entry (corresponding to No.1 roll) to the exit (corresponding to No.5 roll). However, compared to the surface characteristic point temperatures of the ingot ($T_c^{1/8}$ and $T_c^{1/8}$), the temperature decreases at the center position characteristic points of the ingot ($T_c^{1/8}$ and $T_c^{1/2}$) is significantly faster. As a result, the trend of temperature differences ($\Delta T^{1/8}$ and $\Delta T^{1/2}$) in Fig. 4.18c is similar to the trend of $T_c^{1/8}$ and $T_c^{1/2}$ in Fig. 4.18b. At the solidification end of the wide-thick slab (i.e., $f_s = 1.0$) and under the influence of non-uniform cooling, there exists a certain amount of unsolidified two-phase region in the central region near the 1/8 width position of the ingot. During the subsequent cooling and temperature decrease process, the remaining molten steel in the central region of the 1/8 width position continuously releases latent heat of solidification, which to some extent slows down the decrease in $T_c^{1/8}$ and $\Delta T^{1/8}$. As a result, the temperature difference ($\Delta T^{1/8}$) at the 1/8 width

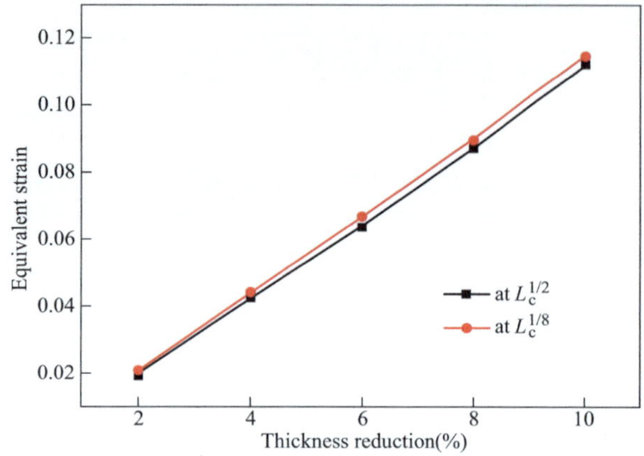

Fig. 4.19 Equivalent strain at the center of slab thickness at 1/8 and 1/2 positions after applying different pressure amounts at the solidification end

position in most of the segment in Fig. 4.18c is significantly greater than that at the 1/2 width position. By combining this with the difference in the closure degree of porosities $P_{1/2}$ and $P_{1/8}$ in Fig. 4.16d, it can be deduced that due to the large temperature difference between the inside and outside of the ingot at the 1/8 width position, the transmission of reduction deformation from the surface to the center of the ingot is more efficient in this region during the reduction process at the solidification end, resulting in more significant improvement of porosities in this region.

Figure 4.19 compares the equivalent plastic strains at the $L_c^{1/8}$ and $L_c^{1/2}$ positions after different reduction deformations. As the reduction amount increases, the equivalent plastic strains at both positions continuously increase, indicating a continuous increase in the reduction deformation degree at the center position of the ingot. Therefore, the closure of porosities at the 1/8 and 1/2 width positions of the ingot can be continuously improved, corresponding to the continuous increase in the closure degree (η_s) of porosities in Fig. 4.16d. It should be noted that the overall equivalent plastic strain at the $L_c^{1/8}$ position in Fig. 4.19 is higher than that at the $L_c^{1/2}$ position. This further indicates that, under the influence of the large temperature difference at the 1/8 width position of the ingot, the reduction deformation at this region during the reduction process at the solidification end can be more efficiently transmitted from the surface to the center, and thus more significantly improve the porosities in this region.

4.2.1.3 The Influence of Reduction Positions on the Evolution of Porosity

For a wide-thick slab continuous casting machine, reduction can be carried out through one or multiple segments, and the reduction position can be flexibly adjusted by changing the gap value of the corresponding segment shrinkage roll. In order to reveal the effect of the reduction position on the closure law of the porosity in the wide-thick slab, the evolution law of the porosity after applying a 6% reduction deformation at different casting positions was calculated.

Figure 4.20 shows the axial deformation and closure change of the porosity after reduction at different reduction positions. The *x*-axis in the figure represents the distance from the entrance of the segment (i.e., the position of the first casting roll) to the solidification end. As the reduction position moves away from the solidification end of the casting, the deformation along the thickness direction in Fig. 4.20a and along the width direction in Fig. 4.20c continuously decreases, while the deformation along the casting direction in Fig. 4.20b continuously increases. This indicates that as the reduction position moves away from the solidification end of the casting, the three axial dimensions of the porosity after reduction all increase. Figure 4.20d shows the closure degree of the porosity after reduction at different casting positions. In comparison with the reduction deformation at the solidification end, when the reduction start position is deviated 3 m along the casting direction from the solidification end, the closure degrees of $P_{1/2}$ and $P_{1/8}$ decrease by 6.3% and 9.3%, respectively. This means that as the casting position at the solidification end increases, the effectiveness of reduction deformation in improving the porosity continuously decreases.

Figure 4.21a and b, respectively, shows the internal and external temperature differences at different positions in the width direction of the casting within the segment, and the central equivalent strain at different casting positions after applying a 6% reduction deformation at different casting positions. As the starting position of

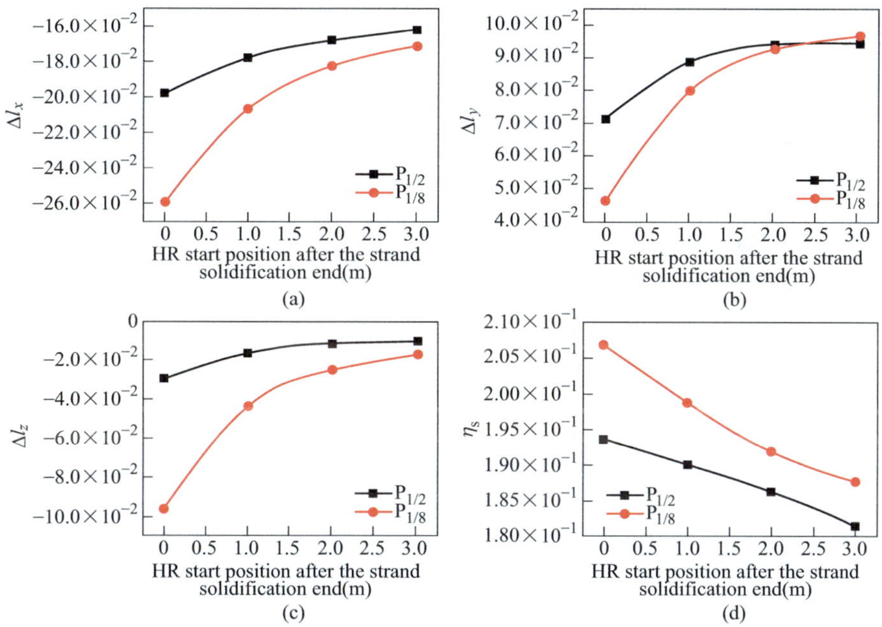

Fig. 4.20 Axial deformation and closure degree of shrinkage pores after applying 6% reduction deformation at different strand positions. **a** Δl_x; **b** Δl_y; **c** Δl_z; **d** η_s

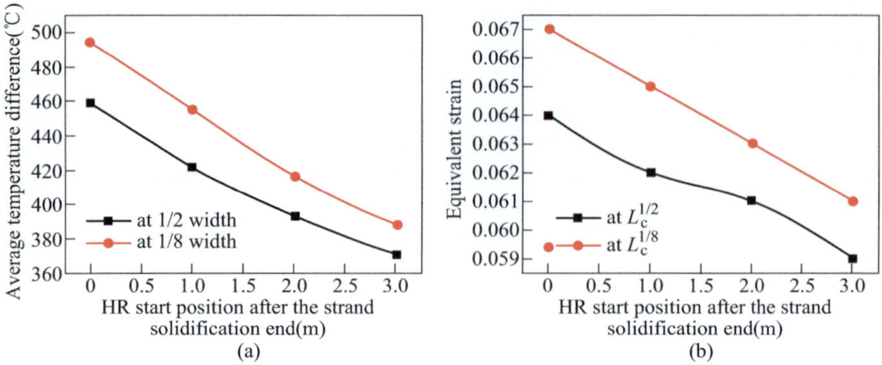

Fig. 4.21 **a** Mean inner and outer temperature differences at 1/8 and 1/2 lengthwise positions within the segment of cast ingots; **b** equivalent strain variations at the 1/8 and 1/2 thickness center positions of slab after 6% reduction applied at different strand locations

the reduction moves away from the solidification end of the casting, the temperature gradient between the inside and outside of the slab, which promotes the transfer of reduction deformation toward the center of the billet, continues to decrease. As depicted in Fig. 4.21b, the equivalent strain also keeps on decreasing as the reduction start position moves further away from the solidification end. Therefore, the closure degree η_s shown in Fig. 4.20d continuously decreases as the reduction start position moves away from the solidification end.

As the reduction start position moves away from the solidification end, the resistance deformation force of the casting to the casting roller during the reduction process continuously increases due to the decrease in the casting temperature. Therefore, the reduction force required for a single segment to apply the same reduction amount significantly increases as the reduction position moves away from the solidification end, as shown in Fig. 4.22. When the reduction start position deviates 3 m from the solidification end, the reduction force required for a single segment to apply a 6% reduction deformation at the solidification end increases by about 20%. This means that as the reduction position moves away from the solidification end, the reduction capacity of the segment for the casting significantly decreases.

In summary, if the reduction start position gradually moves away from the solidification end along the casting direction, the closure degree of the porosity and the reduction capacity of the segment for the casting both continuously decrease, and the improvement effect of reduction on the center porosity of the slab gradually weakens.

4.2.1.4 The Influence of Reduction Modes on the Evolution of Porosity

To enhance the efficiency of reduction for wide-thick slab solidification end, this section investigates the influence of reduction modes on the evolution of porosity in wide-thick slab under the condition of the same reduction amount in a single segment.

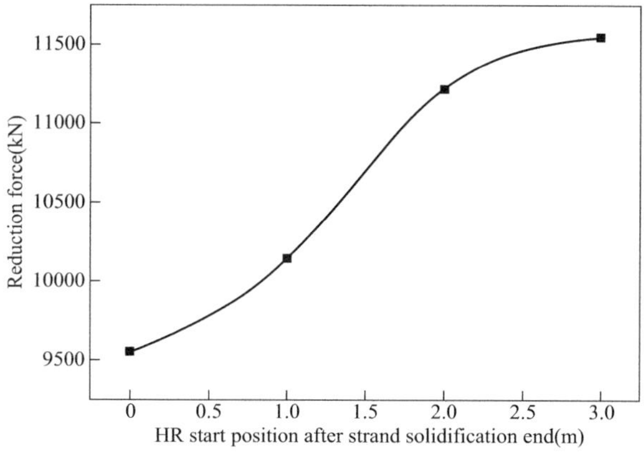

Fig. 4.22 Required reduction force for applying 6% reduction deformation at different strand positions in the segment

Table 4.1 presents the distribution of the amount of reduction within a single segment for the five reduction schemes under study. The total amount of reduction for each scheme is 6% of the thickness of the slab. The variation in slab thickness within the reduction segment for the five schemes is depicted in Fig. 4.23. In Case 1, the reduction deformation is uniformly applied by all rollers of the reduction segment, with each roll applying a reduction deformation of 1.2% of the thickness of the slab, representing the traditional uniform heavy reduction (UHR) reduction mode.

The segment of the slab generally relies on the flexible adjustment of the opening width of the roller gap at the inlet and outlet positions by two pairs of hydraulic cylinders. Based on this, the reduction deformation process in the thickness direction of the slab (the opening width of the roller gap at the outlet of the segment is less than that at the inlet) is achieved. During the reduction process, due to the mechanical structural limitations of the segment, the upper frame (inside the curved side of the caster) rolls along the casting direction, forming a wedge-shaped distribution. Taking the example of a heavy reduction segment with five pairs of rolls, the amount of reduction for No.1 Roll at the inlet of the segment can be adjusted flexibly. The total reduction amount for No.2 Roll to No.5 Roll is the disparity in the opening width of the roller gap between the inlet and outlet of the segment. Since the rolls

Table 4.1 HR deformation distribution within the HR segment in five schemes

Case	Reduction mode	Thickness reduction at each roller of the HR segment (%)					Total thickness reduction (%)
		No.1	No.2	No.3	No.4	No.5	
1	UHR	1.20	1.20	1.20	1.20	1.20	6.0
2		1.80	1.05	1.05	1.05	1.05	6.0
3	SPUHR	2.40	0.90	0.90	0.90	0.90	6.0
4		3.00	0.75	0.75	0.75	0.75	6.0
5		3.60	0.60	0.60	0.60	0.60	6.0

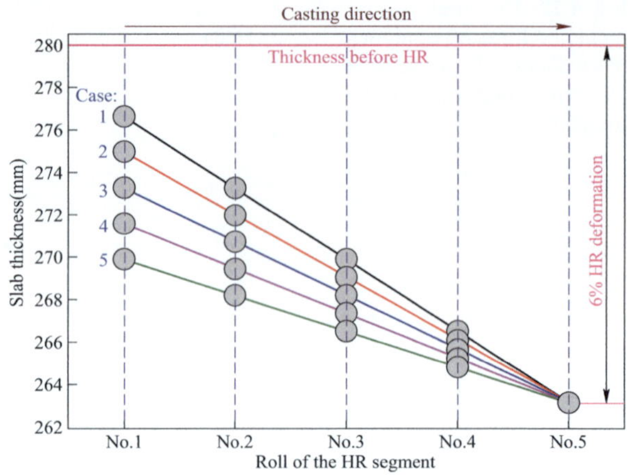

Fig. 4.23 Thickness changes of the slab in the segment under five reduction schemes

on the inner arc can only form a wedge-shaped distribution during the reduction process, the amount of reduction for No.2 Roll to No.5 Roll at each position is the same. Based on the above mechanical structure characteristics of the segment, the "single point + continuous" reduction mode (SPUHR, Single Point and Uniform Heavy Reduction) is proposed on the basis of the uniform reduction mode. In the single point uniform reduction mode (corresponding to Case 2 to Case 5 in Table 4.1), No.1 Roll of the segment applies a larger reduction deformation to the slab, while the remaining reduction deformation is uniformly applied by No.2 Roll to No.5 Roll. Obviously, under the new reduction mode, the individual reduction amount for No.2 Roll to No.5 Roll is less than that for No.1 Roll.

After implementing a single segment 6% reduction deformation on the solidification end using different reduction modes, the evolution of porosity is shown in Fig. 4.24. Compared with the uniform reduction mode of Scheme 1, using the "single point + continuous" reduction mode, as the reduction amount at the inlet of the segment increases, the deformation along the thickness direction in Fig. 4.24a and the deformation along the width of the slab in Fig. 4.24c continuously increase, while the deformation along the casting direction in Fig. 4.24b decreases. This suggests that when the reduction mode is employed, as the reduction amount at the inlet of the segment goes up, the size of the porosity along the three axes diminishes.

The closure degree of the porosity corresponding to the five reduction schemes is shown in Fig. 4.24d. As the reduction amount at the inlet of the segment rises, the corresponding closure degree of the porosity keeps on increasing. Compared with Case 1 of the uniform reduction mode, when the reduction amount at the inlet position of the segment in the "single point + continuous" reduction mode increases to 3.6% in Scheme 5, the closure degrees of $P_{1/8}$ and $P_{1/2}$ increase by 6.2% and 8.2%, respectively. This indicates that the newly proposed "single point + continuous" reduction mode can enhance the efficiency of reduction for improving porosities. Furthermore, as the reduction amount for No.1 Roll at the inlet position of the segment increases, the enhancement effect on the reduction efficiency becomes more apparent.

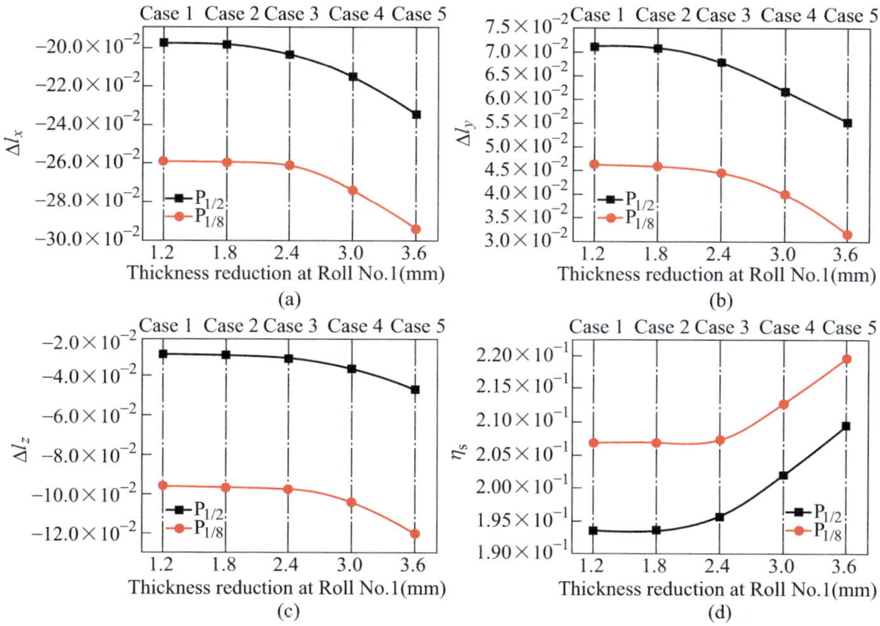

Fig. 4.24 Axial deformation and closure degree of porosity after 6% reduction deformation under different reduction modes. **a** Δl_x; **b** Δl_y; **c** Δl_z; **d** η_s

After implementing reduction for the five reduction schemes, the equivalent strain at the center position of 1/2 and 1/8 of the thickness of the slab in the width direction is shown in Fig. 4.25. The equivalent strain increases continuously from Case 1 to Case 5, indicating that as the reduction amount for No.1 Roll at the inlet position of the segment increases, the efficiency of transmitting the reduction deformation from the surface to the center of the slab continuously increases. Therefore, when using the "single point + continuous" reduction mode, as the reduction amount for No.1 Roll at the inlet position increases, the porosity can be improved more efficiently. This mechanistically explains the trend of the closure degree of the porosities from Case 1 to Case 5 in Fig. 4.24d, proving that the newly proposed "single point + continuous" reduction mode can effectively improve the reduction efficiency.

As analyzed in Sect. 4.2.1.3, the greater the temperature difference between the inside and outside of the slab, the higher the efficiency of transmitting the reduction deformation to the center, and the more obvious the improvement effect on the porosity. During the process of reduction at the solidification end, the temperature difference between the interior and exterior of the slab within the segment gradually diminishes from No.1 Roll to No.5 Roll, and as a result, the reduction efficiency declines. Under the "single point + continuous" reduction mode, as the reduction amount for No.1 Roll at the inlet position of the segment increases, the reduction efficiency inevitably increases. However, in addition to the temperature

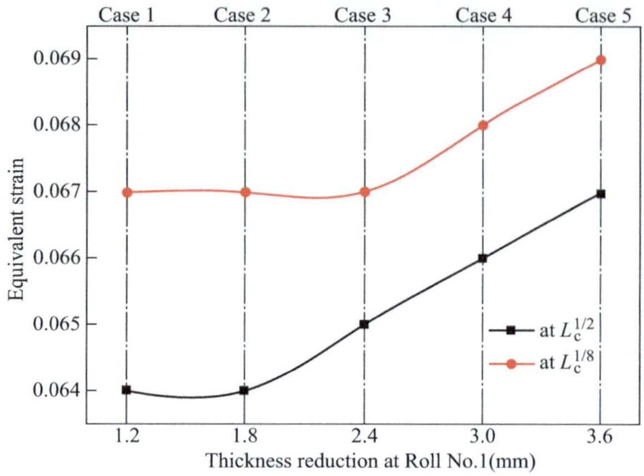

Fig. 4.25 Equivalent strain at the center of the thickness at 1/8 and 1/2 of the slab width after 6% reduction deformation under different reduction schemes

difference between the inside and outside of the slab, the difference in the distribution of the reduction amount within the segment of the reduction mode may also be another potential factor affecting the reduction efficiency. Although the total amount of reduction is the same for the five reduction schemes, even if the temperature change of the slab in the casting direction is ignored, the reduction deformation behavior of the slab at each roll position will still change due to the different distribution of the reduction amount within the segment, which will affect the final deformation characteristics of the slab and the evolution of the porosity.

To study the influence of the distribution of the reduction amount within the segment on the reduction efficiency, the closure degree of the porosity for Case 1 to Case 5 was calculated under the condition of ignoring the temperature field variation of the slab within the segment (i.e., assuming that the temperature field of the slab within the segment is fixed and equal to that at the inlet position of the segment), as shown in Fig. 4.26. It can be seen that, although the temperature field variation of the slab within the segment has been ignored, the closure degrees of $P_{1/8}$ and $P_{1/2}$ still increase by 5.9% and 5.2%, respectively, from Case 1 to Case 5. This indicates that the distribution of the reduction amount within the segment is an important factor affecting the reduction efficiency, and as the reduction amount for No.1 Roll at the inlet position of the segment increases, the reduction efficiency will continuously increase.

4.2.1.5 The Influence of Cast Speed on the Evolution of Porosity

The influence of casting speed on the evolution of porosity. In the actual casting process, the casting speed significantly affects the production rhythm of the casting machine and the solidification process of the slab. To study the influence of casting speed on the solidification end reduction of wide-thick slab, the evolution of porosity

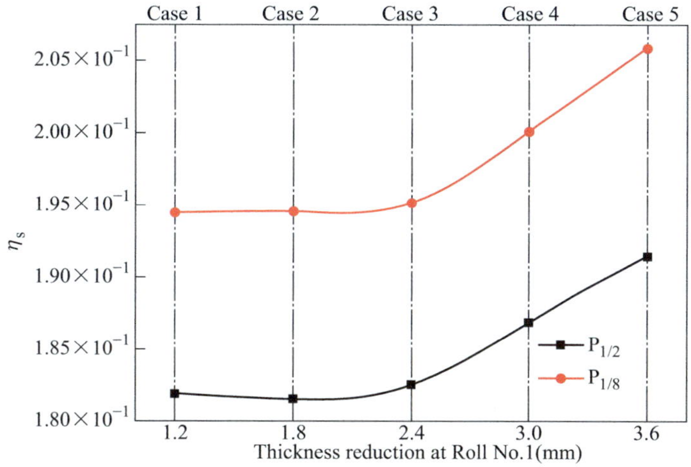

Fig. 4.26 Porosity closure degree of Scheme 1 to Scheme 5 when ignoring the temperature change in the direction of casting direction

after applying 6% reduction to thc solidification end ($f_s = 1.0$) at different casting speeds was calculated and compared. The calculation results are shown in Fig. 4.27.

It can be observed from Fig. 4.27a–c that, with the increase of casting speed, the variation trends of deformation degree of $P_{1/2}$ and $P_{1/8}$ along the three axial directions are opposite. Specifically, the overall deformation degree in the thickness direction of $P_{1/2}$ (see Fig. 4.27a) and the width direction (see Fig. 4.27c) shows a decreasing trend, while the closure degree of porosity in the casting direction (see Fig. 4.27b) shows an overall increasing trend. Figure 4.27d presents the closure degree of porosity following reduction. From it, it can be observed that although there are certain alterations in the closure degree of porosity for $P_{1/2}$ and $P_{1/8}$ as the casting speed increases, the extent of these changes is not substantial. The maximum change amplitudes of the closure degrees for $P_{1/2}$ and $P_{1/8}$ are approximately 0.40% and 1.08%, respectively. This indicates that when reduction is applied at the solidification end, the influence of casting speed on the improvement of porosity closure is not significant.

From the analysis in Sects. 4.2.1.2 and 4.2.1.3, it is known that the temperature difference between the inner and outer sides in the thickness direction of the slab is a key factor affecting the efficiency of transmitting the reduction deformation from the slab surface to its interior. In order to further elucidate the mechanism of the effect of casting speed on the reduction process at the solidification end, the thickness direction temperature distribution at the 1/2 and 1/8 positions in the width direction of the solidification end of the slab at different casting speeds was extracted, as shown in Fig. 4.28. It can be observed that the trends of temperature distribution in the thickness direction of the wide—thick slab under different casting speed conditions are essentially the same. Therefore, the similar temperature difference between the inner and outer sides at the solidification end in the thickness direction of the slab

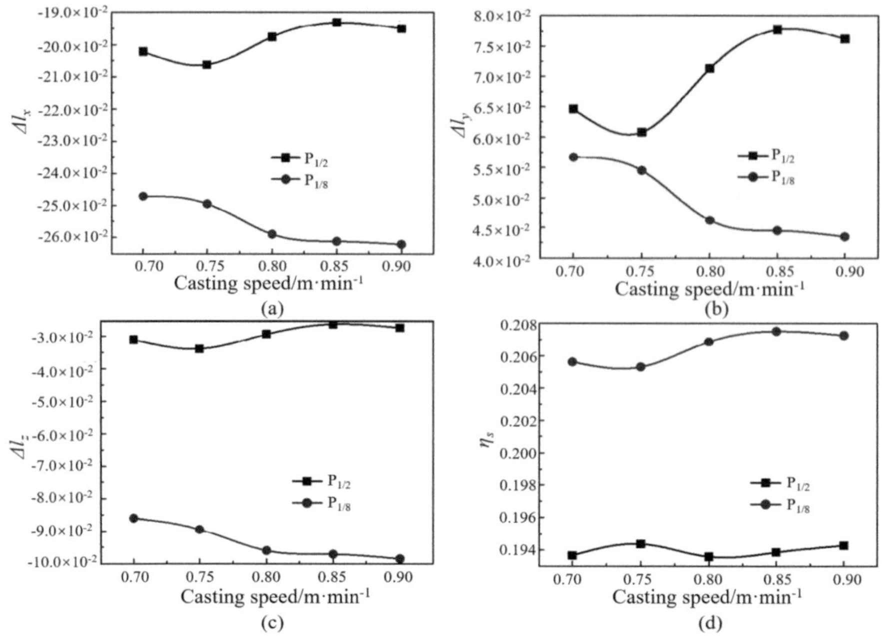

Fig. 4.27 Axial deformation and closure degree of porosity after applying 6% reduction down deformation at the solidification end under different secondary cooling zone water distribution conditions axial deformation and closure ratio of porosity after 6% reduction applied to the solidification end under different casting speed. **a** Δl_x; **b** Δl_y; **c** Δl_z; **d** η_s

makes the change in the closure degree of porosity in Fig. 4.27d not significant with the increase of casting speed.

Figure 4.29 shows the trend of the solidification end of the strand with the change of casting speed. It can be seen that the position of the solidification end of the strand

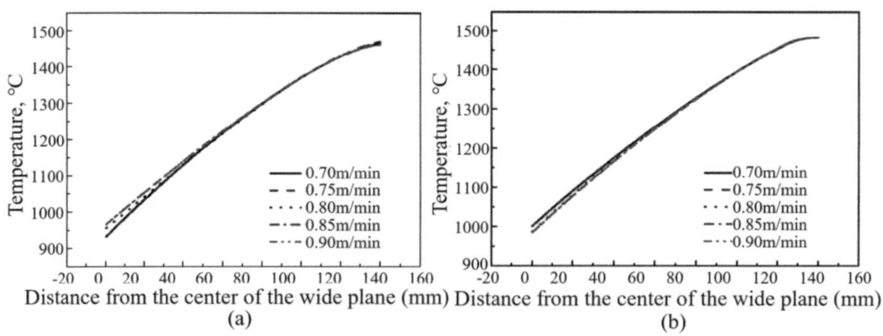

Fig. 4.28 Transverse temperature distribution of solidification end in slab at different casting speed. **a** Position at 1/2 of the width; **b** position at 1/8 of the width

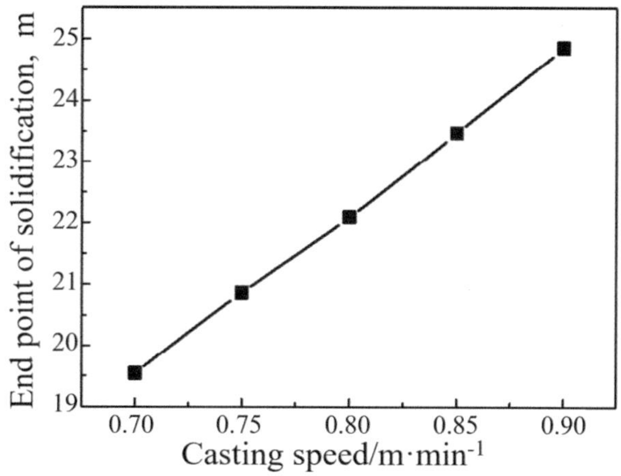

Fig. 4.29 Strand position at the solidification end during different secondary cooling water distribution

changes significantly with the increase of casting speed. When the casting speed increases from 0.70 m/min to 0.90 m/min, the solidification end of the wide-thick slab moves approximately 5.3 m along the casting direction.

In the actual continuous casting process of slabs, the reduction process can be implemented by one or multiple segments. As the solidification end of the strand advances, the efficiency of reduction in enhancing the central porosity diminishes significantly on account of the continuously decreasing temperature difference within the slab. This means that an inappropriate position of the solidification end of the strand (unreasonable casting speed control) will limit the efficient implementation of the reduction process at the solidification end. Therefore, it is necessary to optimize the position of the solidification end of the strand by adjusting the casting speed to fully utilize the advantage of the larger temperature difference near the solidification end of the strand and to ensure effective implementation of the reduction process. Meanwhile, in combination with the crack confinement principle in Chap. 3, it is necessary to avoid applying large deformation reduction in the low solid-phase rate range.

4.2.1.6 The Influence of Secondary Cooling on the Evolution of Porosity

The water distribution in the secondary cooling zone of the continuous cast machine significantly affects the cooling and solidification process of the slab, and the water distribution in the secondary cooling zone can be flexibly adjusted within a certain range. In order to study the effect of the water distribution in the secondary cooling zone on the solidification end reduction of wide-thick slab, the evolution law of porosity during the solidification end reduction process under different water distribution in the secondary cooling zone was calculated and compared.

Figure 4.30a–c show the deformation degree of porosity along three axial directions after applying a 6% reduction deformation at the solidification end of wide-thick slabs under different cooling water conditions. It can be seen that as the water distribution in the secondary cooling zone increases, the deformation degree of the porosity along the thickness direction (see Fig. 4.30a) and along the width direction (see Fig. 4.30c) shows a gradually increasing trend. The degree of deformation of the porosity along the lengthwise direction (refer to Fig. 4.30b) generally exhibits a decreasing trend. Figure 4.30d shows the closure degree of the porosity under different water distribution conditions. Although the closure degree of the porosity shows a gradual decrease as the water distribution increases, the degree of change is not significant. When the water distribution in the secondary cooling zone increases from 0.80 L/kg to 1.00 L/kg, the closure degrees $P_{1/2}$ and $P_{1/8}$ of the porosity only decreased by approximately 1.8% and 2.1%, respectively. This indicates that the water distribution in the secondary cooling zone has no significant impact on improving the porosity closure at the end of slab solidification under reduction.

Under different water distribution conditions in the secondary cooling zone, the temperature distributions in the widthwise 1/2 and 1/8 positions of the slab thickness direction at the solidification end are shown in Fig. 4.31. Although there are significant changes in the water distribution in the secondary cooling zone, there is no significant change in the temperature distribution in the thickness direction at the

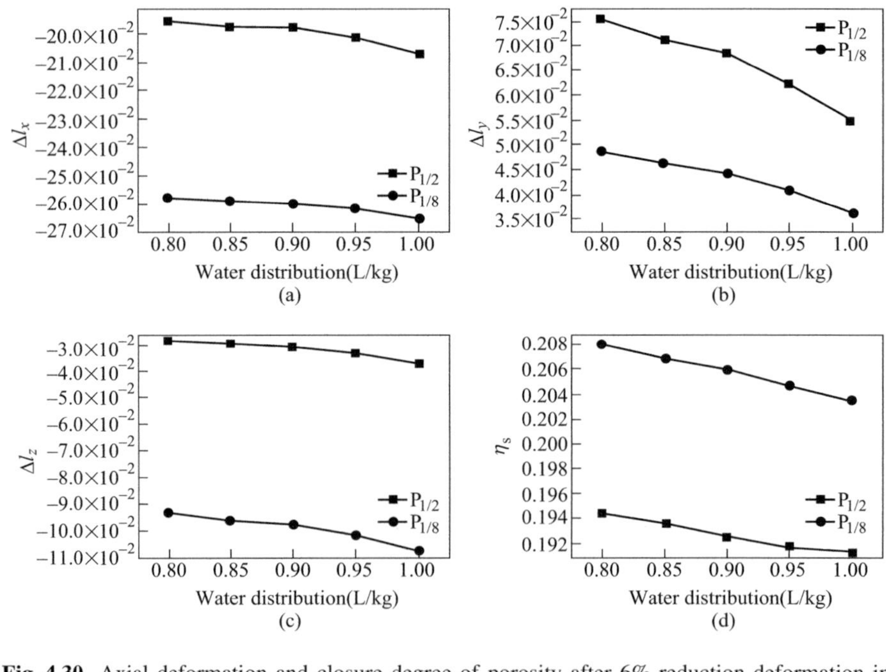

Fig. 4.30 Axial deformation and closure degree of porosity after 6% reduction deformation in single segment of solidification end with different secondary cooling water distribution. **a** Δl_x; **b** Δl_y; **c** Δl_z; **d** η_s

end of the slab solidification. Therefore, the porosity closure degree in Fig. 4.30d is not significantly affected by the water volume in the secondary cooling zone.

Figure 4.32 shows the trend of the location of the strand at the solidification end with the change in water distribution in the secondary cooling zone. It can be observed that the water distribution within the secondary cooling zone has a significant impact on the location of the strand at the solidification end. As the water distribution in the secondary cooling zone increases from 0.80 L/kg to 1.00 L/kg, the location of the solidification end flow moves forward by about 1.3 m. This means that the solidification end of the slab and the appropriate strand position can be adjusted through the appropriate changes in the water distribution in the secondary cooling zone to facilitate the efficient implementation of the solidification end reduction of wide-thick slab.

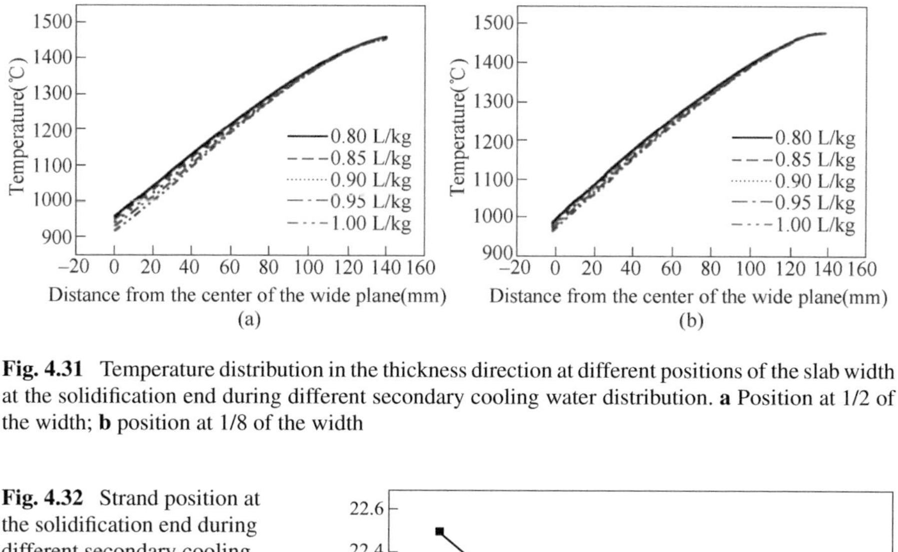

Fig. 4.31 Temperature distribution in the thickness direction at different positions of the slab width at the solidification end during different secondary cooling water distribution. **a** Position at 1/2 of the width; **b** position at 1/8 of the width

Fig. 4.32 Strand position at the solidification end during different secondary cooling water distribution

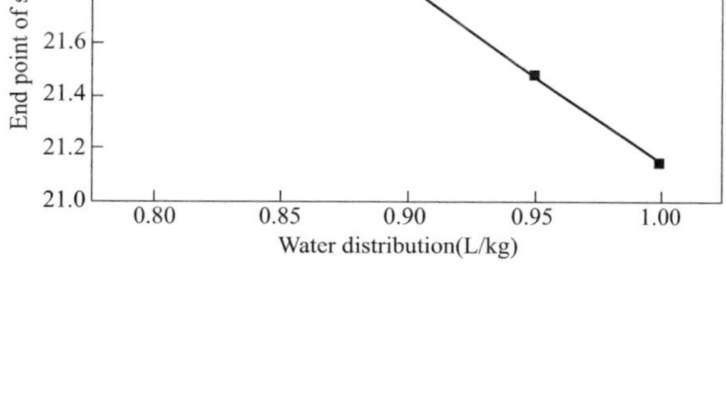

4.2.1.7 Prediction Equation of Macroporosities Closure Under Wide-Thick Slab Reduction

Researchers commonly quantitatively assess the internal closure of porosities in metal materials during the hot working process by establishing a porosity closure theory model [25–28]. Typically, two identical thermal/mechanical coupling models are required to be established for conditions with and without preset porosities. The porosity closure behavior of the model with porosity during the deformation process is related to the stress–strain deformation behavior of the model without porosity (at the corresponding porosity position), thereby establishing a "porosity closure deformation characteristic" relationship.

The degree of deformation at different positions of the material during the deformation process can be quantitatively characterized by the corresponding position's equivalent strain (ε_{eq}). Therefore, some scholars [22, 29–34] use the equivalent strain of the base material at various positions after processing deformation as an index parameter to evaluate the corresponding position's porosity closure behavior. Additionally, some scholars [35–38] also use the static hydrostatic stress integral proposed by Tanaka as an index parameter to evaluate the porosity closure behavior during the processing deformation, where the equation for the static hydrostatic stress integral is:

$$Q = \int_0^{\varepsilon f} (-\sigma_m/\sigma_{eq}) d\varepsilon_{eq} \tag{4.5}$$

where Q is the hydrostatic integration; ε_{eq} is the equivalent strain; ε_f is the final equivalent strain; σ_m and σ_{eq} are hydrostatic stress and equivalent stress, respectively.

In Eq. (4.5), the values of ε_{eq}, ε_f, $\sigma_{m,}$ and σ_{eq} at the porosity position can be readily determined through synchronous no porosity model calculations.

The fundamental deformation outcomes of stress and strain at diverse positions within the ingot during the reduction process can be ascertained by constructing an identical thermal/mechanical coupling model devoid of porosity for the very same deformation process. Based on the different macroporosities evolution laws under different reduction process conditions in Sects. 4.2.1.2– 4.2.1.6, this section attempts to establish quantitative relationship equations that describe the closure behavior of porosity with the corresponding position's equivalent strain (ε_{eq}) or static hydrostatic stress integral (Q). In addition to Eq. (4.3) which defines the porosity closure degree η_s based on the deviation ratio in the shrinkage thickness direction to characterize the closure degree of porosity during the reduction process, another porosity closure degree η_v is defined based on the change in porosity volume before and after reduction:

$$\eta_v = (V_0 - V)/V_0 \tag{4.6}$$

where η_v is the closure degree of porosity defined based on the change of porosity volume; V_0 and V are porosity volume before and after reduction, respectively.

It is approximately assumed that the porosity after reduction takes on an ellipsoidal shape. The post-reduction porosity volume in Eq. (4.6) can be computed based on the deformations of the three axial directions. In addition, the range of η_v is from 0 to 1.0, where a larger η_v value indicates a better improvement in porosity after reduction.

Figure 4.33a–d, respectively, show the scatter plots for the relationships between η_s-ε_{eq}, η_s-Q, η_v-ε_{eq}, and η_v-Q under different reduction conditions. To quantitatively evaluate the relationship between η_s, η_v, and ε_{eq}, Q, Eq. (4.6) is used to calculate the Pearson correlation coefficients of the scatter data in Fig. 4.33a–d, and the corresponding Pearson correlation coefficient values (r) are provided in each figure:

$$r = \frac{\sum (X_i - \overline{X})(Y_i - \overline{Y})}{\left[\sum (X_i - \overline{X})^2 \sum (Y_i - \overline{Y})^2\right]^{1/2}} \tag{4.7}$$

where r is Pierce correlation coefficient; X_i represents ε_{eq} or Q in Fig. 4.36; Y_i represents η_s or η_v in Fig. 4.36; $\overline{X}$, $\overline{Y}$ are the average value of scatter data X_i and Y_i, respectively.

The theoretical range of r is $-$ 1.0 to 1.0. When $r > 0$, it indicates a positive correlation between Y_i and X_i, while $r < 0$ indicates a negative correlation between

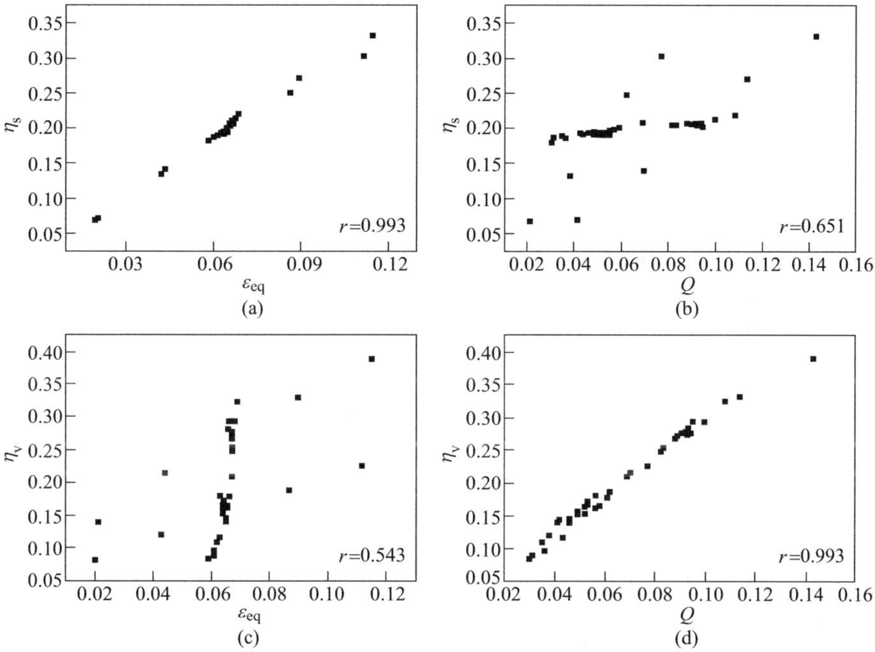

Fig. 4.33 The effect of equivalent strain and hydrostatic stress integral on pore closure degree. **a** η_s-ε_{eq}; **b** η_s-Q; **c** η_v-ε_{eq}; **d** η_v-Q

Y_i and X_i. The larger the absolute value of r, the higher the correlation between Y_i and X_i.

Comparing the Pearson correlation coefficients in Fig. 4.33, it is evident that the Pearson correlation coefficient values in Fig. 4.33a and d are significantly higher than those in Fig. 4.33b and c. This indicates that η_s in Fig. 4.33a has a strong positive correlation with ε_{eq}, and η_v in Fig. 4.33d has a strong positive correlation with Q. Therefore, attempts are made to establish mathematical equations that quantitatively describe the relationship between the porosity closure degree (i.e., η_s or η_v) and the basic deformation law of the ingot after reduction (i.e., Q or ε_{eq}) by selecting η_s-Q in Fig. 4.33a and η_v-ε_{eq} in Fig. 4.33d, aiming to achieve a rapid and quantitative assessment of the process effect of improving porosity in the reduction of wide-thick slab ingots.

Based on the trends of the scatter data in Fig. 4.33a and d, a polynomial form was selected to fit the data in the two figures. Figure 4.34 compares the results of the first to fourth-order polynomial fitting of the scatter data for η_s-ε_{eq} in Fig. 4.33a with the results of the three-dimensional thermal/mechanical coupling model calculation. Additionally, Fig. 4.34 also provides the adjusted coefficient of determination (R^2) for the first-to fourth-order polynomial fitting. A larger R^2 indicates a better fit for the corresponding polynomial. Comparing the R^2 values for the different order polynomial fittings in Fig. 4.34, it is evident that the R^2 value in Fig. 4.34b is relatively large, indicating that the second-order polynomial can accurately describe the quantitative relationship between η_s and ε_{eq}.

Based on the second-order polynomial obtained from the fitting of the scatter data in Fig. 4.34b, the predictive equation for the porosity closure degree in the reduction of the solidification end of wide-thick slab ingots based on the equivalent strain is:

$$\eta_s = -3.08 \times \varepsilon_{eq}^2 + 2.90 \times \varepsilon_{eq} + 4.48 \times 10^{-2} \tag{4.8}$$

Figure 4.35 compares the results of the first-to fourth-order polynomial fitting of the scatter data for η_v-Q in Fig. 4.34d with the results of the original three-dimensional thermal/mechanical coupling model calculation. It also provides the adjusted coefficient of determination (R^2) for the first-to fourth-order polynomial fitting. Comparing the R^2 values for the different order polynomial fittings in Fig. 4.35, it is evident that the R^2 value in Fig. 4.35d is relatively large, indicating that the fourth-order polynomial can accurately describe the quantitative relationship between η_v and Q.

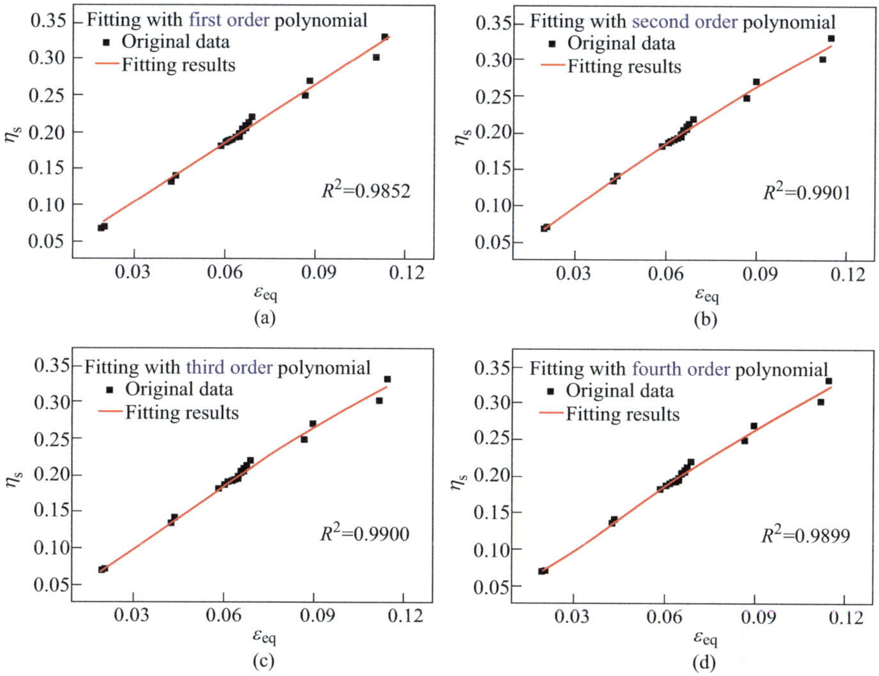

Fig. 4.34 Comparison of η_s-ε_{eq} with different order polynomial fitting results obtained from model calculations. **a** First-order polynomial fitting; **b** second-order polynomial fitting; **c** third-order polynomial fitting; **d** fourth-order polynomial fitting

Based on the fourth-order polynomial obtained from the fitting of the scatter data in Fig. 4.35d, the predictive equation for the porosity closure degree in the reduction of the solidification end of wide-thick slab ingots based on the static hydrostatic stress integral is:

$$\eta_v = -3.61 \times Q^2 + 2.30 \times Q + 5.77 \times 10^{-2} \tag{4.9}$$

4.2.2 *Study on the Evolution Law of Microporosities*

4.2.2.1 Thermal/Mechanical Coupling Simulation Model of Reduction Process with Microporosities

Through ultrasonic scanning and 3D reconstruction in Sect. 4.1, the majority of micro-scale porosities with dimensions ranging from 200 μm to 900 μm can be approximately regarded as ellipsoidal in shape. In order to simulate the evolution

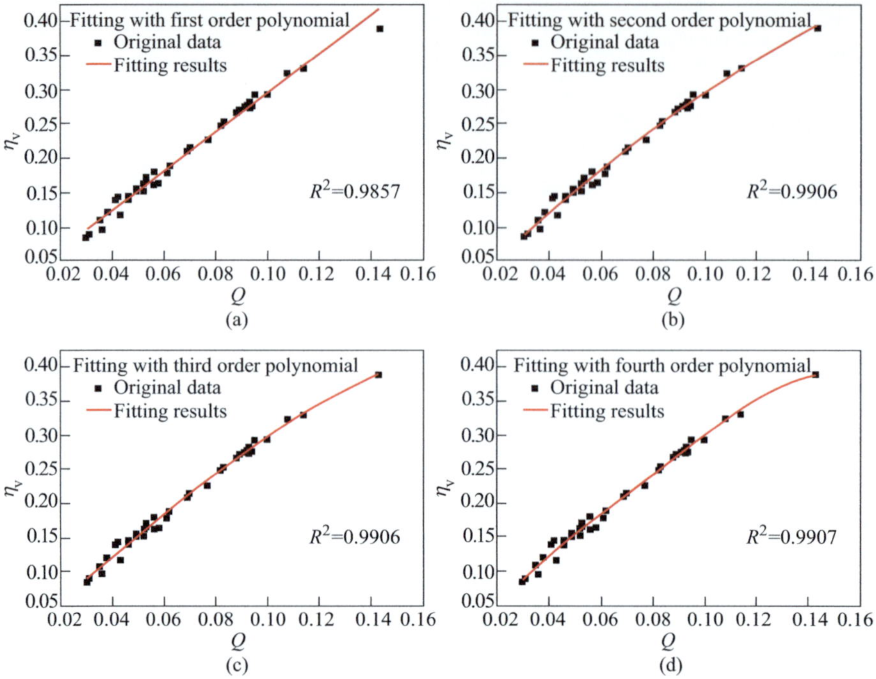

Fig. 4.35 Comparison of η_V-Q with different order polynomial fitting results obtained from model calculations. **a** First-order polynomial fitting; **b** second-order polynomial fitting; **c** third-order polynomial fitting; **d** fourth-order polynomial fitting

process of microporosities in the reduction of continuous casting billets, ellipsoidal artificial microporosities were preset in the plate billet of the 3D thermal/mechanical coupling simulation model. The major axis length of the artificial ellipsoidal microporosity was set at 300 μm, and the minor axis length was set at 200 μm, with an aspect ratio of 1.5. As shown in Fig. 4.36, each microporosity has a different major axis direction. Therefore, in the 3D thermal/mechanical coupling simulation model, different models of cavities with different major axis orientations were established on characteristic planes, including the transverse section of the billet (*XOZ* plane), the middle width longitudinal section (*XOY* plane), and the middle thickness longitudinal section (*YOZ* plane), as shown in Fig. 4.36a–c. These models simulate the evolution process of ellipsoidal microporosities with different major axis orientations during the reduction.

Within each of the planes depicted in Fig. 4.36a–c, there are four microporosities with different major axis orientations, with the angles between the major axis of the microporosity and the reference direction being 0°, 30°, 60°, and 90°. The reference directions for the *XOZ*, *XOY*, and *YOZ* planes in Fig. 4.36a–c are the *X*-axis (reduction direction), *Y*-axis (billet casting direction), and *Z*-axis (width direction), respectively. The names of the microporosities are based on the angles between their

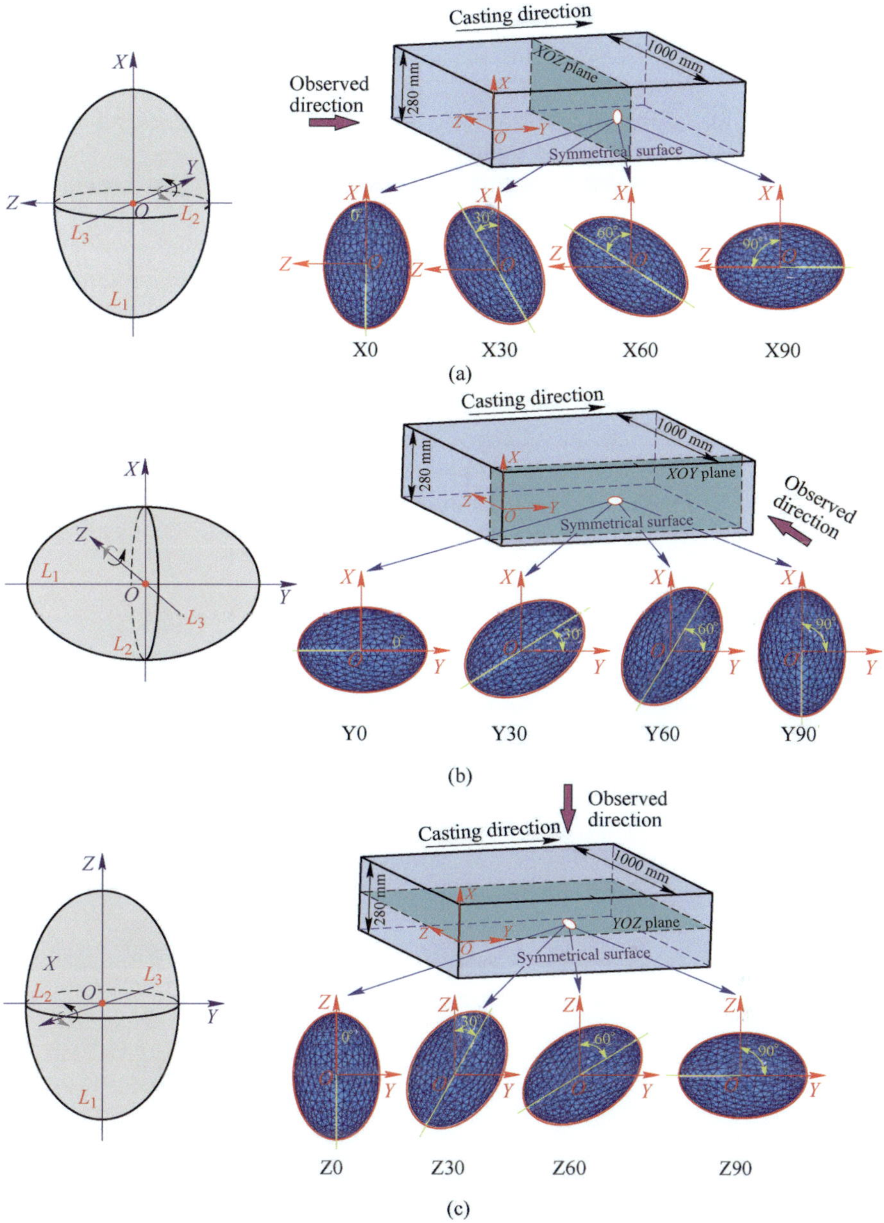

Fig. 4.36 Orientation and position of microporosities on different planes in the three-dimensional thermal-force coupling model. **a** *XOZ* plane; **b** *XOY* plane, and **c** *YOZ* plane

major axis orientations and the reference direction. For example, in Fig. 4.36a, X0–X90 represent the angles between the major axis orientation and the reference X-axis direction, from 0° to 90°. The illustrations in Fig. 4.36 are intended for a clear demonstration of the orientations of porosities and the actual centers of the microporosities are located within the heart of the finite element model.

The observations derived from Fig. 4.5 in Sect. 4.1 suggest that certain local regions display a clustered distribution of microporosities, with relatively short distances between them. During the reduction process, the evolution of a single microporosity in these regions may be influenced by the surrounding microporosities. Consequently, several cluster-type ellipsoidal microporosities with major axes located on the XOZ plane were created in the 3D thermal/mechanical coupling simulation model, as shown in Fig. 4.37a. The dimensions of a single microporosity in the Aggregate model in Fig. 4.37a are the same as those of a single microporosity in Fig. 4.36. The central microporosity (X90') in Fig. 4.37a is located on the center line of the billet. The evolution process of X90 during reduction was simulated and compared with X90, investigating the influence of clustered microporosity distribution on the evolution of microporosities during reduction.

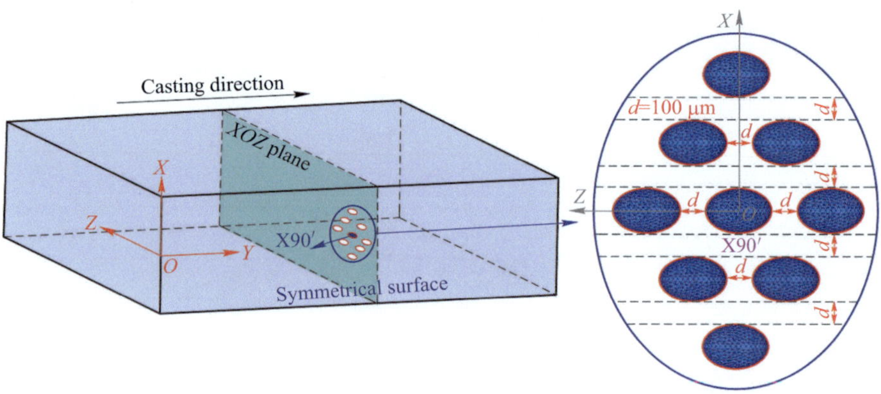

(a)

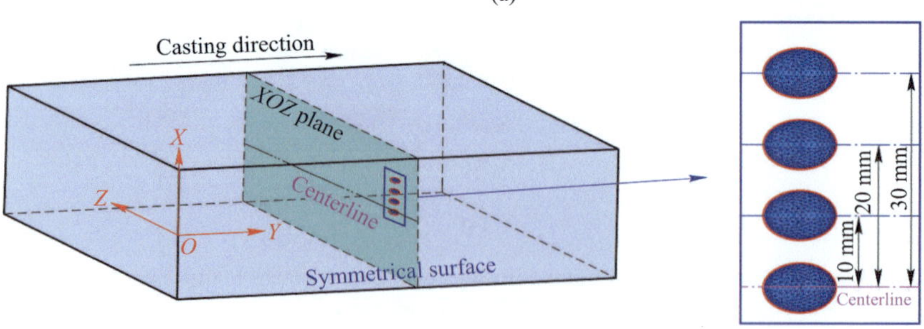

(b)

Fig. 4.37 **a** Distribution of porosities in the thermal finite element model of wide-thick slab; **b** porosities with different thickness distributions

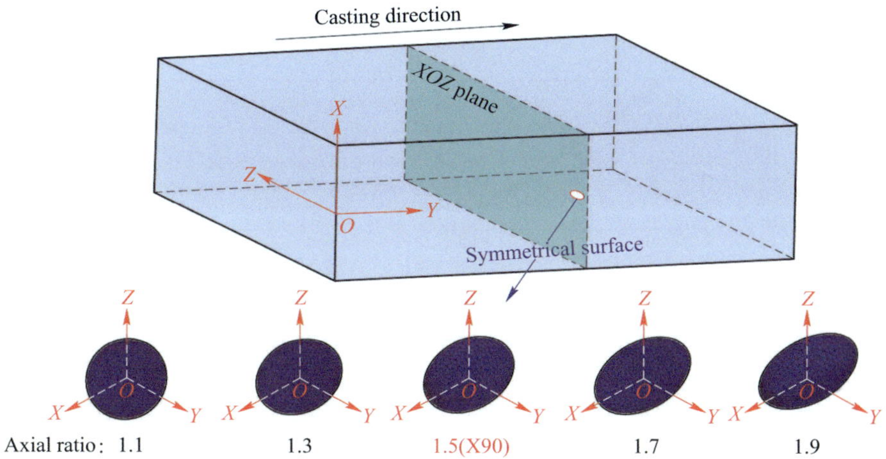

Fig. 4.38 Three-dimensional thermal coupling model of porosities with the same volume and different axial ratios

Figure 4.36 illustrates that the aspect ratios of the microporosities vary significantly, which might exert diverse effects on the final closure evolution of the microporosities. Therefore, as shown in Fig. 4.38, five ellipsoidal microporosities with different aspect ratios (having the same volume but different aspect ratios) were constructed on the *XOZ* plane, named 1.1-X90 to 1.9-X90.

4.2.2.2 The Evolution Law of Microporosities in Transverse Section (*XOZ* Plane)

The comparison in Fig. 4.39 is made between the evolution laws of microporosities in the center position of Fig. 4.36a X90 microporosities calculated based on the single porosity and the multi-pore model. It can be observed that the evolution laws of microporosities calculated by the single and multi-pore models show similar trends in deformation changes under reduction. As the reduction induced deformation at the solidification end increases, the axial deformations of the microporosities in Fig. 4.39a–c continuously increase. Among them, the axial deformation Δl_1 in Fig. 4.39a and the short-axis deformation Δl_3 in Fig. 4.39c are less than 0, while the short-axis deformation Δl_2 in Fig. 4.39b is greater than 0. This indicates that after reduction, the length of the long axis and one of the short axes of the microporosities decrease, while the length of the other short axis increases. Figure 4.39d illustrates the closure degree η_v of the microporosities after different levels of reduction. As the reduction increases, η_v also increases, indicating a continuous improvement of the microporosities with increasing pressure.

Comparing the results of the single and multi-pore models in Fig. 4.39 reveals that under the same reduction conditions, although there are certain differences in the

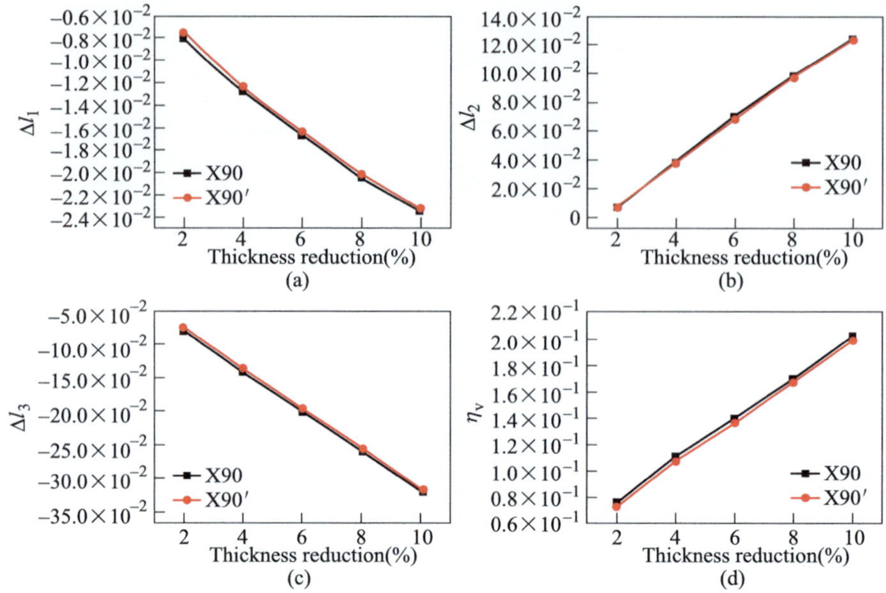

Fig. 4.39 Evolution of porosities calculated by single porosity and multi- porosity models after different amounts of reduction. **a** Long axis deformation Δl_1; **b** short axis deformation Δl_2; **c** short axis deformation Δl_3; **d** microporosities closure degree η_v

evolution laws of the microporosities predicted by the two models, the differences are not significant, and the degree of difference gradually decreases with increasing reduction. This implies that the aggregation of multiple microporosities in local positions does not significantly affect their closure behavior after reduction. Therefore, an ellipsoidal microporosities model including single different positions and orientations can be adopted to systematically reveal the evolution laws of shrinkage within the porosity zone in wide-thick slab after reduction.

After applying 6% reduction to a single segment, the closure laws of X0–X90 microporosities in Fig. 4.36a are shown in Fig. 4.40. The microporosities in the transverse section, with their different orientations, display varying deformations. For example, the axial deformation Δl_1 in Fig. 4.40a and the short-axis deformation Δl_2 in Fig. 4.40c are less than 0, while the short-axis deformation Δl_3 in Fig. 4.40b is greater than 0. This indicates that after reduction, the long axis and the short axis in Fig. 4.40c of the ellipsoidal microporosities tend to decrease, while the short axis in Fig. 4.40b tends to increase. Additionally, as the angle (α_1) between the long axis of the ellipsoidal microporosities and the thickness direction of the slab (*X*-axis) increases, Δl_x and Δl_2 decrease, while Δl_3 increases. This suggests that with the increase of α_1, the trend of the reduction in the long-axis size of the ellipsoidal microporosities after reduction and the increase in size in Fig. 4.40b weakens, while the reducing trend in Fig. 4.40c becomes more pronounced. Figure 4.40d represents the closure degree η_v of the porosity after reduction. As α_1 increases, the corresponding

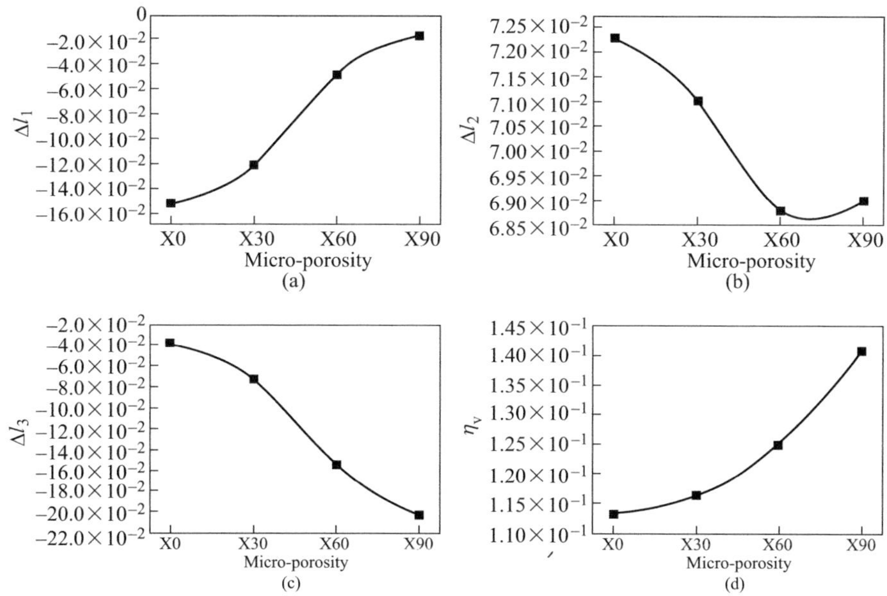

Fig. 4.40 Evolution rule of microporosities in different orientations in the transverse section after 6% heavy reduction. **a** Long axis deformation degree Δl_1; **b**, **c** short-axis deformation degree Δl_2, Δl_3; **d** porosity closure degree η_v

closure degree also increases, indicating that the efficiency of porosity improvement after reduction increases. Furthermore, when α_1 is less than 30° (corresponding to X30 microporosities), the closure degree in Fig. 4.40d increases at a slower rate with α_1. Therefore, the improvement effect of reduction on the microporosities with an angle less than 30° to the thickness direction in the transverse section is relatively poor.

From the SEM scan results in Fig. 4.1, it can be observed that the microporosities in the wide-thick slab primarily distributes within a banded region approximately 30 mm away from the center of the slab thickness. To study the influence of the distribution position of the porosity in the thickness direction on its evolution laws during reduction, the behavior of the closure of the microporosities at different thickness positions within the porosity zone during reduction is calculated by moving the X0–X90 microporosities along the thickness direction (*X*-axis) as shown in Fig. 4.41. It can be seen that the axial deformations Δl_1, Δl_2, Δl_3, and the closure degree η_v at different positions in the thickness direction show similar trends with the change in their long-axis orientation. With the increase of α_1, the ellipsoidal microporosities at different thickness positions exhibit a decrease in Δl_1 and Δl_2, while Δl_3 and η_v continuously increase. For the ellipsoidal microporosities with the same long-axis orientation, as its distance from the center of the slab thickness increases, Δl_1 and Δl_3 decrease, while Δl_2 increases. Further combined with the positive and negative characteristics of Δl_1, Δl_2, and Δl_3, it can be concluded that as the distance of the

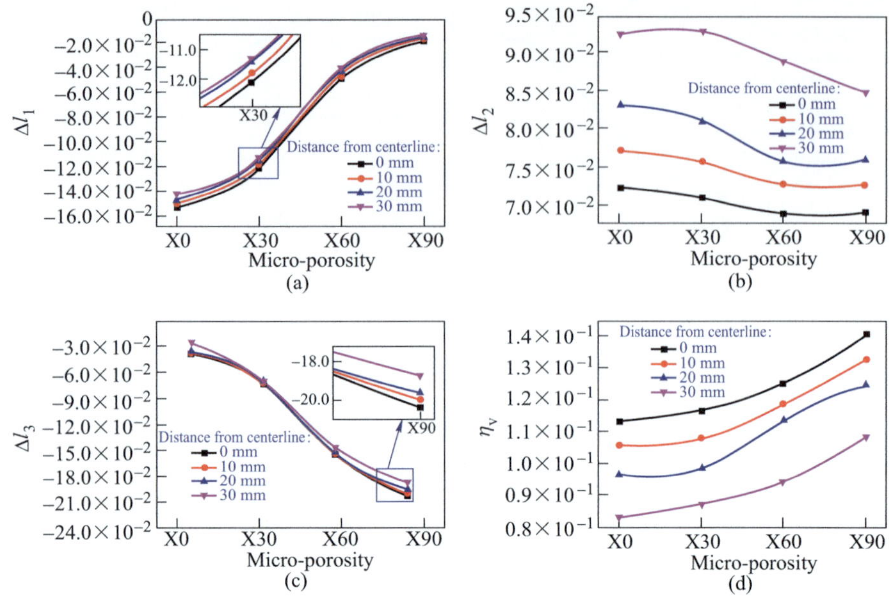

Fig. 4.41 Evolution of microporosities in different orientations and thickness positions in the transverse section after 6% heavy reduction. **a** Δl_1; **b** Δl_2; **c** Δl_3; **d** η_v

microporosities in the transverse section from the center of the slab increases, both the long axis and the two short axis sizes increase after reduction.

Since the axial sizes of the microporosities increase with their distance from the center of the slab thickness after reduction, the closure degree of the ellipsoidal microporosities with the same long-axis orientation in Fig. 4.41d continuously decreases. For the microporosities with long-axis orientations of 0° and 90° in Fig. 4.41 (d) (i.e., X0 and X90 microporosities), when their distribution positions deviate from the center of the slab thickness by 30 mm, the corresponding closure degree decreases by approximately 27% and 23%, respectively. This indicates that within the distribution area of microporosities in the wide—thick slab, the efficiency of improvement brought about by reduction diminishes as the porosity moves away from the center of the slab thickness.

4.2.2.3 The Evolution Law of Microporosities in Longitudinal Section (*XOY* Plane)

To reveal the evolution law of ellipsoidal microporosities in the transverse central longitudinal section under reduction, three-dimensional thermal/mechanical coupling models corresponding to microporosities Y0–Y90 in Fig. 4.36b were established, and their evolution after applying 6% reduction at the solidification end of a single segment were calculated, as shown in Fig. 4.42.

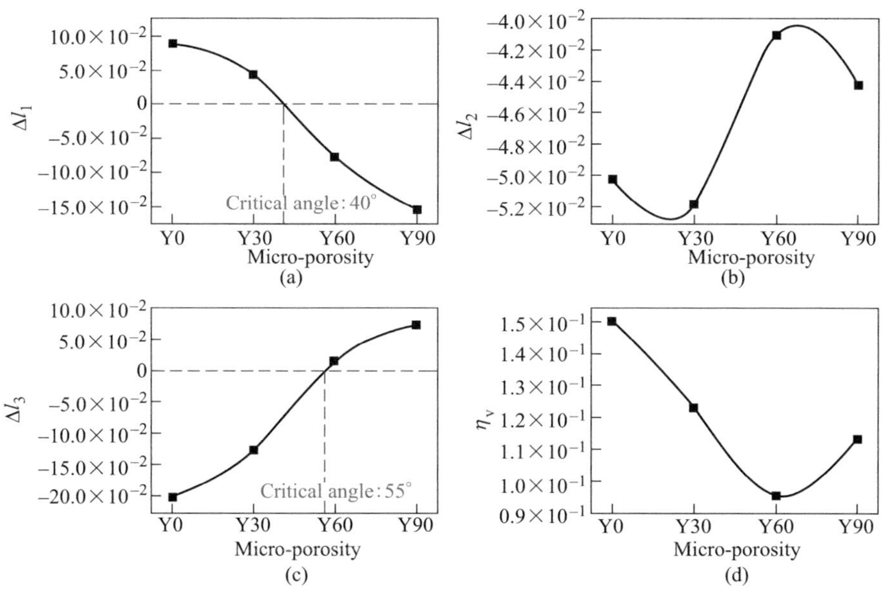

Fig. 4.42 Evolution law of porosities in different directions in the center section of the width after 6% heavy reduction. **a** Long axis deformation Δl_1; **b, c** short-axis deformation Δl_2, Δl_3; **d** porosity closure degree η_v

As the angle (α_2) between the long axis of the ellipsoidal microporosities in the transverse central longitudinal section and the direction of the billet (Y-axis) increases, the deformities of the long axis Δl_1 in Fig. 4.42a and the short axis Δl_2 in Fig. 4.42b after reduction continuously decrease to 0, and then their deformities continuously increase in the opposite direction. Therefore, as α_2 increases, the trend of the increase in the long axis of the ellipsoidal microporosities and the reduction trend of the short axis in Fig. 4.42b after reduction weakens continuously, and when α_2 exceeds a certain critical value (the critical angles in Fig. 4.42a and b are approximately 40° and 55°, respectively), the increase in the long axis of the ellipsoidal microporosities after reduction and the reduction trend of the short axis in Fig. 4.42b weaken. With further increase in α_2, the long axis of the compressed microporosities continuously decreases, while the corresponding short axis in Fig. 4.42b continuously increases. Figure 4.42c shows the deformation Δl_3 of the other short axis of the compressed microporosities, indicating that the change trend of Δl_3 is not fixed with the increase in the angle α_2, but it is consistently less than 0, and the Δl_3 of the microporosities Y0 and Y30 when $\alpha_2 < 45°$ is significantly larger. This indicates that the short axis of the ellipsoidal compressed microporosities decreases in size, and the reduction in the size of the short axis of the microporosities is more significant when $\alpha_2 < 45°$.

Figure 4.42d shows the closure degree of microporosities in different orientations after reduction in the longitudinal section. As α_2 increases, the corresponding closure degree of the microporosities shows a decreasing and then increasing trend. Among them, the microporosities with α_2 of 60° (i.e., Y60) has the smallest closure degree. Therefore, the efficiency of improvement for the microporosities with an angle of approximately 60° between the long axis and the direction of the billet in the longitudinal section is relatively low.

4.2.2.4 The Evolution Law of Microporosities in the Longitudinal Section of Thickness Center (*YOZ* Plane)

In order to investigate the influence of the orientation of microporosities within the longitudinal section at the center of thickness on the deformation under reduction, a three-dimensional thermal/mechanical coupling model for individual microporosities was established based on Fig. 4.36c, and the evolution law after applying a 6% reduction deformation to the solidification end was calculated. The results are shown in Fig. 4.43.

Within the longitudinal section at the center of thickness, as the angle (α_3) between the long axis of the microporosity and the width direction of the ingot (*Z*-axis) increases to a certain critical value (the critical angles corresponding to Fig. 4.43a and c are 20° and 50°, respectively), the long axis deformation Δl_1 in Fig. 4.43a and the short-axis deformation Δl_3 in Fig. 4.43c continue to decrease to 0. As α_3

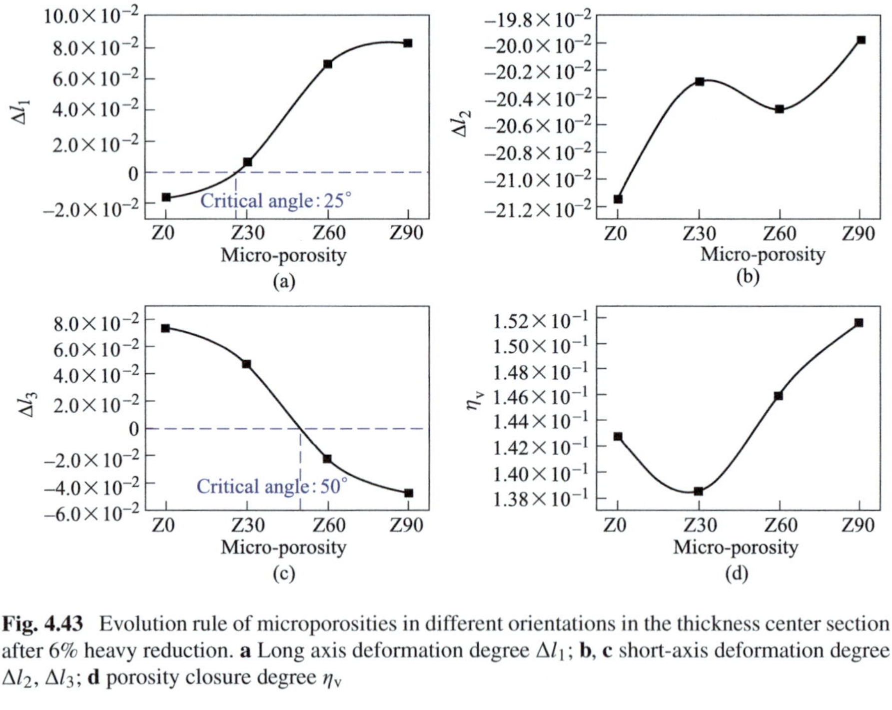

Fig. 4.43 Evolution rule of microporosities in different orientations in the thickness center section after 6% heavy reduction. **a** Long axis deformation degree Δl_1; **b**, **c** short-axis deformation degree Δl_2, Δl_3; **d** porosity closure degree η_v

further increases, Δl_1 and Δl_3 continue to increase in their respective opposite directions. Therefore, when α_3 is less than the corresponding critical angle, the long axis and the short-axis dimensions of the microporosity after reduction exhibit a tendency of reduction and increase, respectively, in Fig. 4.43a and c. Furthermore, with the increase of α_3, the tendency for the long axis to increase and the short axis to decrease in Fig. 4.43c continues to abate. When α_3 exceeds the critical value, the long axis and the short-axis dimensions in Fig. 4.43c exhibit tendencies of increase and decrease, respectively, and as α_3 continues to increase, the tendencies become more pronounced. Figure 4.43b shows the trend of the short-axis deformation Δl_2 of the microporosity after reduction. Within the longitudinal section at the center of thickness, Δl_2 after reduction for the microporosities are all less than 0, and Δl_2 shows an overall increasing trend as α_3 increases, indicating a reduction in the short-axis dimension in Fig. 4.43b, with the reduction becoming more pronounced as α_3 increases.

As shown in Fig. 4.43d, the closure ratio η_v of the microporosities within the longitudinal section at the center of thickness of the wide-thick slab billet decreases and then increases with the increase of α_3. Among them, the microporosity at 30° (i.e., Z30) exhibits the smallest closure, indicating lower efficiency in improving the microporosity closure when the long axis is at an angle of approximately 30° to the width direction of the ingot within the longitudinal section at the center of thickness.

4.2.2.5 The Evolution Law of Microporosities in the Longitudinal Section of Thickness Center (*YOZ* Plane)

The evolution law of microporosities with the same volume but different axis ratios after 6% reduction is shown in Fig. 4.44. When the axis ratio is between 1.1 and 1.9, Δl_1 generally decreases and then increases with the change of the axis ratio. Δl_2 usually decreases with the change of the axis ratio, while the trend of Δl_3 is to increase, then decrease, and then increase again with the change of the axis ratio. Conversely, η_v shows the opposite trend. The results indicate that the porosity closure degree is not positively correlated with the axis ratio. Within the range of axis ratio spanning from 1.1 to 1.9, based on the fitting curve of the porosity closure axis ratio, it is observed that when the axis ratio is 1.43, the closure of microporosities reaches its maximum value. Moreover, the closer the axis ratio gets to either 1.1 or 1.7, the smaller the closure degree of microporosities becomes.

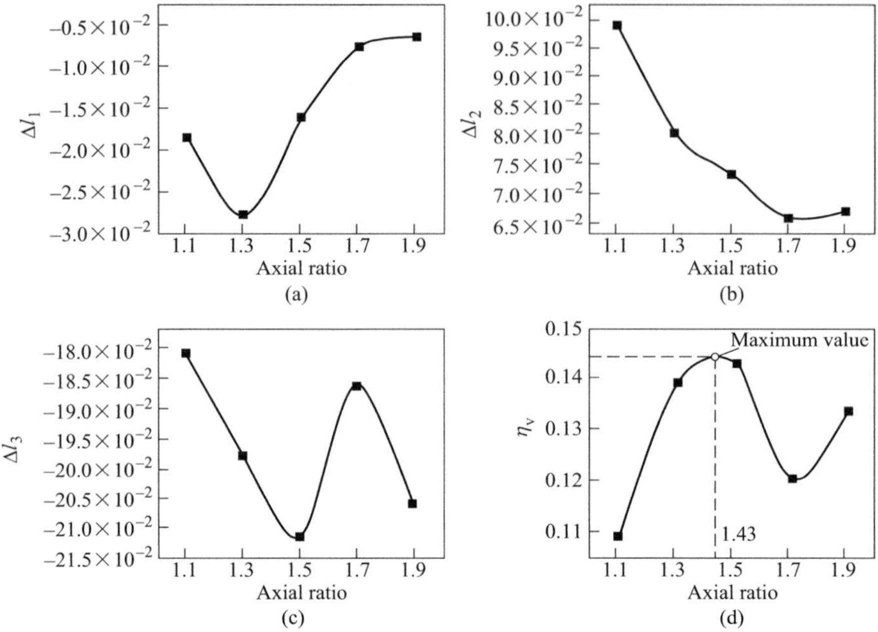

Fig. 4.44 Evolution of different axis ratio microporosities on the transverse section of the slab after implementing 6%HR deformation. **a** Δl_1; **b** Δl_2; **c** Δl_3; **d** η_v

4.3 Evolution Law of Porosities During Heavy Reduction of Blooms

4.3.1 Thermal/Mechanical Coupling Simulation Model of Bloom Solidification End Reduction Process

A 3D thermal/mechanical coupling simulation model for the end reduction process of bloom was established within the calculation domain extending in the width direction for half and in the length direction for 2200 mm along the direction of slab casting speed (adjacent spacing between two casting speed straightening machines is 2200 mm) as shown in Fig. 4.45. The model includes the adjacent two casting speed straightening machines and is used to simulate the reduction deformation behavior of the bloom during the end reduction process, where the reduction roller is used to apply a large reduction deformation to the casting blank, simulating the reduction process at the solidification end of the bloom, and the reduction deformation applied by the reduction roller is denoted as R_r, which is equal to the ratio of the reduction amount of the reduction roller to the thickness of the bloom. The front roller in the model is positioned upstream of the reduction roller in the direction of the cast flow, used to apply a smaller reduction deformation to the bloom for clamping purposes,

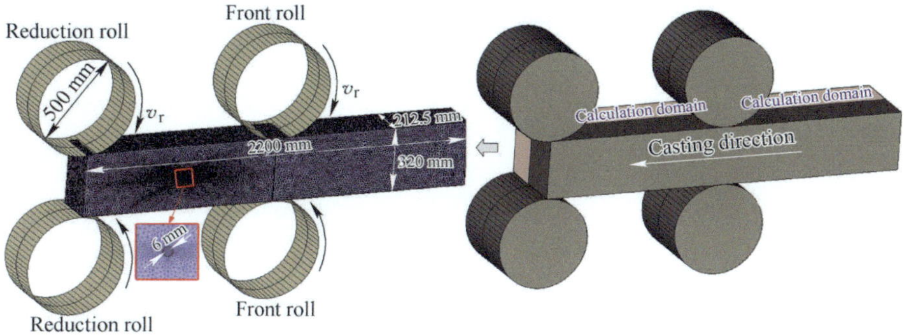

Fig. 4.45 Three-dimensional thermal/force coupling model of the end reduction process during bloom solidification

and also for improving the central segregation of the bloom, and the small reduction deformation applied by the front roller is denoted as r_f, which is equal to the reduction amount applied by the front roller to the thickness direction of the bloom. During the calculation process, the rotational speed of the reduction roller (denoted as v_r) is set to the common casting speed of 0.62 m/min in the onsite production process, while different rotational speeds (denoted as v_f) are set for the front roller to study the evolution law of the preset spherical porosity in the model under the differential speed drive of the casting speed straightening machines. In addition, the differential speed ratio $R_v = v_f/v_r$ between the front and back casting speed straightening machines is defined to quantitatively describe the difference in the rotational speed of the casting rollers between the two straightening machines.

The onsite low power inspection results of the bloom (see Fig. 4.46) show that the porosities in the continuous casting bloom are concentrated near the center of the casting and the size of the porosities is usually less than 10 mm. Therefore, artificial spherical holes of 3–9 mm are preset at the center of the casting blank in the three-dimensional thermal/mechanical coupling model of the end reduction process of the bloom to simulate the porosities at the center of the bloom. Taking a 6 mm diameter artificial spherical porosity as an example, Fig. 4.46 provides the corresponding three-dimensional thermal/mechanical coupling finite element model of the end reduction process of the bloom.

Based on the established three-dimensional thermal/mechanical coupling model of the end reduction process of the bloom, a systematic study was first conducted on the evolution law of the center porosities in the bloom after reduction under different process conditions for a single casting speed straightening machine. To quantitatively describe the different directions and overall deformation behavior of the porosities in the bloom after reduction, the axial deformation (i.e., deformation in the thickness direction Δl_x, in the casting direction Δl_y, and in the width direction Δl_z as defined in Sect. 4.2.1), closure degree (i.e., porosity closure degree η_S defined based on the deviation ratio in the thickness direction as in Eq. (4.4) and porosity closure degree η_V defined based on the volume change of the porosity as in Eq. (4.6)) were adopted.

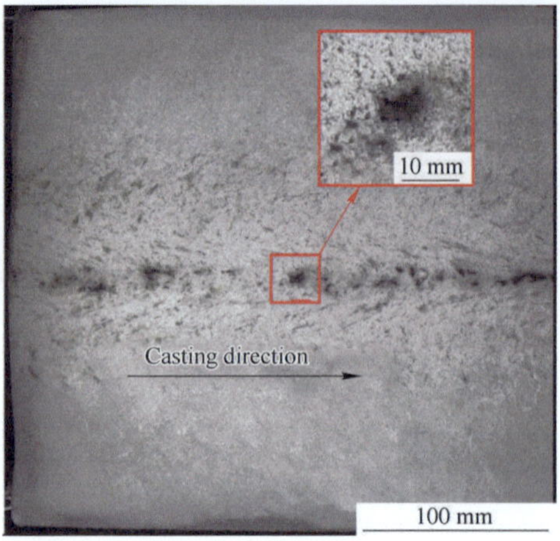

Fig. 4.46 Low magnification of the longitudinal section of the bloom

4.3.2 Porosity Evolution Under Equal Roll Rotation Speed Condition

4.3.2.1 The Influence of Porosity Size on the Evolution Law of Porosity

After being pressed by a No.7 straightening machine, the porosity evolution of large ingot with different initial dimensions (3 mm, 6 mm, and 9 mm) is shown in Fig. 4.47. Following the reduction of the large ingot, the deformation Δl_x of the central shrinkage along the ingot thickness direction (see Fig. 4.47a) is less than 0, while the deformations Δl_y along the ingot casting direction (see Fig. 4.47b) and Δl_z along the width direction of the shrinkage (see Fig. 4.47c) are greater than 0. This indicates that after reduction, the thickness of the central shrinkage of the large ingot decreases, while the dimensions along the casting direction and the width direction increase. Furthermore, since the absolute value of Δl_x is significantly greater than Δl_y and Δl_z, it indicates that the main deformation direction of the central shrinkage of the large ingot after reduction is its thickness direction.

Figure 4.47d shows the closure degree η_s of the shrinkage after reduction. With the increase in reduction amount (expressed as a percentage of the ingot thickness reduction), the central porosity continues to improve, and the corresponding closure degree of the porosity continuously increases. However, the rate of closure degree increase slows down with the increase in reduction amount. This might be because of the increase in the deformation rate of the large ingot during the reduction process as the reduction amount applied by a single straightening machine grows. The enhanced resistance to deformation in the lower temperature regions on both sides of the large ingot, and the hindering effect on the transfer of reduction deformation from the surface of the ingot to the center of the ingot become more prominent, thus affecting the efficiency of the transfer of reduction deformation to the center of the ingot,

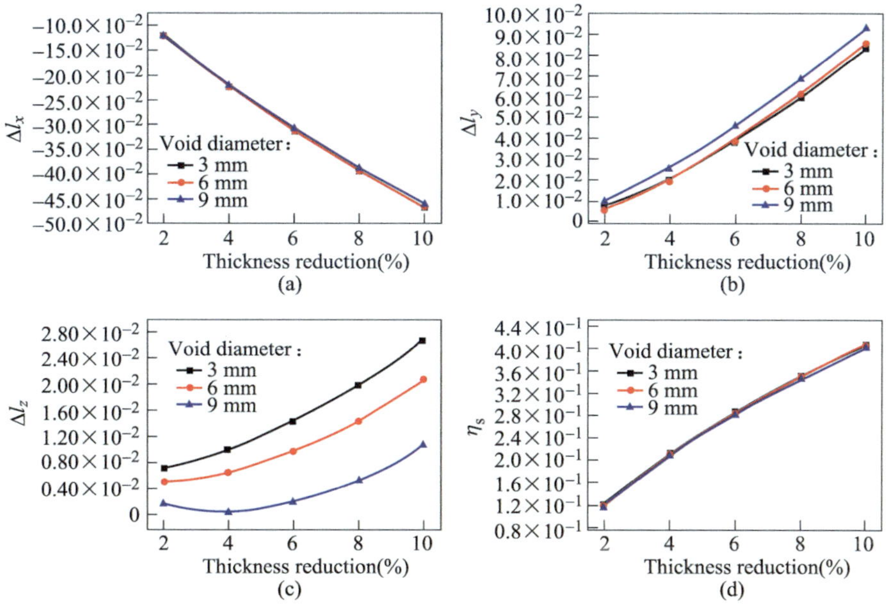

Fig. 4.47 Axial deformation degree and closure degree of porosities of different initial sizes after reduction. **a** Deformation in thickness direction Δl_x; **b** deformation in drawing direction Δl_y; **c** deformation in width direction Δl_z; **d** porosity closure degree η_s

and consequently causing the closure degree of the porosity to slow down with the increase in reduction amount.

Comparing the axial deformations of the shrinkage after reduction of different initial size shrinkages in Fig. 4.47a–c, it is found that there are significant differences in the deformation Δl_y along the casting direction and the deformation Δl_z along the width direction for shrinkages of different sizes. However, the axial deformation Δl_x along the main deformation direction (i.e., the thickness direction) is basically the same for shrinkages of different sizes. In fact, after a 10% reduction, the difference between Δl_y and Δl_z for shrinkages of different sizes is only 2.6% and 1.7% of Δl_x, respectively. This indicates that the evolutionary characteristics of shrinkages of different initial sizes are essentially the same after reduction. Therefore, the closure degrees of shrinkages of different sizes after the same reduction deformation are approximately equal, as shown in Fig. 4.47d. Given this, only the 6 mm diameter shrinkage was studied in the subsequent research to calculate the evolution characteristics of the shrinkage under different reduction process conditions, in order to investigate the effects of various factors on the improvement of the central shrinkage of the large ingot.

To elucidate the differences in the axial deformation of the central shrinkage of the large ingot after reduction from a mechanistic perspective, the axial stresses at the corresponding positions of the shrinkage after reduction were calculated and extracted based on a synchronous three-dimensional thermal/mechanical coupling

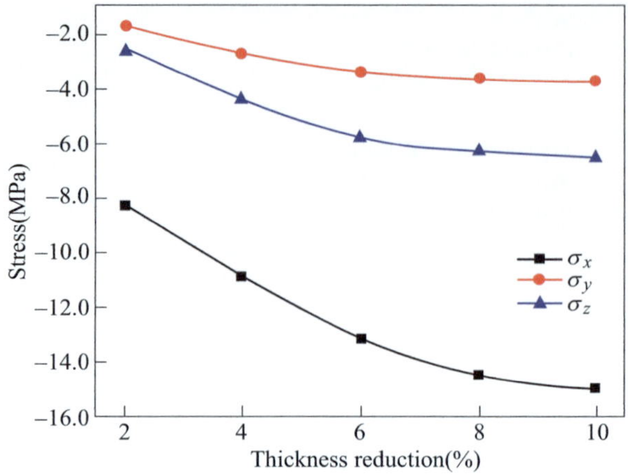

Fig. 4.48 Axial stress in *x*, *y*, and *z* directions at the center of the slab

model without shrinkage, as shown in Fig. 4.48. It can be seen that the three axial stresses at the corresponding positions of the shrinkage are all in a state of reduction (σ_x, σ_y, and σ_z are less than 0), and all three axial stresses increase with the increase in reduction amount. Among them, σ_x is significantly greater than σ_y and σ_z.

Based on the extracted state and magnitude of the axial stresses shown in Fig. 4.48, the axial stress state of the central shrinkage of the large ingot after reduction is depicted in Fig. 4.49. Although all three axial stresses are compressive, the axial stress in the thickness direction (σ_x) is significantly higher than the two axial stresses in the casting direction (σ_y) and the width direction (σ_z). As a result, under the influence of the large axial stress σ_x, the size of the shrinkage along the ingot thickness direction decreases significantly during the reduction process. Under the influence of the decrease in size along the thickness direction of the shrinkage, the sizes of the shrinkage along the casting direction and the width direction exhibit an increasing trend. However, since σ_y and σ_z are significantly smaller than σ_x, they cannot effectively suppress the increasing trend of the shrinkage size along the pulling and width directions after reduction, resulting in a slight increase in the size of the shrinkage along the casting direction compared to the width direction, as depicted in Fig. 4.49.

4.3.2.2 The Influence of Temperature Difference on the Evolution Law of Porosity

Near the solidification end of the bloom, the temperature variations at the center of the broad face (corresponding to point P_1 in Fig. 4.50a) and at the slab center (corresponding to point O in Fig. 4.50a) are shown in Fig. 4.50b. Due to the lack of solidification latent heat released during the solidification process in the central region of the bloom after the solidification end, the temperature at point O in Fig. 4.50b decreases rapidly after the solidification end (i.e., when the temperature falls below the solidus line temperature). As a result, the temperature difference between the

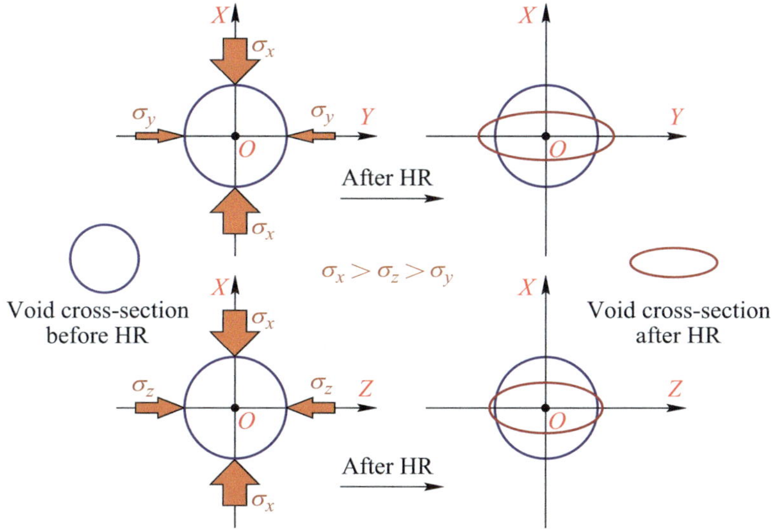

Fig. 4.49 Schematic diagram of the axial force state and deformation of the central porosity in the bloom in the *XOY* plane and *XOZ* plane

surface and the center of the bloom decreases rapidly after the solidification end. To study the effect of temperature differences inside and outside the bloom on the evolution of porosities during the reduction process, the evolution of porosities after 6% reduction at different casting positions after the solidification end was calculated. Table 4.2 provides the temperature at characteristic points and the temperature difference inside and outside the bloom for the studied casting positions. Position 1 to Position 3 represent the casting positions for the No.7 to No.9 straightening machines, respectively. At Position 4, the bloom is at a uniform temperature of 1100°C, representing the temperature characteristics of the rough rolling process.

The evolution of porosities at the center of the bloom after 6% reduction at four different positions is shown in Fig. 4.51. As the temperature difference inside and outside the slab decreases from position 1 to position 4, the deformation of the porosity at the center of the bloom along the thickness direction (Δl_x) continuously decreases, while the deformation along the casting direction (Δl_y) and the width direction (Δl_z) increases.

Figure 4.51d shows the closure degree η_s of porosities after reduction at different positions. As the temperature difference inside and outside the slab continuously decreases from Position 1 to Position 4, η_s decreases significantly. The rate of decrease from Position 1 to Position 2 is faster than that from Position 2 to Position 3. This is mainly due to the similar decreasing trend in the temperature difference inside and outside the slab shown in Fig. 4.48b, indicating that the improvement efficiency of reduction on central porosities decreases as the temperature difference inside and outside the bloom decreases. In addition, as the overall temperature of the bloom

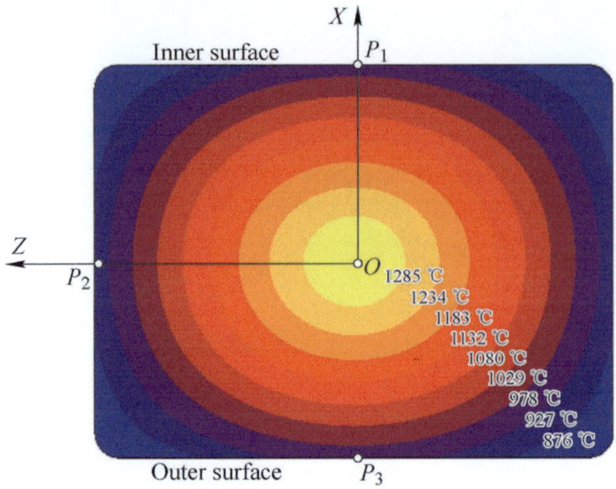

(a)

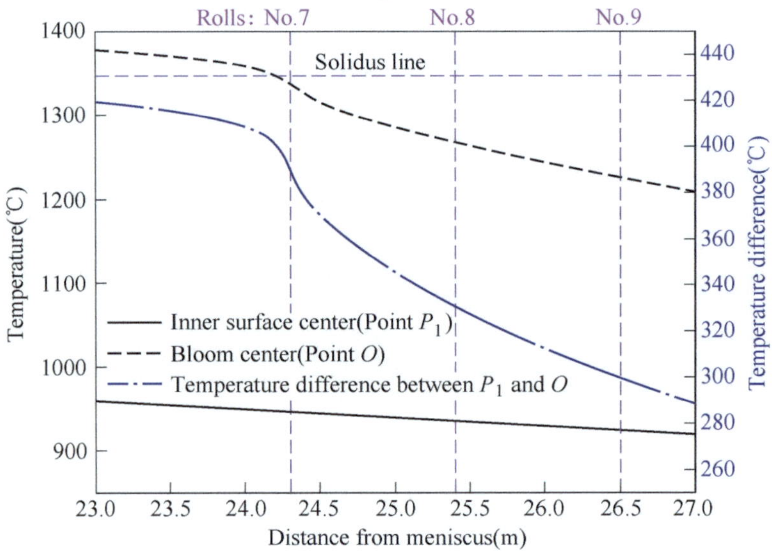

(b)

Fig. 4.50 Temperature field distribution and characteristic point temperature (**a**) and temperature difference inside and outside the slab (**b**) at the solidification end of the bloom

decreases from Position 1 to Position 3, the resistance of the slab to reduction deformation increases. Therefore, the calculated model of the reduction force applied by the casting roll indicates that the reduction force required for a 6% reduction deformation at Position 1 increases significantly from 1860 to 2142 kN, an increase of approximately 15%. This implies that as the casting position advances subsequent to the solidification end, the maximum reduction applicable by the straightening machine will keep decreasing. Therefore, in the production process, appropriate

Table 4.2 Casting position and temperature difference inside and outside the bloom at each position

Position	Distance from the bend meniscus surface (m)	Straightening machine	Temperature (°C)		Temperature difference (°C)
			Center of wide surface	Casting center	
1	24.30	No.7	947.2	1335.1	387.9
2	25.40	No.8	936.7	1267.6	330.9
3	26.50	No.9	925.8	1226.4	300.6
4	–	–	1100.0	1100.0	0.0

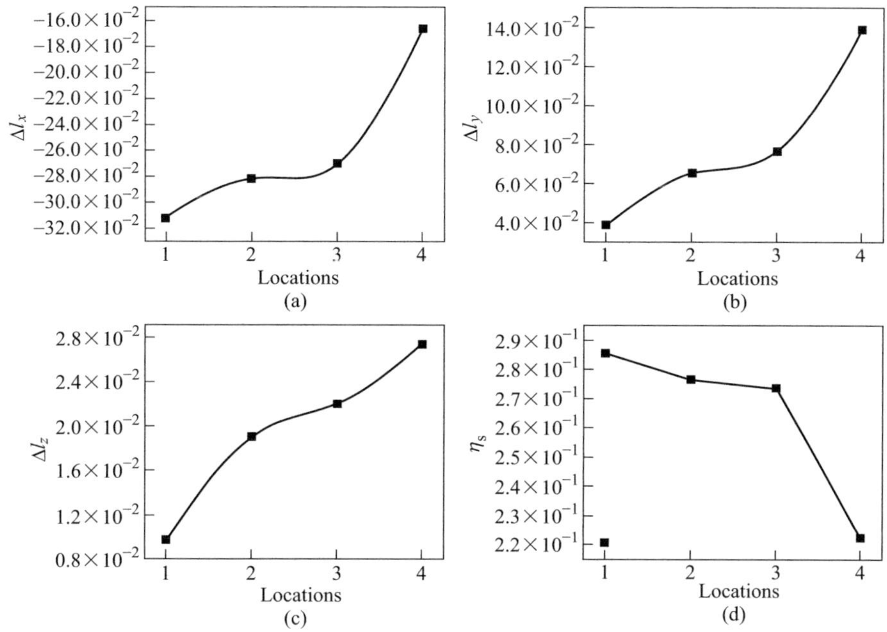

Fig. 4.51 Axial deformation and closure degree of the porosity in the center of the bloom under different positions. **a** Δl_x; **b** Δl_y; **c** Δl_z; **d** η_s

production techniques should be used to position the solidification end of the casting close to the straightening machine in order to utilize the larger temperature difference inside and outside the slab after the solidification end to maximize the efficiency of reduction.

Compared to the bloom at Position 4, which is at a uniform temperature, the blooms at Position 1 to Position 3 have a certain temperature advantage, resulting in significantly higher closure degree of the porosities compared to Position 4. Specifically, the closure degree at Position 1 (with a temperature difference of approximately 390 °C) is approximately 25% higher than that at Position 4.

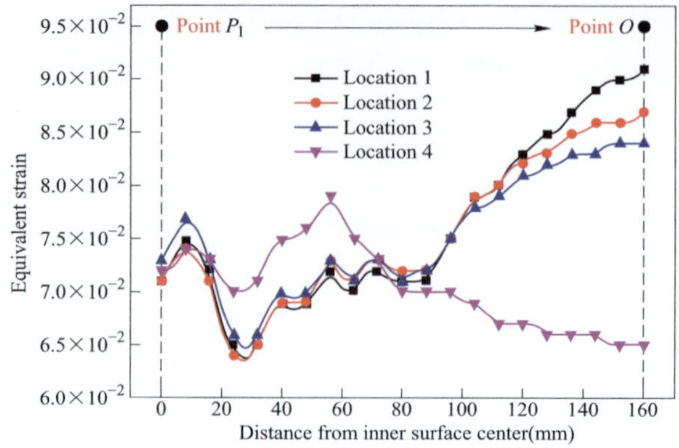

Fig. 4.52 Equivalent strain distribution in the thickness direction of the billet after 6% heavy reduction deformation

Similar to the analysis method for wide-thick slab, to further illustrate the influence of temperature differences on the closure of central porosities in the bloom after reduction, the equivalent strain distribution in the thickness direction of the bloom after 6% reduction at Position 1 to Position 4 was calculated and extracted using the synchronous no porosity, as shown in Fig. 4.52. Due to the uniform temperature at Position 4, the trend of equivalent strain variation from point P_1 to point O after reduction is substantially smaller than that at the other three positions. At Position 4, the overall equivalent strain in the outer region of the bloom after reduction (within a range of about 80 mm below the inner arc surface) is higher than that in the inner region (within a range of about 80–160 mm below the inner arc surface). In contrast, at Position 1 to Position 3, the trend of equivalent strain distribution in the thickness direction of the bloom after reduction is opposite to that at Position 4, with the overall equivalent strain in the outer region being lower than that in the inner region. Based on the above analysis, it can be inferred that the larger temperature difference inside and outside the bloom near the solidification end during the casting process effectively promotes the transfer of reduction deformation from the surface of the casting to its central region, thereby more efficiently improving the central porosity defect. In addition, the decrease in equivalent strain at the central position of the bloom (i.e., position O) from Position 1 to Position 3 after reduction indicates a reduction in the effective transfer of reduction deformation to the central position of the casting. Therefore, the closure degree of porosities shown in Fig. 4.51d gradually decreases from Position 1 to Position 3.

4.3.2.3 The Influence of Roller Shape on the Evolution Law of Porosity

Due to the lower temperature near the narrow faces of the bloom, the corresponding resistance to deformation is stronger in those regions. As shown in Fig. 4.53a, the conventional flat rollers primarily encounter resistance forces from the narrow face regions of the billet, greatly limiting the reduction capacity of the straightening machine, especially for heavy reduction and large deformation. However, the porosity

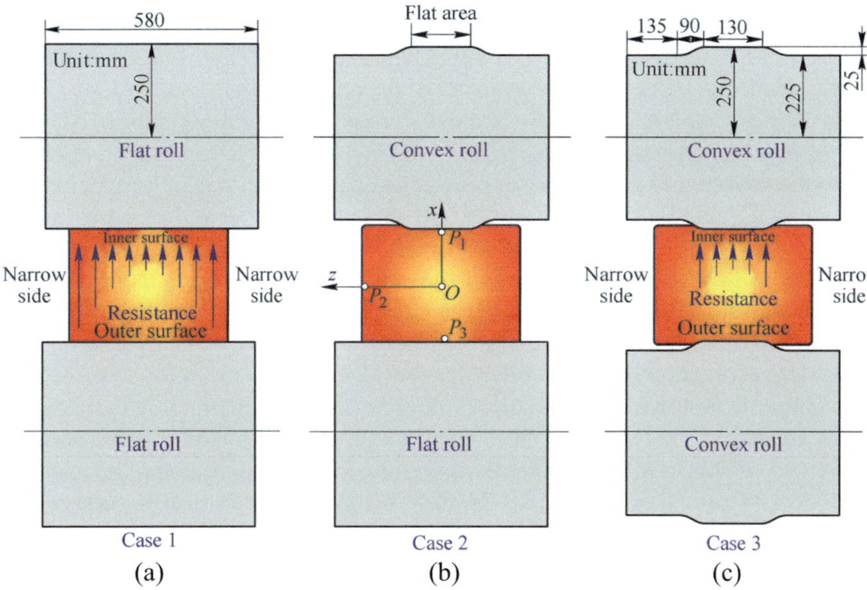

Fig. 4.53 Cases of three roll shape schemes. **a** Case 1 (both inner and outer arcs are flat rolls); **b** Case 2 (convex roll on the inner arc side, flat roll on the outer arc side); **c** Case 3 (both inner and outer arcs are convex rolls)

in the center of the bloom is mainly concentrated in the central region of the billet. This indicates that the application of the convex roller for reduction can significantly suppress the tendency of the porosity in the central part of the bloom to increase in the casting direction, it can effectively enhance the reduction capacity of the straightening machine. Based on this point of view, researchers have proposed and applied convex rollers for billet reduction processes. Our team has also proposed a gradient curvature convex roller for continuous casting of blooms, as detailed in Sect. 7.3 [38].

To study the influence of the roller type on the billet reduction process, the evolution of the porosity in the center of the bloom and the system pressure after reduction using three different roller schemes, as shown in Fig. 4.53, were calculated and compared. Case 1 and Case 3 represent the use of conventional flat rollers and convex rollers for both the inner and outer arc casting rollers of the straightening machine, while Case 2 represents the use of a convex roller for the inner arc and a conventional flat roller for the outer arc of the straightening machine. The roller diameter for both the flat roller and the maximum diameter of the convex roller is 500 mm. The convex roller is a gradient curvature convex roller, with a convex length (determined by the width of the loose distribution region of the porosity in the transverse section of the bloom) of 130 mm.

After applying a 6% reduction deformation to the bloom using the three roller schemes shown in Fig. 4.53, the evolution of the porosity in the center of the bloom is shown in Fig. 4.54. As observed in Fig. 4.54a–c, compared to the traditional flat roller in Case 1, the use of convex roller schemes in Case 2 and Case 3 result in a significant change in the axial deformation of the porosity in the center of the bloom after reduction. In Case 2 and Case 3, the thickness direction deformation (Δl_x) and width direction deformation (Δl_z) of the porosity after reduction are significantly increased compared to Case 1. When using the conventional flat roller scheme, the billet casting direction deformation (Δl_y) of the porosity after reduction is greater than 0. However, when using the convex roller, the corresponding Δl_y in Case 2 and Case 3 becomes negative. This indicates that the application of the convex roller for reduction can significantly suppress the tendency of the porosity in the central part of the bloom to increase in the casting direction.

From the above analysis, it is evident that the thickness direction deformation (Δl_x) and width direction deformation (Δl_z) of the porosity are significantly increased in Case 2 and Case 3 when using the convex roller, thereby increasing the porosity closure (η_s) as shown in Fig. 4.54d. Furthermore, the casting direction deformation of the porosity is reduced, leading to a decrease in the porosity closure η_s. As shown in Fig. 4.55, when the convex roller is employed, the porosity volume reduction in Case 2 and Case 3 is significantly enhanced by 51.8% and 51.5%, respectively, in contrast to that with the traditional flat roller in Case 1. Therefore, it can be concluded that compared to the conventional flat roller reduction, the convex roller can effectively enhance the reduction efficiency at the solidification end of the bloom, contributing to a more pronounced improvement in the porosity defect under the same reduction conditions.

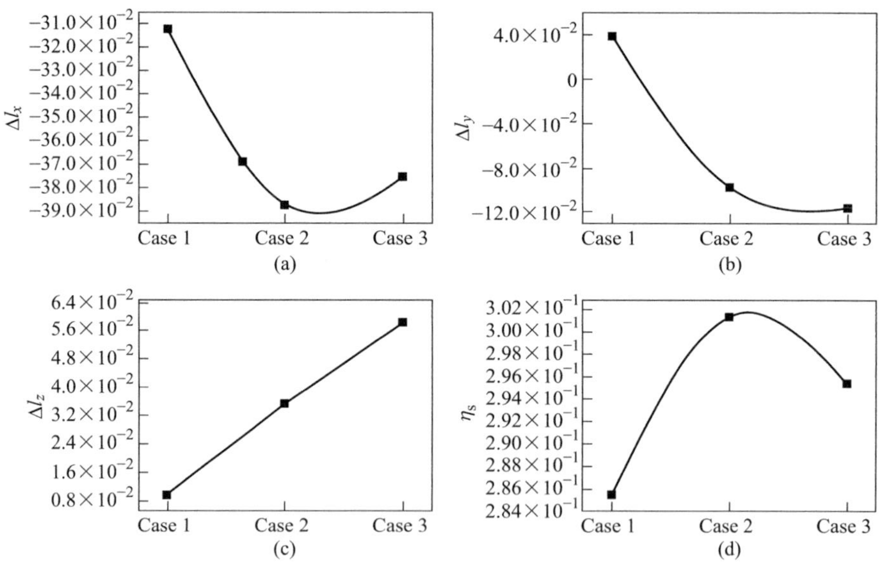

Fig. 4.54 Axial deformation and porosity closure in the center of bloom after reduction with three roll type schemes. **a** Δl_x; **b** Δl_y; **c** Δl_z; **d** η_s

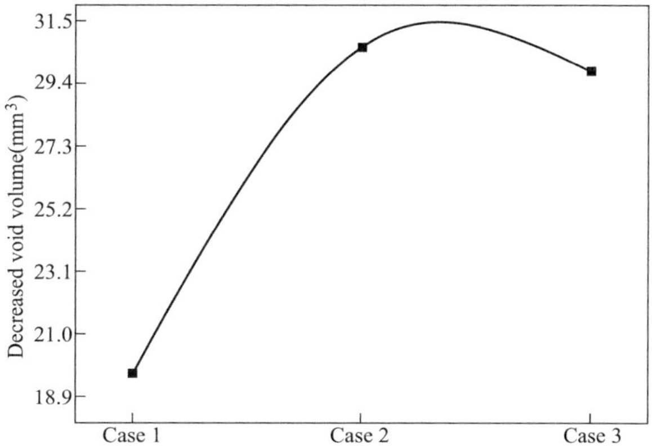

Fig. 4.55 Volume reduction of central porosity in bloom after reduction with three schemes

After reduction using the three roller schemes, the equivalent strain distribution from points P_1/P_3 to O in the thickness direction and from point P_2 to O in the width direction of the bloom (as shown in Fig. 4.53) is shown in Fig. 4.56a and b. It can be observed that the equivalent strain for all three schemes gradually increases from the surface of the bloom toward its center. However, the rate of change in equivalent strain within the width and thickness directions in Case 2 and Case 3 is significantly higher than that in Case 1. Therefore, the internal equivalent strain in the bloom for Case 2 and Case 3 with the convex roller is higher than that of Case 1. This indicates that compared to the conventional flat roller, reduction with the convex roller effectively promotes the transmission of reduction deformation from the surface to the inner region of the bloom, thus more efficiently improving the porosity and loose defects in the center of the bloom.

From Fig. 4.56b, it is evident that the equivalent strain in the width direction in the outer region of the bloom for Case 2 and Case 3 is significantly lower than that of Case 1, while it is higher near the center of the cast billet. This indicates that with the adoption of Case 2 or Case 3, in which the convex roller is utilized for reduction, the reduction deformation is mainly concentrated in the central region of the bloom. This effectively averts resistance to reduction deformation in the lower—temperature regions near the narrow faces of the bloom, thus achieving greater reduction deformation under lower system pressure. Figure 4.56c compares the reduction forces required for the three roller schemes when applying 6% reduction deformation at the solidification end of the bloom. It can be seen that compared to the conventional flat roller in Case 1, the reduction forces required for Case 2 and Case 3, using single and double convex rollers respectively, are reduced by approximately 41.9% and 56.1%, respectively. This demonstrates that the convex roller can effectively reduce the reduction force, thereby significantly enhancing the reduction capacity of the straightening machine.

Comparing the distribution of equivalent strain in the thickness and width directions for Case 2 and Case 3 in Fig. 4.56a and b reveals significant differences. In Fig. 4.56a, the equivalent strain in the thickness direction for Case 3 is higher than that

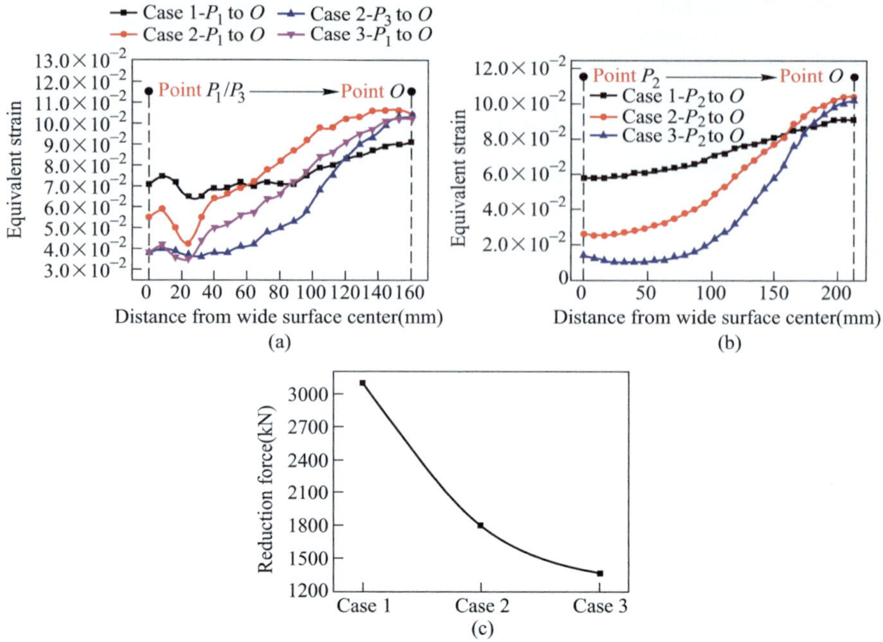

Fig. 4.56 Distribution of equivalent strain in slab thickness and width direction and required reduction force for reduction roll. **a** Thickness direction; **b** width direction; **c** reduction force

of Case 2 from the outer arc surface (point P_3) to the center (point O), while it is less than that of Case 2 from the inner arc surface (point P_1) to the center (point O). From Fig. 4.56b, it can be noted that the equivalent strain in the width direction for Case 3, especially in the lower temperature region near the narrow face of the cast billet, is less than that of Case 2. Therefore, when using double convex rollers for both inner and outer arcs in Case 3, the reduction force required for the straightening machine to perform 6% reduction deformation can be further reduced by 24.3% compared to the use of single convex rollers on the inner arc in Case 2. It can be concluded that, in comparison with the traditional flat roller mode, the application of convex rollers for reduction can effectively decrease the system pressure required for the straightening machine to conduct reduction, thereby augmenting the reduction capacity of the straightening machine. In addition, when both inner and outer arcs of the straightening machine are equipped with double convex rollers, the reduction capacity of the straightening machine can be further improved. Considering factors such as the transfer of guide rods, equipment maintenance, and the safety of continuous casting production, it is generally preferable to use single convex rollers, as shown in Case 2, only installing convex rollers on the inner arc of the casting machine.

4.3.2.4 The Influence of Casting Speed on the Evolution Law of Porosity

To investigate the influence of pulling speed on the end depressurization effect of bloom solidification, the evolution law of porosity after 6% depressurization deformation at the solidification end of bloom under different pulling speeds was calculated using a single-strand straightener, as shown in Fig. 4.57. As the pulling speed increases from 0.52 m/min to 0.62 m/min, the axial deformations of the porosity shown in Fig. 4.57a–c do not exhibit a fixed trend with pulling speed, and the axial deformations along the three directions of the porosity show minor fluctuations with the change of pulling speed. Figure 4.57d shows the closure degree of the porosity after depressurization at the solidification end under different pulling speeds. It can be observed that the closure degree of the porosity gradually increases with the increase of pulling speed, but the overall increase is relatively small. When the pulling speed increases by 19.2% from 0.52 m/min to 0.62 m/min, the corresponding closure degree of the porosity increases by only about 0.7%. The analysis above indicates that the pulling speed has no significant influence on the evolution law of the central porosity after depressurization at the solidification end of bloom.

Figure 4.58 shows the calculated temperature and the solidification end casting location of the bloom at different pulling speeds using a two-dimensional solidification heat transfer model. With the increase of pulling speed, the effective cooling time of the bloom in each cooling zone decreases. At the same casting position in the

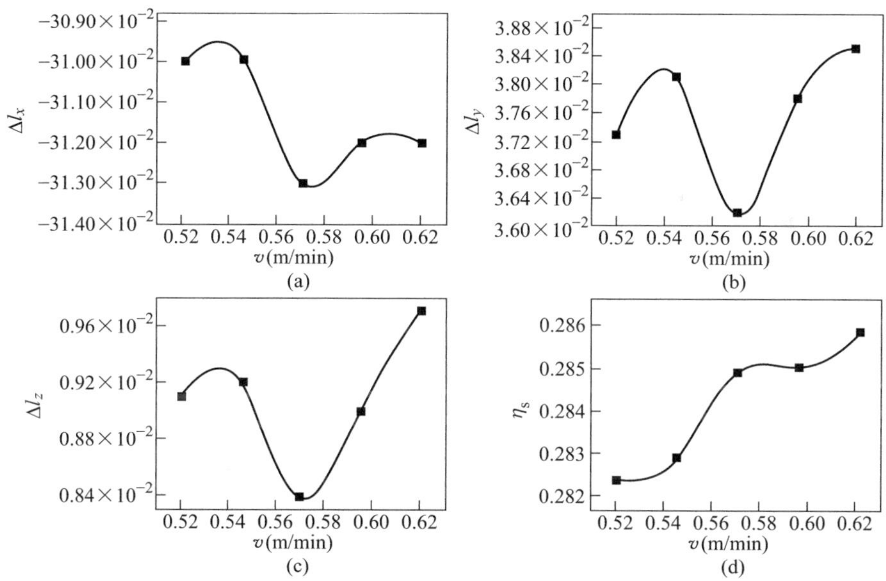

Fig. 4.57 Axial deformation degree and relative closure degree of the central porosity after heavy reduction at the solidification end of bloom at different casting speeds. **a** Δl_x; **b** Δl_y; **c** Δl_z; **d** η_s

air cooling zone, the center temperature of the wide surface of the bloom in Fig. 4.58a continues to rise. In addition, the casting location at the solidification end of the bloom also undergoes significant changes with the increase of pulling speed, as shown in Fig. 4.58b. During the process of increasing the pulling speed from 0.52 m/min to 0.62 m/min, with every 0.1 m/min increase in the pulling speed, the casting location at the solidification end can shift a distance of approximately 4 m along the direction of the slab. However, as shown in Fig. 4.58c, the temperature distribution in the thickness direction at the solidification end of the bloom under different pulling speeds is basically the same, which means that the temperature difference between the inside and outside is also basically equal. Therefore, as shown in Fig. 4.57, the pulling speed has no significant influence on the axial deformation and closure degree of the central porosity after depressurization at the solidification end of the bloom. This implies that, similar to wide-thick slab, in the actual production process, by appropriately adjusting the pulling speed, the casting location at the solidification end can be adjusted to fully utilize the temperature difference advantage at the solidification end of the continuous casting slab to achieve efficient transfer of depressurization deformation toward the central part of the slab.

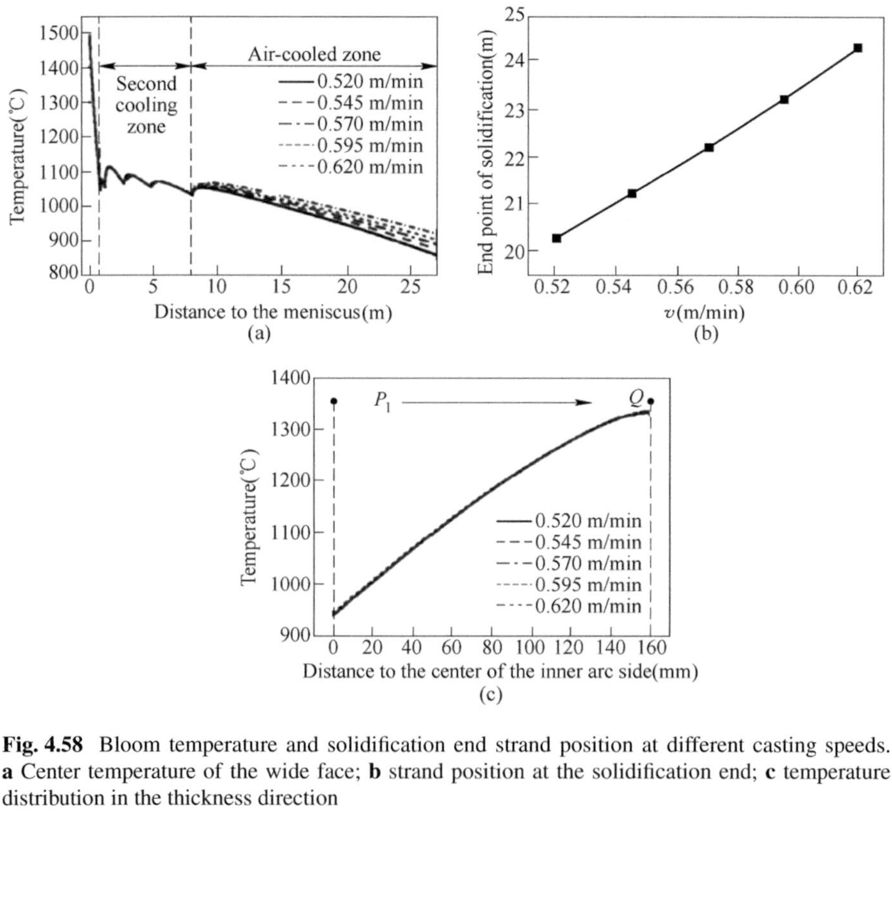

Fig. 4.58 Bloom temperature and solidification end strand position at different casting speeds. **a** Center temperature of the wide face; **b** strand position at the solidification end; **c** temperature distribution in the thickness direction

4.3.2.5 The Influence of Secondary Cooling on the Evolution Law of Porosity

The temperature field distribution of the bloom can significantly affect the evolution of porosity during the solidification end reduction process, and the cooling water distribution in the secondary cooling zone is one of the main factors affecting the cooling and solidification process of the bloom. Therefore, the evolution of porosity after applying a 6% reduction deformation at the solidification end of the bloom was computed and compared under different secondary cooling water volumes, aiming to reveal the impact of the secondary cooling water distribution on the solidification—end reduction of the bloom.

Figure 4.59 shows the temperature and solidification endpoint mold flow position of the bloom under different cooling water distributions. As the ratio of secondary cooling water volume increases from 0.22 L/kg to 0.34 L/kg, the surface temperature of the bloom at the same mold flow position in the secondary cooling zone shown in Fig. 4.59a continuously decreases, and the solidification endpoint mold flow position in Fig. 4.59b moves approximately 1.8 m in the opposite direction of the casting direction. However, the thickness-wise temperature distribution of the solidification endpoint position of the bloom shown in Fig. 4.59c does not undergo significant changes with the ratio of secondary cooling water volume. Therefore, the secondary cooling water distribution has no significant effect on the axial deformation and closure degree of the porosity after solidification end reduction of the bloom depicted in Fig. 4.60. For example, the maximum difference in porosity closure degree under different secondary cooling water volumes in Fig. 4.60d is only about 0.2% of the maximum closure degree of the porosity.

From the above analysis, it can be concluded that, in the continuous casting process of the bloom, the solidification endpoint mold flow position can be appropriately adjusted by the secondary cooling water volume in addition to the casting speed, so as to fully utilize the advantage of the larger temperature difference at the solidification endpoint of the bloom and ensure the efficient implementation of the reduction process.

4.3.3 Porosity Evolution Under Equal Roll Rotation Speed Condition

4.3.3.1 The Influence of Differential Drive on the Evolution Law of Porosity

During the solidification process of continuous casting billets, in addition to the downward deformation, there are also significant extension and width deformations, which affect the efficiency of deformation transmission to the center of the billet. Therefore, apart from controlling the downward pressure amount, the control of the

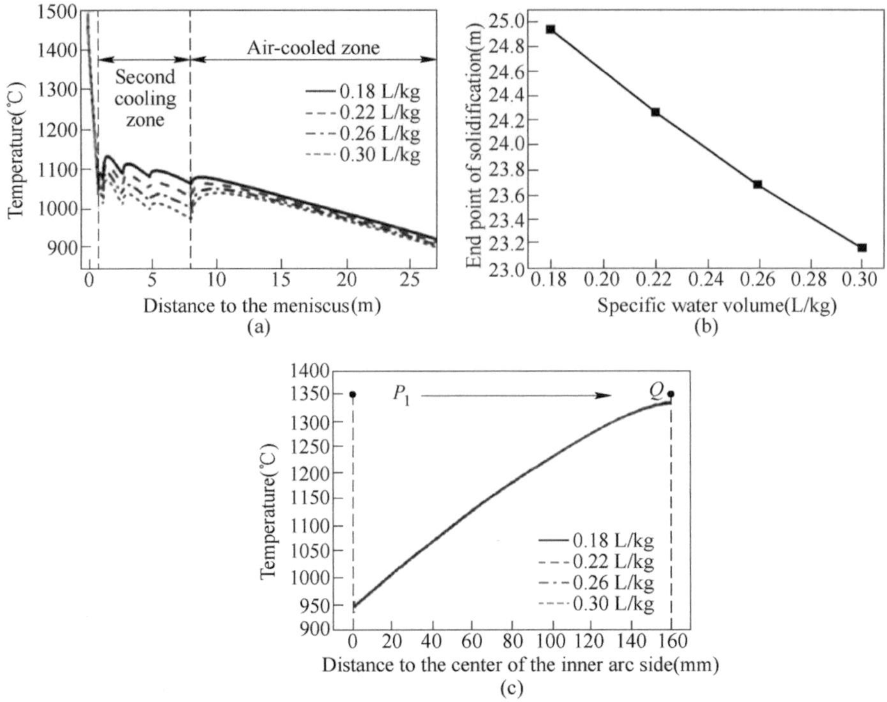

Fig. 4.59 Temperature and strand position at the solidification end of the bloom at different water distributions. **a** Center temperature of the wide face; **b** strand position at the solidification end; **c** temperature distribution in the thickness direction

straightening machine's speed and the adjustment of the directional forces acting on the billet during casting can be utilized to suppress the billet's extension deformation. This leads to an enhancement in the effectiveness of downward pressure by reducing the deformations in both the pulling and pressure directions.

Taking the continuous casting bloom as an example, the variation in the central shrinkage deformation of the billet under different speed ratio of the front and rear reduction rollers is explored. Specifically, by setting the speed of the reduction rollers ($v_r = 0.62$ m/min) and the amount of downward pressure (6%) as fixed conditions, different values of front roller speed (v_f) were applied to calculate the evolution of the central shrinkage deformation of the bloom after downward pressure under different speed ratio ($R_v = v_f/v_r$) conditions, as shown in Fig. 4.61.

As the speed ratio increases from 0.8 to 1.2, the deformation in the casting direction Δl_y (as shown in Fig. 4.61b) and the width direction deformation Δl_z continuously decrease (as shown in Fig. 4.61c). This indicates that with the increase in the speed ratio of the front and rear reduction rollers during the downward pressure process, the extension deformation of the central shrinkage of the bloom in the pulling and width directions is effectively suppressed. As a result of the suppression of the deformation in the pulling and width directions, the dissipation effect of the extension and

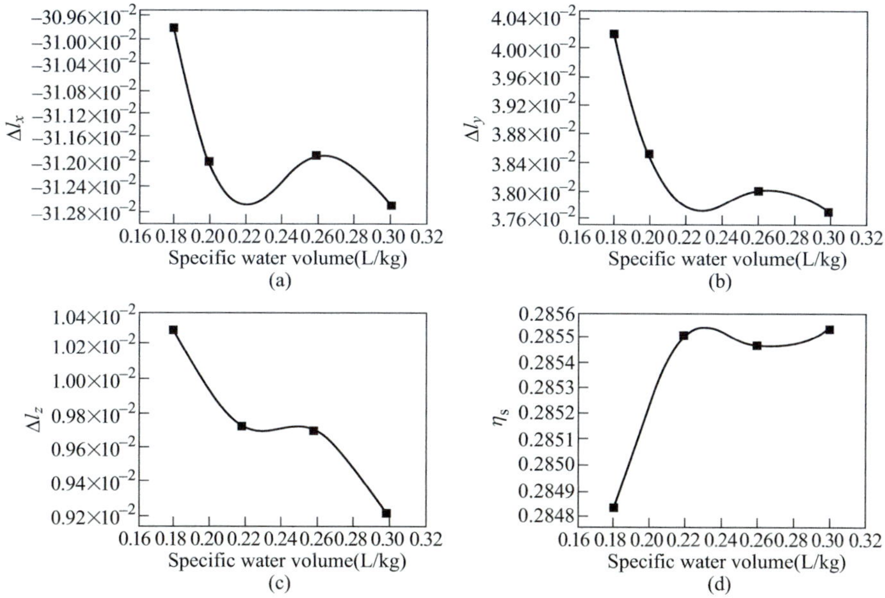

Fig. 4.60 Axial deformation degree of the central porosity after heavy reduction at the solidification end of the bloom under different secondary cooling water distributions. **a** *X*-axis; **b** *Y*-axis; **c** *Z*-axis; **d** porosity closure degree

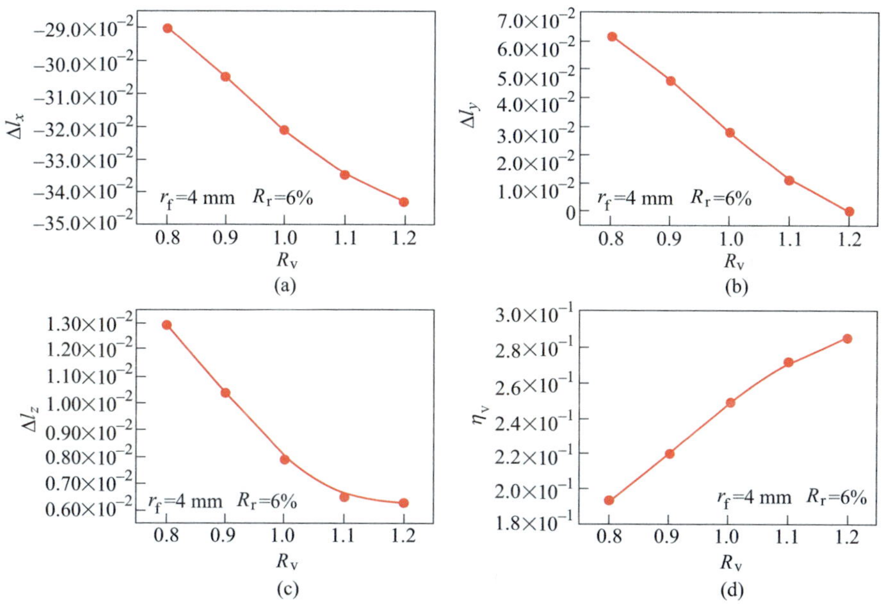

Fig. 4.61 Axial deformation and closure degree of porosity after heavy reduction at the solidification end of bloom with different casting speed

width deformations on the amount of downward pressure is also suppressed, thereby promoting the efficiency of the transmission of downward deformation to the center of the billet. Consequently, the thickness direction closure of the shrinkage Δl_x (as shown in Fig. 4.61a) also continuously increases with the increase in the speed ratio, indicating a reduction in the thickness direction size of the shrinkage due to the increase in the speed ratio. Additionally, the closure degree of the porosity η_v (as shown in Fig. 4.61d) also steadily increases with the increase in the speed ratio, with a 47% increase when the speed ratio increases from 0.8 to 1.2.

Based on the above analysis, it can be inferred that during the downward pressure process at the end of the solidification of the bloom, effectively enhancing the process effect of the downward pressure on the central shrinkage of the bloom can be achieved by appropriately increasing the speed of the casting roller in front of the reduction roller (v_f) so that it exceeds the speed of the reduction roller (v_r). From the perspective of force analysis, when the speed of the front roller is higher than that of the reduction roller, the overall deformation of the bloom in the casting direction tends to be in an extruded state. This is similar to the results obtained by Wu et al. in the study of the improvement mechanism of slight downward pressure on the center segregation of slab under light downward pressure, where it was found that if the slab is under slight extrusion within the fan segment under light downward pressure, it helps to improve the process effect of the improvement of center segregation of the slab under the same amount of downward pressure conditions.

The static water stress integral proposed by Tanaka (see Eq. (4.3)) can quantitatively reflect the improvement of internal shrinkage in the material due to metal processing deformation. To further illustrate the effect of the speed ratio on the closure law of shrinkage after the downward pressure process, the static water stress integral value (Q) at the corresponding position of the shrinkage for different speed ratios was calculated based on the equivalent strain (ε_{eq}) and static water stress (σ_m) determined by the synchronous shrinkage-free model, as shown in Fig. 4.62. It can be observed that the static water stress integral value (Q) continuously increases with the increase in the speed ratio (R_v), and its increasing trend with R_v is similar to the increasing trend of the closure degree of the porosity in Fig. 4.61d. This indicates that the greater the speed ratio R_v, the more effective the improvement in the closure efficiency of the shrinkage after the downward pressure at the end of the solidification of the bloom.

To further reveal the influence of the speed ratio on the closure law of shrinkage after the downward pressure process, the axial stress at the position of the shrinkage when the reduction roller is directly above the shrinkage during the downward pressure process, as shown in Fig. 4.63a, was extracted based on the synchronous shrinkage-free model. The extracted results are presented in Fig. 4.63b, σ_x, σ_y, and σ_z represent the axial stresses in the thickness direction, casting direction, and width direction of the billet, respectively.

It can be observed from Fig. 4.63b that all three axial stresses at the corresponding position of the shrinkage are in reduction (σ_x, σ_y, and σ_z are all less than 0), and σ_x is significantly greater than σ_y, and σ_z. Based on the features of the three axial stresses, a schematic diagram of the axial stress at the shrinkage position and the

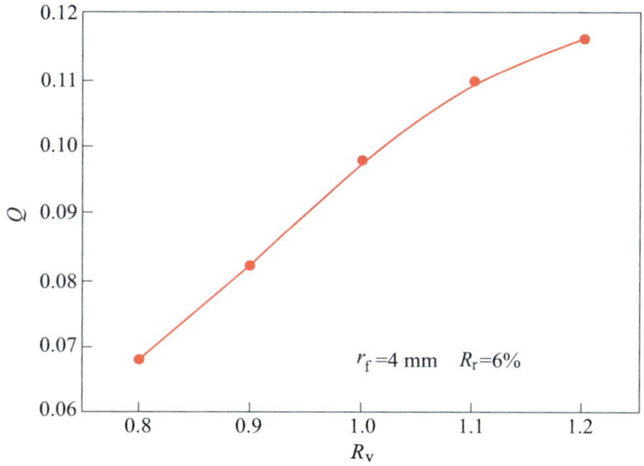

Fig. 4.62 Trend of hydrostatic stress integral (Q) with casting speed

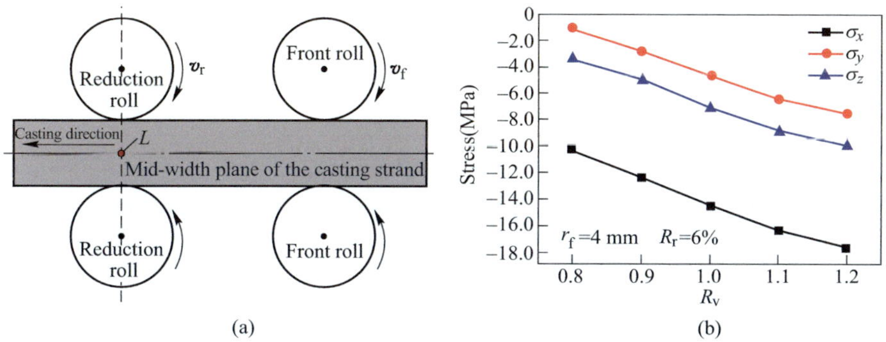

Fig. 4.63 Schematic diagram of the position of the slab axial stress during the reduction process with different casting speed (**a**) and extraction results (**b**)

evolution of the shrink during the bloom's downward pressure process is drawn, as presented in Fig. 4.64. During the downward pressure process, under the action of the larger axial compressive stress σ_x in the thickness direction, the thickness of the shrinkage significantly decreases. As the thickness of the shrinkage, it tends to increase in the pulling and width directions, while the smaller axial stresses σ_y and σ_z can to some extent suppress the trend of increasing the size of the shrinkage in the pulling and width directions. As the speed ratio R_v in Fig. 4.63b, the increasing σ_x further promotes the reduction in the thickness of the shrinkage during the downward pressure process, and the gradually increasing σ_y and σ_z exert a more significant inhibitory effect on the trend of increasing the size of the shrinkage in the pulling and width directions. In conclusion, the axial size the central shrinkage of the bloom after the downward pressure continuously decreases with the increase in the speed ratio R_v, indicating that increasing R_v can significantly enhance the closure efficiency of the downward pressure process on the shrinkage of the bloom after solidification.

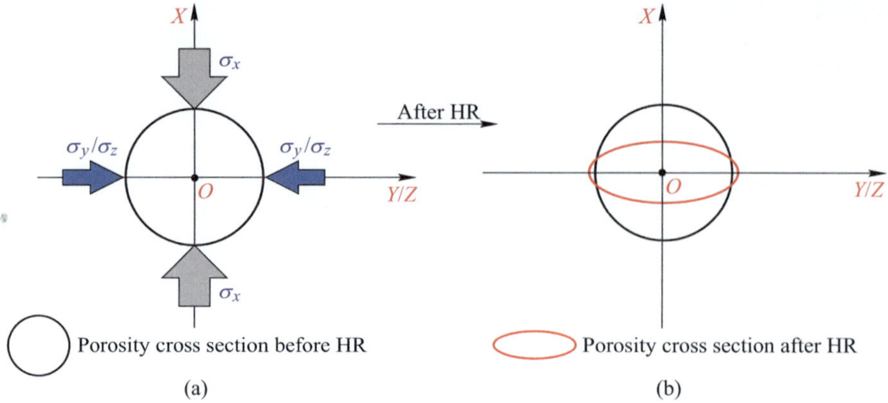

Fig. 4.64 Schematic diagram of axial stress at the porosity position during the reduction process at the solidification end of the bloom in the *XOY* and *XOZ* planes

4.3.3.2 Effect of Differential Drive on Reduction to Improve Porosity Efficiency

Based on the analysis in Sect. 4.3.3.1, it is evident that increasing the front roller speed to be greater than the reduction roller speed (i.e., $R_v > 1.0$) during the downward pressure process at the solidification end of blooms contributes to improving the porosity closure efficiency. This section further investigates the enhancement effect of differential speed drive on the efficiency of the downward pressure at the solidification end of blooms when $R_v > 1.0$. To quantitatively evaluate the improvement effect of the differential speed drive on the efficiency of the downward pressure at the solidification end of blooms, an improvement index is defined based on the porosity closure degree at different differential speed ratios:

$$\text{In} = \left(\eta_v - \eta_v^0\right)/\eta_v^0 \tag{4.10}$$

where In is the improvement index of differential roll rotation speed, and a larger value of In implies a more prominent promotional effect of the differential roll rotation speed on HR; η_v is the porosity closure index after HR under the condition of $R_v \geq 1.0$; η_v^0 is the porosity closure index after HR under condition of $R_v = 1.0$.

Figure 4.65a and b, respectively, present the porosity closure degree at different differential speed ratios and the improvement index of the efficiency of the downward pressure at the solidification end of blooms. As R_v increases from 1.0 to 2.0, the porosity closure degree η_v in Fig. 4.65a and the improvement index In in Fig. 4.65b continue to increase. When R_v increases from 1.0 to 2.0, In increases to 21.8%. This indicates that compared to the traditional equal speed drive mode (i.e., $R_v = 1.0$), implementing the differential speed drive mode with a differential speed ratio of $R_v = 2.0$ for the downward pressure at the solidification end of blooms can significantly increase the efficiency of improving the central shrinkage by approximately 21.8%.

However, it is noteworthy that the rate of increase of In in Fig. 4.65b slows down with the increase of R_v, indicating a decrease in the efficiency improvement of the downward pressure at the solidification end of blooms as the speed difference between the front roller and the reduction roller continues to increase. To explain the above-mentioned trend of In with R_v, the relative error between the front roller speed and the billet forward speed at different differential speed ratios R_v is extracted, and the results are shown in Fig. 4.66. As the differential speed ratio R_v increases, the relative error between the front roller speed and the billet forward speed continues to increase, indicating a more significant relative sliding between the front roller and the billet. As a, the efficiency improvement of the downward pressure with the differential speed drive decreases with the increase of R_v, as shown by the slowing down of the increase in In with R_v in Fig. 4.65b.

By varying the reduction roller reduction amount R_r in the model, the enhancement effect of the differential speed drive on the efficiency of the downward pressure at the solidification end of blooms at different R_r is studied. Figure 4.67a and b, respectively

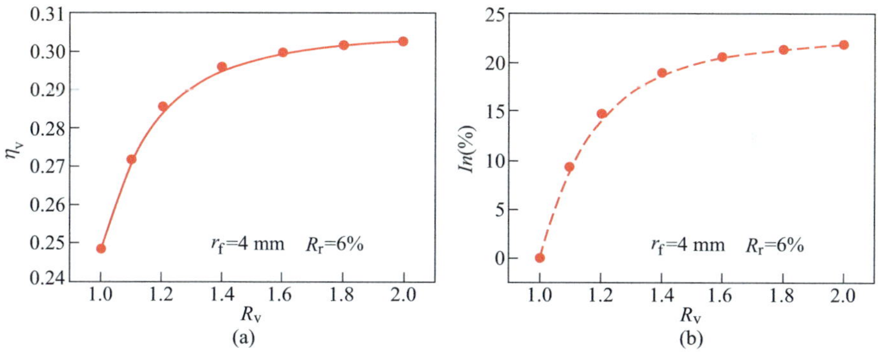

Fig. 4.65 Porosity closure degree (**a**) and reduction efficiency improvement index (**b**) at different differential ratios

Fig. 4.66 Front roll speed and billet advance speed and their relative error at different differential ratios

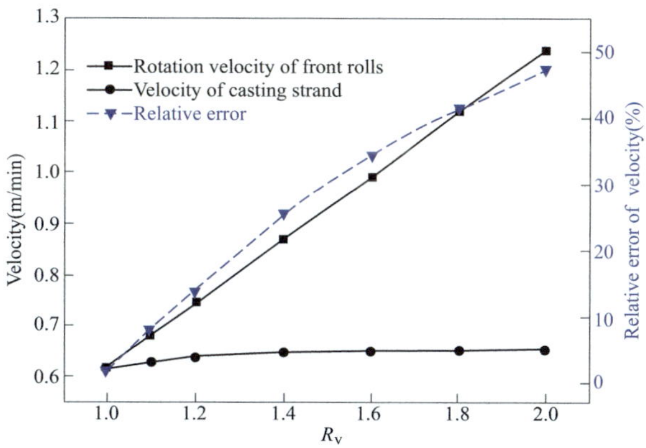

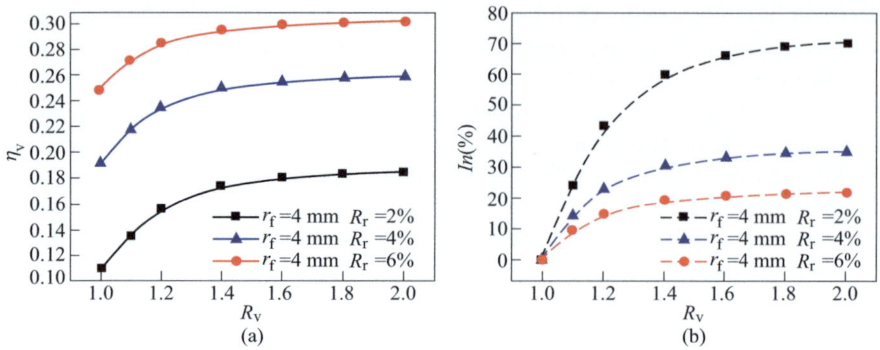

Fig. 4.67 Trend of porosity closure degree and increase index with speed difference ratio at different reduction amounts of the front roller

show the porosity closure degree η_v and the improvement index In with the change of the differential ratio R_v at different R_r. From Fig. 4.67a, it can be observed that the efficiency of downward pressure at different R_r can be further improved under the differential speed drive mode. However, Fig. 4.67b shows that the improvement effect of the differential speed on the efficiency of downward pressure at different R_r is significantly different, and under the same differential speed ratio R_v conditions, the improvement effect of the differential speed drive on the efficiency of downward pressure becomes more pronounced as the reduction roller reduction amount R_r decreases. For example, when $R_v = 0.0$ and $R_r = 6\%$, the corresponding improvement index In is approximately 21.8%. However, under the same differential speed ratio (i.e., $R_v = 2.0$) condition, as the reduction roller reduction amount R_r decreases to 2%, the corresponding In value increases to approximately 70%.

By varying the reduction amount r_f of the front roller, the enhancement effect of the differential speed drive on the efficiency of the downward pressure at the solidification end of blooms at different r_f is studied. Figure 4.68a and b, respectively, the porosity closure degree η_v and the improvement index r_f with the change of the differential speed ratio R_v at different r_f. Figure 4.68a shows that when $R_v = 1.0$, the porosity closure degree η_v corresponding to different reduction of the front roller is basically the same, indicating that the reduction amount of the front roller does not significantly affect the improvement effect of the downward pressure on the central shrinkage under the conventional equal speed drive mode. As the differential speed ratio R_v between the front roller and the reduction roller increases, shrinkage closure degree η_v in Fig. 4.68a and the corresponding improvement index In in Fig. 4.68b continue to increase, and the differences in η_v and In corresponding to different r_f become more pronounced. Under the same differential speed ratio R_v condition the porosity closure degree η_v in Fig. 4.68a and the improvement index In in Fig. 4.68b show an increasing trend with the increase of r_f. This indicates that with the increase of the reduction amount r_f of the front roller, the enhancement effect of the differential drive on the efficiency of the downward pressure can be further strengthened. However, it should be noted that as the reduction amount r_f of the front

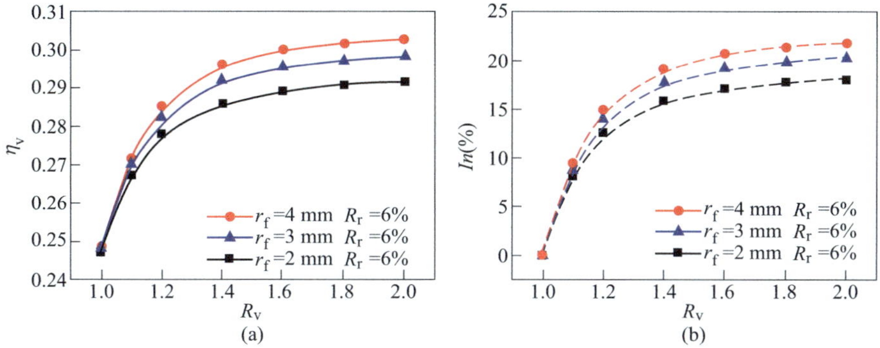

Fig. 4.68 Trend of pore closure degree (**a**) and improvement index (**b**) with differential speed ratio at different front roller reduction amounts

roller increases from 2 to 4 mm, the improvement index In at $R_v = 2.0$ in Fig. 4.68b only increases from 18.1 to 21.8%. In contrast, under the same differential speed ratio condition, as the reduction roller reduction amount R_r decreases from 6 to 2% as shown in Fig. 4.67b, the In value significantly increases from 21.8 to 70%. This indicates that compared to the reduction amount r_f of the front roller, the reduction roller reduction amount R_r will have a more significant impact on the enhancement effect of the differential speed drive mode on the efficiency of the downward pressure.

4.3.4 Prediction Equation of Porosity Closure for Bloom Reduction

Similar to the research method of continuous casting slab, for the quantitative evaluation of the process effect of shrinkage improvement by the press-down process at the solidification end of bloom, and to provide theoretical guidance for the formulation of the press-down process at the solidification end of bloom, this section establishes the shrinkage evolution law after the press-down process under different process conditions calculated previously, and develops the prediction equation for the closure degree of shrinkage after the press-down of the bloom. Scatter plots of η_s-ε_{eq}, η_s-Q, η_v-ε_{eq}, and η_v-Q under different process conditions are shown in Fig. 4.69. Here, η_s and η_v represent the closure degree based on the thickness direction deviation ratio of shrinkage (see Eq. (4.3)) and the closure degree based on the volume change of shrinkage (see Eq. (4.5)), while ε_{eq} and Q, respectively, represent the equivalent strain at the corresponding position of shrinkage calculated by the synchronous no-porosity and the static water stress integral (see Eq. (4.4)). In addition, Eq. (4.6) is used to calculate the Pearson correlation coefficient (r) of the scatter data in Fig. 4.69a–d, and the r values are labeled in the corresponding figures. By comparing the r values

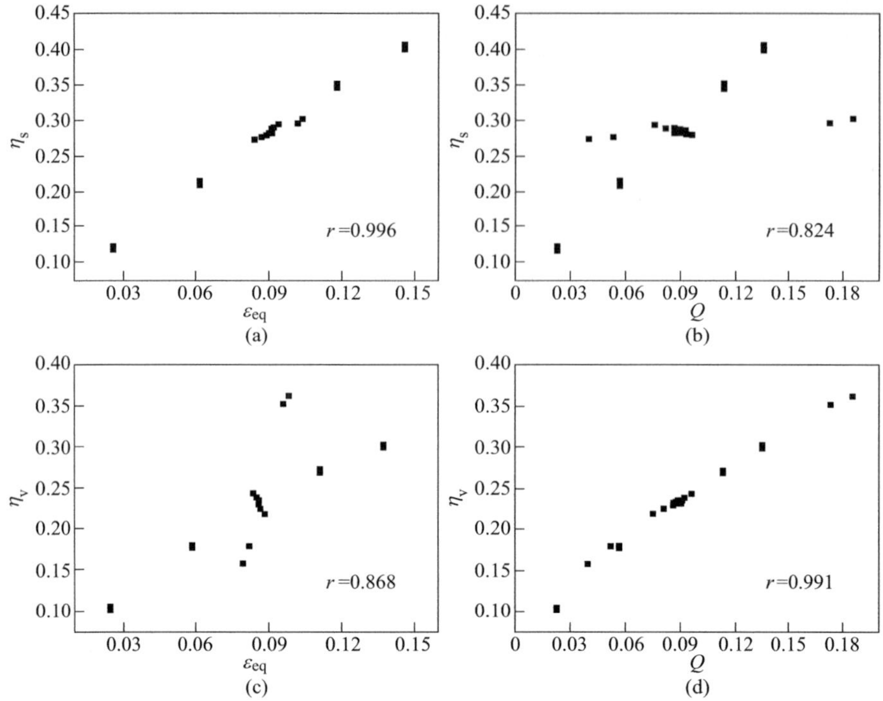

Fig. 4.69 Effect of equivalent strain and hydrostatic stress integral on porosity closure. **a** η_s-ε_{eq}; **b** η_s-Q; **c** η_v-ε_{eq}; **d** η_v-Q relation

of the scatter data in each figure, it is found that there is a close positive correlation between η_s-ε_{eq} in Fig. 4.69a and η_v-Q in Fig. 4.69d.

Given the close positive correlation in Fig. 4.69a and d, the scatter data from these two figures were selected to establish the prediction equation for the closure degree of shrinkage after the press-down process of the bloom. Based on the distribution characteristics of the scatter data in Fig. 4.69a and d, a polynomial form was adopted in deriving the closure degree prediction equation. Figure 4.70a–d compares the first to fourth-order polynomial fitting results of the η_s-ε_{eq} scatter data in Fig. 4.69a with the results of the three-dimensional thermal/mechanical coupling model calculation. In addition, Fig. 4.70a–d also provide the adjusted determination coefficient (R^2) for the first-to fourth-order polynomial fitting. Obviously, the R^2 value in Fig. 4.70b is relatively large, indicating that the second-order polynomial can accurately describe the quantitative relationship between η_s and ε_{eq}.

Based on the second-order polynomial obtained from the scatter data in Fig. 4.69a, the closure degree prediction equation for the shrinkage after the press-down process of the bloom based on the equivalent strain is:

$$\eta_s = -3.08 \times \varepsilon_{eq}^2 + 2.90 \times \varepsilon_{eq} + 4.48 \times 10^{-2} \quad (4.11)$$

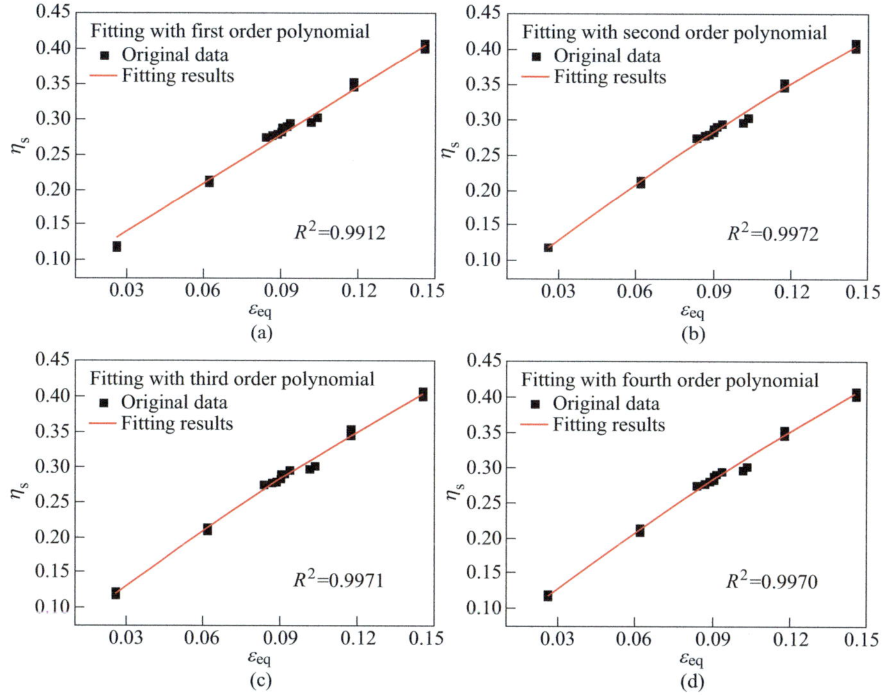

Fig. 4.70 Comparison between η_s-ε_{eq} calculated by model and fitting results of polynomials of different orders. **a** First-order polynomial fitting; **b** quadratic polynomial fitting; **c** cubic polynomial fitting; **d** quadratic polynomial fitting

Figure 4.71a–d compares the first-to fourth-order polynomial fitting results of the η_v-Q scatter data in Fig. 4.69d with the results of the original three-dimensional thermal/mechanical coupling model calculation, and provides the adjusted determination coefficient (R^2) for the first-to fourth-order polynomial fitting. Obviously, the R^2 value in Fig. 4.71d is relatively large, indicating that the fourth-order polynomial can accurately describe the quantitative relationship between η_v and Q.

Based on the fourth-order polynomial obtained from the scatter data in Fig. 4.71d, the closure degree prediction equation for the shrinkage after the press-down process of the bloom based on the static water stress integral is:

$$\eta_v = -3.61 \times Q^2 + 2.30 \times Q + 5.77 \times 10^{-2} \quad (4.12)$$

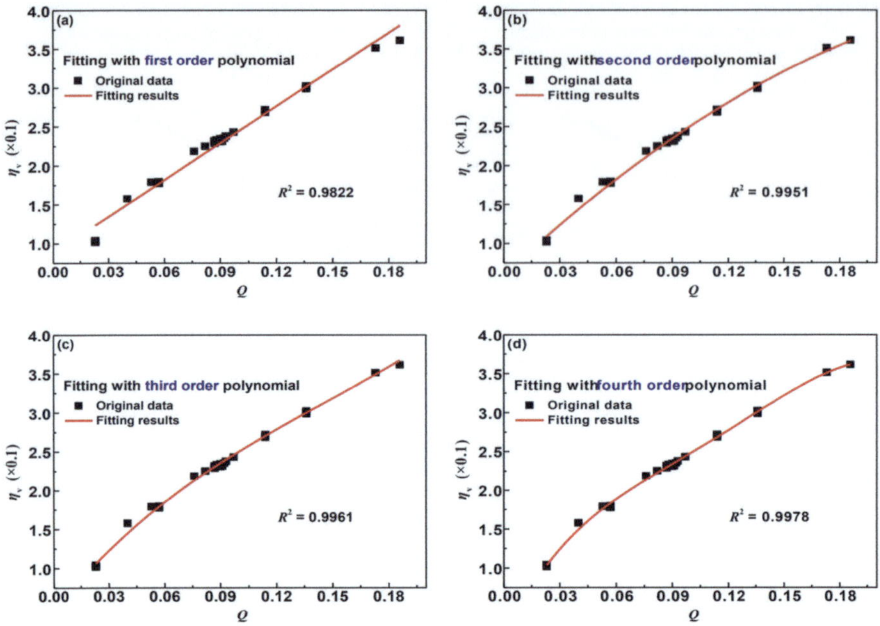

Fig. 4.71 Comparison of the first-to fourth-order polynomial fitting results of the η_V-Q scatter data

References

1. Cáceres CH, Djurdjevic MB, Stockwell TJ et al (1999) The effect of Cu content on the level of microporosity in Al–Si–Cu–Mg casting alloys. Scr Mater 40(5):631–637
2. Murakami Y (1985) Analysis of stress intensity factors of modes I, II and III for inclined surface cracks of arbitrary shape. Eng Fract Mech 22(1):101–114
3. Zhang XX, Cui ZS, Wen C et al (2009) A criterion for void closure in large ingots during hot forging. J Mater Process Technol 209(4):1950–1959
4. Chen J, Chandrashekhara K, Mahimkar C et al (2011) Void closure prediction in cold rolling using finite element analysis and neural network. J Mater Process Technol 211(2):245–255
5. Dong QP, Zhang JM et al (2016) Shrinkage porosity and its alleviation by heavy reduction in continuously cast strand. J Mater Process Technol 238:81–88
6. Cheng R, Zhang J, Zhang L et al (2018) Comparison of porosity alleviation with the multi-roll and single-roll reduction modes during continuous casting. J Mater Process Technol 266:96–104
7. Wu C, Ji C, Zhu M (2019) Influence of differential roll rotation speed on evolution of internal porosity in continuous casting bloom during heavy reduction. J Mater Process Technol 271:651–659
8. Wu C, Ji C, Zhu M (2019) Closure of internal porosity in continuous casting bloom during heavy reduction process. Metall Mater Trans B 50(6):2867–2883
9. Dahle AK, John D, Thevik HJ et al (1999) Modeling the fluid-flow-induced stress and collapse in a dendritic grid. Metall Mater Trans B 30(2):287–293
10. Boileau JM, Allison JE (2003) The effect of solidification time and heat treatment on the fatigue properties of a cast 319 aluminum alloy. Metall Mater Trans A 34(9):1807–1820
11. Roy N, Samuel AM, Samuel FH (1996) Porosity formation in AI-9 Wt Pct Si-3 Wt Pct Cu alloy systems: metallographic observations. Metall Mater Trans A 27(2):415–429

12. Lee YW, Chang E, Chieu CF (1990) Modeling of feeding behavior of solidifying Al-7Si-0.3 Mg alloy plate casting. Metall Trans B 21(4):715–722
13. Pang D (1998) Modeling of casting solidification stochastic or deterministic? Can Metall Q 229–239
14. Stefanescu DM (2005) Computer simulation of shrinkage related defects in metal castings–a review. Int J Cast Met Res 18(3):129–143
15. Lee PD, Hunt JD (2001) Hydrogen porosity in directionally solidified aluminium-copper alloys: a mathematical model. Acta Mater 49(8):1383–1398
16. Carlson KD, Beckermann C (2009) Prediction of shrinkage pore volume fraction using a dimensionless niyama criterion. Metall Mater Trans A 40(1):163–175
17. Xia M et al (2017) Four-phase dendritic model for the prediction of macrosegregation, shrinkage cavity, and porosity in a 55-ton ingot. Metall Mater Trans A 48(3):1139–1150
18. Ji C, Luo S, Zhu M et al (2014) Uneven solidification during wide-thick slab continuous casting process and its influence on soft reduction zone. ISIJ Int 54(1):103–111
19. Plancher E, Gravier P, Chauvet E et al (2019) Tracking pores during solidification of a Ni-based superalloy using 4D synchrotron microtomography. Acta Mater 181:1–9
20. Takahashi T et al (1980) An analysis of fluidity, strength formation and cracking susceptibility on the basis of the solid-liquid coexisting zone in al-base binary alloys. J Jpn Inst Met 44(10):1089–1097
21. Kakimoto H, Arikawa T, Takahashi Y et al (2010) Development of forging process design to close internal voids. J Mater Process Technol 210(3):415–422
22. Chen MS, Lin YC (2013) Numerical simulation and experimental verification of void evolution inside large forgings during hot working. Int J Plast 49:53–70
23. Zhao X, Zhang J, Lei S et al (2014) Finite-element analysis of porosity closure by heavy reduction process combined with ultra-heavy plates rolling. Steel Res Int 85(11):1533–1543
24. Chen DC (2006) Rigid-plastic finite element analysis of plastic deformation of porous metal sheets containing internal void defects. J Mater Process Technol 180(1–3):193–200
25. Dudra SP, Im YT (1990) Analysis of void closure in open-die forging. Int J Mach Tool Manu 30(1):65–75
26. Park JJ (2013) Finite-element analysis of cylindrical-void closure by flat-die forging. ISIJ Int 53(8):1420–1426
27. Park JJ (2013) Prediction of void closure in steel slabs by finite element analysis. Met Mater Int 19(2):259–265
28. Lee YS, Lee SU, Tyne CJV et al (2011) Internal void closure during the forging of large cast ingots using a simulation approach. J Mater Process Technol 211(6):1136–1145
29. Park JJ (2016) Effect of shear deformation on closure of a central void in thin-strip rolling. Metall Mater Trans A 47(1):479–487
30. Ståhlberg U, Keife H (1992) A study of hole closure in hot rolling as influenced by forced cooling. J Mech Work Technol 30(1):131–135
31. Wang B, Zhang J, Xiao C et al (2015) Analysis of the evolution behavior of voids during the hot rolling process of medium plates. J Mater Process Technol 221:121–127
32. Nakasaki M, Takasu I, Utsunomiya H (2006) Application of hydrostatic integration parameter for free-forging and rolling. J Mater Process Technol 177(1–3):521–524
33. Tanaka M, Ono S, Tsuneno M (1987) A numerical analysis on void crushing during side reduction of round bar by flat dies. J Jpn Soc Technol Plast 28:238–244
34. Rodgers JL, Nicewander WA (1988) Thirteen ways to look at the correlation coefficient. Am Stat 42(1):59–66
35. Ji C, Wu CH, Zhu MY (2016) Influence of heavy reduction (HR) on the internal quality of the bearing steel GCr15 bloom. In: CFD Modeling and Simulation in Materials Processing 2016. Wiley, pp 247–254
36. Moon CH, Oh KS, Lee JD et al (2012) Effect of the roll surface profile on centerline segregation in soft reduction process. ISIJ Int 52(7):1266–1272
37. Wu M, Domitner J, Ludwig A (2012) Using a two-phase columnar solidification model to study the principle of mechanical soft reduction in slab casting. Metall Mater Trans A 43(3):945–964

38. Ji C, Li G, Wu C et al (2019) Design and application of CSC-roll for heavy reduction of the bloom continuous casting process. Metall Mater Trans B 50(1):110–122

Chapter 5
Solute Transport Law of Continuous Casting Billet During Reduction Process

The solidification heat transfer, melt flow, microstructure evolution, and solute transport behavior under the action of external fields such as electromagnetic stirring [1] and solidification end reduction [2] are complex and changeable in the continuous casting process. At the same time, factors such as thermal buoyancy, solute buoyancy, thermal shrinkage, and solidification shrinkage will also significantly affect the solute transport behavior inside the continuous casting billet [3]. Finally, grave macrosegregation defects are formed inside the continuous casting billet [4], which are difficult to be effectively eliminated during the subsequent loading and rolling process [5], thus affecting the yield ratio and mechanical properties of the rolled products [6]. Because it is difficult to directly observe the internal macro and micro phenomena in the solidification process of the billet, the traditional mold ingot experiment [7] and the continuous casting industrial experiment [8] widely used at present cannot accurately reveal the solute transport behavior of continuous casting billet. Therefore, most rely on combining numerical simulation with physical experiments to study the solute transport behavior in the continuous casting process. In recent decades, the continuous medium model [9], CA-FE model [10], and multiphase solidification model [11–13] have become the main research methods for the numerical simulation of solute transport in the solidification process in sequence. However, due to the limitation of the discretization method of the model calculation process, the continuous medium cannot effectively track the solidification interface front, resulting in the inability to accurately simulate the solute transport behavior in the two-phase region. The CA-FE model can accurately describe the solidification dendrite evolution behavior in the micro-scale range, but it is difficult to effectively simulate the solute transport behavior in the macro-scale range due to the huge demand for computing resources. The multiphase solidification model, which can effectively consider the microstructure evolution of the solidification process and describe the microscopic transport phenomena such as mass, momentum, energy, and solute between the solid-liquid interface, is the most commonly used research method.

C. Ji et al., *Solidification End Reduction of Continuous Casting Process*, Key Steel Manufacturing Technologies by Collaborative Innovation,
https://doi.org/10.1007/978-981-96-8544-8_5

To further disclose the interaction between solidification heat transfer, solute microsegregation, and solute macrosegregation at the solidification end of continuous casting billet, a multiphase solidification model of the continuous casting process is established in this chapter. This model realizes the coupled calculation of mass, momentum, energy, and solute transport behavior under macro and micro conditions and systematically reveals the liquid-phase flow, solidification structure evolution, and solute transport behavior in continuous casting billet under multiple external fields such as electromagnetic stirring and solidification end reduction. On this basis, the genetic evolution law of solute segregation in the subsequent hot charging process of continuous casting was further studied, and the associated mechanism between solute segregation and shrinkage cavity defects of continuous casting billet and the improvement law of solute segregation under different heating conditions were revealed. This provides a theoretical basis for resolving the solute segregation defects of continuous casting billet from the source.

5.1 Research on Solute Transport Law of Bloom Under Multiple External Field Regulation

The multiphase solidification model first proposed by Professor Beckermann [14, 15] from the University of Iowa in the United States calculates the solid-liquid two phases in the solidification process separately and strictly defines the mass, momentum, energy, and solute transport behavior between the two phases under microscopic grain growth conditions. Professor Ludwig and Meng-Huai Wu [16, 17] from the University of Leoben, in Austria, have also done a lot of work in this field, including handling the liquid phase, inter-dendritic melt phase of equiaxed grains, solid phase of equiaxed grains, inter-dendritic melt phase of columnar dendrites, and solid phase of columnar dendrites separately and systematically studying the mass and energy transfer between different phases, the interphase drag force, and the microstructure evolution behavior. In addition, Rappaz [18], Combeau [19], Ciobanas [20], and other scholars have analyzed scientifically the influence of the growth of columnar grain and equiaxed grain and the CET transition behavior on solute diffusion behavior. Professor Hou-Fa Shen and Bai-Cheng Liu from Tsinghua University [21] and Professor Dian-Zhong Li from the Institute of Metal Research, Chinese Academy of Sciences [22], also expound on the causes of solute segregation defects through multiphase models in the field of ingot casting. In recent years, our team [23, 24] combined the multiple external field characteristics of the continuous casting process, used the multiphase solidification model to couple the macroscopic melt flow heat transfer and microscopic grain nucleation growth, and systematically clarified the effects of multiple external field factors such as electromagnetic stirring and solidification end mechanical reduction on the microstructure evolution and solute segregation behavior of continuous casting billets.

5.1.1 *Multiphase Solidification Model of Continuous Casting Bloom Under Multiple External Fields*

To accurately describe the solidification heat transfer, melt flow, dendritic nucleation growth, and solute transport of continuous casting billet under the coordinated control of multiple external fields such as reduction, deformation, and cooling, a coupled model of multiphase solidification was established (as shown in Fig. 5.1), considering five phases: the liquid phase, inter-dendritic melt phase of equiaxed grains, solid phase of equiaxed grains, inter-dendritic melt phase of columnar dendrites, and solid phase of columnar dendrites, and explored the characteristics of solidification structure such as dynamic tracking of columnar dendrite tip and equiaxed grains nucleation growth.

The microscopic change process between each phase is fully considered in the five-phase model, and the coupling calculation workload is huge. Therefore, the corresponding assumption is made, and the relationship of the five phases in the evolution of solidification structure is described as follows:

$$f_l + f_e + f_c = 1 \tag{5.1}$$

$$f_e = f_e^i + f_e^s, f_c = f_c^i + f_c^s \tag{5.2}$$

$$f_e c_e = f_e^i c_e^i + f_e^s c_e^s, f_c c_c = f_c^i c_c^i + f_c^s c_c^s \tag{5.3}$$

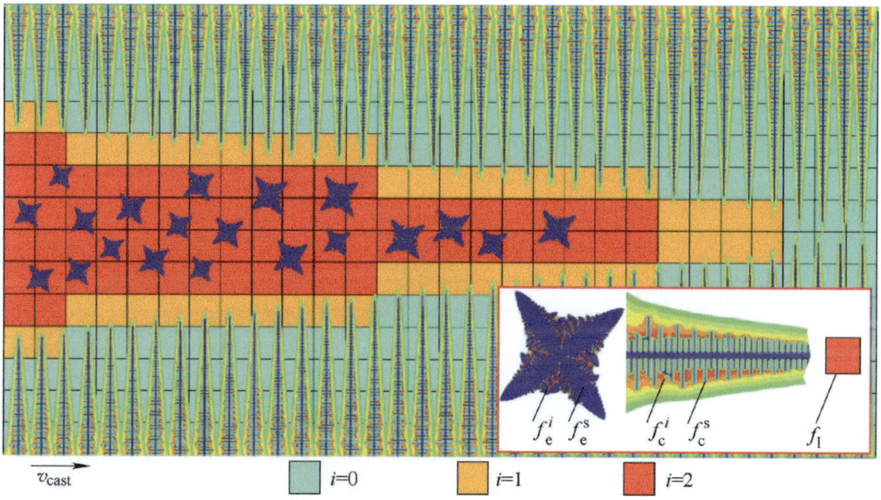

Fig. 5.1 Coupled model of multiphase solidification in continuous casting

where f_l is the volume fraction of liquid phase; f_e is the volume fraction of equiaxed grains phase; f_c is the volume fraction of columnar dendrites phase; c_e is the solute concentration of equiaxed grains phase, %; c_c is the solute concentration of columnar dendrites' phase, %; f_e^i is the volume fraction of the inter-dendritic melt phase of equiaxed grains; c_e^i is the solute concentrations of the inter-dendritic melt phase of equiaxed grains, %; f_c^i is the volume fraction of the inter-dendritic melt phase of columnar dendrites; c_c^i is the solute concentrations of the inter-dendritic melt phase of columnar dendrites, %.

According to the characteristics of solidification structure evolution in the continuous casting process, the casting flow region is meshed first, so that the subsequent computer can identify the regions of different phases correspondingly. For the columnar dendrite, different values of the element index denote different dendrite morphologies, thus helping the computer choose an action to take for dendritic growth. When (*i*) is 0, the element represents the columnar trunk. When (*i*) is 1, the element represents the columnar tip. When (*i*) is 2, the element represents the equiaxed and liquid phases simultaneously. In the process of microstructure evolution calculation, for the equiaxed grains phase and liquid phase, the model will set that the equiaxed grains phase and the liquid phase have the same speed, in other words, the equiaxed grains phase is free in the liquid phase.

As shown in Fig. 5.2, the multiphase solidification coupling model based on the volume averaging method described in this section is mainly composed of four parts: melt flow, mass transfer, solute transfer, and energy transfer conservation equations, its momentum transfer equation mainly includes the mixed k–ε turbulence model and the Kozeny-Carman model.

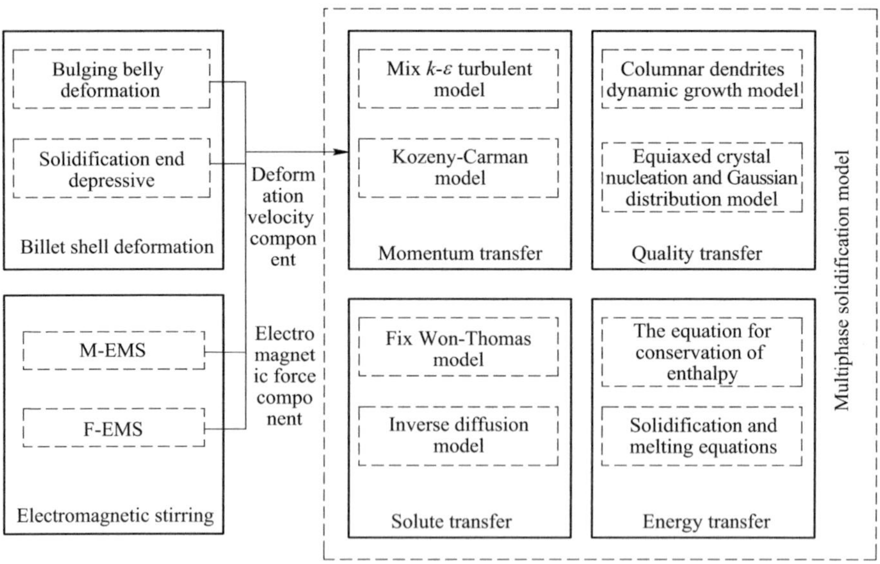

Fig. 5.2 Structure of multiphase solidification coupling model in continuous casting process

5.1.1.1 The Mass Transfer Model

According to Eqs. (5.1)–(5.3), the mass transfer model of the solidification process in continuous casting can be expressed by the following equations:

$$\frac{\partial(\rho_l f_l)}{\partial t} + \nabla(f_l \rho_l \boldsymbol{v}_l) = -S_{le} - S_{lc} \tag{5.4}$$

$$\frac{\partial(\rho_c f_c)}{\partial t} + \nabla(f_c \rho_c \boldsymbol{v}_c) = S_{lc} + S_{ce} \tag{5.5}$$

$$\frac{\partial(\rho_e f_e)}{\partial t} + \nabla(f_e \rho_e \boldsymbol{v}_e) = S_{le} - S_{ce} \tag{5.6}$$

$$\frac{\partial\left(\rho_c f_c^s\right)}{\partial t} + \nabla\left(f_c^s \rho_c \boldsymbol{v}_c\right) = S_c^s \tag{5.7}$$

$$\frac{\partial\left(\rho_e f_e^s\right)}{\partial t} + \nabla\left(f_e^s \rho_e \boldsymbol{v}_e\right) = S_e^s \tag{5.8}$$

where ρ_l is the density of liquid phase, kg/m^3; ρ_e is the density of equiaxed grains phase, kg/m^3; ρ_c is the density of columnar dendrites phase, kg/m^3; $\boldsymbol{v}_l$ is the velocity vectors of the liquid phase, m/s; $\boldsymbol{v}_e$ is the velocity vectors of equiaxed grains phase, m/s; $\boldsymbol{v}_c$ is the velocity vectors of columnar dendrites phase, m/s; S_{le} is the liquid-equiaxed net mass transfer rate, kg/(m s^3); S_{lc} is the liquid-columnar net mass transfer rate, kg/(m s^3); S_{ce} is the columnar-equiaxed net mass transfer rate, kg/(m s^3); S_c^s is the inter-dendritic net mass transfer rate in the columnar phase, kg/(m s^3); S_e^s is the inter-dendritic net mass transfer rate in the equiaxed phase, kg/(m s^3).

In the calculation process of the multiphase model, to simplify the calculation of the model, the equiaxed grain and the columnar dendrite are assumed to be spherical and cylindrical respectively, as shown in Fig. 5.3. This assumption helps to reduce the complexity of the model and make the subsequent calculations more manageable.

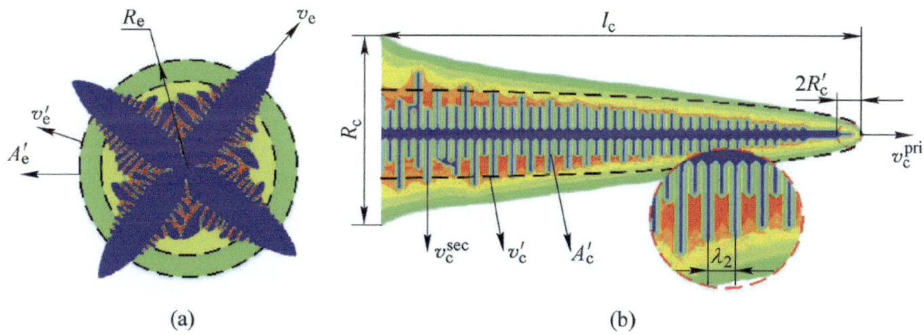

Fig. 5.3 Schematic diagrams of the growth morphology of equiaxed (**a**) and columnar (**b**) dendrites

The net mass transfer rate between phases will greatly affect the calculation of the multiphase solidification model. The equations of net mass transfer rate for each phase are shown as follows:

$$S_{\mathrm{le}} = v'_{\mathrm{e}} A'_{\mathrm{e}} \rho_{\mathrm{e}} = \left(I_{\mathrm{s}}^{\mathrm{e}} v_{\mathrm{e}}\right)\left(n_{\mathrm{e}} 4\pi R_{\mathrm{e}}^{2} f_{\mathrm{l}}\right)\rho_{\mathrm{e}} \tag{5.9}$$

where v'_{e} is the interfacial growth velocity of an equiaxed grain, m/s; v_{e} is the growth velocity of the dendritic tip of an equiaxed grain, which can be calculated by the Lipton-Glicksman-Kurz (LGK) [25] model, m/s; A'_{e} is the concentration of the surface area, m/s; ρ_{e} is the density of equiaxed grains phase, kg/m^3; $I_{\mathrm{s}}^{\mathrm{e}}$ is he influence factor of the grain outline and is equal to 0.68 [26]; n_{e} is the nucleation density, m^{-3}; R_{e} is the radius of the equiaxed grain, m.

$$S_{\mathrm{lc}} = v'_{\mathrm{c}} A'_{\mathrm{c}} \rho_{\mathrm{c}} + iS'_{\mathrm{c}} = \left(I_{\mathrm{s}}^{\mathrm{c}} v_{\mathrm{c}}^{\mathrm{sec}}\right)\left(2\pi R_{\mathrm{c}} f_{\mathrm{l}}/\lambda_{1}^{2}\right)\rho_{\mathrm{c}} + i v_{\mathrm{c}}^{\mathrm{pri}} n_{\mathrm{c}} \pi R'_{\mathrm{c}} \rho_{\mathrm{c}} f_{\mathrm{l}} \tag{5.10}$$

where $v_{\mathrm{c}}^{\mathrm{pri}}$ is the growth velocities of the dendritic tip of the primary dendrite, m/s; $v_{\mathrm{c}}^{\mathrm{sec}}$ is the growth velocities of the dendritic tip of secondary dendrite, m/s; ρ_{c} is the density of columnar dendrites phase [27, 28], kg/m^3; R_{c} is the radius of the columnar dendritic trunk, m; $R_{\mathrm{c}'}$ is the radius of the columnar dendritic tip, m; λ_1 is the primary dendrite arm spacing [29], m.

With a change in the element where the dendritic tip is located, the element indexes (i) will change from 0 to1. $I_{\mathrm{s}}^{\mathrm{c}}$ is similar to $I_{\mathrm{s}}^{\mathrm{e}}$ and is equal to 0.7979 [26].

$$S_{\mathrm{e}}^{\mathrm{s}} = v_{\mathrm{e}}^{\mathrm{s}} A_{\mathrm{e}}^{\mathrm{s}} \rho_{\mathrm{e}} = \frac{2D_{\mathrm{l}}\Omega}{\beta\lambda_{2} f_{\mathrm{e}}^{\mathrm{s}}} \times \frac{2f_{\mathrm{e}}^{\mathrm{s}} f_{\mathrm{e}}}{\lambda_{2}} \rho_{\mathrm{e}} \tag{5.11}$$

where $v_{\mathrm{e}}^{\mathrm{s}}$ is the interfacial growth velocity between the inter-dendritic melt and the solid phase of the equiaxed grain, m/s; $A_{\mathrm{e}}^{\mathrm{s}}$ is the concentration of the interfacial surface area, %; D_{l} is the diffusion coefficient in the liquid, m^2/s; β is the constant; Ω is the dimensionless solute supersaturation $\left(c_{\mathrm{l}}^{*} - c_{\mathrm{s}}^{\mathrm{e}}\right)/\left(c_{\mathrm{l}}^{*} - c_{\mathrm{s}}^{*}\right)$.

The secondary dendritic arm spacing λ_2 is calculated according to an equation provided in previous work [5].

$$S_{\mathrm{c}}^{\mathrm{s}} = v_{\mathrm{c}}^{\mathrm{s}} A_{\mathrm{c}}^{\mathrm{s}} \rho_{\mathrm{c}} + iS'_{\mathrm{c}} = \frac{2D_{\mathrm{l}}\left(c_{\mathrm{l}}^{*} - c_{\mathrm{s}}^{\mathrm{c}}\right)}{\beta\lambda_{2} f_{\mathrm{c}}^{\mathrm{s}}\left(c_{\mathrm{l}}^{*} - c_{\mathrm{s}}^{*}\right)} \times \frac{2f_{\mathrm{c}}^{\mathrm{s}} f_{\mathrm{c}}}{\lambda_{2}} \rho_{\mathrm{c}} + i v'_{\mathrm{c}} n_{\mathrm{c}} \pi R'_{\mathrm{c}} \rho_{\mathrm{c}} f_{\mathrm{l}} \tag{5.12}$$

where $v_{\mathrm{c}}^{\mathrm{s}}$ is the interfacial growth velocity between the inter-dendritic melt and the solid phase of the columnar dendrite, m/s; $A_{\mathrm{e}}^{\mathrm{c}}$ is the concentration of the interfacial surface area in the columnar dendrite, %; n_{c} is nucleation density of the columnar dendritic trunk, m^{-3}.

5.1.1.2 The Momentum Transfer Model

Because equiaxed grains move freely in a liquid phase, the equiaxed and liquid phases are only considered in the transport of momentum. According to the momentum conservation law, the Navier-Stokes equations are expressed as follows:

$$\frac{\partial}{\partial t}(f_1\rho_1\boldsymbol{v}_1) + \nabla\cdot(f_1\rho_1\boldsymbol{v}_1\boldsymbol{v}_1) = -f_1\nabla P + \nabla\cdot\left(f_1\left(\mu_1 + \mu_{\mathrm{t,k}}\right)\left(\nabla\boldsymbol{v}_1 + (\nabla\boldsymbol{v}_1)^{\mathrm{T}}\right)\right) + \boldsymbol{F}_{\mathrm{T}}^{1} + \boldsymbol{F}_{\mathrm{C}}^{1} + \boldsymbol{V}_{\mathrm{cl}} + \boldsymbol{V}_{\mathrm{el}} \tag{5.13}$$

$$\frac{\partial}{\partial t}(f_{\mathrm{e}}\rho_{\mathrm{e}}\boldsymbol{v}_{\mathrm{e}}) + \nabla\cdot(f_{\mathrm{e}}\rho_{\mathrm{e}}\boldsymbol{v}_{\mathrm{e}}\boldsymbol{v}_{\mathrm{e}}) = -f_{\mathrm{e}}\nabla P + \nabla\cdot\left(f_{\mathrm{e}}\left(\mu_{\mathrm{e}} + \mu_{\mathrm{t,k}}\right)\left(\nabla\boldsymbol{v}_{\mathrm{e}} + (\nabla\boldsymbol{v}_{\mathrm{e}})^{\mathrm{T}}\right)\right) + \boldsymbol{F}_{\mathrm{T}}^{\mathrm{e}} + \boldsymbol{F}_{\mathrm{C}}^{\mathrm{e}} + \boldsymbol{F}_{\mathrm{u}} + \boldsymbol{V}_{\mathrm{le}} + \boldsymbol{V}_{\mathrm{ce}} \tag{5.14}$$

where P is pressure, Pa; μ_1 is the viscosity of the liquid phase, 0.006 Pa s; μ_{e} is the viscosity of the equiaxed phase [13, 31], Pa s; $\mu_{t,k}$ is turbulent viscosity that is calculated by the k–ε mixture turbulence model [5], Pa s; $\boldsymbol{F}_{\mathrm{T}}^{1}$, $\boldsymbol{F}_{\mathrm{C}}^{1}$, $\boldsymbol{F}_{\mathrm{T}}^{\mathrm{e}}$, $\boldsymbol{F}_{\mathrm{C}}^{\mathrm{e}}$ is determined the effects of thermal buoyancy and solute buoyancy in the liquid and equiaxed phases[5], kg/(m s^2); $\boldsymbol{F}_{\mathrm{u}}$ is a discriminatory function applied to judge the computing method of the mushy zone [5].

The momentum transfer source terms of $\boldsymbol{V}_{\mathrm{cl}}$, $\boldsymbol{V}_{\mathrm{ce}}$ and $\boldsymbol{V}_{\mathrm{el}}$ between each phase are used for the treatment of the mushy zone:

$$\boldsymbol{V}_{\mathrm{cl}} = K_{\mathrm{cl}}(\boldsymbol{v}_1 - \boldsymbol{v}_{\mathrm{c}}) + v_{\mathrm{cl}}^{*}S_{\mathrm{cl}} \tag{5.15}$$

$$\boldsymbol{V}_{\mathrm{ce}} = K_{\mathrm{ce}}(\boldsymbol{v}_{\mathrm{c}} - \boldsymbol{v}_{\mathrm{e}}) + v_{\mathrm{ce}}^{*}S_{\mathrm{ce}} \tag{5.16}$$

$$\boldsymbol{V}_{\mathrm{el}} = K_{\mathrm{el}}(\boldsymbol{v}_{\mathrm{e}} - \boldsymbol{v}_1) + v_{\mathrm{el}}^{*}S_{\mathrm{el}} \tag{5.17}$$

where K_{cl} is the liquid-columnar momentum exchange coefficient [30, 31, 32]; K_{ce} is the columnar-equiaxed momentum exchange coefficient [17, 33]; K_{el} is the liquid-equiaxed momentum exchange coefficient in the Kozeny-Carman model [16, 34]; v_{cl}^{*} is the momentum source term at the interfaces between the liquid and columnar phases, kg/(m^2 s^2); v_{ce}^{*} is the momentum source term at the interfaces between the columnar and equiaxed phases, kg/(m^2 s^2); v_{el}^{*} is the momentum source term at the interfaces between the liquid and equiaxed phases, kg/(m^2 s^2).

5.1.1.3 The Heat Transfer Model

Similar to the momentum conservation equations, those for the heat transfer model are simplified to mainly study the energy transfer between the liquid, equiaxed, and columnar phases. The energy conservation equations are described below.

$$\frac{\partial}{\partial t}(f_l \rho_l H_l) + \nabla(f_l \rho_l \boldsymbol{v}_l H_l) = \nabla\big(f_l k^*(\nabla T_l)\big) - H(S_{le} + S_{lc}) - H^*(2T_l - T_e - T_c) \tag{5.18}$$

$$\frac{\partial}{\partial t}(f_e \rho_e H_e) + \nabla(f_e \rho_e \boldsymbol{v}_e H_e) = \nabla\big(f_e k^*(\nabla T_e)\big) + HS_{le} + H^*(T_l - T_e) \tag{5.19}$$

$$\frac{\partial}{\partial t}(f_c \rho_c H_c) + \nabla(f_c \rho_c \boldsymbol{v}_c H_c) = \nabla\big(f_c k^*(\nabla T_c)\big) + HS_{lc} + H^*(T_l - T_c) \tag{5.20}$$

where k^* is the effective thermal conductivity considering the influence of turbulence, W/(m K); T_l is the temperatures of the liquid phase, K; T_e is the temperatures of the equiaxed phase, K; T_c is the temperatures of the columnar phase, K; H_l is the enthalpies of the liquid phase, J/kg; H_e is the enthalpies of the equiaxed phase, J/kg; H_c is the enthalpies of the columnar phase, J/kg; H is the phase transition enthalpy, J/kg.

The above calculation parameters and heat transfer boundary conditions are described in Chap. 3 [31, 35]. Taking into account the effects of the solidification and re-melting of dendrite, H is equal to H_l.

To satisfy the assumption of local thermal equilibrium (low fluid velocity and sufficient heat exchange between each phase), a large diffusional heat exchange coefficient ($H^* = 108$) is set to balance the temperature between each phase [17, 36]. Therefore, the effective thermal conductivity can be simplified and expressed as follows [16, 32, 37, 38]:

$$k^* = \begin{cases} k_l + \frac{\mu_{t,k} c_{p(l)}}{P_r} & T \geq T_l \\ k_l f_l + k_e f_e + k_c f_c & T_e = T_c = T_s < T < T_l \\ k_e \text{ or } k_c & T < T_s = T_e = T_c \end{cases} \tag{5.21}$$

where k_l is the temperature-dependent conductivities of the liquid phases, W/(m K); k_e is the temperature-dependent conductivities of the equiaxed phases, W/(m K); k_c is the temperature-dependent conductivities of the columnar phases, W/(m K); P_r is the turbulent Prandtl number, which is 0.9 here.

In Eqs. (5.18)–(5.20), the enthalpies of H_l, H_e, and H_c are given by [33, 39]:

$$H_l = h_l^{\text{ref}} + \int_{T_{\text{ref}}}^{T_l} c_{p(l)} \mathrm{d}T + f_l L \tag{5.22}$$

$$H_e = H_c = h_s^{\text{ref}} + \int_{T_{\text{ref}}}^{T_e} c_{p(s)} \mathrm{d}T + f_l L \tag{5.23}$$

where h_l^{ref} is the reference enthalpies of liquid phases, J/kg; h_s^{ref} is the reference enthalpies of solid phase (equiaxed and columnar phases), J/kg; $c_{p(l)}$ is the specific heats of the liquid phase, J/(kg K); $c_{p(s)}$ is the specific heats of the solid phase (equiaxed and columnar phases), J/(kg K); T_{ref} is the reference temperature for enthalpy definition, K; L is the latent heat of fusion, W/(m^3 K).

The liquid fraction, f_l, can be defined as [34, 40]:

$$f_l = \begin{cases} 1 & T \geqslant T_l \\ (T - T_s)/(T_l - T_s) & T_e = T_c = T_s < T < T_l \\ 0 & T < T_s = T_e = T_c \end{cases} \tag{5.24}$$

where T_s is the temperature of the solid phase, K.

Considering the effect of the back-diffusion model [5], the temperatures of each phase (T_l, T_e, and T_c) can be defined as:

$$T_l = T_f + mc_0\left[1 - (1 - \gamma k)(1 - f_l)\right]^{(k-1)/(1-\gamma k)} \tag{5.25}$$

$$T_e = T_c = T_s = T_f + \frac{1}{k} mc_0\left[1 - (1 - \gamma k)(1 - f_l)\right]^{(k-1)/(1-\gamma k)} \tag{5.26}$$

where T_f is the fusion heat of pure iron and is 1808 K; m is the liquidus slope, K-1; c_0 is the initial carbon concentration, %; γ is a nondimensional diffusion parameter of the back-diffusion model between 0 (as in the Scheil rule) and 1 (as in the Lever rule); k is the solute partition coefficient at the liquid-solid interface.

5.1.1.4 The Solute Transfer Model

Considering the effects of phase change and diffusion on solute transfer, the solute conservation equations of the multiphase model are as follows [11, 26, 41, 42]:

$$\frac{\partial(\rho_l f_l c_l)}{\partial t} + \nabla \cdot (f_l \rho_l \boldsymbol{v}_l c_l) = -C_{le}^{P} - C_{le}^{D} - C_{lc}^{P} - C_{lc}^{D} \tag{5.27}$$

$$\frac{\partial(\rho_e f_e c_e)}{\partial t} + \nabla \cdot (f_e \rho_e \boldsymbol{v}_e c_e) = C_{le}^{P} + C_{le}^{D} - C_{ec}^{P} - C_{ec}^{D} \tag{5.28}$$

$$\frac{\partial(\rho_c f_c c_c)}{\partial t} + \nabla \cdot (f_c \rho_c \boldsymbol{v}_c c_c) = C_{lc}^{P} + C_{lc}^{D} + C_{le}^{P} + C_{le}^{D} \tag{5.29}$$

$$\frac{\partial\left(\rho_e f_e^{s} c_e^{s}\right)}{\partial t} + \nabla \cdot \left(f_e^{s} \rho_e \boldsymbol{v}_e c_e^{s}\right) = C_e^{sP} + C_e^{sD} \tag{5.30}$$

$$\frac{\partial\left(\rho_c f_c^{s} c_c^{s}\right)}{\partial t} + \nabla \cdot \left(f_c^{s} \rho_c \boldsymbol{v}_c c_c^{s}\right) = C_c^{sP} + C_c^{sD} \tag{5.31}$$

where $C_{\mathrm{le}}^{\mathrm{P}}$, $C_{\mathrm{lc}}^{\mathrm{P}}$, $C_{\mathrm{ec}}^{\mathrm{P}}$, $C_{\mathrm{c}}^{\mathrm{sP}}$, $C_{\mathrm{e}}^{\mathrm{sP}}$ is the solute transfer source terms with respect to phase change;$C_{\mathrm{le}}^{\mathrm{D}}$, $C_{\mathrm{lc}}^{\mathrm{D}}$, $C_{\mathrm{ec}}^{\mathrm{D}}$, $C_{\mathrm{c}}^{\mathrm{sD}}$, C_{e}^{sD} is the solute transfer source terms with respect to diffusion.

Similarly, $C_{\mathrm{ce}}^{\mathrm{P}} = -C_{\mathrm{ec}}^{\mathrm{P}}$ and $C_{\mathrm{ce}}^{\mathrm{D}} = -C_{\mathrm{ec}}^{\mathrm{D}}$ are not considered in this model. In addition, by using a cellular automaton model to modify the Won-Thomas model [35, 43], the effects of solute distribution and back diffusion can be taken into account. The modified solute transfer source terms can be replaced with the following:

$$C_{\mathrm{le}}^{\mathrm{P}} + C_{\mathrm{le}}^{\mathrm{D}} = S_{\mathrm{le}} c_{\mathrm{le}}^{*} - S_{\mathrm{h}} \rho_{\mathrm{e}} A_{\mathrm{e}}' D_{\mathrm{l}} \left(c_{\mathrm{le}}^{*} - c_{\mathrm{l}}\right) / l_{\mathrm{l}}^{\mathrm{e}} - \gamma \rho_{\mathrm{e}} k f_{\mathrm{e}} \partial (S_{\mathrm{le}}) / \partial t \tag{5.32}$$

$$C_{\mathrm{lc}}^{\mathrm{P}} + C_{\mathrm{lc}}^{\mathrm{D}} = S_{\mathrm{lc}} c_{\mathrm{lc}}^{*} - S_{\mathrm{h}} \rho_{\mathrm{c}} A_{\mathrm{c}}' D_{\mathrm{l}} \left(c_{\mathrm{lc}}^{*} - c_{\mathrm{l}}\right) / l_{\mathrm{l}}^{\mathrm{c}} - \gamma \rho_{\mathrm{c}} k f_{\mathrm{c}} \partial (S_{\mathrm{lc}}) / \partial t \tag{5.33}$$

$$C_{\mathrm{e}}^{\mathrm{sP}} + C_{\mathrm{e}}^{\mathrm{sD}} = S_{\mathrm{e}}^{\mathrm{s}} c_{\mathrm{e}}^{i*} - \gamma \rho_{\mathrm{e}} k f_{\mathrm{e}}^{\mathrm{s}} \partial \left(S_{\mathrm{e}}^{\mathrm{s}}\right) / \partial t \tag{5.34}$$

$$C_{\mathrm{c}}^{\mathrm{sP}} + C_{\mathrm{c}}^{\mathrm{sD}} = S_{\mathrm{c}}^{\mathrm{s}} c_{\mathrm{c}}^{i*} - \gamma \rho_{\mathrm{c}} k f_{\mathrm{c}}^{\mathrm{s}} \partial \left(S_{\mathrm{c}}^{\mathrm{s}}\right) / \partial t \tag{5.35}$$

where c_{le}^{*} is the equilibrium concentrations at the interface between the liquid and equiaxed phases, %; c_{lc}^{*} is the equilibrium concentrations at the interface between the liquid and columnar phases, %; c_{e}^{i*} is the equilibrium concentrations of equiaxed phases, %; c_{c}^{i*} is the equilibrium concentrations of columnar phases, %; S_{h} is the Sherwood number; $l_{\mathrm{l}}^{\mathrm{e}}$ is solute diffusion lengths from the liquid to equiaxed phases, m; $l_{\mathrm{l}}^{\mathrm{c}}$ is solute diffusion lengths from the liquid to columnar phases, m.

The macrosegregation behaviors of other elements (Si, Mn, P, and S) are similar to those of carbon [36, 37, 44, 45], and only carbon segregation is considered to simplify the calculation. The segregation degree of carbon (c_{mix}/c_0) is one of the main parameters for quality evaluation and can be calculated by Eq. (5.36):

$$\frac{c_{\mathrm{mix}}}{c_0} = \frac{\rho_{\mathrm{l}} f_{\mathrm{l}} c_{\mathrm{l}} + \rho_{\mathrm{e}} f_{\mathrm{e}} c_{\mathrm{e}} + \rho_{\mathrm{c}} f_{\mathrm{c}} c_{\mathrm{c}}}{(\rho_{\mathrm{l}} f_{\mathrm{l}} + \rho_{\mathrm{e}} f_{\mathrm{e}} + \rho_{\mathrm{c}} f_{\mathrm{c}}) c_0} \tag{5.36}$$

5.1.1.5 The Nucleation Model of Equiaxed Grains

Under the cooling effect of the mold copper plate, the columnar dendrite grows vertically from the surface of the billet to the center, and the melt temperature in the liquid pool gradually decreases. When a certain undercooling degree is reached, the grains nucleate in the liquid phase and migrate with the liquid-phase flow. Among them, the nucleation of equiaxed grains is obtained by Eq. (5.37).

$$\frac{\partial n_{\mathrm{e}}}{\partial t} + \nabla (\boldsymbol{u}_{\mathrm{e}} n_{\mathrm{e}}) = N \tag{5.37}$$

where $\boldsymbol{u}_e$ is the velocity of the equiaxed grain phase, m/s; n_e is the nucleation density of the equiaxed grain, m^{-3}; N is the nucleation rate of the equiaxed grain, m^{-3}/s; N is determined by the relationship $N = \frac{dn}{d(\Delta T)} \times \frac{d(\Delta T)}{dt}$.

To describe the distribution of nucleation grains, the continuous nucleation model is adopted. It is considered that the nucleation density obeys the Gaussian distribution, which is expressed by Eq. (5.38):

$$\frac{dn}{d(\Delta T)} = \frac{n_{max}}{\sqrt{2\pi}\Delta T_\sigma} e^{-\frac{1}{2}\left(\frac{\Delta T - \Delta T_N}{\Delta T_\sigma}\right)^2} \tag{5.38}$$

where n_{max} is the maximum nucleation density of the equiaxed grain, m^{-3}; ΔT_N is the average nucleation undercooling degree, K; ΔT_σ is the standard deviation, K; ΔT is the undercooling degree of melt, K [38, 46].

In the process of grain nucleation, the nucleation rate of the grain is calculated by the relationship that temperature changes with time. In the traditional ingot solidification process, the undercooling degree can be directly obtained by using the relationship $\Delta T = T_f + mc_l - T_l$, according to the solute concentration and temperature of the current volume element. However, during the solidification process of continuous casting, the high-temperature molten steel continuously enters the mold from the submerged nozzle, and the surface solidified billet is gradually pulled out and enters the secondary cooling zone. When the solidification process of continuous casting enters a stable state, the volume element temperature and solute concentration in the model no longer change with time, so the traditional method $d(\Delta T)/dt$ to obtain equiaxed grain nucleation will no longer be applicable. Consequently, for steady-state casting, the liquid-phase temperature T_l and solute concentration c_l of the fluid microelement continuously change with the liquid-phase flow during the movement of the fluid microelement from position P1 to positions P2 and P3, as shown in Fig. 5.4.

Therefore, this section proposes an equiaxed grain nucleation model suitable for the continuous casting solidification process based on the element temperature gradient and solute concentration.

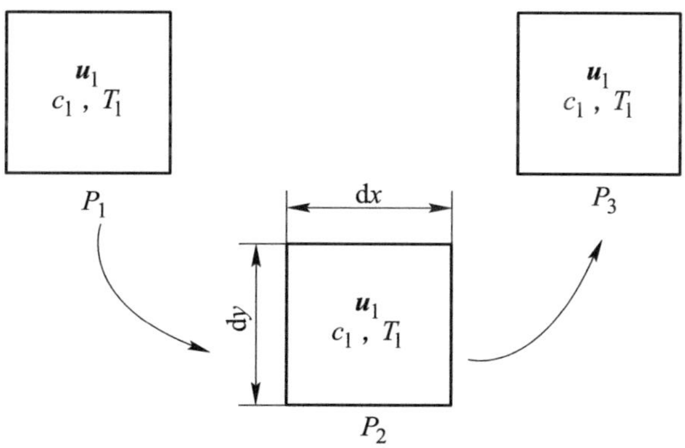

Fig. 5.4 Microscopic elements of fluid

$$\frac{d(\Delta T)}{dt} = m\boldsymbol{G}_{c,l}\boldsymbol{u}_l - \boldsymbol{G}_{T,l}\boldsymbol{u}_l \tag{5.39}$$

where $\boldsymbol{G}_{c,l}$ is the concentration gradient of liquid solute, m^{-1}; $\boldsymbol{G}_{T,l}$ is the temperature gradient of liquid phase, K/m.

When the grain density reaches the maximum nucleation density, the melt stops nucleation. When the volume fraction of equiaxed grain phase is less than 0.001, the grains melt with the increase of melt temperature.

5.1.1.6 The Dynamic Tracking Model of Columnar Dendrite Tip

Under the cooling effect of mold copper plate, the surface temperature of molten steel decreases rapidly, which causes the columnar dendrite grow toward the liquid pool in a manner perpendicular to the surface. As the temperature gradually decreases, when the melt reaches the supercooled state, the equiaxed grains nucleate and grow in the liquid pool. Since the equiaxed grain zone and the columnar dendrite zone have completely different effects on the flow of the melt, it is necessary to track the tip of the columnar dendrite to divide the columnar dendrite zone, the equiaxed grain zone, and the liquid-phase zone.

In the multiphase solidification model, the state parameters are assigned to each volume element to mark the state of the current volume element. It is considered that the initial stage of the whole model is liquid ($i_c = 0$), and the columnar dendrite grows perpendicular to the surface of the billet under the action of mold cooling. Therefore, the surface state parameter of the billet near the meniscus is the columnar dendrite tip ($i_c = 1$). As the billet temperature decreases, the columnar dendrite tip advances to the liquid poor at a certain velocity (v_{tip}^{c}), as shown in Fig. 5.5. During the solidification process of the ingot, the relative position between the dendrites is fixed, and the columnar dendrite is unidirectionally toward the liquid phase. However, for continuous casting, the billet is continuously draw out of the mold into the secondary cooling zone, and the growth length and the tip position of the columnar dendrite need to be advanced along the casting direction. Therefore, this study proposes a dynamic tracking model of the columnar dendrite tip based on the growth rate of the dendrite tip and the casting speed.

$$l_{y+\Delta y} = l_y + \int_{\tau}^{\tau+\Delta\tau} v_{tip}^{c} d\tau \tag{5.40}$$

where l_y is the dendrite tip growth length at y, m; $l_{y+\Delta y}$ is the dendrite tip growth length at $y + \Delta y$, m; $\Delta\tau$ is the time interval based on the characteristic length and billet moving velocity, which is obtained by the equation $\Delta\tau = \Delta y / u_{c,y}$, s; Δy is the characteristic length of the casting direction, m.

When the growth length l_y of the columnar dendrite tip reaches the characteristic length l_{ref}, the columnar dendrite tip passes through the element and converts it into

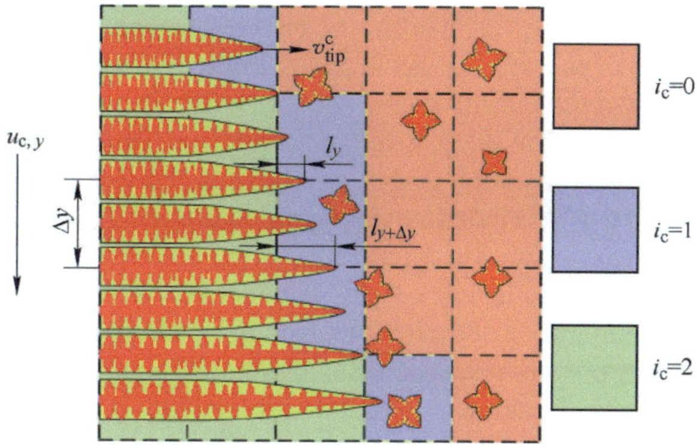

Fig. 5.5 Dynamic tracking diagram of columnar dendrite tip

the columnar dendrite trunk, and the state parameter (i_c) changes from 1 to 2. At the same time, the state parameter of the liquid-phase element adjacent to the columnar dendrite changes from 0 to 1, becoming a new columnar dendrite tip. Currently, the growth length of the columnar dendrite is $l_{y,\text{new}} = l_y - l_{\text{ref}}$, and the integral operation is performed in the new time step. When the growth time of the columnar dendrite tip reaches Δ τ, the tip position and growth length of the columnar dendrite are transmitted along the casting direction, entering a new calculation period.

During the growth of columnar dendrites, heat is lost from the surface of the billet, which makes the melt temperature gradually decrease. When the undercooling reaches a critical value, equiaxed grains nucleate in the melt and gradually grow. When the equiaxed grain volume of the columnar dendrite tip element reaches a critical fraction ($f_{\text{scr}} = 0.49$), the columnar dendrite tip stops growing into the liquid phase, but the equiaxed grain continues to nucleate and grow during the subsequent solidification process.

5.1.1.7 Calculation Model of Electromagnetic Stirring for Continuous Casting

Continuous casting electromagnetic stirring technology is an electromagnetic metallurgical technology that aims to enhance the quality of the billet. It achieves this by installing various types of electromagnetic stirring devices at different positions within the continuous casting machine. Through the electromagnetic force generated, it intensifies the flow of molten metal within the billet and modifies the flow, heat transfer, and mass transfer conditions during the solidification process. Since continuous casting demands different flow states at different locations and different processes, the electromagnetic stirring process is also different. For example, in the mold, the high-temperature molten steel enters the mold from the tundish through the submerged nozzle. Under electromagnetic stirring, the molten steel is driven to rotate and flow. Therefore, a reasonable electromagnetic stirring process can benefit the removal of inclusions, increase the equiaxed grain nucleus, uniform the temperature of the molten steel, and promote the uniform growth of the shell. However, excessive electromagnetic stirring will excessively wash the front edge of the initial solidified shell, increasing the white band. Because the flow velocity of molten steel at the solidification end is significantly reduced, the application of electromagnetic stirring at this position can significantly increase the flow velocity of molten steel, which is conducive to the mixing of segregated molten steel. However, too much electromagnetic stirring will also increase the swirling effect of molten steel, resulting in the segregation of molten steel to the center of the billet and aggravating segregation. Given this, this chapter takes the bloom continuous casting process as an example and uses the numerical simulation method to establish the three-dimensional electromagnetic field model of mold electromagnetic stirring and final electromagnetic stirring, analyzing the electromagnetic field distribution under different electromagnetic parameters, which provides a theoretical basis for the formulation of reasonable process parameters.

The operating principle of the electromagnetic stirrer is analogous to that of the asynchronous motor. The coil is equivalent to the stator, while the molten steel within the mold is equivalent to the rotor. The coil is equivalent to the stator, and the molten steel inside the mold is equivalent to the rotor. After the three-phase alternating current is introduced into the coil winding, a rotating magnetic field is generated in the stirrer. The molten steel in the mold is cut by the magnetic field line of the rotating magnetic field, just like the conductor cutting the magnetic field line, and the current is induced in the molten steel. The electromagnetic force generated by the action of current and magnetic field acts upon the molten steel, creating a rotational torque at different positions relative to the center to prompt the molten steel to rotate

The generation of electromagnetic force in electromagnetic stirring molten steel mainly follows two laws:

(1) The law of electromagnetic induction. The three-phase alternating current introduced into the coil excites the rotating magnetic field. The magnetic field is unevenly distributed inside the mold and has a certain rotation speed. The magnetic field line is cut by the molten steel to generate an induced current.

$$\boldsymbol{j} = \sigma \boldsymbol{v} \times \boldsymbol{B} \tag{5.41}$$

where $\boldsymbol{j}$ is the induced current density, A/m^2; σ is electrical conductivity of molten steel, S/m; $\boldsymbol{v}$ is magnetic field rotation velocity, m/s; B is magnetic induction intensity, T.

The steel is a ferromagnetic material at room temperature, and the Curie point is 760 °C. When the temperature exceeds the Curie point, it loses its magnetism. The temperature of molten steel in the mold reaches about 1500 °C. At this time, the molten steel is conductive but not magnetically conductive, that is, it is transformed into a paramagnet, which can move with the rotation of the electromagnetic field.

(2) The electromagnetic interaction. The electromagnetic force is generated by the interaction between the induced current and the changing magnetic field in the

molten steel:

$$F = j \times B \tag{5.42}$$

where $\boldsymbol{F}$ is the electromagnetic force, N/m^3.

The electromagnetic force generated by induction is the volume force, which acts on the molten steel, generates the electromagnetic torque around the central axis of the billet, and drives the molten steel to rotate, which is the principle of electromagnetic stirring. Electromagnetic stirring is to use the electromagnetic force generated in the molten steel to change the flow state of the molten steel in the mold and change the heat transfer and solidification process of the molten steel through the effect of force and heat.

Taking the electromagnetic stirrer of the bloom continuous casting mold and final—stage electromagnetic stirring as an example, a finite element model is established based on the actual data of the field equipment, and the following simplifications are made according to the actual requirements:

(1) In order to save computing resources, only the billet of the stirrer part is modeled.
(2) Ignoring the formation of the initial solidified shell in the mold, it is considered that the billet is all flowing high-temperature molten steel.
(3) There is a cooling water tank on the copper plate of the mold. When the model is established, the influence of the cooling water is disregarded.
(4) There is cooling water in the electromagnetic stirrer, and the electromagnetic characteristics of water and air are the same, so this part is treated as air.
(5) To ensure the accuracy of the calculation, the simulation of the electromagnetic field needs to model the air at infinity, which is impossible for the calculation model. In addition, due to the action of the yoke and the core, the magnetic field outside a certain range is almost zero, so the finite volume air can be modeled. According to experience, air equivalent to 3–5 times the volume of the magnet is generally modeled.
(6) The stirrer winding coil is wound by a plurality of closely arranged copper wires, which simplifies the coil into a conductive area with the same conductive area. Table 5.1 shows the size parameters of the mold and final-stage electromagnetic stirrer model.

The model established on the basis of the above assumptions is shown in Fig. 5.6 (the air part is not shown). The casting direction is the Y-axis, and the cross-section of the slab is the X-Z plane. In the mesh division of the model, the air part uses tetrahedral elements, and the other part uses hexahedral elements. The number of meshes is about 500,000.

In the model, the liquid steel, copper plate, and magnetic yoke are considered to be isotropic materials, and their relative permeability is constant. The physical parameters of different materials are shown in Table 5.2.

The model load is a three-phase alternating current, and the phase difference of each phase current is 120°. The relative two coils have the same phase of the current, and the calculation equation for each phase current density is as follows:

Table 5.1 Size parameters of mold and final electromagnetic stirrer

	Parameters	Value		Parameters	Value
Mold electromagnetic stirrer (concentrated)	Section size (mm × mm)	280 × 380	Mold electromagnetic stirrer (concentrated)	Relative magnetic conductibility of copper sheet (H/m)	1.0
	Length (mm)	850		Relative magnetic conductibility of steel (H/m)	1.0
	Inside diameter (mm)	776		Relative magnetic conductibility of air (H/m)	1.0
	Outside diameter (mm)	1300		Relative magnetic conductibility of magnetic yoke (H/m)	1000
	Heights (mm)	486		Conductivity of mold copper plates (S/m)	1.78 × 107

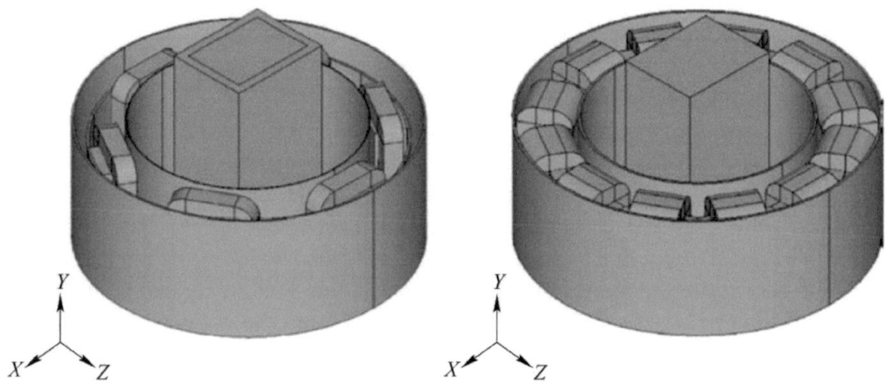

(a) (b)

Fig. 5.6 Finite element model of electromagnetic stirrer. **a** Mold electromagnetic stirring, **b** final electromagnetic stirring

Table 5.2 Material physical parameters used in the calculation

Parameters	Value
Relative magnetic conductibility of copper sheet (H/m)	1.0
Relative magnetic conductibility of steel (H/m)	1.0
Relative magnetic conductibility of air (H/m)	1.0
Relative magnetic conductibility of magnetic yoke (H/m)	1000
Conductivity of mold copper plates (S/m)	1.78 × 107
Conductivity of steel (S/m)	7.14 × 105

$$J_{ax} = j_0 \sin(\omega t) \tag{5.43}$$

$$J_{by} = j_0 \sin(\omega t - 2\pi/3) \tag{5.44}$$

$$J_{cz} = j_0 \sin(\omega t + 2\pi/3) \tag{5.45}$$

where ω is the angular velocity, $\omega = 2\pi f$; t is time, s; f is the current frequency, Hz; J_0 is the amplitude of the coil current density is determined by the current value, the number of turns, and the cross-sectional area of the coil.

5.1.1.8 Solidification End Reduction Model of Continuous Casting

To systematically analyze the effect of solidification end reduction on the liquid-phase flow and solute segregation in the two-phase region, the shell deformation factor is further considered in the multiphase solidification model, and the shell deformation during the reduction process is added as a boundary condition to the multiphase solidification model. Taking the continuous casting bloom as an example, Fig. 5.7 is a mathematical modeling diagram for the reduction interval. Due to the symmetry of the billet along the thickness direction (X-axis) after the solidification end reduction, the model calculation area only considers the deformation behavior in the 1/2 region of the billet thickness direction. In addition, to describe the process of the solidification end reduction of continuous casting more clearly, the influence of the bending section of the casting machine is not considered.

In Zone 1 ($y_0 \leq y < y_1$), the billet is pulled out from the mold area into the secondary cooling zone, and the thickness size (ω) of the billet is considered to be unchanged. In this region, the transverse velocity of the solid phase is set to zero ($u_{s,x}^{z1} = 0$), and the longitudinal velocity is set to the casting speed ($u_{s,y}^{z1} = u_{\text{cast}}$).

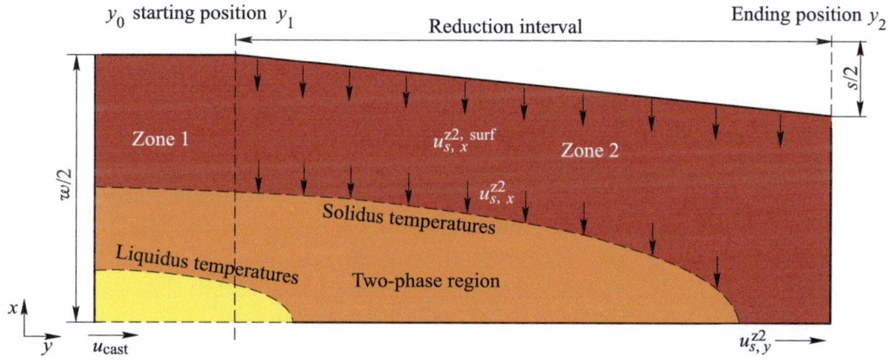

Fig. 5.7 Mathematical simplified diagram of the calculation area

In Zone 2 ($y_1 \leq y < y_2$), the billet gradually enters the withdrawal and straightening machine in the solidification end reduction zone of the bloom. Under the driving of the hydraulic cylinder, the reduction roller forces the solidified shell to squeeze into the center of the billet. It can be considered that the thickness of the billet decreases linearly, which is expressed by Eq. (5.46).

The mathematical simplified equation in the solidification end reduction interval of the whole bloom is as follows:

$$x^{z2,\ \mathrm{suf}} = \omega - \frac{s(y - y_1)}{2(y_2 - y_1)}(y_1 < y \leq y_2) \tag{5.46}$$

$$u_{s,y}^{z2} = \frac{u_{\mathrm{cast}}\omega}{\omega - \int_{y1}^{Y}\left(\frac{s}{2(y_2-y_1)}\eta\right)\mathrm{d}y} \tag{5.47}$$

$$\begin{aligned} u_{s,x}^{z2} &= u_{s,x}^{z2,\mathrm{suf}} + \frac{\partial u_{s,y}^{z2}}{\partial y}\left(x^{\mathrm{suf}} - x\right) \\ &= \frac{s}{2(y_2 - y_1)}u_{s,y}^{z2} + \frac{u_{\mathrm{cast}}\omega\frac{s}{2(y_2-y_1)}\eta}{\left[\omega - \int_{y1}^{Y}\left(\frac{s}{2(y_2-y_1)}\eta\right)\mathrm{d}y\right]^2}\left(x^{\mathrm{suf}} - x\right) \end{aligned} \tag{5.48}$$

where $x^{z2,\ \mathrm{suf}}$ is surface position of billet, m; s is the reduction of the fan-shaped segment, mm; η is center solid fraction of billet, m; y_1 is the starting position of mechanical reduction, m; y_2 is the ending position of mechanical reduction, m.

5.1.2 *Boundary Conditions of Multiphase Solidification Model*

According to the casting machine parameters provided in the field, the geometric model is established and the mesh is divided by taking the bloom as an example. In the mathematical model, different boundary conditions are applied to calculate the melt flow, solidification heat transfer, grain nucleation, and solute transport in the solidification process, analyzing the evolution of solidification structure and solute segregation.

For the multiphase solidification model, it is necessary to set reasonable velocity boundary conditions. Firstly, the whole calculation area and boundary conditions of the continuous casting process are divided. As shown in Fig. 5.8, the calculation area of the continuous casting process can be divided into four areas: inlet (mold nozzle), mold liquid level, billet surface, and outlet (casting machine outlet). The slab surface can be subdivided into mold, secondary cooling zone, and air-cooling zone. For the inlet velocity boundary condition, the molten steel is considered incompressible fluid in the multiphase solidification model, so the constant inlet velocity can be calculated according to the casting speed. For the liquid level of the mold, the flow behavior of the mold flux and its meniscus is not considered, so the liquid level of the mold is assumed to flow along the tangential direction without a velocity gradient. The surface velocity boundary condition of each region on the billet surface is set as “wall”, that is, the no-slip boundary condition is applied in each region, which is suitable for large-scale multiphase transport simulation of the continuous casting process. The outlet of the casting machine adopts the pressure outlet boundary condition. For the columnar dendrite phase, equiaxed grain phase, and liquid phase in the multiphase solidification model, the velocity boundary conditions of each phase in the above regions are consistent. In addition, for the external fields such as electromagnetic stirring and solidification end reduction, the electromagnetic force and deformation speed are comprehensively considered by adding the source term of the fluid domain.

In addition, the heat transfer boundary conditions such as the mold and secondary cooling zone involved in the multiphase solidification model have been introduced in Chap. 3.

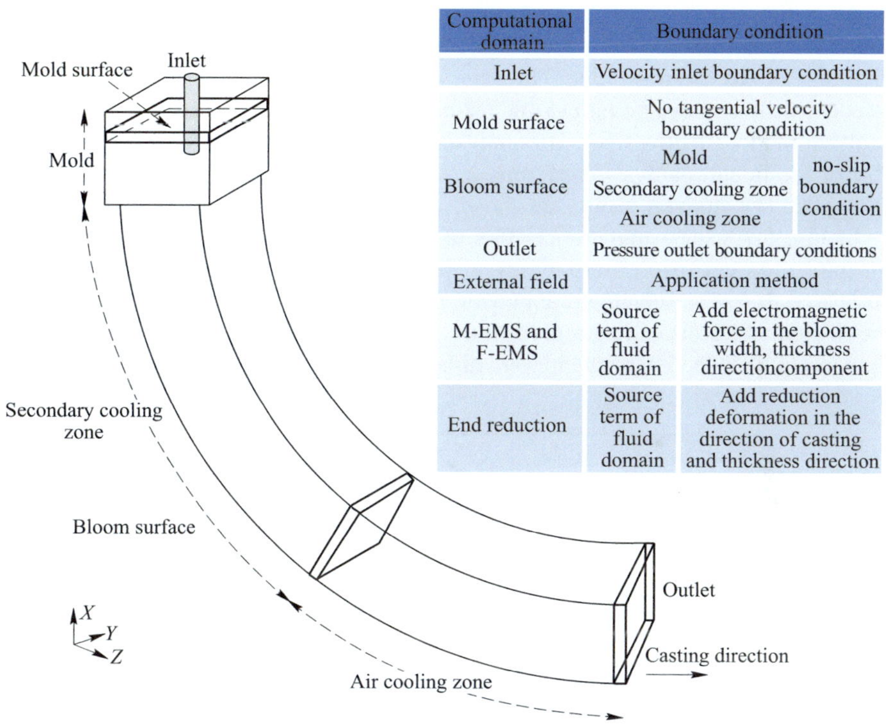

Computational domain	Boundary condition	
Inlet	Velocity inlet boundary condition	
Mold surface	No tangential velocity boundary condition	
Bloom surface	Mold	no-slip boundary condition
	Secondary cooling zone	
	Air cooling zone	
Outlet	Pressure outlet boundary conditions	
External field	Application method	
M-EMS and F-EMS	Source term of fluid domain	Add electromagnetic force in the bloom width, thickness directioncomponent
End reduction	Source term of fluid domain	Add reduction deformation in the direction of casting and thickness direction

Fig. 5.8 Calculation area division diagram of the whole process of continuous casting

5.1.3 Verification of Multiphase Solidification Model

Although the bloom at the end of solidification in the simulation analysis is in the high-temperature state, the macrosegregation difference of carbon element between the cold state (25 °C) and the hot state (< 900 °C) is very small. The carbon and sulfur analyzer and metal in situ statistical analyzer (OPA-200) were used to analyze the carbon segregation in different parts of the continuous casting billet. At the same time, the simulation results of carbon element distribution and macrostructure at the solidification end of continuous casting billet were systematically studied by combining solidification structure analysis with element analysis and in situ statistical distribution analysis.

To obtain the macroscopic morphology of the solidification defects of the slab, as shown in Fig. 5.9a, a 300 mm long longitudinal sample was taken from the cross-section of 320 mm × 410 mm heavy-haul railway's rail steel bloom (carbon mass fraction 0.78%) for hot acid corrosion, and the low-magnification results are shown in Fig. 5.10b. Further, drill cuttings sampling (carbon and sulfur analysis) and in situ analysis were performed on the longitudinal profile samples along the thickness direction of the casting billet. The sampling position is shown in Fig. 5.9b. The selected drill cuttings sampling bit diameter is 5mm, and the carbon and sulfur analyzer model are Eltra CS-i. The analysis results are shown in Fig. 5.11; the sample size of the in situ analysis is 45 mm × 75 mm, and the in situ analyzer model is OPA-200. The analysis results are shown in Fig. 5.12.

In Fig. 5.10, the simulation results of carbon solute distribution and the low-magnification corrosion morphology of the longitudinal section of the bloom are compared. It can be seen that the two are in good agreement. Figure 5.11 compares the calculated results and measured values of the transverse segregation behavior on the longitudinal section of the bloom, although there is a certain deviation between

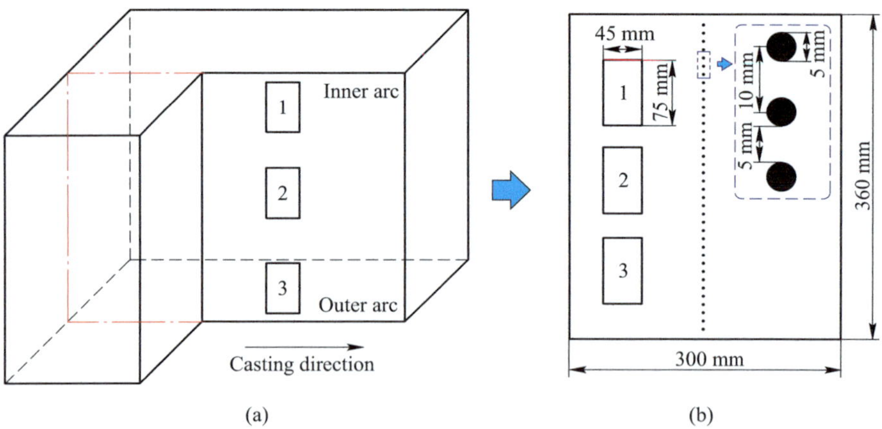

Fig. 5.9 Bloom longitudinal sampling diagram. **a** Bloom sampling diagram; **b** drilling cuttings sampling and in situ analysis sampling location

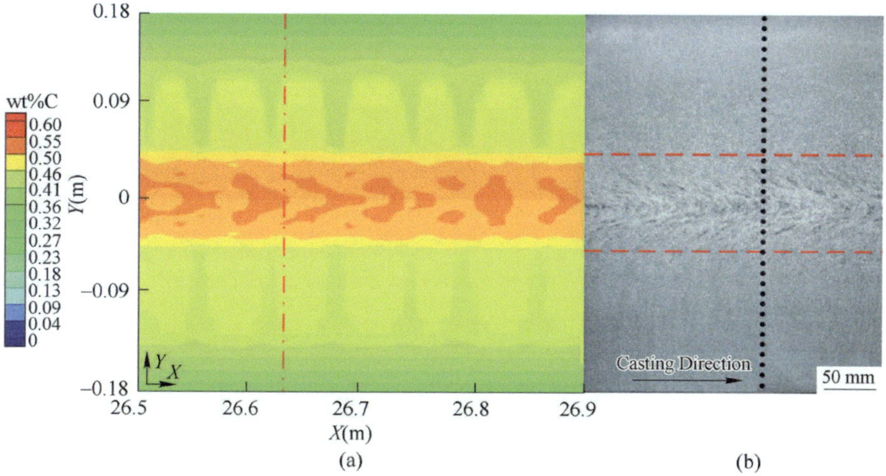

Fig. 5.10 Heavy-haul railway's rail steel bloom longitudinal section macrosegregation. **a** Simulation results; **b** low power photo of hot acid corrosion

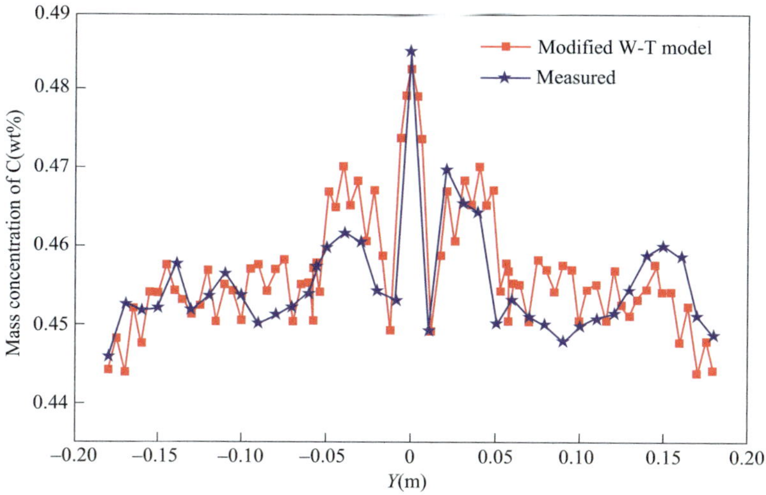

Fig. 5.11 Comparison of simulated and measured results of solute segregation in the thickness direction of heavy-haul railway's rail steel bloom

the calculated results and the measured results in the transverse direction, the trend of the two results is consistent. The results show that the center positive segregation of the inner arc side is slightly higher than that of the outer arc side due to the sinking of the equiaxed grain to the outer arc side. As shown in Fig. 5.12, the OPA-200 in situ analyzer was used to obtain the measurement results of carbon element distribution at different locations. Both positive carbon segregation and negative carbon segregation

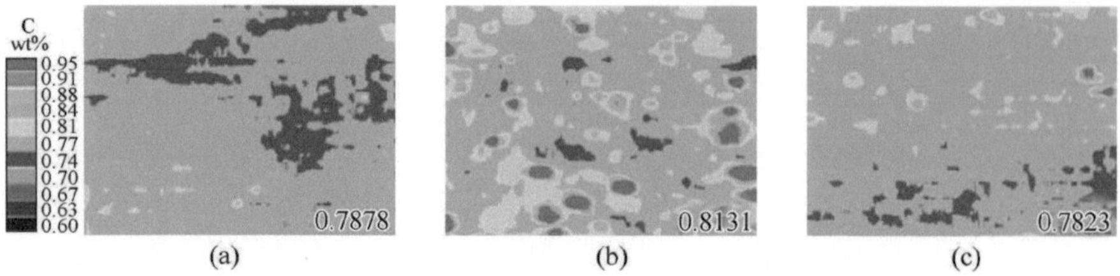

Fig. 5.12 C In situ analysis results of heavy rail steel in different positions in the vertical section of large billet (sampling location shown in Fig. 5.9). **a** Position a, inner arc; **b** location b, center; **c** position c, outer arc side

were observed in the center of the bloom, and negative carbon segregation was observed in the edge region of the bloom. The average carbon concentration (mass fraction) of samples 1, 2, and 3 was 0.788%, 0.813%, and 0.782%, respectively. The experimental results of Figs. 5.11 and 5.12 show that the measurement results of the in situ analyzer are consistent with the numerical simulation results, indicating that the calculation of the mathematical model is correct and feasible.

To verify the rationality and accuracy of the established model, it is necessary to compare the measured data with the calculated value of the model. At present, there are two methods to verify the electromagnetic field model in practical application. The common method is to measure the magnetic induction intensity; however, some scholars believe that the magnetic induction intensity value is only the intermediate parameter between the electromagnetic force or the electromagnetic torque and the excitation current, which cannot represent the stirring intensity scientifically and reasonably. It is recommended to use the electromagnetic torque as the performance index of the electromagnetic stirrer. However, at present, the standard of electromagnetic torque measuring instruments has not been unified, and the results of different measuring instruments have large errors. Therefore, this paper still verifies the numerical model according to the magnetic induction intensity.

The instrument used for the field magnetic induction intensity test is the CT3 tesla meter. Its working principle is to use the hall effect to convert the magnetic induction intensity into the electromotive force for measurement. The principle of the hall effect is as follows: if the current direction is perpendicular to the magnetic field, an additional transverse electric field will be generated in the direction perpendicular to the current and magnetic field, and a weak electromotive force will appear at both ends of the carrier fluid. The measurement of the magnetic field can be experimentally measured by measuring the electromotive force.

When a thin current-carrying conductor plate is placed in a magnetic field, the hall potential difference between the upper and lower ends of the plate is proportional to the current intensity I and the magnetic induction intensity B and inversely proportional to the thickness d of the plate along the B direction.

$$U = R_{\mathrm{H}} \frac{I_{\mathrm{H}} B}{d} \tag{5.49}$$

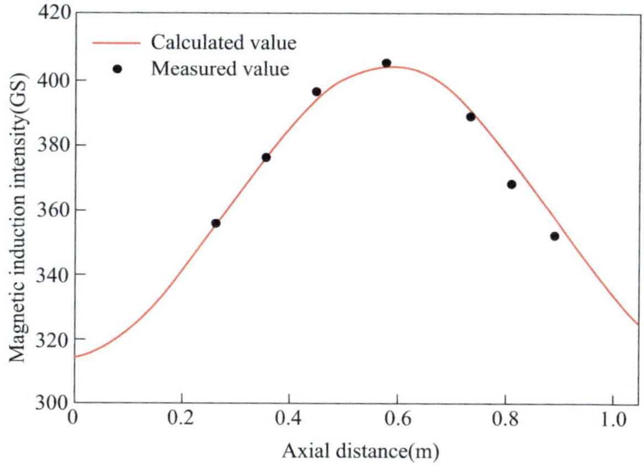

Fig. 5.13 Comparison between the calculated value and the measured value of the magnetic induction intensity on the centerline of the billet

where R_H is the hall coefficient, which is only related to the material of the conductor, is a constant; I_H is hall working current, A; B is magnetic induction intensity, Gs; d is the thickness of the conductor along B, m.

Taking the mold electromagnetic stirrer as an instance, the magnetic induction intensity at the center of the stirrer under different electromagnetic parameters was gauged by a tesla meter, and contrasted with the computed value of the model to verify the precision of the calculation outcomes of the magnetic induction intensity of the model. The measurement was conducted under no-load circumstances, and there is no molten steel in the mold. Some scholars have proposed that the measured value can represent the normal value when there is no load. Figure 5.13 is the comparison between the calculated value and the measured value of the magnetic induction intensity on the centerline of the casting billet in the stirrer. From the figure, it can be seen that the magnetic induction distribution on the centerline of the billet is "large in the middle and small at both ends". The maximum magnetic induction intensity appears in the middle of the stirrer, about 0.6m away from the upper mouth of the mold. There is a certain divergence between the top and bottom of the mold, yet the two are in good accordance, signifying that the established model is rational and accurate.

5.1.4 Research on Solute Transport Behavior of Bloom

Current intensity and frequency are the two most important process control parameters of electromagnetic stirring. However, the frequency of electromagnetic stirring has little effect on the magnetic induction intensity. In particular, the thickness of the shell in the mold is thin, the penetration effect of electromagnetic stirring is good, and the difference in magnetic induction intensity caused by stirring frequency can be neglected. According to practice, mold electromagnetic stirring generally uses low-frequency current, a value of 2.0–3.5 Hz. Therefore, this section only compares and analyzes the current intensity of the mold electromagnetic stirring.

5.1.4.1 Effect of Mold Electromagnetic Stirring on the Solute Transport Behavior of Bloom

The simulation object is the continuous casting bloom heavy-haul railway's rail steel (carbon mass fraction is 0.78%), and the bloom section is 320 mm × 410 mm. Using the established multiphase solidification model, the electromagnetic field distribution characteristics in the casting billet were calculated when the current parameters were 350 A and 2.4 Hz. It can be seen from Fig. 5.14 the magnetic field distribution within the mold is non-uniform, particularly in the axial direction of the billet. The magnetic induction at the center of the electromagnetic stirrer is relatively large, while the magnetic field intensity at a distance from the stirrer is relatively small. The magnetic induction intensity at the edge of the billet near the mold wall is slightly higher than that at the surface center.

Figure 5.15 is the distribution cloud diagram of electromagnetic force in millet. It can be seen from the diagram that the distribution trend of electromagnetic force in millet is roughly the same as that of magnetic induction intensity. The electromagnetic force at the position of the stirrer is large, and another peak value of the electromagnetic force appears at the lower mouth of the mold. The main reason is that the mold copper plate has a strong shielding effect on the magnetic field. After leaving the mold, the shielding effect of the mold copper plate is reduced, and the electromagnetic force peaks.

Figure 5.16 illustrates the electromagnetic force distribution at the middle position of the electromagnetic stirrer along the thickness direction of the billet. It can be seen

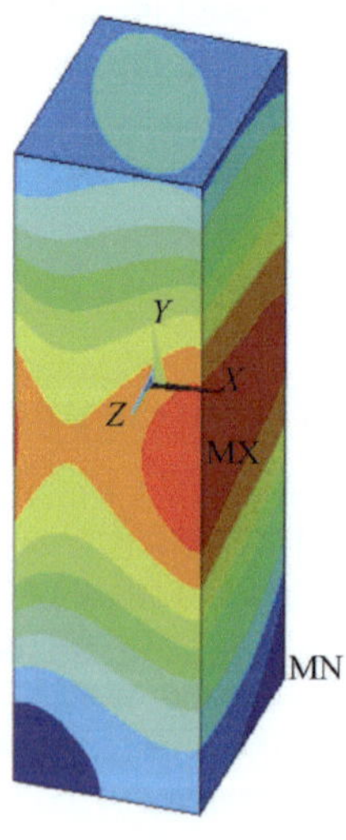

Fig. 5.14 Distribution of magnetic induction intensity in billet

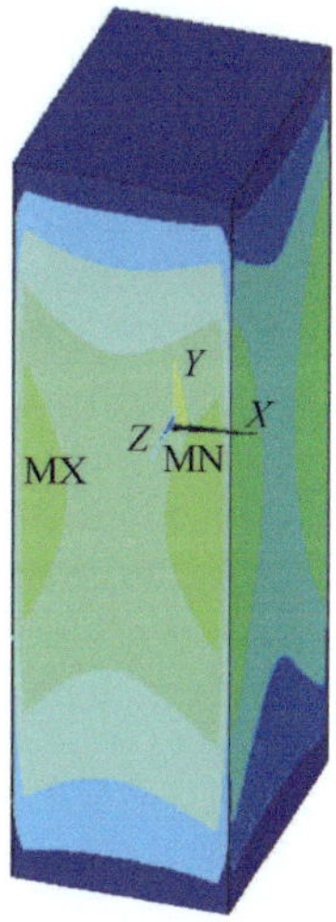

Fig. 5.15 Distribution of electromagnetic force in billet

that the electromagnetic force is the largest at the edge of the billet and decays to the center. With the increase of radial distance, the negative value in the figure indicates that the direction is opposite to the positive value. This distribution law makes the torque formed under the action of electromagnetic stirring, and the velocity of molten steel at the edge of the billet is larger, which also reflects the process characteristics of electromagnetic stirring.

Figure 5.17 presents the tangential electromagnetic force distribution of the cross-section at the middle position of the stirrer. It can be seen that the electromagnetic force is circumferentially distributed on the cross-section, and a rotating torque is generated in general. Under the action of this torque, the molten steel in the mold rotates horizontally, which accelerates the flow of molten steel. This is precisely the objective of employing the mold electromagnetic stirring technology to enhance the

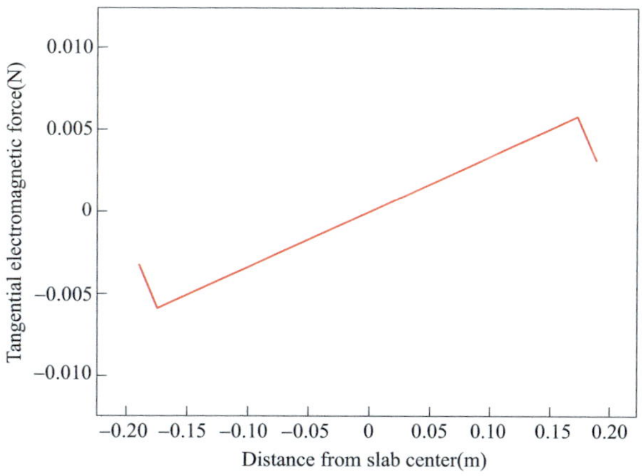

Fig. 5.16 Distribution of tangential electromagnetic force in the middle of the stirrer along the thickness direction of the billet

flow of the molten steel in the mold, boost the equiaxed grains rate, and promote the uniform distribution of solute elements and temperature fields.

Figure 5.18 shows the distribution of axial magnetic induction intensity under different stirring current intensities when the stirring current frequency of the mold is 2.4 Hz. It can be seen that the magnetic induction intensity of the stirrer center is "parabolic" along the centerline of the casting billet, that is, the closer to the stirrer, the greater the magnetic induction intensity of the centerline of the casting billet. On the other hand, with the increase of stirring current intensity, the central magnetic induction intensity increases significantly. When the current increases from 200 to 500 A, the central magnetic induction intensity of the stirrer increases from 178.44 Gs to 446.10 Gs.

Figure 5.19 presents the tangential electromagnetic force distribution along the thickness direction of the slab under different current intensities. It can be observed

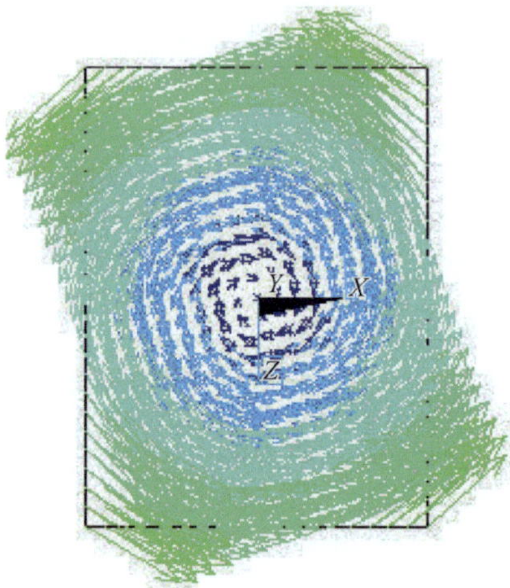

Fig. 5.17 Tangential electromagnetic force distribution of the cross-section in the middle of the stirrer

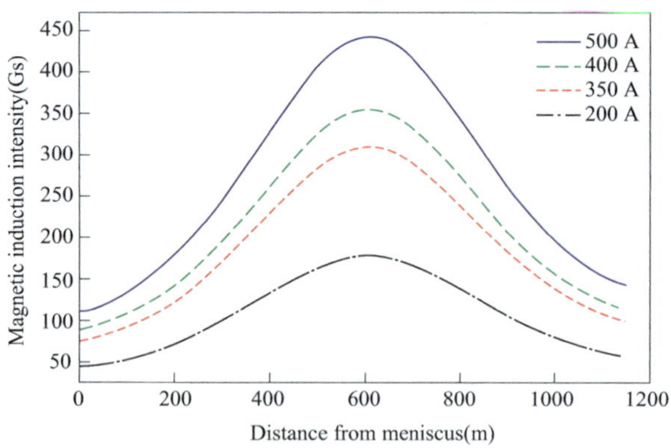

Fig. 5.18 Effect of current intensity on magnetic induction intensity

from the diagram that the greater the current intensity, the greater the magnetic induction intensity, the greater the induced current, and the greater the electromagnetic force. In addition, due to the skin effect of molten steel, the closer to the center of the billet, the weaker the magnetic field, the smaller the induced current, and the smaller the electromagnetic force.

Figure 5.20 is the effect of different current intensities on the equiaxed grain density. It can be seen that with the increase of the mold stirring current intensity, the equiaxed grain rate in the center of the billet increases significantly. However, after the current intensity reaches 400 A, with the increase of the stirring current intensity continues the equiaxed grain rate in the center of the billet increases to a limited extent.

It can be seen from Fig. 5.21 that increasing the current intensity can promote the long-distance transmission of solute elements enriched in the liquid phase at the solidification interface, resulting in the formation of negative segregation near the edge of the billet. The greater the current intensity, the more obvious the negative segregation at the edge of the billet, but the current intensity has a limited effect on the solute concentration of the molten steel in the center of the liquid cavity. On

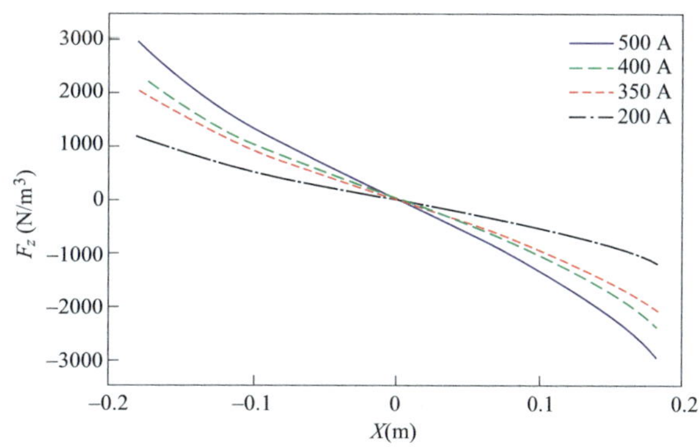

Fig. 5.19 Tangential electromagnetic force distribution along the thickness direction of the billet under different current intensities

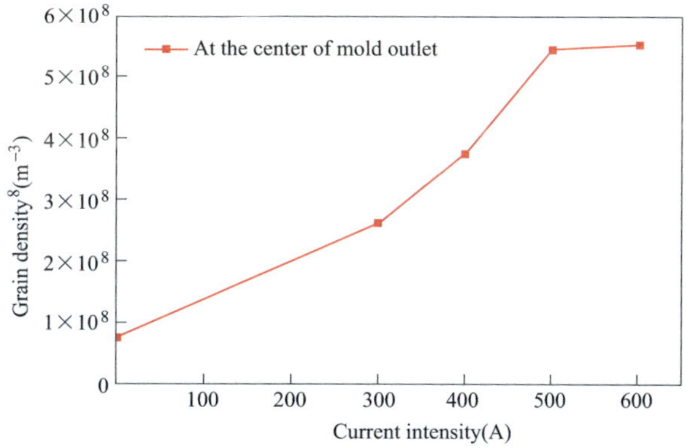

Fig. 5.20 Effect of different current intensities on equiaxed grain density

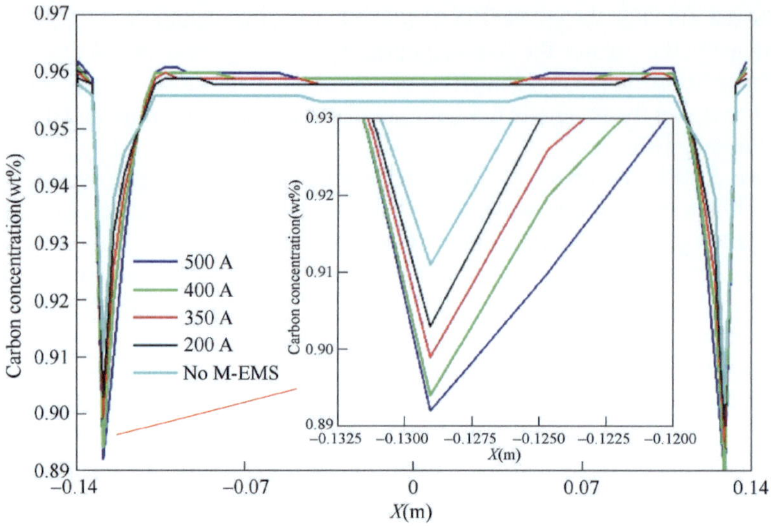

Fig. 5.21 Carbon element distribution in the thickness direction of the billet at 1.2 m from the meniscus under different current intensities

the other hand, increasing the current intensity can effectively promote the increase of the center grain density, especially when the current intensity reaches 400 A, the proportion of equiaxed grains increases substantially, which is beneficial to the improvement of the center segregation of the final solidified billet.

Therefore, for mold electromagnetic stirring, the optimal current intensity should be greater than 350 A, and the current greater than 400 A has a smaller equiaxed grain density of the billet, which leads to an increase in negative segregation at the edge of the billet.

5.1.4.2 Effect of Final Electromagnetic Stirring on the Solute Transport Behavior of Bloom

Compared with mold electromagnetic stirring, the flow rate of molten steel in the center of the slab at the end of solidification is significantly reduced, and a stronger electromagnetic force is needed to regulate the movement of molten steel, that is, the intensity of final electromagnetic stirring is higher than that of mold electromagnetic stirring. At the same time, due to the thickening of the solidified shell, the electromagnetic force penetration effect is weakened. Therefore, the influence of different stirring frequencies on the magnetic induction intensity of the stirrer is first analyzed. The simulation object of final electromagnetic stirring is heavy-haul railway's rail steel bloom with a section of 320 mm × 410 mm (carbon mass fraction is 0.78%). Figure 5.22 shows the distribution of the magnetic induction intensity of the billet in the center of the stirrer along the central axis under different stirring frequencies

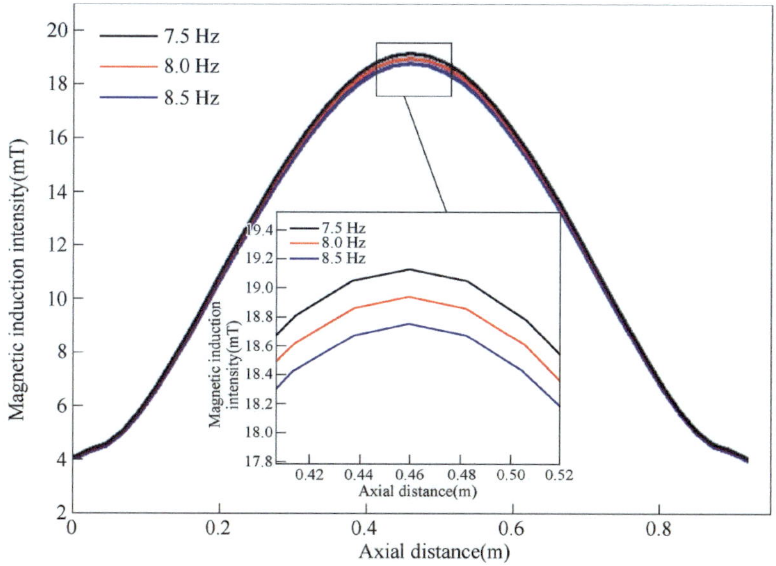

Fig. 5.22 Influence of the current frequency of the final electromagnetic stirring on the magnetic induction intensity

(7.5, 8.0, 8.5 Hz) when the current intensity of the final electromagnetic stirring is 300 A.

The distribution of axial magnetic induction intensity in the final electromagnetic stirring center is similar to that of the mold electromagnetic stirring. Both of them are parabolically distributed along the centerline of the billet (Figs. 5.18 and 5.22). On the other hand, when the current frequency is reduced from 8.5 to 7.5 Hz, the magnetic induction intensity in the center of the stirrer (at an axial distance of 0.5 m) increases from 18.75 to 19.13 mT, and the change is not large. It can be seen that adjusting the current intensity of the final electromagnetic stirrer is also the main means to adjust its magnetic induction intensity. The following simulation experiments will also focus on the adjustment of current intensity.

Figure 5.23 is the tangential electromagnetic force in the middle of the final electromagnetic stirrer along the narrow direction of the billet. The current frequency of F-EMS is 8.0 Hz, the casting speed is 0.68 m/min, and the final electromagnetic stirrer is located at 14.0 m from the meniscus. When the current intensity of F-EMS increases from 250 to 400 A, the maximum tangential electromagnetic force in the z direction of the bloom grows from 1938.33 to 10272.72 N/m^3. Under the influence of the skin effect, the electromagnetic force diminishes linearly from the edge of the billet to its center along the narrow surface direction of the billet.

Figure 5.24 shows the tangential stirring speed distribution along the narrow direction of the billet in the middle of the final electromagnetic stirrer. Because the solidified shell of 62.16 mm has been formed at this position, the stirring speed at the edge of the billet is 0 m/s. The tangential stirring speed of molten steel reaches the

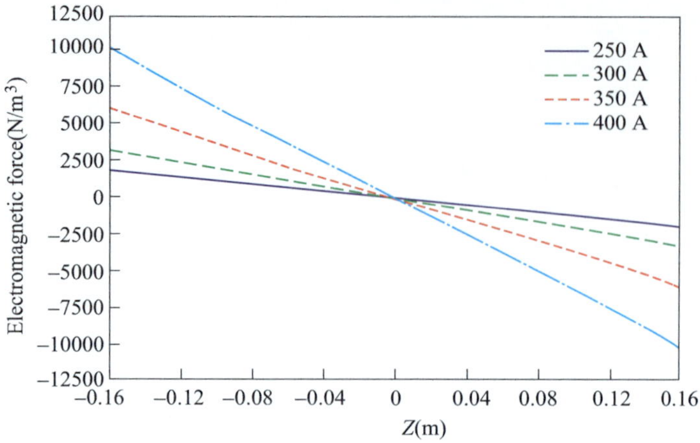

Fig. 5.23 Effect of the current intensity of the final electromagnetic stirring on the tangential electromagnetic force in the narrow face of the center of the billet

maximum at 72.32 mm from the center of the gear steel billet and gradually decreases to the center of the billet under the action of the skin effect. Similar to M-EMS, the stirring speed of molten steel inside the billet under the action of F-EMS increases with the increase of current intensity.

Figure 5.25 shows the phase fraction along the narrow direction of the bloom center at 14.0 m from the mold meniscus under different F-EMS current intensities. As the current intensity of F-EMS increases from 250 to 400 A, the distribution range of the columnar dendrite phase decreases gradually, the distribution range of

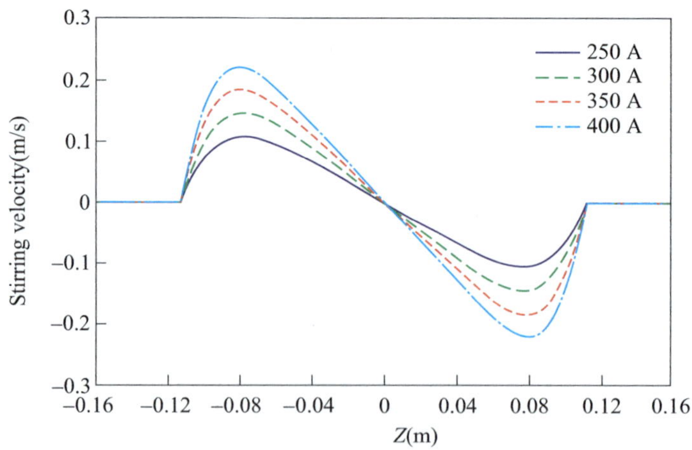

Fig. 5.24 Effect of the current intensity of the final electromagnetic stirring on the tangential stirring speed in the narrow face of the center of the billet

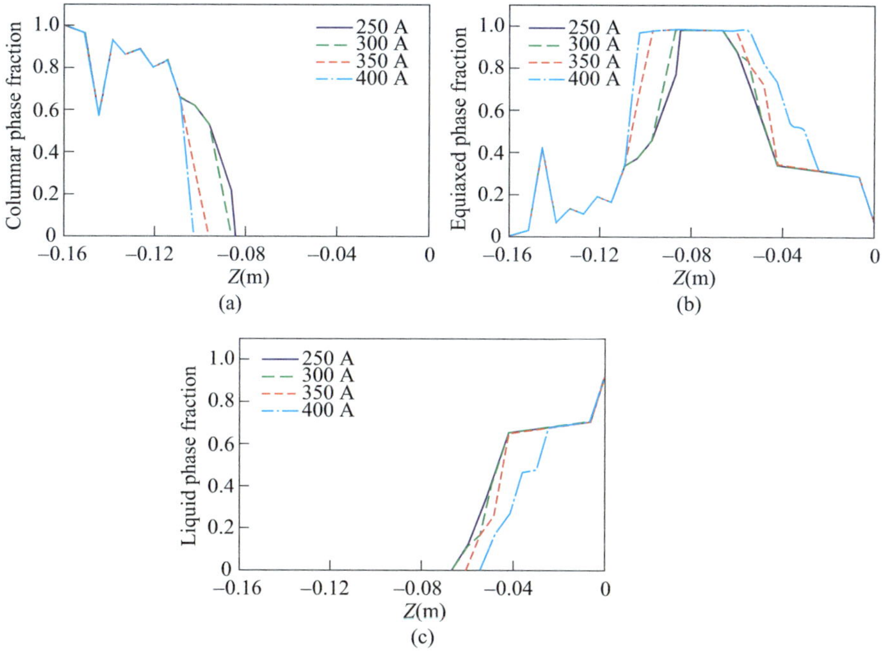

Fig. 5.25 Phase fraction along the narrow surface direction at the center of the bloom at 14.0 m from the mold meniscus under different F-EMS current intensities. **a** Columnar dendrite phase, **b** equiaxed grain phase; **c** liquid phase

the equiaxed grain phase increases continuously, and the distribution range of the liquid phase in the center of bloom decreases continuously.

Figure 5.26 gives a schematic diagram of melt flow, solute transport, and equiaxed grain movement under the action of F-EMS. With the increase of F-EMS current intensity, the electromagnetic force continuously promotes the rotational flow of molten steel inside the billet, and the columnar dendrite tip at the front of the solidification interface is continuously disrupted or remelted by the solute erosion mechanism, which reduces the columnar dendrite region and correspondingly increases the equiaxed grain region. In addition, due to the rotating stirring effect of F-EMS, the temperature of molten steel decreases rapidly with the increase of current intensity, resulting in the acceleration of the solidification rate and the formation of freer equiaxed grains. Under the action of rotating centrifugal force, the free equiaxed grain is far away from the center of the bloom, and deposits at the front of the solidification interface, so that the equiaxed grain region becomes larger and the liquid-phase region becomes smaller.

Figure 5.27a shows the degree of carbon segregation in the thickness direction of the bloom at 14.0 m from the meniscus with different F-EMS current intensities. With the increase of F-EMS current intensity from 250 A to 400 A, the center positive segregation degree of bloom decreases from 1.100 to 1.074. When the current

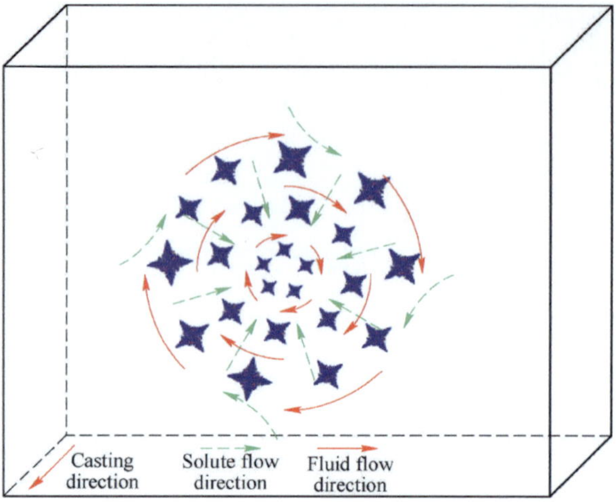

Fig. 5.26 Schematic diagram of melt flow, solute transport, and equiaxed grain movement under the action of F-EMS

intensity of F-EMS increases from 250 to 400 A, the negative segregation degree near the center of the bloom along the z direction decreases from 0.988 to 0.976.

Combined with Fig. 5.26, the rotating flow of molten steel is driven by electromagnetic force, which leads to an increase in solidification rate and the formation of more solute-poor equiaxed grains. Under the action of rotating centrifugal force, a large number of solute-poor equiaxed grains are accumulated in front of the solidification boundary, forming a local negative segregation zone, which is the white band. Therefore, the larger F-EMS current intensity will enhance the solute scouring effect at the front of the solidification interface and aggravate the negative segregation defects near the center of the billet. Figure 5.27b shows the degree of carbon segregation in the centerline of the billet along the casting direction under different

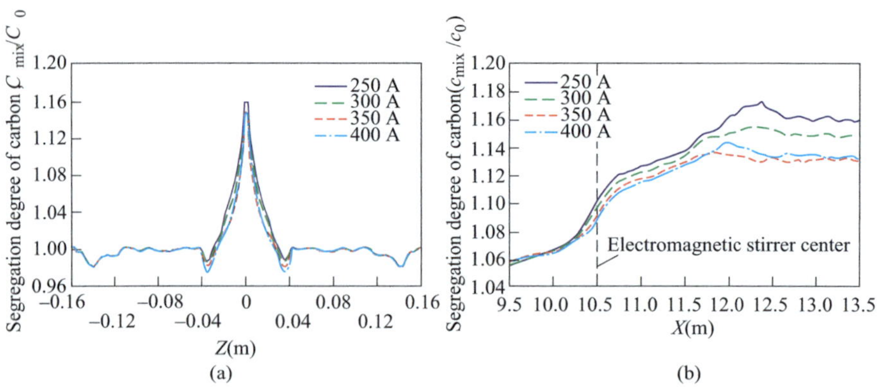

Fig. 5.27 Carbon segregation degree at different positions of bloom under different F-EMS current intensities. **a** The thickness direction of the bloom at 14.0 m from the meniscus of the mold; **b** bloom centerline

F-EMS current intensities. With the increase of current intensity, the carbon concentration in the center of the bloom increases significantly after the molten steel enters the final electromagnetic stirrer, and the carbon concentration further increases after the molten steel flows out of the electromagnetic stirrer. The reason is that equiaxed grains are deposited at the front of the solidification interface under the action of rotating centrifugal force, which forces the enriched solute steel liquid at the front of the solidification interface to transfer to the center of the bloom, increasing the central carbon concentration of the bloom. When the billet is pulled out from final the electromagnetic stirrer, the carbon concentration in the center of the billet will decrease to a certain extent after reaching the maximum value under the action of high-temperature diffusion.

However, compared with the carbon concentration at the center of the bloom when the current intensity is 400 A, the carbon concentration at the center of the bloom is the lowest when the current intensity is 350 A. The results show that the higher current intensity of final electromagnetic stirring does not necessarily regulate the macrosegregation of the billet well. This is due to the dual effect of final electromagnetic stirring on segregation. On the one hand, under the influence of F-EMS, the temperature of molten steel at the center of the bloom decreases rapidly, the solidification rate increases continuously, and the dendrite arm spacing decreases significantly, which effectively controls the macrosegregation at the center of the bloom. On the other hand, when the F-EMS current intensity is too large, the rotation speed of the molten steel increases, so that the solute-rich inter-dendritic melt in the columnar dendrite flows to the center of the bloom, thereby increasing the segregation degree at the center of the bloom. Therefore, for this specific production condition, the F-EMS current intensity is set to 350 A, which can better reduce the positive segregation and negative segregation defects near the center of the heavy-haul railway's rail steel bloom.

5.1.4.3 Effect of Superheat on the Solute Transport Behavior of Bloom

When the casting speed is 0.70 m/min, the multiphase solidification model is used to simulate the solidification end segregation of heavy-haul railway's rail steel bloom ($C_0 = 0.78\%$) with 320 mm × 410 mm cross-section at different superheats. The simulation results of solidification end segregation at superheat of 10, 20, and 30 °C are shown in Fig. 5.28.

It can be seen from Fig. 5.27 that the solute segregation in the longitudinal section of continuous casting bloom is a discontinuous "V"-shaped distribution, and with the increase of casting superheat, the central macrosegregation defect at the solidification end of continuous casting bloom becomes more serious. Figure 5.28 shows the distribution of carbon segregation at the solidification end section along the thickness direction of the billet under different casting superheats. It can be seen from the diagram that when the casting superheats are reduced from 30 to 10 °C, the carbon mass fraction at the center of the continuous casting billet is reduced from 0.95 to 0.85%. As positive segregation emerges in the center, the coexistence of positive and

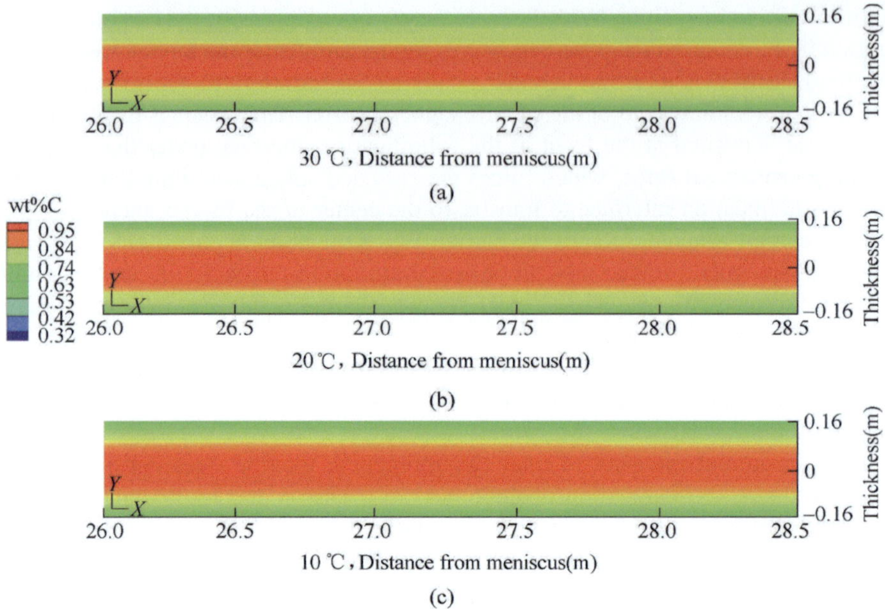

Fig. 5.28 Segregation morphology of solidification end of continuous casting billet at different superheats

negative segregation occurs in the vicinity of the center area, and there is a mild negative segregation near the surface area. In addition, due to the V-shaped distribution of segregation defects in the longitudinal section, the carbon mass fraction distribution curve shown in Fig. 5.29 does not have a peak value in the center of the billet, which is slightly different from the conventional solute segregation distribution curve.

5.1.4.4 Effect of Solidification End Reduction on the Solute Transport Behavior of Bloom

Based on the thermomechanical coupling model of the solidification end reduction process of the continuous casting billet which was developed in Chap. 3, the deformation law of the shell during the reduction process can be ascertained and linked to the multiphase solidification model as a boundary condition. Using this model, the liquid-phase flow and solute segregation behavior in the two-phase region under the action of shell deformation were studied and analyzed, and the influence of different reductions on the central segregation of the billet was revealed.

Figure 5.30 compares the solute transport behavior of the billet with a total reduction of 9 mm and 11 mm. It can be seen that under the action of solid-phase shrinkage and mechanical pressure, the solid-phase deformation behavior in the two-phase

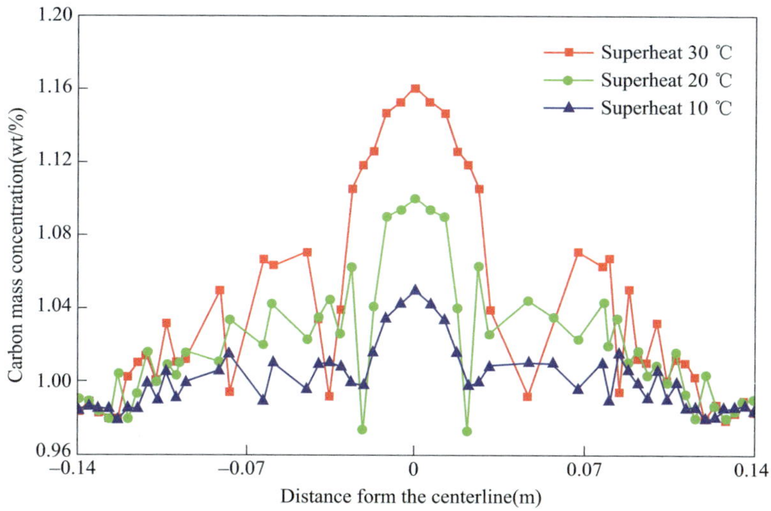

Fig. 5.29 Distribution law of carbon segregation along the thickness direction at the solidification end of heavy-haul railway's rail steel billet under different superheat degrees

region is significantly affected, thus changing the relative flow of the solute-rich liquid phase in the center of the billet.

In the vicinity of the solidification end point, the volume of the central area of the billet shrinks rapidly under the combined action of heat shrinkage and solidification shrinkage, which promotes the accelerated pumping flow of the central liquid phase to

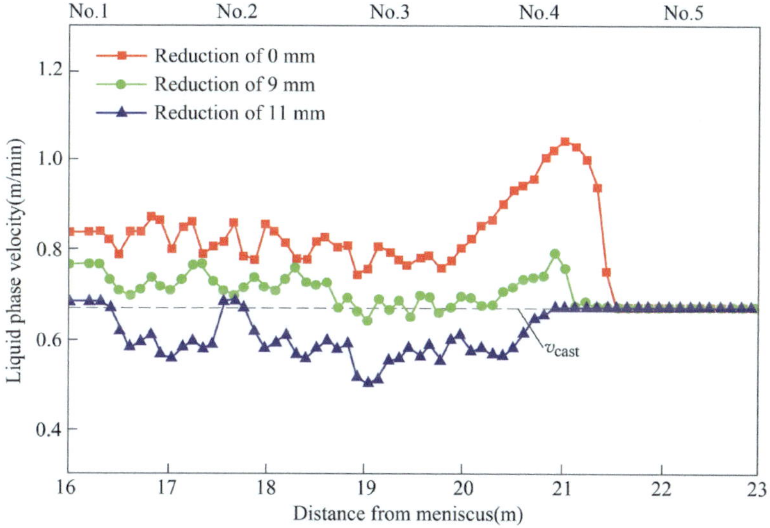

Fig. 5.30 Liquid-phase velocity on the centerline of the billet

the solidification end point. As shown in Fig. 5.30, with the gradual increase of reduction, the flow velocity of the solute-rich liquid phase to the solidification end point (21.56 m) gradually decreases, which is the main reason for the gradual reduction of central segregation of billet. Specifically, as shown in Fig. 5.30, without reduction, under the shrinkage of the central region of the billet, the liquid-phase velocity in the center of the billet in the reduction range (No.1–No.5 withdrawal and straightening machine) increases continuously, and because the closer to the solidification end, the stronger the shrinkage effect, the phenomenon of accelerated liquid-phase flow occurs in the area before the solidification end (No.4–No.5 withdrawal and straightening machine). When the reduction is 8mm, the liquid velocity in the center of the billet in the reduction interval (No.1–No.5 withdrawal and straightening machine) decreases as a whole, especially the accelerated flow behavior of the liquid phase before the solidification end point is inhibited to a certain extent, which indicates that the reduction effect compensates the shrinkage of the central area of the billet. However, due to the insufficient reduction, the solute-rich liquid phase will still flow to the solidification end point of the billet. When the reduction is increased to 16 mm, the central liquid-phase velocity of the billet in the reduction range (No.1-No.5 withdrawal and straightening machine) is less than the casting speed, indicating that the reduction of 16 mm can fully compensate for the shrinkage demand of the central region of the billet and promote the reverse flow of the central liquid phase, especially the accelerated flow behavior of the liquid phase before the solidification end point disappears, which is the key to solve the central segregation of the billet. Figure 5.31 shows the change of central segregation along the casting direction under different reduction conditions. It can be seen from the figure that the center segregation of the billet shows a trend of decreasing first and then increasing. With the increase of reduction, the degree of negative segregation in the initial solidification stage increases obviously, which is due to the reverse flow trend of the central liquid phase in the initial stage of reduction, resulting in the decrease of local solute concentration. As the billet continuously solidifies, under the combined influence of heat shrinkage and solidification shrinkage at the solidification end, the solute-rich liquid phase is drawn and flows toward the solidification end, thereby promoting the formation of central segregation within the billet. Therefore, on the whole, with the increase of the solidification end reduction, the thermal shrinkage and solidification shrinkage are compensated, and the liquid phase flowing to the solidification end is obviously reduced, so the increase of the central segregation of the billet is significantly slowed down. In addition, with the increase of the solidification end reduction, the billet is continuously thinned and the solidification end point is moved forward.

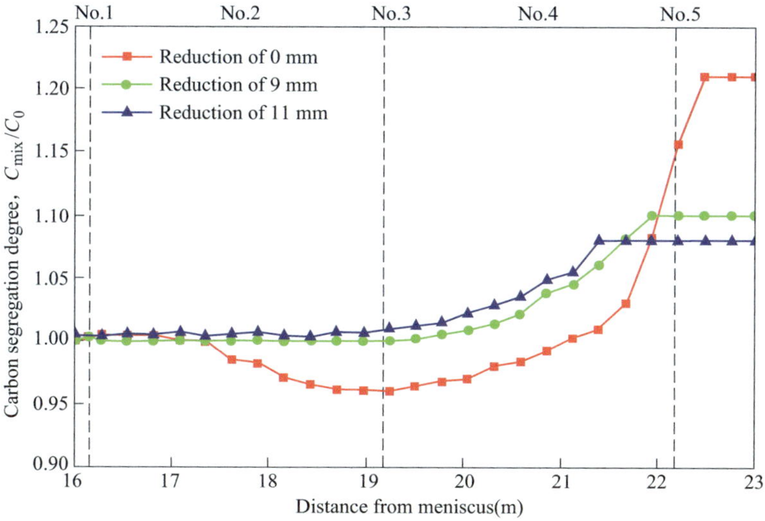

Fig. 5.31 Evolution law of carbon segregation in the centerline of continuous casting billet under different reduction conditions

5.2 Research on Solute Transport Law of Wide-Thick Slab Under the Control of Multiple External Fields

5.2.1 *Shell Deformation Model Based on Multiphase Solidification Model*

In this section, based on the deformation law of the wide-thick slab during the reduction process in Chap. 3, a multiphase (liquid-phase-columnar dendrite phase-equiaxed grain phase) solidification model of the reduction process was developed to study the liquid-phase flow and solute segregation behavior in the two-phase region of the wide-thick slab under the action of shell deformation and the influence of different reductions on the central segregation of the slab. Figure 5.32 is a schematic diagram of the improvement of segregation behavior by solidification end reduction of wide-thick slabs.

As shown in Fig. 5.33a, according to the thermodynamic analysis of the reduction process in Chap. 3, it is found that the position of the maximum deflection point of the slab bulging is not in the middle of the distance between the two adjacent supporting rollers, but along the direction of the casting speed to the downstream of the casting flow, which will greatly affect the molten steel flow and macrosegregation behavior inside the slab.

As shown in Fig. 5.33b, the asymmetric bulging morphology of the solidification end reduction process of the thick slab is fitted by a piecewise sine function curve. Because the sine and cosine function curves are only different in phase difference, to

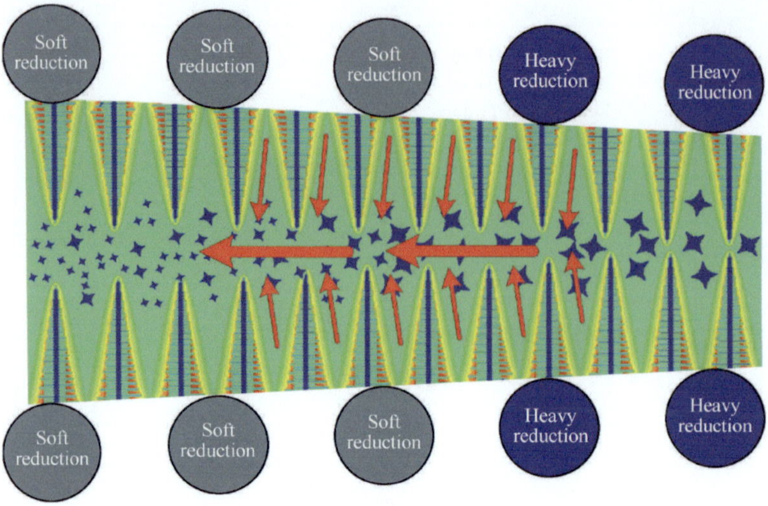

Fig. 5.32 Schematic diagram of the improvement of segregation at the solidification end reduction of the wide-thick slab

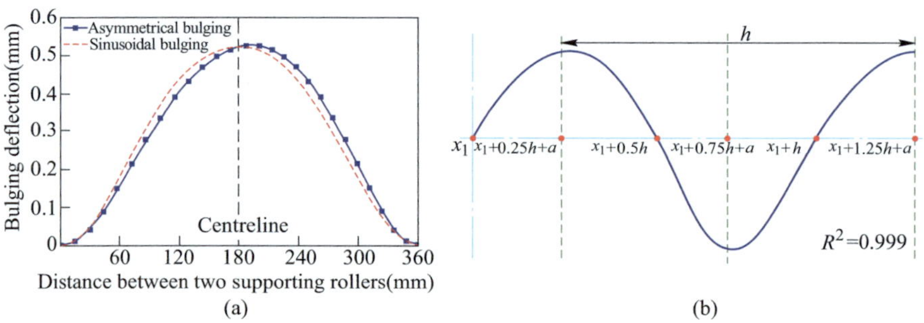

Fig. 5.33 Slab bulging morphology calculated according to the three-dimensional bulging model (**a**) and asymmetric drum bulging morphology fitting curve (**b**)

ensure that the coordinate starting point is bulging, the height is 0mm, so this section is based on the sine function curve to form an asymmetric bulging morphology. The asymmetric bulging morphology fitting calculation equation is as follows:

$$y_{\mathrm{bul}}^{i}(x)=\begin{cases}\dfrac{Y_i}{2}-\dfrac{d_i}{2}\left(1-\dfrac{x}{n_ih_i}\right)\sin\left(\dfrac{2\pi}{h_i+4a_i}\right),\\ x=x_i\sim x_i+0.25h_i+a_i;\\ \dfrac{Y_i}{2}-\dfrac{d_i}{2}\left(1-\dfrac{x}{n_ih_i}\right)\sin\left(\dfrac{2\pi}{h_i-4a_i}\left(x-2\left(k_n^i+1\right)a_i\right)\right),\\ x=x_i+0.25\,h_i+a_i+k_n^ih_i\sim x_i+0.75h_i-a_i+k_n^ih_i,\,k_n^i\in[0,n_i];\\ \dfrac{Y_i}{2}-\dfrac{d_i}{2}\left(1-\dfrac{x}{n_ih_i}\right)\sin\left(\dfrac{2\pi}{h_i+4a_i}\left(x+4\left(k_n^i+1\right)a_i\right)\right),\\ x=x_i+0.75h_i-a_i+k_n^ih_i\sim x_i+1.25h_i+a_i+k_n^ih_i,\,k_n^i\in[0,n_i]\end{cases},x=x_i\sim x_{i+1} \tag{5.50}$$

where $y_{\mathrm{bul}}^{i}(x)$ is Slab thickness only under the influence of asymmetric bulging, mm; k_n^i is the number of cycles of the asymmetric bulging morphology curve in each calculation area.

For continuously cast thick slabs, taking into account the asymmetric bulging morphology and the solidification end reduction process during the continuous casting of the thick slabs, the deformation behavior of the thick slab shell is calculated by disregarding the spreading behavior of the slab shell. In Fig. 5.34, half of the thickness of the longitudinal section of the thick slab is selected to establish a two-dimensional calculation model. The continuous casting thick slab is divided into five calculation regions along the casting flow length, namely Zone 0, Zone 1, Zone 2, Zone 3, and Zone 4. Among them, Zone 0 is the mold section, Zone 1 is the asymmetric bulging section, Zone 2 and Zone 3 are the asymmetric bulging and mechanical reduction section, and Zone 4 is the horizontal section.

Other calculation parameters used to describe the shape of the asymmetric bulging are shown in Table 5.3.

Combined with the comprehensive influence of asymmetric bulging and solidification end reduction, the surface morphology equation of continuous casting thick slab shell is as follows:

$$y_0(x)=\frac{T}{2},x=0\sim x_1 \tag{5.51}$$

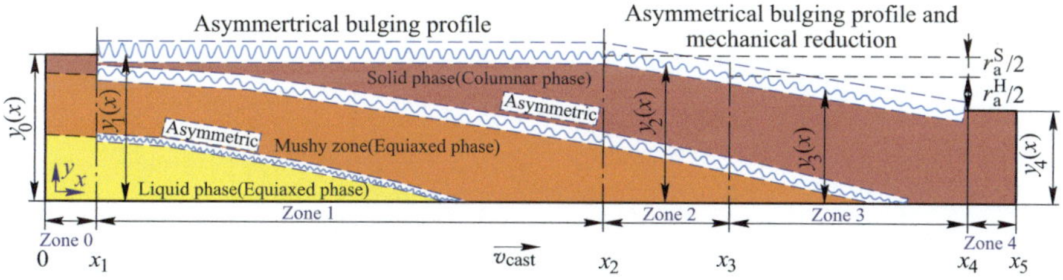

Fig. 5.34 Schematic diagram of different calculation area divisions of the longitudinal section of wide-thick slab

Table 5.3 Calculation parameters used to describe the shape of the asymmetric bulging

Item	Unit	Zone i		
		$i = 1$	$i = 2$	$i = 3$
The slab thickness at the inlet of each zone	Y_i (mm)	w	y_{Z2}, bul (x_2)	y_{Z2}, bul (x_3)
The initial bulging height at the inlet of each zone	d_i (mm)	0.80	0.35	0.54
The total number of rollers in each zone	n_i	58	18	15
The distance between two rollers in each zone	h_i (mm)	268	360	410
The offset of the point of maximum bulging deflection	α_i (mm)	11.17	14.40	16.45

$$y_1(x) = y_{\text{bul}}^1(x), x = x_1 \sim x_2 \tag{5.52}$$

$$y_2(x) = y_{\text{bul}}^2(x) - \frac{r_{\text{a}}^{\text{S}}}{2}\left(\frac{x - x_2}{x_3 - x_2}\right), x = x_2 \sim x_3 \tag{5.53}$$

$$y_3(x) = y_{\text{bul}}^3 - \frac{r_{\text{a}}^{\text{S}}}{2} - \frac{r_{\text{a}}^{\text{H}}}{2}\left(\frac{x - x_3}{x_4 - x_3}\right), x = x_3 \sim x_4 \tag{5.54}$$

$$y_4(x) = \frac{T}{2} - \frac{r_{\text{a}}^{\text{S}} + r_{\text{a}}^{\text{H}}}{2}, x = x_4 \sim x_5 \tag{5.55}$$

where $y_0(x)$ is the thickness of slab in Zone 0, mm; $y_1(x)$ is the thickness of the slab in Zone 1, mm; $y_2(x)$ is the thickness of the slab in Zone 2, mm; $y_3(x)$ is the thickness of the slab in Zone 3, mm; $y_4(x)$ is the thickness of the slab in Zone 4, mm; T is the initial thickness of the thick slab, mm; r_{a}^{S} is the soft reduction in Zone 2, mm; r_{a}^{H} is the soft reduction in Zone 3, mm.

The surface velocity equation of the slab shell parallel to the casting speed direction caused by asymmetric bulging morphology in Zone 0, Zone 1, and Zone 2 is as follows:

$$v_{\text{bul}}^{i,x}(x) = v_{\text{cast}}$$

$$v_{\text{bul}}^{i,y}(x) = \begin{cases} v_{\text{bul}}^{i,x}(x) \cdot \left[\frac{d_i}{2n_i h_i}\sin\left(\frac{2\pi}{h_i + 4a_i}x\right) - \frac{\pi d_i}{h_i + 4a_i}\left(1 - \frac{x}{n_i h_i}\right)\cos\left(\frac{2\pi}{h_i + 4a_i}x\right)\right] \\ v_{\text{bul}}^{i,x}(x)\left\{\frac{d_i}{2n_i h_i}\sin\left[\frac{2\pi}{h_i - 4a_i}\left(x - 2\left(k_n^i + 1\right)a_i\right)\right]\right. \\ \left. - \frac{\pi d_i}{h_i - 4a_i}\left(1 - \frac{x}{n_i h_i}\right)\cos\left[\frac{2\pi}{h_i - 4a_i}\left(x - 2\left(k_n^i + 1\right)a_i\right)\right]\right\} \\ v_{\text{bul}}^{i,x}(x)\left\{\frac{di}{2ni_i h_i}\sin\left[\frac{2\pi}{h_i + 4a_i}\left(x + 4\left(k_n^i + 1\right)a_i\right)\right]\right. \\ \left. - \frac{di}{2ni_i h_i}\sin\left[\frac{2\pi}{h_i + 4a_i}\left(x + 4\left(k_n^i + 1\right)a_i\right)\right]\right\} \end{cases} \tag{5.56}$$

where $\boldsymbol{v}_{\text{bul}}^{i,x}(x)$ is parallelling the casting speed, the surface velocity of the slab shell direction caused by asymmetric bulging morphology in Zone 0, Zone 1, and Zone 2, mm; $\boldsymbol{v}_{\text{bul}}^{i,y}(x)$ is perpendicular to the casting speed, the surface velocity of the slab shell direction caused by asymmetric bulging morphology in Zone 0, Zone 1, and Zone 2, mm; v_{cast} is the casting speed, m/s.

Considering the influence of solidification end reduction, the velocity equation of the shell surface of the whole casting flow is expressed as follows:

$$\boldsymbol{v}_0^x(x) = v_{\text{cast}},\ \boldsymbol{v}_0^y(x) = 0 \tag{5.57}$$

$$\boldsymbol{v}_1^x(x) = \boldsymbol{v}_{\text{bul}}^{1,x}(x),\ \boldsymbol{v}_1^y(x) = \boldsymbol{v}_{\text{bul}}^{1,y}(x) \tag{5.58}$$

$$\begin{aligned} \boldsymbol{v}_2^x(x) &= \boldsymbol{v}_{\text{bul}}^{2,x}(x) + \frac{x - x_2}{x_3 - x_2} \times \frac{v_{\text{cast}} r_{\text{a}}^{\text{S}}}{y_1(x_2) - r_{\text{a}}^{\text{S}}}, \\ \boldsymbol{v}_2^y(x) &= \boldsymbol{v}_2^x(x)\left[\boldsymbol{v}_{\text{bul}}^{2,y}(x) - \frac{r_{\text{a}}^{\text{S}}}{2(x_3 - x_2)}\right] \end{aligned} \tag{5.59}$$

$$\begin{aligned} \boldsymbol{v}_3^x(x) = \boldsymbol{v}_{\text{bul}}^{3,x}(x) &+ \frac{v_{\text{cast}} y_1(x_2)}{\left[y_1(x_2) - r_{\text{a}}^{\text{S}}\right]} + \frac{x - x_3}{x_4 - x_3} \\ &\times \frac{v_{\text{cast}}\left(r_{\text{a}}^{\text{S}} y_2(x_3) + r_{\text{a}}^{\text{H}} y_1(x_2) + r_{\text{a}}^{\text{S}} r_{\text{a}}^{\text{H}}\right)}{\left[y_2(x_3) - r_{\text{a}}^{\text{H}}\right]\left[y_1(x_2) - r_{\text{a}}^{\text{S}}\right]} \end{aligned} \tag{5.60}$$

$$\boldsymbol{v}_3^y(x) = \boldsymbol{v}_3^x(x)\left[\boldsymbol{v}_{\text{bul}}^{3,y}(x) - \frac{r_{\text{a}}^{\text{H}}}{2(x_4 - x_3)}\right] \tag{5.61}$$

$$\boldsymbol{v}_4^x(x) = \frac{y_2(x_3) y_1(x_2) v_{\text{cast}}}{\left[y_1(x_2) - r_{\text{a}}^{\text{S}}\right]\left[y_2(x_3) - r_{\text{a}}^{\text{H}}\right]},\ \boldsymbol{v}_4^y(x) = 0 \tag{5.62}$$

where $\boldsymbol{v}_0^x(x), \boldsymbol{v}_1^x(x), \boldsymbol{v}_2^x(x), \boldsymbol{v}_3^x(x), \boldsymbol{v}_4^x(x)$ is the surface velocity of thick slab shell parallel to casting speed direction, mm; $\boldsymbol{v}_0^y(x), \boldsymbol{v}_1^y(x), \boldsymbol{v}_2^y(x), \boldsymbol{v}_3^y(x), \boldsymbol{v}_4^y(x)$ is the surface velocity of thick slab perpendicular to casting speed direction, mm.

To accurately describe the velocity change inside the thick slab, the velocity equation inside the entire casting flow is expressed as follow:

$$\boldsymbol{v}_0^{x,\text{in}}(x) = \boldsymbol{v}_0^x(x),\ \boldsymbol{v}_1^{x,\text{in}}(x) = \boldsymbol{v}_1^x(x),\ \boldsymbol{v}_2^{x,\text{in}}(x) = \boldsymbol{v}_2^x(x)$$

$$\boldsymbol{v}_3^{x,\text{in}}(x) = \boldsymbol{v}_3^x(x),\ \boldsymbol{v}_4^{x,\text{in}}(x) = \boldsymbol{v}_4^x(x) \tag{5.63}$$

$$\boldsymbol{v}_0^{y,\text{in}}(x) = \boldsymbol{v}_0^y(x),\ \boldsymbol{v}_1^{y,\text{in}}(x) = \boldsymbol{v}_1^y(x),\ \boldsymbol{v}_4^{y,\text{in}}(x) = \boldsymbol{v}_4^y(x), \tag{5.64}$$

$$\boldsymbol{v}_2^{y,\text{in}}(x) = \boldsymbol{v}_2^y(x) + \frac{v_{\text{cast}} r_{\text{a}}^{\text{S}}\left[y_2(x) - y_2^{\text{in}}(x)\right]}{\left[y_1(x_2) - r_{\text{a}}^{\text{S}}\right](x_3 - x_2)}, \tag{5.65}$$

$$v_3^{y,\mathrm{in}}(x) = v_3^y(x) + \frac{v_{\mathrm{cast}}\left[r_a^S y_2(x_3) + r_a^H y_1(x_2) + r_a^S r_a^H\right]\left[y_3(x) - y_3^{\mathrm{in}}(x)\right]}{\left[y_2(x_3) - r_a^H\right]\left[y_1(x_2) - r_a^S\right](x_4 - x_3)} \tag{5.66}$$

where $v_0^{x,\mathrm{in}}(x)$, $v_1^{x,\mathrm{in}}(x)$, $v_2^{x,\mathrm{in}}(x)$, $v_3^{x,\mathrm{in}}(x)$, $v_4^{x,\mathrm{in}}(x)$ is the internal velocity of thick slab shell parallel to casting speed direction, mm; $v_0^{y,\mathrm{in}}(x)$, $v_1^{y,\mathrm{in}}(x)$, $v_2^{y,\mathrm{in}}(x)$, $v_3^{y,\mathrm{in}}(x)$, $v_4^{y,\mathrm{in}}(x)$ is the internal velocity of thick slab perpendicular to casting speed direction, mm; $y_2^{\mathrm{in}}(x)$, $y_3^{\mathrm{in}}(x)$ is boundary morphology of mushy zone inside the slab.

5.2.2 *Model Parameters and Calculation Process*

Similar to the multiphase solidification coupling model of the bloom continuous casting process established in Sect. 5.1.1, a multiphase solidification coupling model of the wide-thick slab is established under the framework of the volume-averaged Euler method. Because the deformation of the shell during the reduction process of the wide-thick slab is more complicated, the main difference between the wide-thick slab model and the bloom model is the different boundary conditions of the shell deformation. As detailed in Sect. 5.2.1, it is requisite to work out a mathematical model of shell deformation and deduce the deformation velocity component based on the actual deformation of the shell during the bulging deformation as well as the solidification end reduction process of the wide-thick slab continuous casting process. Subsequently, this component should be integrated into the momentum transfer equation by utilizing the source term of the fluid domain. In addition, combined with the production of wide-thick slab continuous casting, the effect of electromagnetic stirring is not considered. The calculation process of the model is shown in Fig. 5.35.

The process parameters and material thermophysical parameters used in the simulation process of the multiphase solidification model of the wide-thick slab described in this section are shown in Table 5.4. This paper selects a 280 mm wide-thick slab continuous casting production line as the research object. The casting machine is shown in Fig. 3.11. A two-dimensional symmetrical model with a longitudinal section size of 280 mm × 30430 mm is established using the preprocessing software GAMBIT (Ansys, Inc, Canonsburg, PA). The whole calculation domain is divided into about 3.4 million elements, and the mesh size is 5 mm × 5 mm. The mold length is 800 mm, the casting speed is 0.90 m/min, and the steel grade is low alloy steel Q420M with a carbon content of 0.16. The other simulation parameters are similar to the multiphase solidification coupling model of the bloom continuous casting process established in Section 5.1.1.

For the initial bulging morphology of the inlet of the wide-thick slab calculation Zone 1 in Fig. 5.34, this section combines the specific parameters of the casting machine shown in Fig. 3.11 with the relevant literature [3] to determine the initial slab thickness of the calculation Zone 1. It is 280 mm, the initial bulging height is 0.8 mm, and the roller spacing in the zone is 268 mm. The remaining relevant parameters are shown in Table 5.3. In addition, to further clarify the initial bulging morphology in

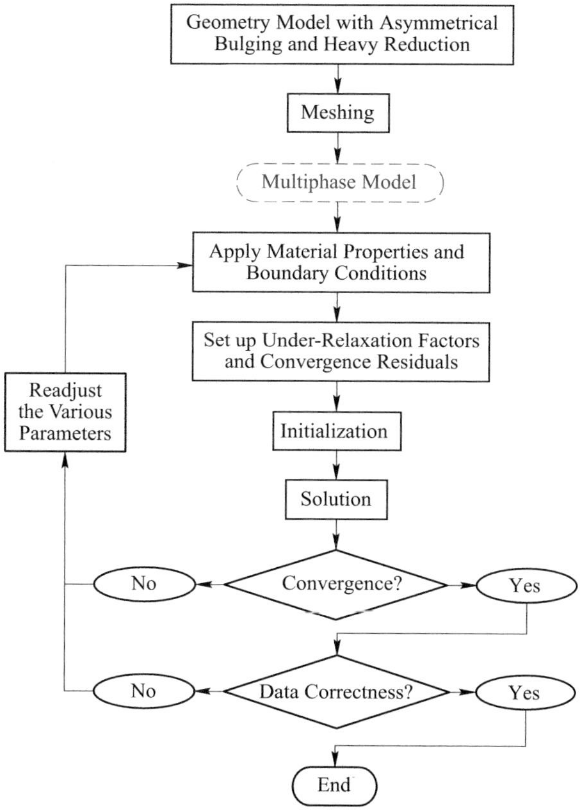

Fig. 5.35 Calculation flow chart of multiphase solidification coupling model of the wide-thick slab under the framework of volume average Euler method

the calculation zone (Zones 2 and 3) of the solidification end reduction in Fig. 5.34, based on the thermal/mechanical model described in Chap. 3, Fig. 5.36 gives the bulging deformation of the shell along the width and thickness direction of the inlet sector section of Zone 2 and Zone 3. Among them, δ_x and δ_z are used to characterize the bulging deformation of the wide surface of the shell along the X-axis direction (that is, perpendicular to the wide surface direction) and the narrow surface of the shell along the Z-axis direction (perpendicular to the narrow surface direction). As depicted in the figure, under the influence of the static pressure of molten steel, the shell undergoes significant bulging deformation along both the thickness direction and the width direction of the billet; among them, due to the inability to obtain the effective support of the casting roller, the bulging deformation (δ_x) of the shell along the thickness direction of the billet is mainly concentrated in the vicinity of the middle position of the roller spacing, while the maximum bulging deformation along the width direction of the billet is located on the narrow surface of the shell.

Figure 5.37 compares the bulging deformation characteristics of the wide surface of the shell along the thickness direction at the entrance position of Zone 2 and Zone 3. From the calculation Zone 2 to Zone 3, the thickness of the solidified shell

Table 5.4 Process parameters and the thermophysical parameters of materials for simulation

Item	Unit	Value
Mold length	x_1 (mm)	800
Casting speed	v_{cast} (m/min)	0.90
Initial thickness of the slab	W (mm)	280
Density (liquid/equiaxed/columnar)	ρ_l, ρ_e, ρ_c (kg/m^3)	7109/7109/7384
Conductivity (liquid/equiaxed/columnar)	k_l, k_e, k_c [W/(m K)]	39/33/33
Specific heat (liquid/solid)	$c_p(l), c_p(s)$ [J/(kg K)]	824.62/660.87
Liquidus slope	M (K_{wt}/p_{ct})	− 78
Latent heat of fusion	ΔHm (J/kg)	272000
Initial carbon concentration	c_0	0.0016
Liquid diffusion coefficient	D_l (m^2/s)	2×10^{-8}
Partition coefficient	k	0.45
Adjustment factor	β	0.8
Maximum nucleation density	n_{max} (m^{-3})	2.224×10^6
Standard deviation	$\Delta T\sigma$ (K)	1.5
Average nucleation undercooling	ΔTN (K)	20

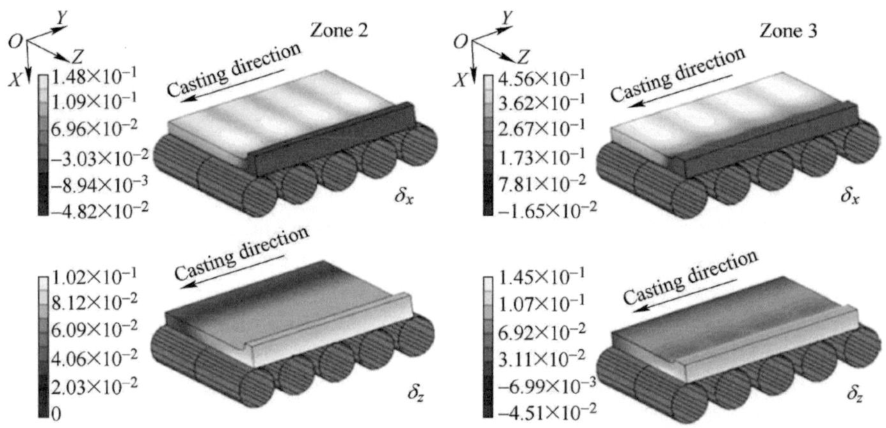

Fig. 5.36 Bulging deformation of the wide-thick slab shell along the width and thickness direction

increases continuously, which indicates that the deformation resistance of the shell is continuously enhanced. However, with the increase of the height difference with the liquid level of the mold, the static pressure of the molten steel and the roll spacing are also increasing. Under the combined action of the two, the bulging deformation of the wide surface shell is increasing, and the increasing speed is more obvious, indicating that the influence of the static pressure of the molten steel and the roll spacing on the bulging deformation is more obvious. In addition, due to the influence of creep

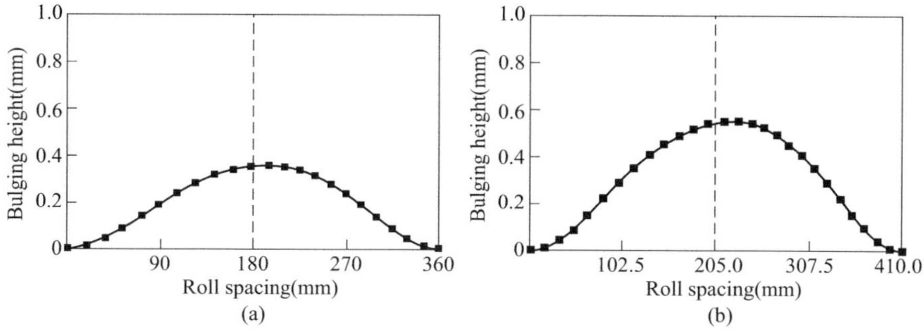

Fig. 5.37 Bulging deformation law of the wide and narrow surface of the shell at the entrance position of calculation Zone 2 (**a**) and Zone 3 (**b**)

deformation, the maximum bulging deformation does not appear in the middle of the two rollers but shifts backward along the casting direction.

According to the calculation results of Fig. 5.37, the detailed calculation parameters (bulging height, number of supporting rollers, roller spacing, etc.) used to describe the asymmetric bulging morphology in the Sect. 5.2.1 shell deformation model can be finally obtained, as shown in Table 5.3.

5.2.3 *Verification of Coupled Calculation Model*

The macrosegregation behavior on the longitudinal section of the slab with a total reduction of 25 mm was investigated through macroscopic low-power corrosion analysis and infrared carbon-sulfur analysis. To analyze the macroscopic low-magnification morphology, a 200 mm long low-magnification sample was taken from the center of the thick slab along the longitudinal section, as shown in Fig. 5.38. Among them, due to the thinning of the slab after reduction, the longitudinal profile height is 255 mm. Similarly, the carbon and sulfur analysis samples were taken on the low-magnification samples, and the position is shown in Fig. 5.38b.

Figure 5.39 compares the numerical simulation results with the low-magnification photos of the longitudinal section of the solidified end of the thick slab obtained by corrosion. The results show that the calculated results of the central macrosegregation of the continuous casting billet in Fig. 5.39a are very similar to the low-magnification corrosion morphology of the macrosegregation of the thick slab in Fig. 5.39b, indicating that the simulation results are in good agreement with the experimental results. Relying on the multiphase solidification model, the extent of carbon segregation in the transverse direction within the longitudinal section, which is located 29.2 m away from the meniscus, was computed, as depicted in Fig. 5.40. According to the position shown in Fig. 5.38, the drill cuttings with a diameter of 5mm were used and the elemental segregation was measured by an infrared carbon-sulfur analyzer. The

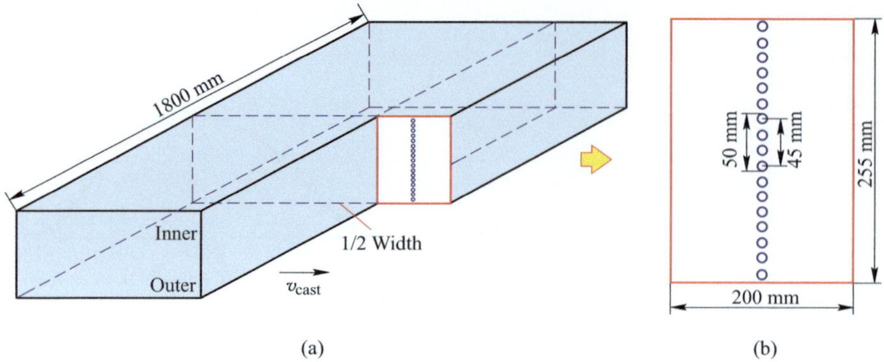

Fig. 5.38 Thick slab longitudinal section of different sampling positions. **a** Schematic diagram and the specific size of different samples (**b**)

calculated value of the carbon segregation index and the measured value are shown in Fig. 5.40.

The experimental results show that the calculation results of the asymmetric bulging morphologies are more in line with the actual measurement values than the sine curve bulging morphologies. The experimental data are consistent with the theoretical calculation, indicating that the theoretical model and numerical simulation are correct and feasible. Since the electromagnetic stirring model is not considered in this study, no negative segregation band ("white band") is found at the 1/4 thickness of the thick slab shown in Fig. 5.39a, b.

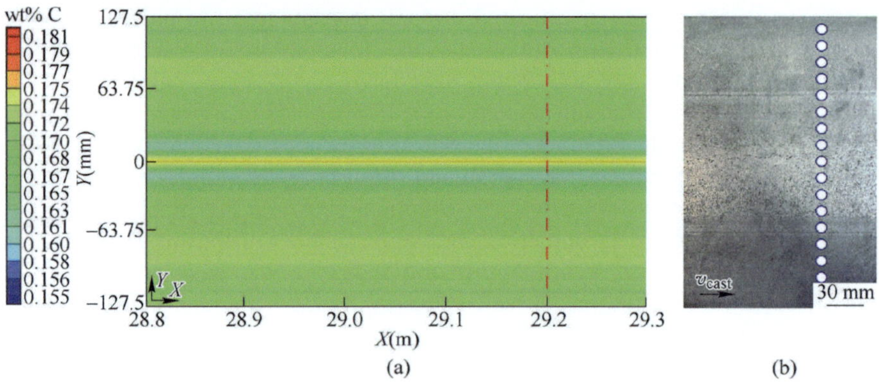

Fig. 5.39 Macrosegregation calculation results of the longitudinal section at the solidification end of the thick slab (**a**) are compared with the low-magnification corrosion morphology (**b**)

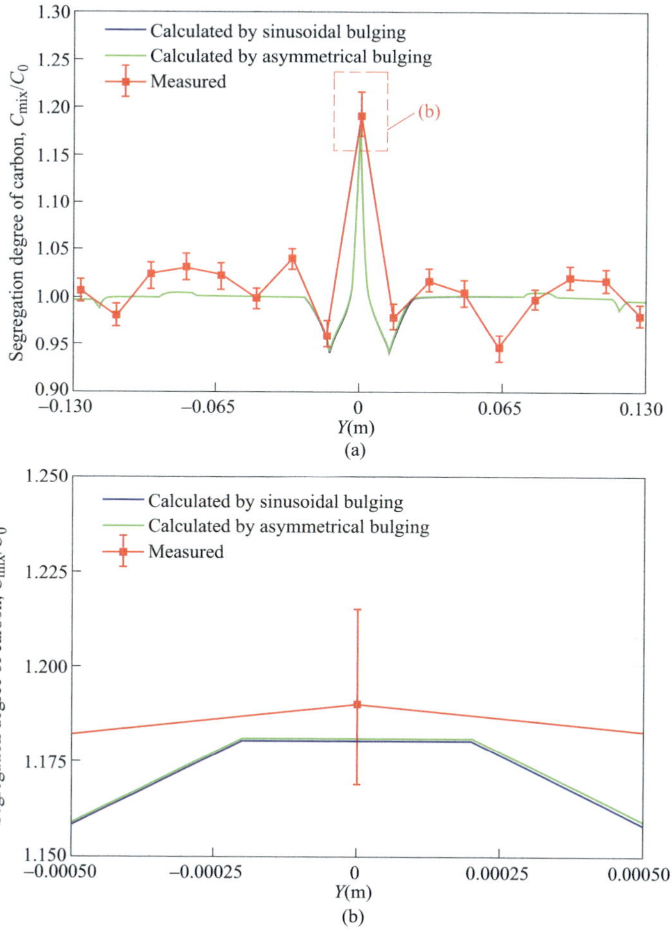

Fig. 5.40 Measured and calculated results of carbon segregation degree (**a**) and the local magnification of segregation degree at the center position (**b**) under different bulging morphologies

5.2.4 *Solute Transport Behavior Law of Wide-Thick Slab*

5.2.4.1 Simulation Scheme

In the continuous casting process of the wide-thick slab, bulging deformation and mechanical reduction will significantly affect the formation of macrosegregation. Therefore, based on the multiphase coupling model of wide-thick slab solidification, two simulation experiments were designed. The first experiment primarily simulated the bulging deformation prior to the reduction interval (15.2–15.8 m from the meniscus). In particular, the influence of sinusoidal bulging and non-sinusoidal bulging on the solute segregation of the slab was compared and examined. The second

Table 5.5 Comparison of simulation parameters under different reductions in the second experiment

Item	Reduction amount (mm)		Total reduction amount before solidification end in Segment 9 and Segment 10, ra (mm)	Total reduction amount for the whole continuous casting strand (mm)	Reduction process
	Segment 9, raS	Segment 10, raH			
Case 1	0	0	0	0	No reduction
Case 2	3	0	3	9	Soft reduction
Case 3	3	6	9	15	Heavy reduction
Case 4	3	12	15	25	Heavy reduction

experiment mainly simulates the solute segregation behavior in the reduction interval (the 9th and the 10th fan-shaped segment, 20.5–24.5 m from the meniscus) without reduction and under the conditions of 3, 9, and 15 mm reduction. The specific four cases are shown in Table 5.5.

5.2.4.2 Effect of Different Bulging Morphologies on Macrosegregation Behavior of Thick Slab (The First Experiment)

To understand the influence of different shell bulging morphologies on the velocity field distribution of continuous casting billet, the simulation of solute transport behavior under sinusoidal bulging and non-sinusoidal bulging was carried out. The results are shown in Fig. 5.41.

The simulation results show that when the bulging morphology between adjacent support rollers changes, the velocity vector also changes. For the sine curve bulging, as shown in Fig. 5.41a, the velocity field on the longitudinal section at 15.20–15.80 m from the meniscus shows the same trend of sinusoidal change. For the asymmetric bulging morphology, as shown in Fig. 5.41b, the velocity field change trend is similar to the former, showing a non-sinusoidal curve change trend. Regarding the two different bulging morphologies, the simulation results demonstrate that the speeds in the first half and the second half of the adjacent support roller possess different trends. However, in the simulation of asymmetric bulging under actual working conditions, the trend of velocity distribution is generally inclined toward the downstream support roller.

Figure 5.42 shows the velocity and volume fraction of the central liquid phase on the longitudinal section of the slab under different bulging morphologies. It can be seen that there is a significant difference in the liquid-phase velocity curve between the sine curve bulging morphology and the asymmetric bulging morphology.

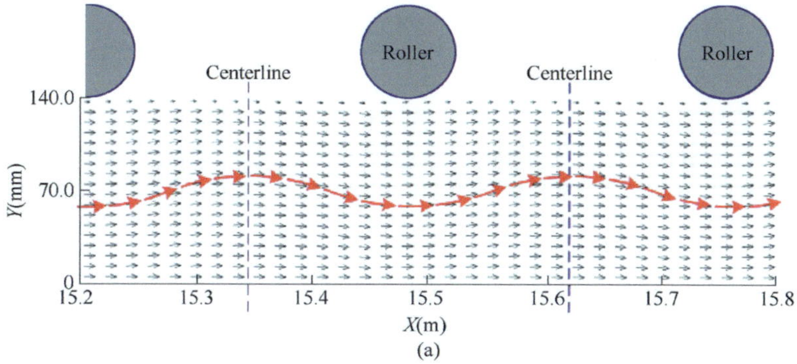

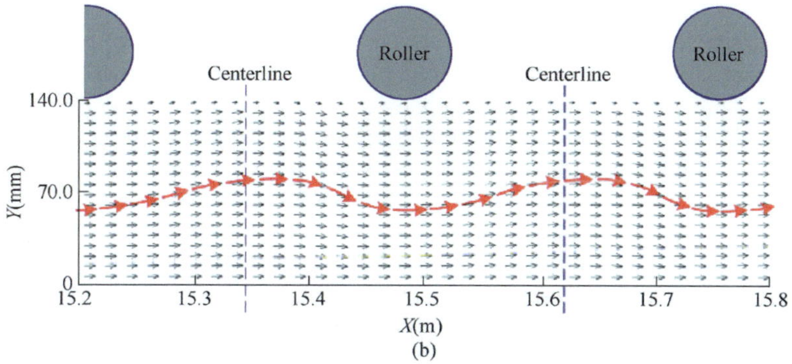

Fig. 5.41 Longitudinal profile velocity field under different bulging morphologies. **a** Sine curve bulging; **b** asymmetric bulging

As the volume fraction of the liquid phase decreases from 0.24 to 0.19, the maximum velocity of the liquid phase at the centerline of the gap between the two adjacent support rollers under the action of sine curve bulging decreases from 936.19×10^{-3} m/ min to 936.11×10^{-3} m/ min, which is greater than the casting speed of 0.90 m / min. In addition, considering the influence of asymmetric bulging morphology, the melt velocity curve shows a non-sinusoidal change trend, and the position of the maximum velocity of the liquid phase is horizontally offset from the centerline of the support roller by about 0.03 m. When the volume fraction of the liquid phase decreases from 0.24 to 0.19, the maximum liquid-phase velocity under the action of asymmetric bulging morphology decreases from 936.24×10^{-3} m/min to 936.14×10^{-3} m/min, which is still greater than the casting speed (0.90 m/min). However, whether the bulging morphology is a conventional sine curve morphology or an asymmetric bulging morphology, as the main driving force of the liquid-phase velocity change, the velocity of the liquid phase in the center of the billet is always greater than the casting speed, and the liquid phase continuously flows to the solidification end. In contrast, the maximum liquid-phase velocity in the center of the slab increases markedly under asymmetric bulging. The reason for this outcome is that when the maximum deflection point of the asymmetric bulging morphology moves backward from the center of the adjacent support roller, the suction effect of the bulging is intensified, thereby boosting the liquid-phase velocity at the center of the slab.

Figure 5.43 shows the segregation degree of carbon elements on the centerline of the billet under different bulging morphologies. It can be seen that the variation trend of the segregation degree of the billet is shifted obviously to the next support roller under the condition of asymmetric bulging morphology. The carbon segregation degree increased from 99.77×10^{-2} and 99.85×10^{-2} to 99.95×10^{-2} and 100.01×10^{-2} under sine curve bulging and asymmetric bulging conditions, respectively. By combining the variation law of segregation degree in Fig. 5.43 with the results of melt flow velocity shown in Fig. 5.42, it can be concluded that the suction effect caused by asymmetric bulging morphology is significantly greater than that caused by sinusoidal bulging. Under the action of asymmetric bulging morphology, the enriched solute liquid phase between columnar dendrites is sucked out of the original position, transmitting to the subsequent support roller at a higher flow rate to form a new accumulation zone, increasing the degree of carbon segregation. The degree of segregation gradually changes from negative segregation to positive segregation, indicating that the bulging of the shell is one of the reasons for the positive segregation. Considering the asymmetric bulging morphology, the degree of central segregation of the billet will be further deteriorated.

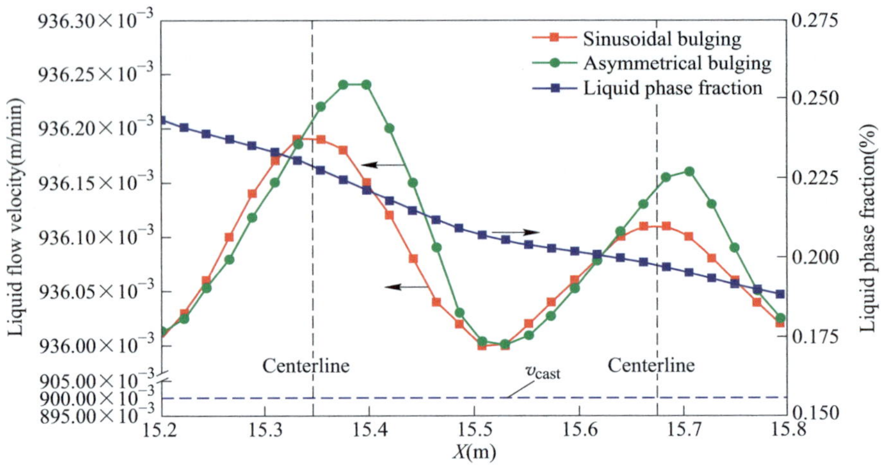

Fig. 5.42 Velocity and volume fraction of the liquid phase on the centerline of the slab under different bulging morphologies

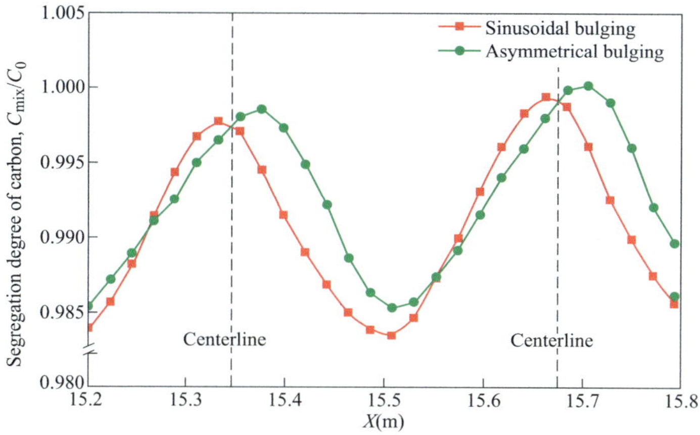

Fig. 5.43 Segregation degree of carbon element on the centerline of the billet under different bulging morphologies

5.2.4.3 Effect of Different Solidification End Reduction Processes on Macrosegregation Behavior of Thick Slab (The Second Experiment)

Figures 5.44, 5.45, 5.46, and 5.47 show the solidification behavior of the longitudinal section of the billet under the conditions of no reduction and a reduction of 3, 9, and 15 mm in the reduction range (the 9th and the 10th fan-shaped segment, 20.5–24.5 m from the meniscus), mainly including the liquid-phase velocity vector and carbon segregation distribution. The reduction of the 9th segment is r_a^S, and the central solid fraction of the billet (f_s) is 0.75-0.84. The reduction of the 10th segment is r_a^H, and the central solid fraction (f_s) is 0.90–0.99.

When no reduction is applied (Case 1, $r_a^S = 0$ mm and $r_a^H = 0$ mm), as shown in Fig. 5.44a, due to the influence of shell suction at the end of solidification, the molten steel in the upstream of the casting flow continuously flows into the 9th segment. Due to the low degree of segregation in the upstream molten steel, no obvious solute segregation defects were observed in the center of the 9th segment (Fig. 5.44b). During the first half of the 10th segment, the heat shrinkage and solidification shrinkage will cause the movement of the dendrite tip toward the surface of the shell, and a local negative pressure zone will be formed in the center of the billet. It can be seen in Fig. 5.44a that the liquid phase in the center of the billet accelerates under the action of suction, which is the main reason for the formation of local negative segregation defects under the mechanism of "Solute flush" (Fig. 5.44b). In the second half of the 10th segment, due to continuous shrinkage of the billet, the severe suction caused by the continuous shrinkage of the slab will cause the solute in the center of the billet to converge before the solidification end point and finally form a positive segregation defect (Fig. 5.44b). In summary, the distribution of carbon segregation along

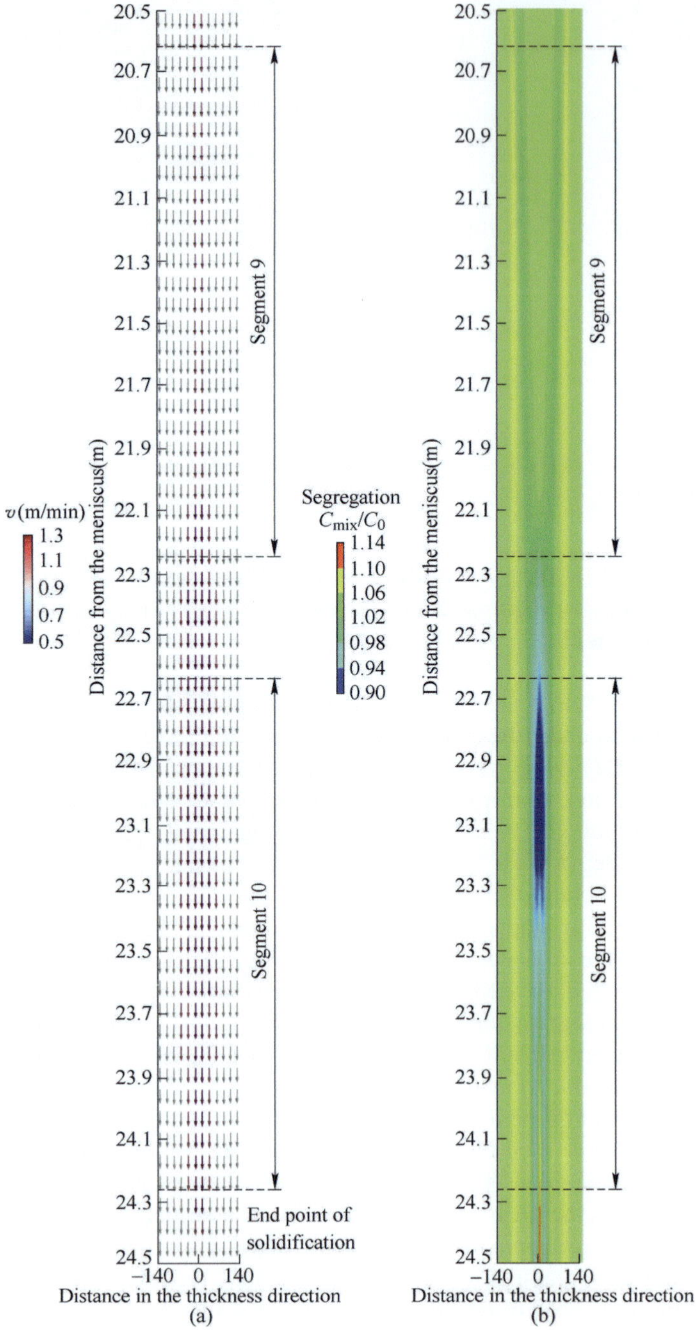

Fig. 5.44 Case 1: solidification behavior of longitudinal section of slab without reduction. **a** Liquid velocity vector; **b** carbon segregation degree

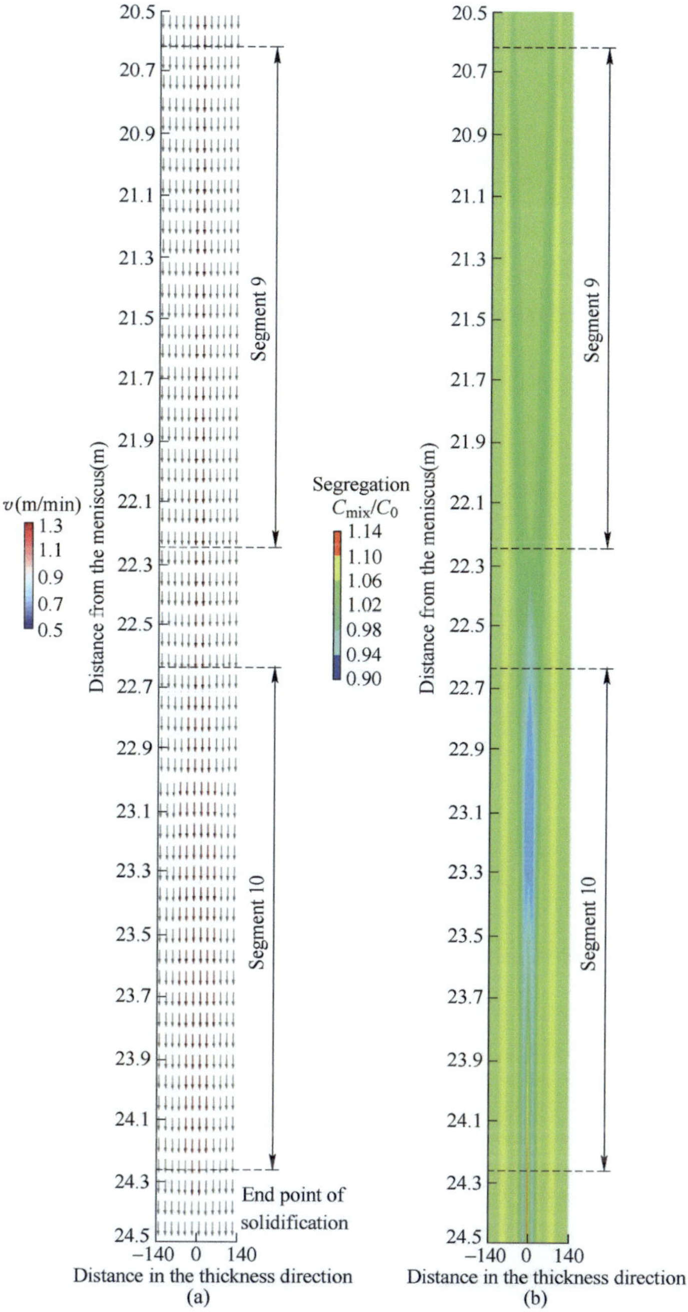

Fig. 5.45 Case 2: the solidification behavior of the longitudinal section of the slab during soft reduction. **a** Liquid velocity vector; **b** carbon segregation degree

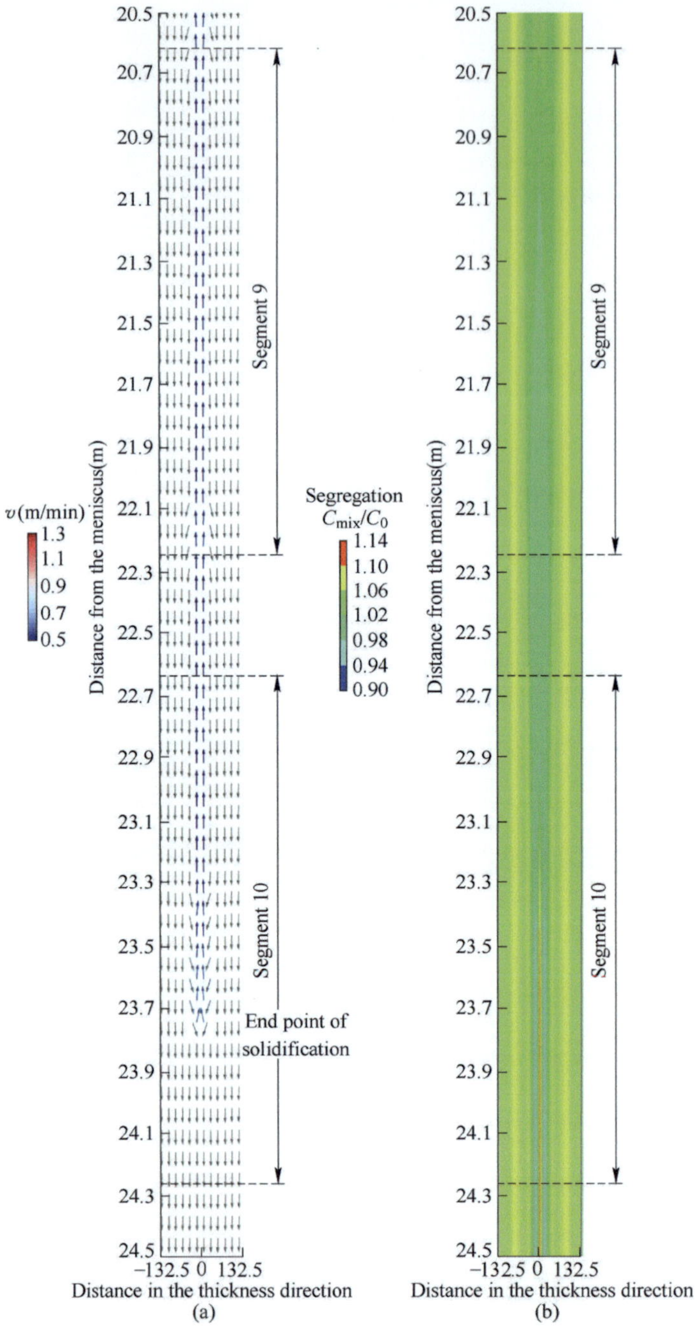

Fig. 5.46 Case 3: the solidification behavior of the longitudinal section of the slab during heavy reduction. **a** Liquid velocity vector; **b** carbon segregation degree

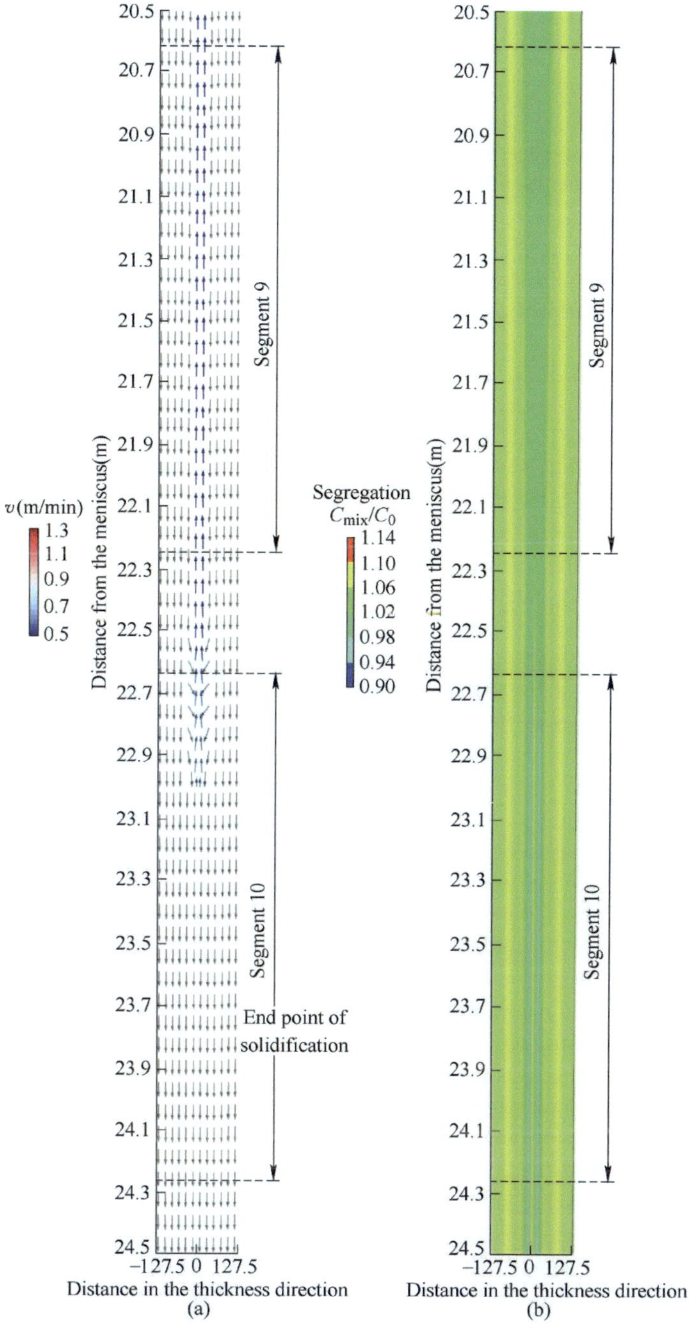

Fig. 5.47 Case 4: the solidification behavior of the longitudinal section of the slab during heavy reduction. **a** Liquid velocity vector; **b** carbon segregation degree

the centerline of the billet within the 10th segment exhibits a tendency of negative segregation first, followed by positive segregation.

When applying soft reduction (Case 2, $r_{\mathrm{a}}^{\mathrm{S}} = 3$ mm and $r_{\mathrm{a}}^{\mathrm{H}} = 0$ mm), as shown in Fig. 5.45a, the liquid-phase flow velocity in the central region of the 9th and 10th segments of the billet decreases as a whole, especially in the first half of the 10th segment. The decrease of the liquid-phase flow velocity in the center of the billet indicates that the soft reduction of the 9th segment can better suppress the liquid-phase flow in the center of the billet. In the 10th segment, the liquid-phase velocity of the centerline of the billet will still increase because the deformation (3 mm) is not enough to compensate for the shrinkage of the central area of the billet. However, the liquid velocity decreases and the increment is insufficient in the 9th segment caused by soft reduction, so the increase of liquid velocity under the action of shell shrinkage in the 10th segment is smaller than that without soft reduction. At the same time, the "solute flush" and solute aggregation in the center of the billet will also be weakened, so the solute enrichment in the 10th segment will also be weakened, but positive segregation defects will still be formed before the solidification end point (Fig. 5.45b). In summary, compared with no reduction, soft reduction has a certain effect on improving the macrosegregation in the central region.

For two kinds of heavy reduction process (Case 3, $r_{\mathrm{a}}^{\mathrm{S}} = 3$ mm and $r_{\mathrm{a}}^{\mathrm{H}} = 6$ mm; Case 4, $r_{\mathrm{a}}^{\mathrm{S}} = 3$ mm and $r_{\mathrm{a}}^{\mathrm{H}} = 12$ mm), as shown in Figs. 5.46a and 5.47a, although the reduction of the 9th segment is the same as that of the soft reduction, the heavy reduction of the 10th segment promotes the obvious reverse flow of the liquid phase in the center of the billet and causes the decrease of the liquid-phase velocity in the center of the billet in the 9th segment. Simultaneously, the significant reduction accelerates the thinning of the shell, causing the solidification process to end prematurely. With the increase of the reduction (the shell thinning increases), the solidification end point moves to the front of the meniscus (Figs. 5.46b and 5.47b). In addition, the shrinkage of the solidified shell can be completely compensated under heavy reduction, the suction effect in the center of the billet disappears, and the strong shell deformation will promote the reverse flow of the liquid phase in the center of the billet, so that the enriched solute liquid phase is squeezed out before the solidification end point, and the solute enrichment in the center of the billet is effectively inhibited. However, since the enriched solute liquid phase cannot be completely discharged, a slight central macrosegregation will be formed, as shown in Figs. 5.46b and 5.47b. It can be seen from the comparison of the two heavy reduction processes that with the increase of the reduction amount, the extrusion effect of the enriched solute liquid phase before the solidification end point will be strengthened, and the improvement effect on the central segregation will be more obvious.

5.3 Research on the Evolution Law of Shrinkage Cavity Segregation During the Hot Charging Process

In the continuous casting process, solute redistribution will lead to the uneven composition of the billet and solute enrichment between dendrites, resulting in microsegregation [49]. Severe microsegregation will form banded structures along the rolling direction during the rolling process, thus destroying the mechanical properties of the steel [50, 51]. At the same time, under the action of solidification shrinkage and dendrite bridging, the center of the billet is easy to forms serious defects such as shrinkage porosity, and the shrinkage position is often accompanied by solute segregation defects [52]. However, most of the existing high-temperature diffusion studies on the improvement of microsegregation focus on the high-temperature diffusion of inter-dendritic segregation elements [53, 54], and there is little research on the improvement process of microsegregation in the shrinkage cavity.

5.3.1 Microscopic Shrinkage Cavity Morphology Characteristics of Heavy-Haul Railway's Rail Steel Billet

Macro/micro shrinkage cavities are easily formed in the center of the billet under the effect of solidification shrinkage at the continuous casting solidification end. The non-destructive layered scanning of slab samples was carried out by an ultrasonic wave, and a three-dimensional model was established based on the scanning results, to obtain the true morphology and distribution of the micro shrinkage cavity in the center of the billet. As shown in Fig. 5.48, most of the micro-shrinkage cavities are ellipsoidal in the heavy-haul railway's rail steel billet, the length of the long axis is not more than 1mm, and a small part is spherical and irregular shape.

To systematically study the influence of shrinkage cavity size, soaking temperature, and soaking time on the microsegregation diffusion inside the shrinkage cavity, the best experimental method is to design an orthogonal experiment. In this study, the shrinkage cavity size can be categorized into three distinct types, and the soaking temperature and soaking time are also selected as three types. Therefore, the orthogonal experiment can choose the orthogonal table of L9 (34), where 9 is the number of experiments, 3 is the number of experimental levels, 4 indicates that the experimental table has 4 columns, and the orthogonal experimental factor level table is shown in Table 5.6.

The orthogonal experiment table is shown in Table 5.7. In the experimental scheme, $R_1B_1C_1$ indicates that the defect size is less than 100 μm, the soaking temperature is 1150 °C, and the soaking time is 60 min.

The electron probe JXA8530 F produced by Japan Electronics Co., Ltd., was used as the experimental instrument. Before the high-temperature diffusion process, the electron probe surface was scanned at the shrinkage cavity of the sample to

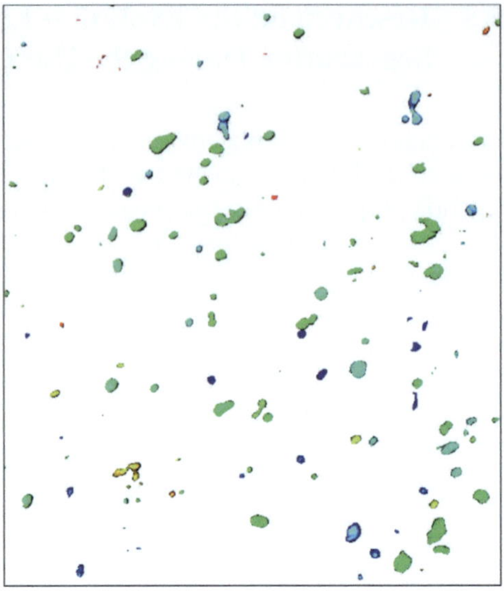

Fig. 5.48 Three-dimensional modeling diagram of micro shrinkage cavity in the center of heavy-haul railway's rail steel billet

Table 5.6 Factor level table

Level	Shrinkage cavity size, R (μm)	Holding temperature, B (°C)	Holding time, C (min)
1	$R \leq 100$	1150	60
2	$100 < R \leq 400$	1200	90
3	$400 < R$	1250	120

Table 5.7 Orthogonal experiment table

No.	Factors			Experiments
	Shrinkage cavity size, *R* (μm)	Holding temperature, *B* (°C)	Holding time, *C* (min)	
1	$R \leq 100$	1150	60	R1B1C1
2	$R \leq 100$	1200	90	R1B2C2
3	$R \leq 100$	1250	120	R1B3C3
4	$100 < R \leq 400$	1200	120	R2B2C3
5	$100 < R \leq 400$	1250	60	R2B3C1
6	$100 < R \leq 400$	1150	90	R2B1C2
7	$400 < R$	1250	90	R3B3C2
8	$400 < R$	1150	120	R3B1C3
9	$400 < R$	1200	60	R3B2C1

observe the shrinkage cavity and the distribution of carbon elements around it. At the same time, the scanning position before the high-temperature diffusion process was recorded. To prevent the oxidation of the sample surface during the soaking process and affect the experimental results, a vacuum glass tube was used to encapsulate the sample. Following the homogenization of the sample in line with the pre-designed experimental scheme, the sample was lightly thrown, and the position before the high-temperature diffusion process was found. The electron probe surface scanning analysis was performed again to observe the changes in the segregation of carbon elements inside and around the shrinkage cavity after homogenizing heat treatment.

5.3.2 *Effect of Soaking Process on Microsegregation at Shrinkage Cavity*

In the previous test results, the electron probe test results can only record the carbon content at different positions on the entire scanning surface, and cannot quantitatively describe the carbon content at different positions on the scanning surface. To quantify the test results of EPMA and facilitate the calculation of orthogonal test data, the concept of Change of Microsegregation Area (CMA) is introduced here. The equation is as follows:

$$\text{CMA} = \frac{|S_2 - S_1|}{S_1} \times \frac{1}{100 - S_1} \times 100\% \tag{5.67}$$

where S_1 is the area ratio of microsegregation before soaking; S_2 is the area ratio of microsegregation after soaking.

Figures 5.49, 5.50, and 5.51 show the distribution of carbon elements Inside and around the microscopic shrinkage cavity of the billet. Obviously, there exists significant carbon segregation both within and in the vicinity of the micro shrinkage cavity. It can be seen from the figure that the microsegregation inside and around the shrinkage cavity has been significantly improved after the soaking process at different temperatures and times. The maximum value of the characteristic X-ray count intensity of EPMA is adjusted to be consistent (adjusted to a maximum of 370), which is convenient to compare the changes of microsegregation of carbon before and after the soaking process of each sample.

Figure 5.52 shows the area ratio of microsegregation of different samples after heat treatment under different conditions from Figs. 5.49, 5.50, and 5.51. It can be seen that the area of microsegregation of carbon elements inside and around the shrinkage cavity decreases after soaking, and the matrix area increases significantly. The dotted line in Fig. 5.52 is the average microsegregation area in the set shrinkage cavity size range. When the shrinkage cavity size is less than 100 μm, the average area of microsegregation is 7.87%. When the micro shrinkage size is greater than 100 μm, the average area of microsegregation is greater than 10%. It is evident that

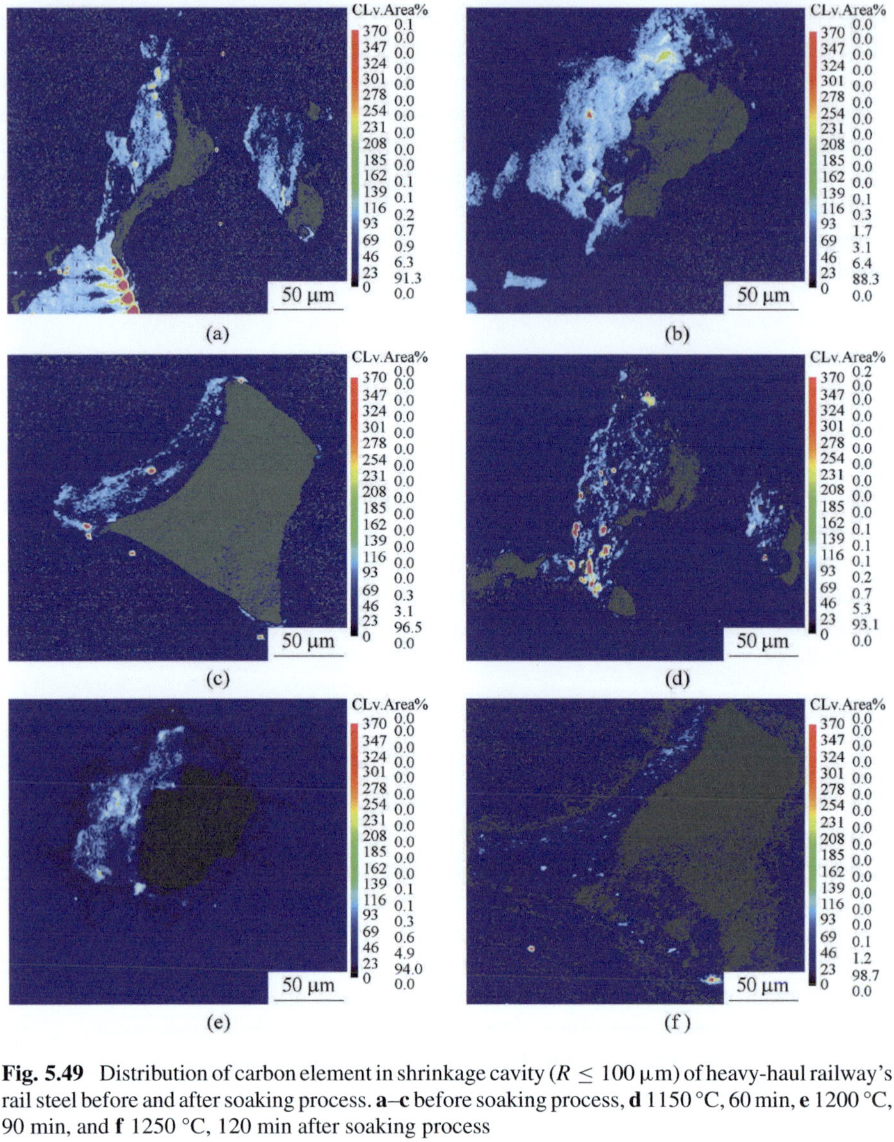

Fig. 5.49 Distribution of carbon element in shrinkage cavity ($R \leq 100$ μm) of heavy-haul railway's rail steel before and after soaking process. **a–c** before soaking process, **d** 1150 °C, 60 min, **e** 1200 °C, 90 min, and **f** 1250 °C, 120 min after soaking process

when the shrinkage cavity size is large, the carbon element segregation inside and around it is large.

The results in Table 5.8 are calculated by Eq. (5.67). The calculation results show that the change of microsegregation area ratio is the most when the soaking temperature is 1250 °C in the same shrinkage cavity size range. The effect of soaking temperature on the experimental results is the most significant, while the effect of

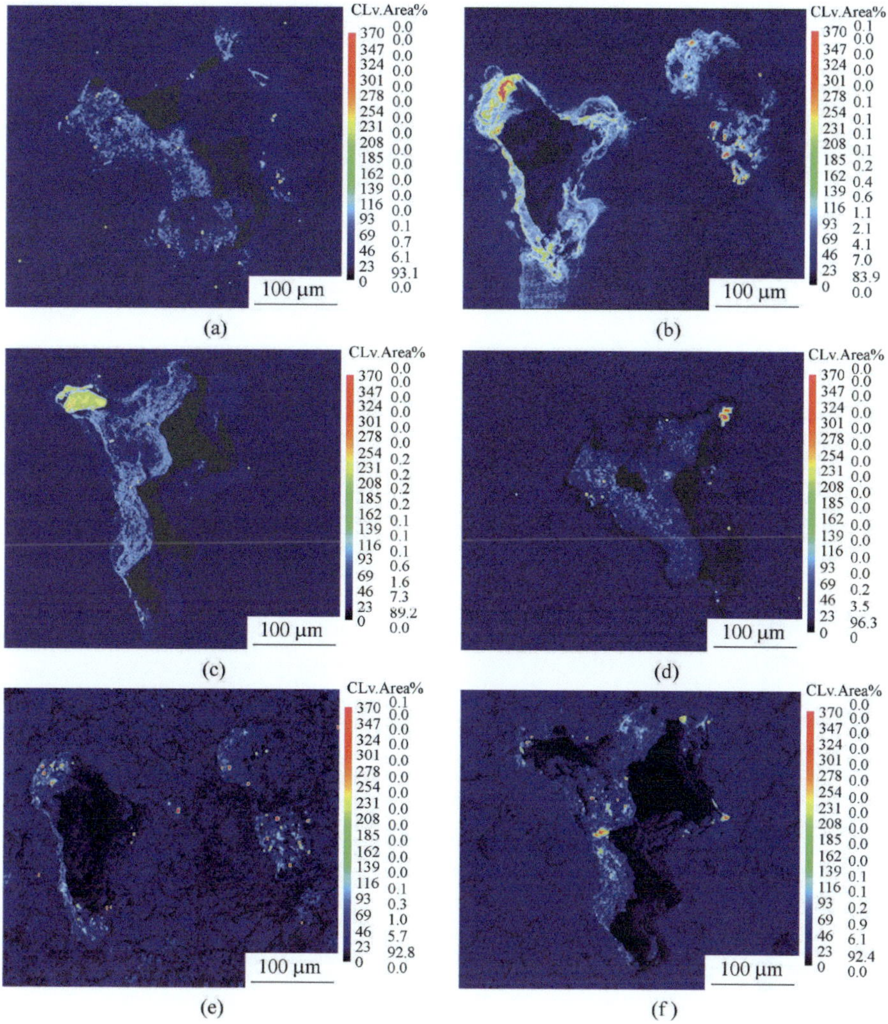

Fig. 5.50 Distribution of carbon element in shrinkage cavity ($100\,\mu m < R \leq 400\,\mu m$) of heavy-haul railway's rail steel before and after soaking process. **a–c** Before soaking process, **d** 1150 °C, 60 min, **e** 1200 °C, 90 min, and **f** 1250 °C, 120 min after soaking process

shrinkage cavity size and soaking time on the improvement of microsegregation is not as obvious as that of soaking temperature.

The results of range analysis and variance analysis are shown in Tables 5.9 and 5.10. The range analysis and variance analysis of the orthogonal experiment can determine the strength of each influencing factor on the experimental results. The range analysis value E is the largest, indicating that this factor is the most important factor affecting the test results. From Table 5.8, it is observable that the maximum value of the soaking temperature E is 0.42, indicating that the soaking temperature

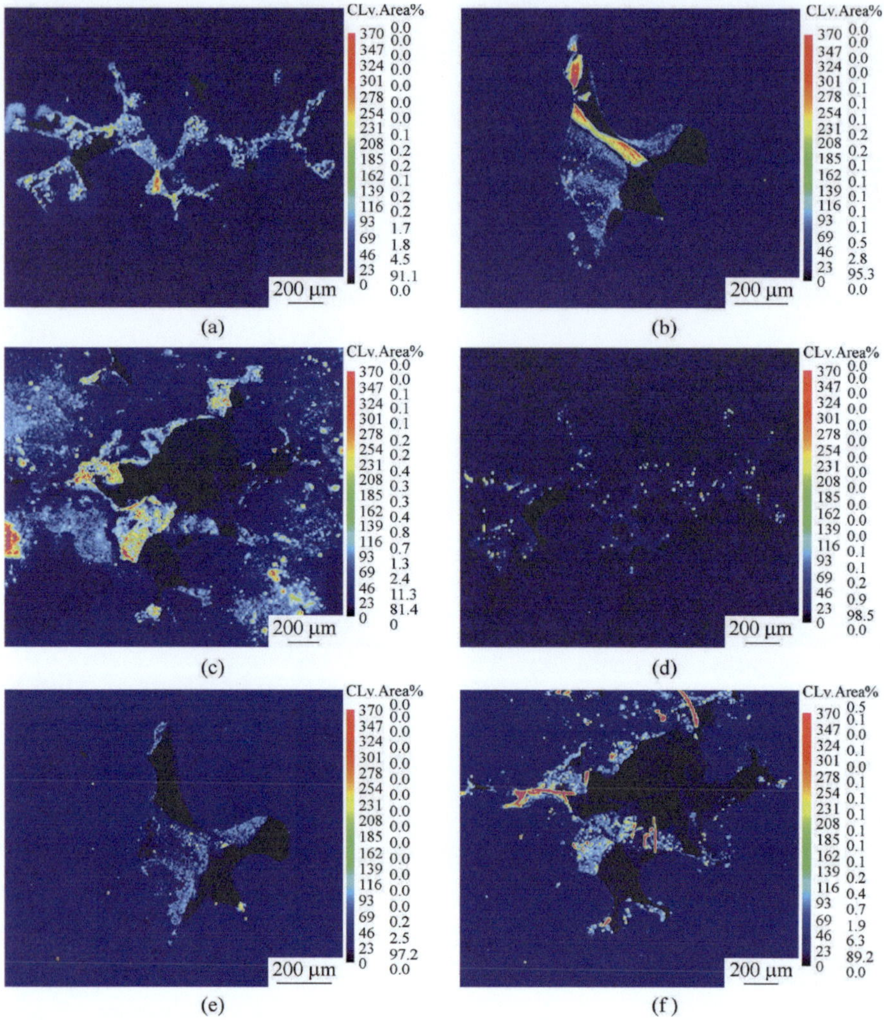

Fig. 5.51 Distribution of carbon element in shrinkage cavity ($R > 400\ \mu m$) of heavy-haul railway's rail steel before and after soaking process. **a–c** Before soaking process, **d** 1150 °C, 60 min, **e** 1200 °C, 90 min, and **f** 1250 °C, 120 min after soaking process

has the greatest influence on the improvement of segregation. The range E value of the microscopic shrinkage cavity size and the soaking time is not much different. It can be seen that the shrinkage cavity size and the soaking time have the same high-temperature diffusion effect on the carbon element. According to the characteristics of the orthogonal test analysis of variance, the test results of Table 5.10 were tested by the F test. The F value of the soaking temperature in the diffusion process of carbon element is the largest ($F_{0.01}(2, 2) = 99 > 83.19 > F_{0.05}(2, 2) = 19$), which shows that the correlation between the soaking temperature and the high-temperature

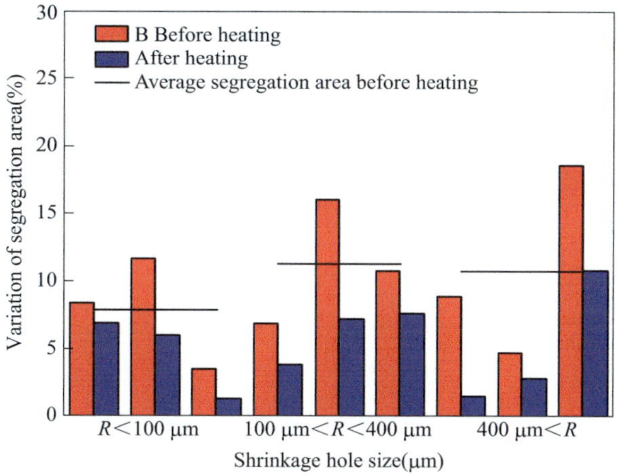

Fig. 5.52 Variation of segregation area in soaking process

Table 5.8 Change of microsegregation area ratio

Factor			Change of microsegregation area ratio
Shrinkage hole size (μm)	Soaking temperature (°C)	Soaking time (min)	
$R \leq 100$	1150	60	0.1949
$R \leq 100$	1200	90	0.5517
$R \leq 100$	1250	120	0.6514
$100 < R \leq 400$	1200	120	0.4826
$100 < R \leq 400$	1250	60	0.6589
$100 < R \leq 400$	1150	90	0.3322
$400 < R$	1250	90	0.9127
$400 < R$	1150	120	0.4242
$400 < R$	1200	60	0.5152

diffusion effect of carbon element is more than 95%. The F value of the soaking time is the smallest ($F_{0.05}(2, 2) = 19 > 9.55 > F_{0.1}(2, 2) = 9$), but its correlation also reaches more than 90%, which still has a significant effect on the high-temperature diffusion of carbon. The results of the F test show that the three factors of soaking temperature, soaking time, and shrinkage cavity size can affect the improvement of microsegregation. Among them, the soaking temperature has the greatest influence on the improvement of microsegregation, and the influence of shrinkage cavity size and soaking time on microsegregation is less than that of soaking temperature. According to the range analysis and variance analysis of carbon elements in Tables 5.9 and 5.10, it is considered that the optimal soaking time is 90 min when the soaking temperature is 1250 °C, and the experimental results in Table 5.7 are consistent.

Table 5.9 Range analysis table

Item	Shrinkage cavity size (μm)	Blank column	Holding temperature (°C)	Holding time (min)
k_1	0.47	0.53	0.32	0.46
k_2	0.49	0.54	0.52	0.60
k_3	0.62	0.50	0.74	0.52
E	0.15	0.05	0.42	0.14

Table 5.10 Variance analysis table

Factor	SS	df	MS	F	Critical value
Shrinkage cavity size (μm)	0.04	2.00	0.02	12.30	$F0.05(2, 2) = 19$
Holding temperature (°C)	0.27	2.00	0.13	83.19	$F0.1(2, 2) = 9$
Holding time (h)	0.03	2.00	0.02	9.55	$F0.01(2, 2) = 99$
Error	0.00	2.00	0.00	–	–

5.3.3 *Microsegregation Diffusion Kinetics at Shrinkage Cavity*

The research shows that the distribution of microsegregation between dendrite arms conforms to the cosine law and has a strong periodic law. However, the distribution of elements not only has a periodic law, but also has the characteristics of inconsistent element content between adjacent dendrite arms. This feature will cause the cosine model to not necessarily conform to the distribution of microsegregation between dendrite arms. For the purpose of resolving the disparity in element content among different dendrite arms, Zhang et al. [55] established a Gaussian diffusion model for the distribution of solute elements between dendrite arms. The Gaussian model solute distribution is shown in Fig. 5.53. The Gaussian model has the following characteristics: (1) the elements between dendrite arms are periodic; (2) The element content between different dendrite arms is different.

The microsegregation Gaussian diffusion model is shown in Eq. (5.68). For the diffusion of alloying elements between dendrite arms, the researchers only considered the diffusion process of a single dendrite in a single direction toward another dendrite. This diffusion direction has only one dimension and does not consider the diffusion of elements in other directions. However, the microsegregation diffusion inside the shrinkage cavity is not only unidirectional diffusion in one-dimensional space but multidirectional diffusion in three-dimensional space, which will make the Gaussian diffusion model unsuitable for the microsegregation diffusion behavior inside the shrinkage cavity.

$$C(x, t) = \frac{S}{\sqrt{4\pi Dt}} \sum_{i=0}^{n} \exp \frac{-(x \pm il)^2}{4Dt} \tag{5.68}$$

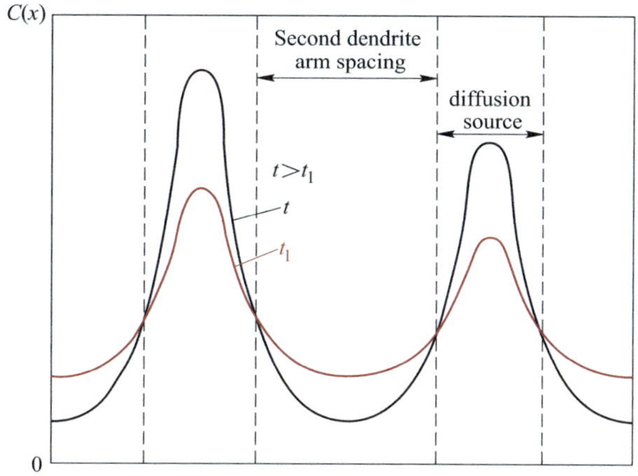

Fig. 5.53 Elements distribution of the Gaussian diffusion curve

Therefore, it is necessary to improve the Gaussian diffusion model between dendrite arms and consider the diffusion behavior in multiple directions. Therefore, the microsegregation diffusion at the shrinkage cavity should be assumed: (1) the diffusion source at the whole shrinkage cavity is regarded as a diffusion point; (2) The microsegregation reaches the maximum value at the center of the shrinkage cavity. Therefore, the diffusion equation of microsegregation at the shrinkage cavity can be expressed as:

$$C(x_1, x_2, t) = \frac{S^2}{\sqrt{4\pi Dt}} \sum_{i=0}^{n} \exp \frac{-(x_1 \pm il)^2 - (x_2 \pm il)^2}{4Dt} \tag{5.69}$$

where l is the shrinkage cavity size, its length is equal to the size of the area of the circle; t is the soaking time, s; S is the total amount of atoms at each point; D is A diffusion coefficient related to temperature.

D can be expressed as by Arrhenius equation diffusion coefficient:

$$D = D_0 \exp \frac{-Q}{R(T + 273)} \tag{5.70}$$

where D_0 is the element diffusion constant; Q is the diffusion activation energy; R is the gas diffusion constant is 8.314 J/(mol K); T is the diffusion temperature.

The diffusion coefficient of carbon element [56] verified by literature is $D_0 = 0.11\ \text{cm}^2/\text{s}$, $Q = 129.09\ \text{kJ/mol}$.

Obviously, when $t = 0$, the initial element distribution at the shrinkage cavity is not consistent with the Eq. (5.68). There is a value of t such that the equation satisfies the initial element distribution law at the shrinkage cavity. The initial element distribution can be expressed as:

$$C(x_1, x_2, t) = \frac{S}{\sqrt{4\pi Dt}} \sum_{i=0}^{n} \exp \frac{-(x_1 \pm il)^2 - (x_2 \pm il)^2}{4Dt} \tag{5.71}$$

After a period of diffusion time (t'), Eq. (5.71) can be expressed as:

$$C(x_1, x_2, t) = \frac{S}{\sqrt{4\pi Dt}} \sum_{i=0}^{n} \exp \frac{-(x_1 \pm il)^2 - (x_2 \pm il)^2}{4D(t' + t)} \tag{5.72}$$

At t time, the microsegregation distribution at the shrinkage cavity is shown in Fig. 5.54a, and the microsegregation is more serious at this time. After the high-temperature diffusion of t′ time, the improvement effect of microsegregation at the shrinkage cavity is shown in Fig. 5.54 (b). It can be seen that the microsegregation ratio SR (SR $= C_{\max}/C_{\min}$) is greatly reduced.

According to the characteristics of the Gaussian diffusion model, the value of $C_{\max}$ is obtained when $x_1 = x_2 = 0$, and the minimum value is obtained at $x_1 = x_2 = r$. The microsegregation ratio SR can be expressed as Eq. (5.73). SRAH is segregation ratio after homogenization.

$$\mathrm{SR}_{\mathrm{AH}} = \frac{C_{\max}}{C_{\min}} = \sum_{i=0}^{n} \exp \frac{-(\pm il)^2}{2D(t' + t)} \Bigg/ \exp \frac{-(r \pm il)^2}{2D(t' + t)} \tag{5.73}$$

It can be found in the test results that the shrinkage cavities are generally far away from each other, showing a single distribution state. Therefore, there is only one diffusion source $n = 1$, and Eq. (5.73) can be expressed as:

$$\mathrm{SR}_{\mathrm{AH}} = \frac{1 + 2\exp\frac{-2r^2}{D(t'+t)}}{2\exp\frac{-r^2}{2D(t'+t)} + \exp\frac{-9r^2}{2D(t'+t)}} \tag{5.74}$$

In the equation, t' can be calculated from the electron probe image before soaking. Different samples have different initial microsegregation ratios, so the t of each sample is also different. To verify the effect of a shrinkage cavity on microsegregation diffusion, the calculation results of the Gaussian diffusion model in a position without a shrinkage cavity are compared. The Gaussian diffusion model of microsegregation of a position without shrinkage cavity can be expressed as:

$$\mathrm{SR}_{\mathrm{AH}} = \frac{1 + 2\exp\frac{-l^2}{4D(t'+t)}}{2\exp\frac{-l^2}{16D(t'+t)} + \exp\frac{-9l^2}{16D(t'+t)}} \tag{5.75}$$

Table 5.11 shows the calculation results of high-temperature diffusion behavior of microsegregation without and with shrinkage cavity, and the calculation errors of the two are solved. The closer the SR value after the soaking process is to 1, the

Fig. 5.54 Schematic diagram of carbon element distribution inside shrinkage cavity. **a** Before heating; **b** after heating

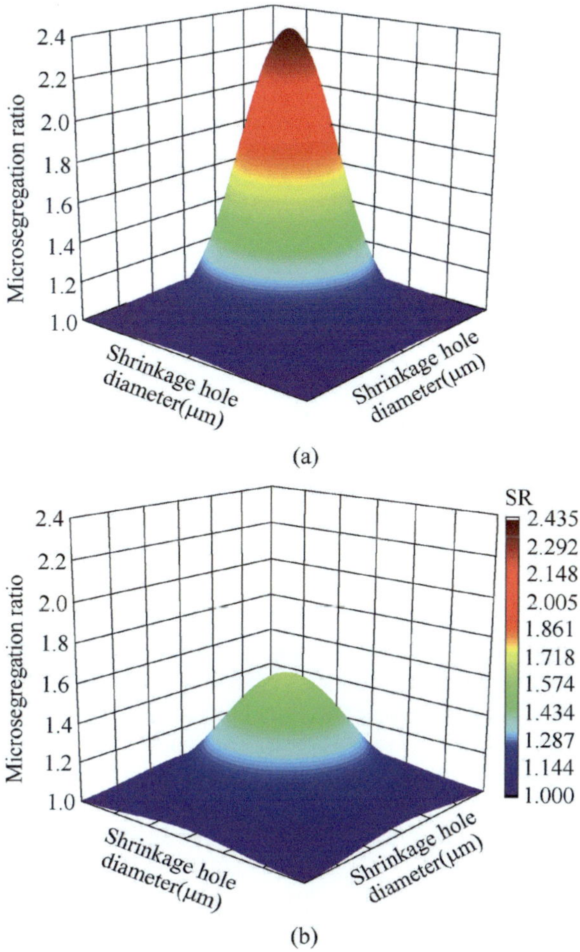

more complete the microsegregation diffusion is. The results show that after high-temperature diffusion treatment with different soaking temperatures and soaking times, the microsegregation ratio SR of carbon element remains between 1.388 and 1.776. Compared with the initial microsegregation ratio of 1.707–2.783, the change in microsegregation ratio shows that the high-temperature diffusion effect is better. It can be seen from Table 5.11 that the error between the experimental value and the calculated value of most samples is within 10%. When the shrinkage cavity size is less than 100μm, the shrinkage cavity has little effect on the microsegregation diffusion. When the shrinkage cavity size is greater than 100 μm, the deviation of the microsegregation calculation outcome in the vicinity of the shrinkage cavity is more limited, indicating that the microsegregation Gaussian diffusion model at the shrinkage cavity has higher accuracy in calculating the microsegregation inside the shrinkage cavity.

Table 5.11 Experiment and result comparison

Heating process	Shrinkage cavity size (μm)								
	$R \le 100\,\mu m$			$100\,\mu m < R \le 300\,\mu m$			$300\,\mu m < R$		
	1150 °C 1.0 h	1200 °C 1.5 h	1250 °C 2.0 h	1200 °C 2.0 h	1250 °C 1.0 h	1150 °C 1.5 h	1250 °C 1.5 h	1150 °C 2.0 h	1200 °C 1.0 h
SRC	1.592	1.536	1.429	1.437	1.433	1.637	1.523	1.496	1.477
SRS	1.656	1.58	1.488	1.504	1.492	1.708	1.596	1.604	1.581
Error (P_{ct})	3.865	3.335	3.965	4.455	3.954	4.157	4.574	6.733	6.578
Avg. error (P_{ct})	3.722			4.189			5.962		

Figure 5.55a shows the change in microsegregation ratio at different soaking temperatures under the condition of keeping the initial SRBH and the size of the shrinkage cavity unchanged. It can be seen from the figure that SRBH gradually decreases with the increase of soaking time, but when the soaking time exceeds 1.5 h, the change curve of the microsegregation ratio at the shrinkage cavity begins to become relatively flat. This shows that when the soaking time is in the range of 1.5 h, the high-temperature diffusion process can still be carried out effectively. When the soaking time exceeds 1.5 h, the high-temperature diffusion has been fully carried out, and the continuous increase of the soaking time has little effect on the improvement of microsegregation, which proves that the optimal high-temperature diffusion process parameters in the above orthogonal test conclusions are reasonable. In Fig. 5.55b, the smaller the shrinkage cavity size is, the shorter the soaking time is, and the larger the shrinkage cavity diameter is, the longer the soaking time is.

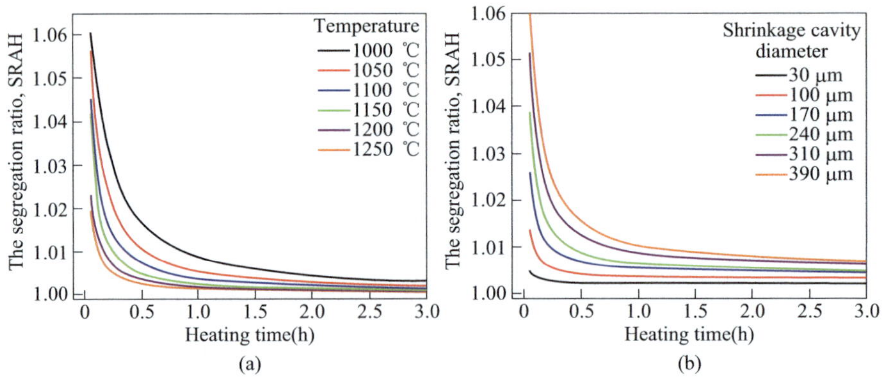

Fig. 5.55 Variation of microsegregation ratio with soaking time. **a** The same shrinkage cavity diameter, $r = 90\,\mu m$; **b** the same soaking temperature, $T = 1150$ °C

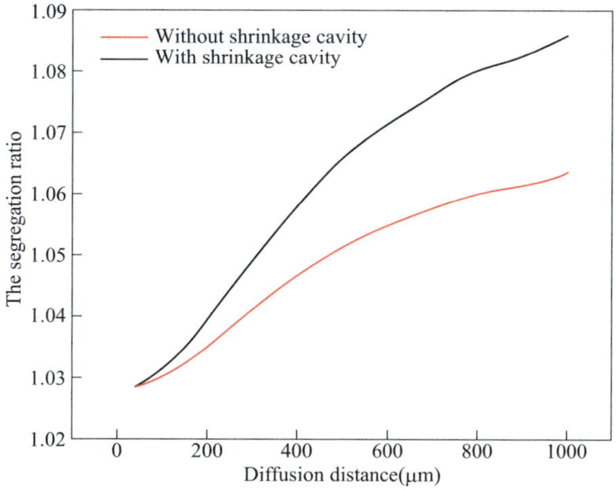

Fig. 5.56 Effect of element diffusion distance on microsegregation ratio with or without shrinkage cavity (calculation conditions SRBH = 2, soaking temperature = 1200 °C, soaking time = 1 h)

As shown in Fig. 5.56, under the same high-temperature diffusion process conditions, when the element diffusion distance and the shrinkage cavity diameter at position without shrinkage cavity are both 50 μm, the improvement effect of microsegregation is the same. When the element diffusion distance is 350 μm, the micro-segregation ratio at the shrinkage cavity is reduced to 1.014, and the micro-segregation at the non-shrinkage cavity is reduced to 1.007. It can be seen that the residual micro-segregation inside the shrinkage cavity is larger. The reason is that when there is a shrinkage cavity, in the same soaking time, most of the segregation elements inside the shrinkage cavity participate in surface diffusion, and a small part of the segregation elements participate in grain boundary diffusion and intragranular diffusion. The intragranular diffusion and grain boundary diffusion rate are less than the surface diffusion rate. After reaching the set soaking time, a part of the microsegregation elements stay inside the shrinkage cavity due to surface diffusion, and the other part of the microsegregation elements diffuse to the outside of the shrinkage cavity. Therefore, the microsegregation left in the shrinkage cavity increases, which also constitutes the principal factor accounting for the obstruction of microsegregation diffusion by the shrinkage cavity. Combined with the above information, it can be found that there are three reasons for hindering the shrinkage cavity diffusion in the process of segregation diffusion: (1) the larger the shrinkage cavity size, the more serious the microsegregation inside the shrinkage cavity; (2) in the same soaking time, surface diffusion mainly occurs inside the shrinkage cavity, and many segregation elements are remaining inside the shrinkage cavity. (3) When there is no shrinkage, the surface diffusion area is large, which further increases the diffusion effect. At the same time, it can be seen from Eq. (5.74) that the most important factor affecting the high-temperature diffusion process is the soaking temperature, followed by the shrinkage cavity size and the soaking time, which not only verifies the conclusion of the above orthogonal experiment but also shows the accuracy of the high-temperature diffusion experiment results in this chapter.

References

1. Yu HQ, Zhu MY (2012) Influence of electromagnetic stirring on transport phenomena in round billet continuous casting mould and macrostructure of high carbon steel billet. Ironmaking Steelmaking 39(8):574–584
2. Ji C, Wu C, Zhu M (2016) Thermo-mechanical behavior of the continuous casting bloom in the heavy reduction process. JOM 68(12):3107–3115
3. Mayer F, Wu M, Ludwig A (2010) On the formation of centre line segregation in continuous slab casting of steel due to bulging and/or feeding. Steel Res Int 81(8):660–667
4. Fachinotti VD, Le Corre S, Triolet N et al (2006) Two-phase thermo-mechanical and macrosegregation modelling of binary alloys solidification with emphasis on the secondary cooling stage of steel slab continuous casting processes. Int J Numer Methods Eng 67(10):1341–1384
5. Guan R, Ji C, Zhu M et al (2018) Numerical simulation of V-shaped segregation in continuous casting blooms based on a microsegregation model. Metall Mater Trans B 49(5):2571–2583
6. Yu W, Zhang YM, He CY et al (2011) Production of heavy-gauge steel plates by clad rolling process. Chin J Eng 33(11):1391–1395
7. Ridder SD, Kou S, Mehrabian R (1981) Effect of fluid flow on macrosegregation in axi-symmetric ingots. Metall Mater Trans B 12(3):435–447
8. Oh KS, Chang YW (1995) Macrosegregation behavior in continuously cast high carbon steel blooms and billets at the final stage of solidification in combination stirring. ISIJ Int 35(7):866–875
9. Bennon WD, Incropera FP (1987) A continuum model for momentum, heat and species transport in binary solid-liquid phase change systems—I. Model formulation. Int J Heat Mass Transfer 30(10):2161–2170
10. Gandin C, Rappaz M (1994) A coupled finite element-cellular automaton model for the prediction of dendritic grain structures in solidification processes. Acta Metall Mater 42(94):2233–2246
11. Wu M, Ludwig A (2009) Modeling equiaxed solidification with melt convection and grain sedimentation—I: Model description. Acta Mater 57(19):5621–5631
12. Guan R, Ji C, Zhu M et al (2020) Modeling the effect of combined electromagnetic stirring modes on macrosegregation in continuous casting blooms. Metall Mater Trans B 51(3):1137–1153
13. Wang CY, Beckermann C (1996) Equiaxed dendritic solidification with convection: Part I. Multiscale/multiphase modeling. Metall Mater Trans A 27(9):2754–2764
14. Beckermann C, Viskanta R (1988) Double-diffusive convection during dendritic solidification of a binary mixture. Phys Chem Hydrodynamics 10(2):195–213
15. Ni J, Beckermann C (1991) A volume-averaged two-phase model for transport phenomena during solidification. Metall Mater Trans B 22(3):349–361
16. Ludwig A, Wu M (2002) Modeling of globular equiaxed solidification with a two-phase approach. Metall Mater Trans A 33(12):3673–3683
17. Wu M, Ludwig A (2006) A three-phase model for mixed columnar-equiaxed solidification. Metall Mater Trans A 37(5):1613–1631
18. Rappaz M, Thevoz P (1987) Solute diffusion model for equiaxed dendritic growth. Acta Metall 35(7):1487–1497
19. Combeau H, Založnik M, Hans S et al (2009) Prediction of macrosegregation in steel ingots: influence of the motion and the morphology of equiaxed grains. Metall Mater Trans B 40(3):289–304
20. Ciobanas AI, Fautrelle Y (2007) Ensemble averaged multiphase Eulerian model for columnar/equiaxed solidification of a binary alloy: I. The mathematical model. J Phys D: Appl Phys 40(12):3733–3762
21. Tu W, Shen HF, Liu BC (2014) Two-phase modeling of macrosegregation in a 231 t steel ingot. ISIJ Int 54(2):351–355
22. Du Q, Li DZ (2000) Quantitative prediction of macrosegregation formation caused by natural convection during solidification of steel casting. Acta Metall Sinica 36(11):1197–1200

23. Jiang D, Zhu M (2017) Center segregation with final electromagnetic stirring in billet continuous casting process. Metall Mater Trans B 48(1):444–455
24. Guan R, Ji C, Wu CH et al (2019) Numerical modelling of fluid flow and macrosegregation in a continuous casting slab with asymmetrical bulging and mechanical reduction. Int J Heat Mass Transfer 141:503–516
25. Lipton J, Glicksman ME, Kurz W (1984) Dendritic growth into undercooled alloy metals. Mater Sci Eng 65(1):57–63
26. Wu M, Ludwig A, Fjeld A (2010) Modelling mixed columnar-equiaxed solidification with melt convection and grain sedimentation—Part II: Illustrative modelling results and parameter studies. Comput Mater Sci 50(1):43–58
27. Hou Z, Jiang F, Cheng G (2012) Solidification structure and compactness degree of central equiaxed grain zone in continuous casting billet using cellular automaton-finite element method. ISIJ Int 52(7):1301–1309
28. Kurz W, Giovanola B, Trivedi R (1986) Theory of microstructural development during rapid solidification. Acta Metall 34(5):823–830
29. Jiang DB, Zhu MY (2016) Solidification structure and macrosegregation of billet continuous casting process with dual electromagnetic stirrings in mold and final stage of solidification: a numerical study. Metall Mater Trans B 47(6):3446–3458
30. Gu JP, Beckermann C (1999) Simulation of convection and macrosegregation in a large steel ingot. Metall Mater Trans A 30(5):1357–1366
31. Wu CH, Ji C, Zhu MY (2018) Numerical simulation of bulging deformation for wide-thick slab under uneven cooling conditions. Metall Mater Trans B 49(3):1346–1359
32. Poole GM, Heyen M, Nastac L, et al (2014) Numerical modeling of macrosegregation in binary alloys solidifying in the presence of electromagnetic stirring. Metall Mater Trans B 45(5):1834–1841
33. Voller VR, Prakash C (1987) A fixed-grid numerical modeling methodology for convection-diffusion mushy region phase-change problems. Int J Heat Mass Transfer 30:1709–1720
34. Voller VR, Swaminathan CR(1991) Generalized source-based method for solidification phase change. Numer Heat Transfer B 19(2):175–189
35. Won YM, Thomas BG (2001) Simple model of microsegregation during solidification of steels. Metall Mater Trans A 32(7):1755–1767
36. Schneider MC, Beckermann C (1995) Simulation of micro-/macrosegregation during the solidification of a low-alloy steel. ISIJ Int 35(6):665–672
37. Nabeshima S, Nakato H, Fujii T, et al (1995) Control of centerline segregation in continuously cast blooms by continuous forging process. ISIJ Int 35(6):673–679
38. Wang WL, Ji C, Luo S, et al (2018) Modeling of dendritic evolution of continuously cast steel billet with cellular automaton. Metall Mater Trans B 49(1):200–212
39. Savage J, Pritchard WH (1954) The problem of rupture of the billet in the continuous casting of steel. J Iron Steel Inst London 178(11):268–277
40. Nozaki T, Matsuno J, Murata K, et al (1976) Secondary cooling pattern for the prevention of surface cracks of continuous casting slab. Trans Iron Steel Inst Jpn 62(12):1503–1512
41. Flemings MC (2000) Our understanding of macrosegregation: past and present. ISIJ Int 40(9):833–841
42. Krauss G (2003) Solidification, segregation, and banding in carbon and alloy steels. Metall Mater Trans B 34(6):781–792
43. Smith R (2018) Microsegregation measurement: methods and applications. Metall Mater Trans B 49(6):3258–3279
44. Lan P, Zhang JQ (2014) Numerical analysis of macrosegregation and shrinkage porosity in large steel ingot. Ironmaking & Steelmaking 41(8):598–606
45. Han Y, Li C, Ren J, et al (2019) Dendrite segregation changes in high temperature homogenization process of as-cast H13 stee. ISIJ International 59(10):1893–1900
46. He S, Li C, Ren J (2018) Investigation on alloying element distribution in Cr8Mo2SiV Cold-work die steel ingot during homogenization. Steel Res Int 89(10):180014

47. Zhang DF, Peng J (2005) Diffusion Models for Solid-State Homogenization of Dendrite Segregation. Journal of Wuhan University of Technology-Materials Science Edition 20(2):111–114
48. Timmerscheid TA, von Appen J, et al (2014) A molecular-dynamics study on carbon diffusion in face-centered cubic iron. Comput Mater Sci 91:235–239

Chapter 6
The Microstructure Evolution of Continuous Casting Strand for Reduction Process

The experimental study shows that the austenite recrystallization occurs in the cast slab with the increase of the reduction deformation. In addition, the pinning effect of microalloyed carbonitrides can be effectively retained before rolling, which is beneficial to the improvement of the microstructure and properties of the rolled metal. Figure 6.1 shows the microstructure of the microalloyed steel (Q345E) continuous casting strand under different reductions. It can be seen that the increase of the reduction deformation significantly changes the structure of the slab. The ferrite grains on the surface layer of the strand were significantly refined, and the central microstructure of the strand was transformed into acicular ferrite. The structure of the strand has the obvious advantages of high plasticity and not easy to grow up during the heating process. Therefore, the systematic study of the recrystallization behavior of austenite grains and the evolution of the microstructure in the subsequent heating and rolling process of the strand under heavy reduction at the end of continuous casting solidification has important research significance and practical value for improving the performance of large-scale steel products.

The researchers have carried out related research, the changes of the dynamic recrystallization volume fraction are affected by the stress, strain and deformation temperature distribution on the wide and thick slab during the reduction process of the strand [1, 2]. This chapter combines the specific deformation characteristics of the microalloyed steel Q345E wide and thick slab continuous casting during heavy reduction. The laws of dynamic recrystallization during reduction [3], sub-dynamic recrystallization in reduction pass interval [4], and static recrystallization [5] were systematically studied. The refinement mechanism of the recrystallized grains by the heavy-pressure reduction process under the multifactor conditions of "stress/strain-strain rate-temperature-pass interval" is expounded. A prediction model of recrystallized grain size was established, and the microstructure evolution of the strand during the subsequent heating process and its influence on the microstructure and properties of the rolled product were explored.

C. Ji et al., *Solidification End Reduction of Continuous Casting Process*, Key Steel Manufacturing Technologies by Collaborative Innovation,
https://doi.org/10.1007/978-981-96-8544-8_6

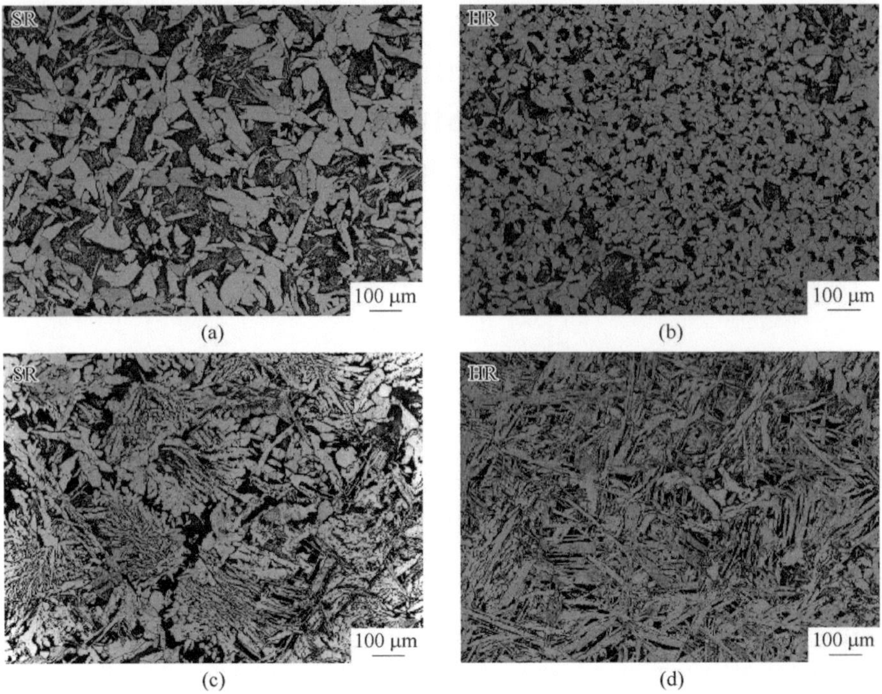

Fig. 6.1 Influence of different reductions on the microstructure of continuous casting slab at room temperature. **a** The amount of reduction is 9 mm, and the surface of the billet is 5 mm under the skin; **b** the amount of reduction is 35 mm, and the surface of the billet is 5 mm under the skin; **c** the reduction is 9 mm, the center of the billet; **d** the reduction is 35 mm, the center of the billet

6.1 Recrystallization Characteristics of Continuous Casting Strand for Reduction Process

Under certain temperature conditions, the process of nucleation and growth of undistorted equiaxed grains inside the deformed metal and gradually replacing the deformed grains is called recrystallization. During the reduction process at the end of solidification, especially the large deformation under heavy reduction, dynamic recrystallization of austenite will occur in the casting strand, and static recrystallization and sub-dynamic recrystallization will occur in the interval of reduction passes.

Due to the widening and thickening of the cross-section size, the cross-section temperature and stress span are significantly expanded, resulting in more complex recrystallization behavior. Figure 6.2 shows the stress–strain curve of the thermal deformation process and the softening effect of the stress when the strain is interrupted during the deformation process. As shown in Fig. 6.2a, when the flow stress is less than the critical strain (εcd) of dynamic recrystallization, a large number of substructures appear inside the metal material during the deformation process

[7], and the initial stress increases rapidly after work hardening [8–10]. When the strain exceeds the critical strain of dynamic recrystallization (εcd), the stress first increases slowly and then decreases gradually, and the metal material undergoes dynamic recrystallization. After the strain reaches the strain (εcp) at which dynamic recrystallization occurs completely, the softening and strain hardening caused by dynamic recrystallization will reach a certain dynamic balance, and the stress will progressively reach a stable state.

Figure 6.2b shows the strain interruption during the reduction process, which is also the three static softening phenomena during the reduction pass interval. When the strain is in Region A, the strain breaks below the critical strain (εcs) below which static recrystallization occurs, and only static recovery occurs within the pass interval. When the strain is interrupted in Region B (εcs < ε < εcd), only static recrystallization occurs during the pass interval. When the strain is interrupted in the Region C (εcd < ε < εcp), three softening mechanisms of static recovery, static recrystallization and sub-dynamic recrystallization can occur in the interval between

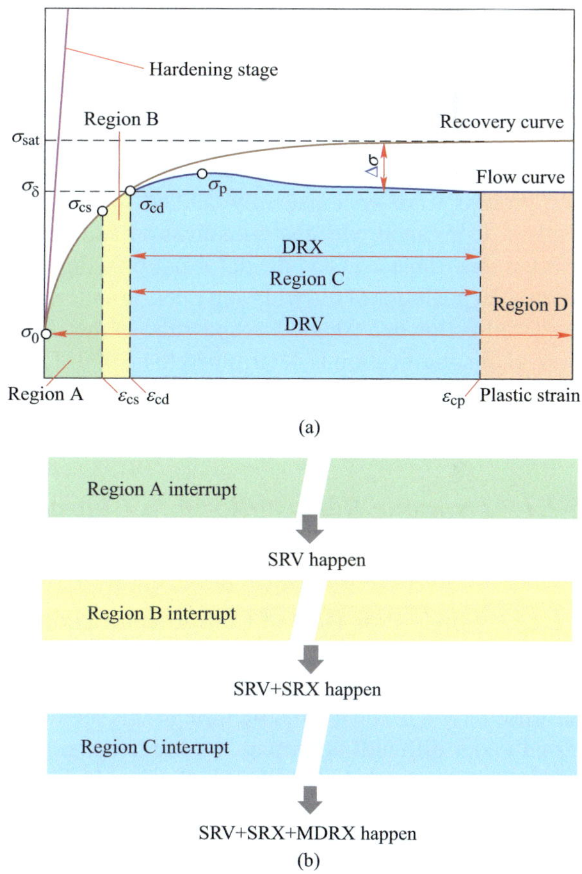

Fig. 6.2 Softening process may occur during thermal deformation [6]. **a** Dynamic; **b** static

Table 6.1 Three conditions for recrystallization

Project	Dynamic recrystallization	Static recrystallization	Sub-dynamic recrystallization
Depression process	Occur	–	–
Pass interval	–	Occur	Occur
Occurrence condition	ε > εcd	εcs < ε < εcd	ε > εcd
Interaction occurs	–	With static reply	εcd < ε < εcp, with static recrystallization

passes. The strain (ε > εcp) is interrupted when reaching the Region D, and only sub-dynamic recrystallization occurs during the pass interval.

During the reduction process at the end of continuous casting strand solidification, multiroller is generally used for continuous reduction. In this process, the conditions such as strain and temperature of the surface layer and the core of the slab are very different, and the temperature, strain rate, strain, initial austenite grain size and pass interval time can affect the type of recrystallization and the occurrence rate of recrystallization. Therefore, the recrystallization behavior during the reduction process at the end of solidification of the slab is complex and changeable. In view of this, taking Q345 microalloyed steel wide and thick continuous casting strand as the research object, the dynamic recrystallization, static recrystallization and sub-dynamic recrystallization of the strand during the solidification end reduction process were systematically studied by high-temperature physical simulation method. Combined with the classification and occurrence conditions of recrystallization (see Table 6.1) and the deformation temperature, the strain and strain rate at the crucial positions of the wide and thick strand, which were acquired in Chap. 3, the recrystallization deformation temperature of the slab during the reduction process at the end of solidification is determined to be 900–1350 °C, the strain rate varies from 0.001 s^{-1} to 0.1 s^{-1}.

6.1.1 *Dynamic Recrystallization Characteristics of Continuous Casting Strand for Reduction Process*

6.1.1.1 Single-Pass Heat of Compression Simulation

The general traditional method to determine the volume fraction of DRX is through the observation of the microstructure. In this method, a large number of samples are tested under different deformation conditions and the recrystallized grains need to be precisely determined, which is tedious and labor-intensive. More researchers use the stress–strain curve during hot compression to study the microstructural evolution of metallic materials. In this section, the DRX mathematical model is established through the stress–strain curve data to calculate the DRX volume fraction, which can

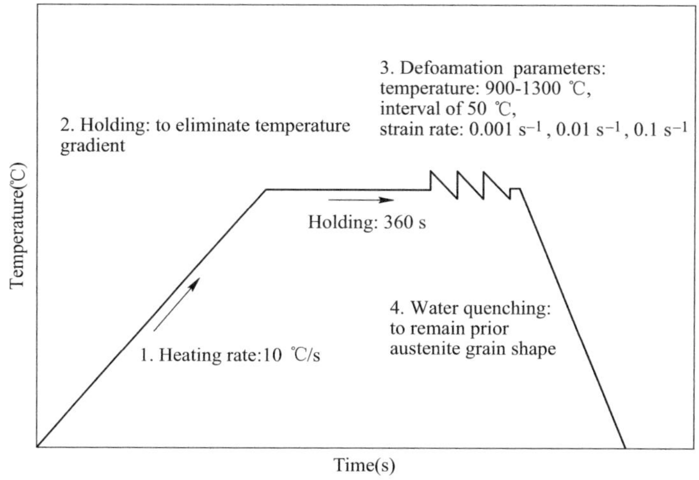

Fig. 6.3 Experimental scheme

greatly reduce the workload. The main deformation direction in the reduction process at the end of solidification is perpendicular to the drawing direction, so samples were taken along the direction of the slab compaction for high-temperature compression tests (see Fig. 6.3). By means of the derivation of the true stress–strain curve, the evolution law of recrystallization and the differential operation of the stress–strain data in the compression process are determined to provide key parameters for the dynamic recrystallization model.

During the procedure of sample preparation for thermal compression experiments, the sample size should be reduced as much as possible. The method of rapid heating is used to reduce the influence of the heating process on the tissue. Rapid quenching after deformation to retain the high-temperature structure is due to the time required for microstructure transformation (recrystallization, phase transformation), and the rapid cooling and heating as much as possible can effectively inhibit the structure transformation. Specifically, the size of the thermal simulation sample used is ϕ8 mm × 12 mm.

The compression test scheme is shown in Fig. 6.3. That is, the experimental compression temperature is set to 900–1350 °C, the interval of each set of compression curves is 50 °C, and the single-pass compressive strain rate is set to 0.001 s^{-1}, 0.01 s^{-1}, and 0.1 s^{-1}, respectively. The quenched samples after deformation were polished and chemically etched with saturated picric acid solution, and then the austenite grain boundaries were observed and the grain size was counted.

Due to the steel composition, the deformation conditions are the same as those in Chap. 2, and the true stress–strain curves under different deformation conditions are shown in Fig. 6.4 in Sect. 2.1.3.2. The stress–strain curves under a single pass can be divided into two categories; one is Dynamic Recrystallization (DRX) [9]. The true stress–strain curve exhibits the typical features of unimodal stress and gradually

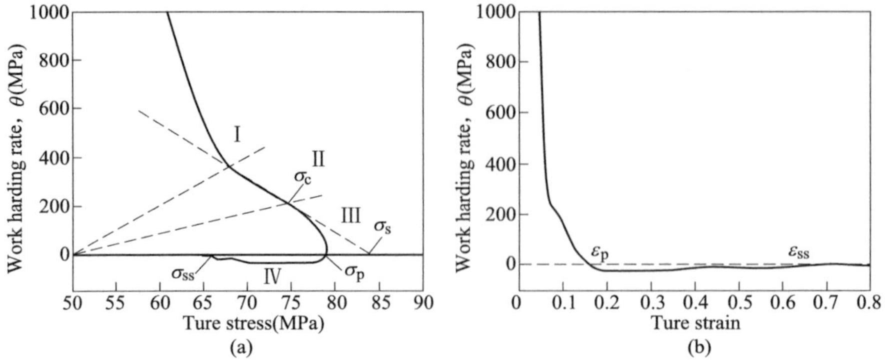

Fig. 6.4 Characteristic curve of work hardening during deformation process. **a** Work hardening rate θ-σ curves corresponding to different stresses; **b** work hardening θ-ε curves corresponding to different strains

decreases to the steady-state stress with the increase of strain, such as the stress–strain curve corresponding to the deformation temperature of 1950 °C. The other type is Dynamic Recovery (DRV) [10], in which the strain of the true stress–strain curve increases with the increase of stress until it reaches saturation and then tends to be stable.

Combined with the reduction characteristics of the solidification end, under the same deformation temperature and strain conditions, the flow stress increases with the increase of the strain rate. Under the same strain rate and strain, the flow stress decreases with the increase of temperature, and the peak strain (εp) and critical strain (εcd) decrease, indicating that dynamic recrystallization is easily induced at high deformation temperature. Higher deformation temperature provides greater nucleation driving force for dynamic recovery, resulting in faster dynamic recrystallization frequency and grain boundary migration rate. Although the low strain rate reduces the nucleation rate of dynamic recrystallization, it takes relatively long strain times for dynamic recrystallization to fully occur. Therefore, the peak strain (εp) and critical strain (εcd) decrease with decreasing strain rate, and dynamic recrystallization occurs more fully. When the deformation condition is low temperature and high strain rate, only DRV can occur, and the deformation energy storage of the material does not reach the critical condition of DRX. The true stress–strain curve shows a steady trend, which is due to the softening mechanism caused by DRV partially offsetting the increase in dislocation density caused by work hardening (WH) [11].

6.1.1.2 Modeling of Dynamic Recrystallization

The condition for dynamic recrystallization is $\varepsilon c > \varepsilon$, where εc is the critical strain. The critical strain εc corresponding to the beginning of dynamic recrystallization is an important parameter of the thermal deformation process of metal materials, and its value can be calculated according to the work hardening curve proposed by Poliak and Jonas [12], as shown in Eq. (6.1).

$$\theta = d\sigma/d\varepsilon \tag{6.1}$$

where σ is the metal deformation stress; ε is the metal deformation strain.

Taking the derivation of the stress and strain of Eq. (6.1), the characteristic curve of work hardening during deformation can be obtained (Fig. 6.4). At the same time, characteristic parameters such as peak strain (εp), peak stress (σp), critical stress (σc), saturation stress (σs), and steady-state stress (σss) can be obtained from this curve. These are the basic characteristic data for establishing the kinetic equation of dynamic recrystallization.

The curve in Fig. 6.4a is divided into four segments. The first and second stages are linear hardening stages. The element diffusion, dislocation entanglement, and substructure formation lead to an increase in dislocation density during the deformation process, which causes dynamic recovery rate weaken and decrease the slope of the curve until the end of the first stage. In segment II, because of the occurrence of DRX, the softening rate suddenly increased, which made the θ-σ curve deviate from a linear segment; the critical point at which the slope falls is σ_c. In segment III, the softening rate gradually decreased until $\theta = 0$ due to the onset of DRX when the flow stress increased to σ_p. If DRX does not occur, the curve will proceed along the dotted straight-line trajectory, the value of θ will decrease linearly to zero, and the corresponding σ will be defined as a saturation stress (σ_{ss}) when a balance between the WH and the DRV is achieved. In segment IV, the value of θ increased to zero again when σ reached σ_{ss} and the deformation reached a new balance between the WH and softening induced by DRV and DRX. The deformation strikes a balance between work hardening, dynamic recovery, and dynamic recrystallization-induced softening, as shown in Fig. 6.4b.

Based on the methodology proposed by Poliak and Jonas, Najafizadeh and Jonas [13] further optimized the solution method of the equation stress, and the eigenvalues calculated by this method are more accurate and reliable. The derived DRX initial stress is shown in Eq. (6.2):

$$\frac{\partial}{\partial\sigma}\left(-\frac{\partial\theta}{\partial\sigma}\right) = 0 \tag{6.2}$$

The work hardening characteristic curve is obtained by calculating the cubic partial derivative of Eq. (6.2), as shown in Fig. 6.5a.

The critical stress point is determined according to the θ-σ curve, and the result is shown in Fig. 6.5b. Based on the above results, the characteristic parameter values under different deformation conditions can be easily calculated, and the values are shown in Table 6.2. These characteristic parameters provide the necessary data support for the calculation of the DRX dynamic model.

At present, most DRX volume fractions (X_{DRX}) can be expressed in terms of stress ε (see Eq. (6.3)), where ε is the strain value at the peak stress. The model has high prediction accuracy for high strain rate DRX in rolling or forging process, but its

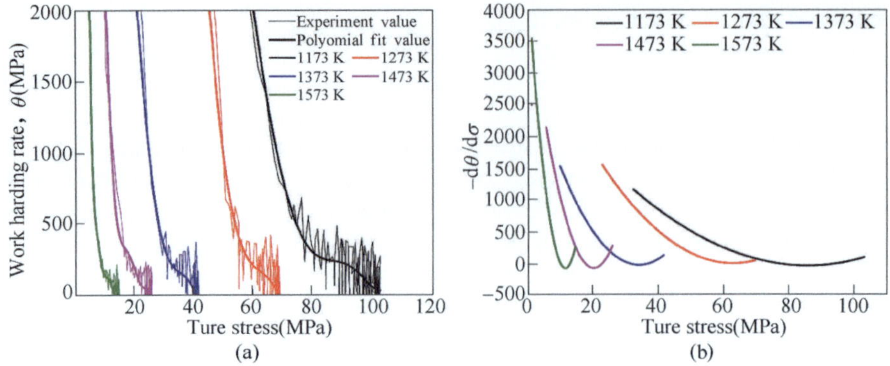

Fig. 6.5 Curves of, **a** Experimental θ-σ values and their polynomial fitting curves and **b** –dθ/dσ–σ

Table 6.2 Data for characteristic points in flow curves under different conditions

Temperature (K)	Strain rate (s^{-1})	ε_c	ε_p	σ_0	σ_c	σ_p	σ_{ss}	σ_s
1173	0.001	0.098	0.179	34.850	79.469	82.857	65.407	92.791
	0.01	0.145	0.213	35.659	108.569	111.903	93.321	123.586
	0.1	–	–	44.025	–	–	–	178.401
1273	0.001	0.073	0.142	21.780	45.668	47.761	38.284	53.025
	0.01	0.104	0.186	25.254	72.067	74.775	61.626	83.650
	0.1	0.135	0.225	26.066	99.032	103.991	84.285	115.557
1373	0.001	0.076	0.120	10.189	27.554	28.796	20.424	32.044
	0.01	0.078	0.138	11.205	40.764	45.559	38.822	49.832
	0.1	0.101	0.162	17.719	63.522	69.189	61.178	77.854
1473	0.001	0.065	0.113	6.369	16.196	17.962	14.322	20.306
	0.01	0.073	0.122	6.639	25.509	27.608	22.180	31.085
	0.1	0.079	0.143	9.551	41.363	45.091	38.758	50.017
1573	0.001	0.058	0.087	2.890	9.509	10.060	8.7522	11.243
	0.01	0.068	0.103	1.838	13.515	14.932	12.523	16.731
	0.1	0.087	0.150	3.381	24.633	28.009	22.115	29.832

prediction accuracy is relatively low due to the low strain rate in continuous casting process.

$$X_{\mathrm{DRX}} = 1 - \exp\left[-k_1\left(\frac{\varepsilon - \varepsilon_c}{\varepsilon**}\right)^{p_1}\right] \tag{6.3}$$

where X_{DRX} is the volume fraction of dynamic recrystallization; k_1, p_1 are the material constant; ε_c is the critical strain; ε^{**} is the strain value of stress peak value.

With the aim of ascertaining the values of k_1 and p_1, it can be expressed in double natural logarithmic form.

The maximum strain due to the stress peak is not necessarily the maximum stress peak during recrystallization. Therefore, in order to improve the accuracy of the model, based on previous work [13], the strain ε^* at the maximum dynamic recrystallization rate is used to replace the strain ε^{**} at the peak stress. And ε^* is determined by derivation of the true stress q by the sixth-order polynomial of V_{DRX} (the velocity of DRX). Therefore, the accuracy of the DRX model can be effectively improved, and the DRX behavior of high temperature and low strain in the continuous casting process can be accurately predicted. The revised model is as follows:

$$X_{DRX} = 1 - \exp\left[-k_1\left(\frac{\varepsilon - \varepsilon_c}{\varepsilon *}\right)^{p_1}\right] \tag{6.4}$$

ε^* represents the strain value corresponding to the maximum recrystallization rate. That is, the strain corresponding to the maximum first-order partial derivative of dynamic recrystallization (X_{DRX}). In order to solve ε^*, the value of X_{DRX} must be determined first, and ε^* is an unknown quantity, so X_{DRX} is solved in the form of stress, as shown in Eq. (6.5) [14]:

$$X_{DRX} = \frac{\sigma_{DEV} - \sigma_{DRX}}{\sigma_S - \sigma_{SS}}(\varepsilon \geq \varepsilon_c) \tag{6.5}$$

where σDRV is the flow stress; if DRV is the only softening mechanism; σs is the saturation stress; σss is the steady-state stress.

σDRV is calculated by plastic flow in different models of continuous deformation with hot work hardening proposed by Kocks and Mecking [15], the equation is as follows:

$$\theta = \theta_0\left(1 - \frac{\sigma}{\sigma_s}\right) \tag{6.6}$$

where $\theta_0 = \alpha$Gbk1/2 has nothing to do with the strain rate; σs is the saturation stress.

The equation for calculating σDRV is as follows:

$$\begin{aligned} \sigma_{DRV} &= \left[\sigma_S^2 + (\sigma_0^2 - \sigma_S^2)e^{-\Omega\varepsilon}\right]^{0.5} \\ \Omega &= 35.382\dot{\varepsilon}^{-0.090}\exp\left(-\frac{16209.040}{RT}\right) \end{aligned} \tag{6.7}$$

where σ0 is the yield stress.

To obtain ε^*, the first-order partial derivative of dynamic recrystallization volume fraction with strain, in particular, the relationship between DRX (V_{DRX}) velocity and true strain is established. As Eq. (6.8):

$$V_{\mathrm{DRX}} = \frac{\partial X_{\mathrm{DRX}}}{\partial \varepsilon} \tag{6.8}$$

where V_{DRX} is the speed of DRX, which reflects the speed of DRX. The calculation results are shown in Fig. 6.6, where the highest point of the curve should be ε^*.

By calculating the derivation of Eq. (6.8), the calculation equation of ε^* can be expressed as:

$$\varepsilon* = 0.015 d_0^{0.136} \dot{\varepsilon}^{0.084} \exp\left(\frac{30808.130}{\mathrm{RT}}\right) \tag{6.9}$$

The modified DRX kinetic model combined with Eqs. (6.3) and (6.5) is as follows:

$$X_{\mathrm{DRX}} = \frac{\sigma_{\mathrm{DRV}} - \sigma_{\mathrm{DRX}}}{\sigma_{\mathrm{S}} - \sigma_{\mathrm{SS}}} = 1 - \exp\left[-k_1 \left(\frac{\varepsilon - \varepsilon_c}{\varepsilon*}\right)^{p1}\right] \tag{6.10}$$

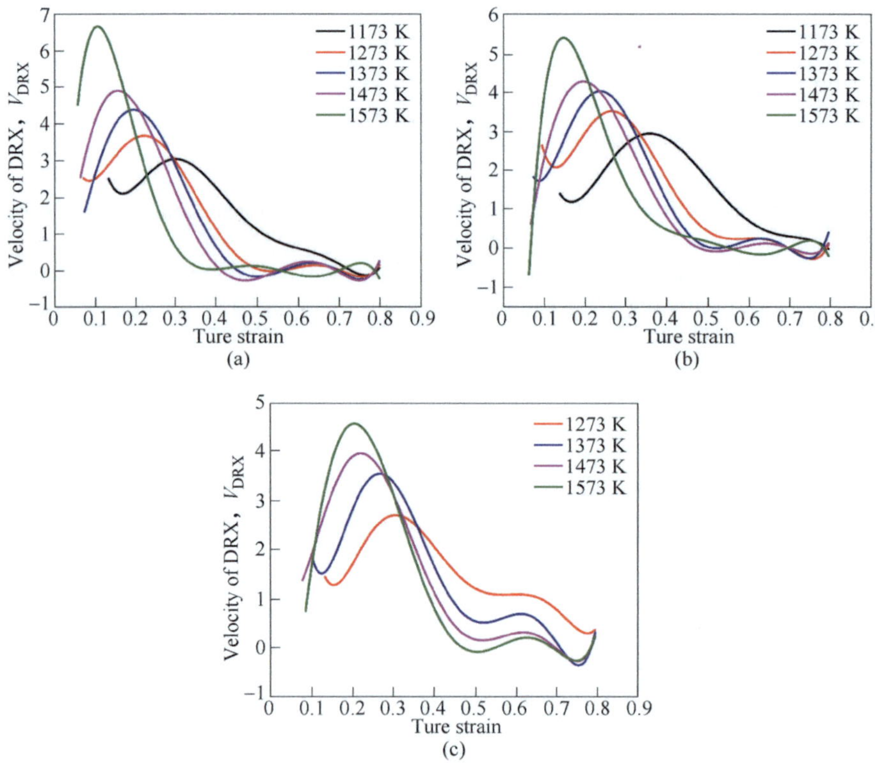

Fig. 6.6 Derivative of the sixth-order polynomial curve of DRX velocity. **a** Strain rate 0.001 s^{-1}; **b** strain rate 0.01 s^{-1}; **c** strain rate 0.1 s^{-1}

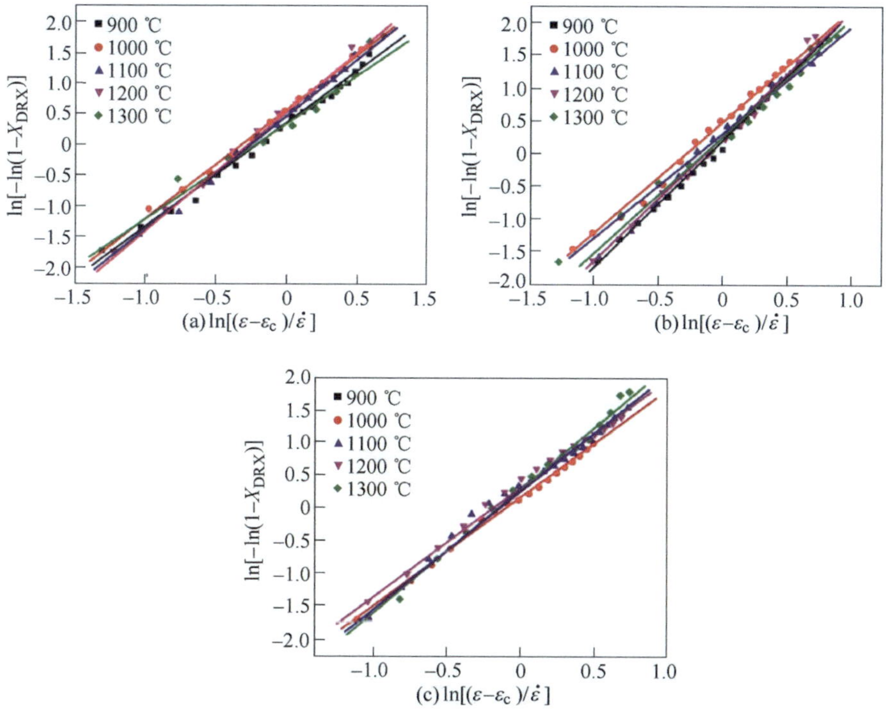

Fig. 6.7 Linear relationship between ln[-ln(1-X_{DRX})] and ln[(ε-εc)/ε*]. **a** 0.001 s^{-1}; **b** 0.01 s^{-1}; **c** 0.1 s^{-1}

where X_{DRX} is the volume fraction of DRX; k_1 and p_1 are material constants; εc is the critical strain; and ε* is the strain at the maximum speed of DRX. According to Eq. (6.10), all the measured values were analyzed by linear regression, and the results are shown in Fig. 6.7. $p_1 = 1.689$, $k_1 = 1.508$, the DRX volume fraction model equation of the slab is shown in Eq. (6.11).

$$\begin{cases} X_{DRX} = 1 - \exp[-1.508(\frac{\varepsilon-\varepsilon_c}{\varepsilon*})^{1.689}] \\ \varepsilon* = 0.015 d_0^{0.136} \dot{\varepsilon}^{0.084} \exp(\frac{30808.110}{RT}) \\ \varepsilon_c = 0.0072 d_0^{0.126} Z^{0.069} \end{cases} \quad (6.11)$$

The comparison between the predicted and experimental values of the improved DRX kinetic model is shown in Fig. 6.8. The predicted value is in good agreement with the experimental value. The volume fraction of DRX increases with increasing temperature and decreases with increasing strain rate. When the strain is 0.001 s^{-1} and 0.01 s^{-1}, the true strain of deformation is 0–0.8, and the temperature range is 900–1300 °C, the X_{DRX} of the slab can reach 100%. When the strain is 0.1 s^{-1}, the X_{DRX} can reach 100% in the temperature range of 1000–1300 °C. When the

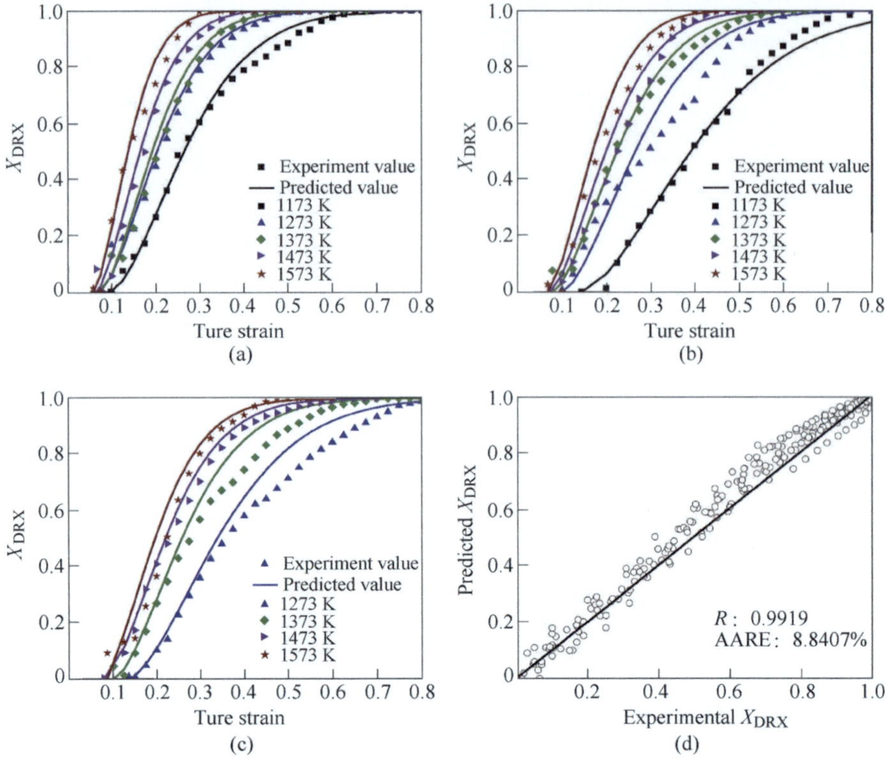

Fig. 6.8 Verification between the predicted X_{DRX} and experimental X_{DRX} of the modified DRX kinetics model under typical strains of **a** 0.001 s^{-1}; **b** 0.01 s^{-1}; and **c** 0.1 s^{-1}; subfigure **d** shows the error analysis

temperature is 900 °C, the flow curve shows the characteristics of DRV because the low temperature and high strain rate are not conducive to the occurrence of DRX.

For the purpose of further assessing the prediction precision of the enhanced model, standard statistical parameters such as correlation coefficient R and average absolute relative error (AARE) were used to quantify the prediction accuracy. The equation is as follows [16, 17]:

$$R = \frac{\sum_{i=1}^{N}(X_{ei} - \overline{X}_e)(X_{ci} - \overline{X}_c)}{\sqrt{\sum_{i=1}^{N}(X_{ei} - \overline{X}_e)^2 \sum_{i=1}^{N}(X_{ci} - \overline{X}_c)^2}} \tag{6.12}$$

$$\text{AARE} = \frac{1}{N}\sum_{i=1}^{N}\left|\frac{X_{ci} - X_{ei}}{X_{ei}}\right| \times 100\% \tag{6.13}$$

where Xc is the calculated volume fraction of DRX; Xe is the experimental DRX volume fraction; $\overline{X}\ c$ is the average value of the calculated DRX volume fraction;

$\overline{X}\ e$ is the average value of the experimental DRX volume fraction; N is the total number of data used in this study.

The predicted and experimental X_{DRX} comparison results are shown in Fig. 6.8. The *R* value and AARE value are 0.9919 and 8.8407%, respectively, and the prediction reliability is high. Among the many DRX dynamic models, the DRX dynamic models established by Laasraoui and Jonas [18] and Serajzadeh and Taheri [19] are the most widely used. Based on this theory, the prediction model of the DRX volume fraction of the slab is shown in Eq. (6.14). The X_{DRX} predicted value of the model established by Laasraoui and Jonas was compared with the experimental value and verified. It is manifest that the predicted value exhibits a significant disparity from the experimental value, and the R and AARE are 0.967 and 24.812%, respectively.

$$\begin{cases} X_{DRX} = 1 - \exp\left[-1.691\left(\frac{\varepsilon-\varepsilon_c}{\varepsilon_p}\right)^{0.649}\right] \\ \varepsilon_p = 0.0082 d_0^{0.162} Z^{0.079} \\ \varepsilon_c = 0.0072 d_0^{0.126} Z^{0.069} \end{cases} \tag{6.14}$$

Later, based on the Laasraoui and Jonas recrystallization model, researchers revised and supplemented it. Yoda et al. [20] proposed a DRX kinetic model based on the Avrami function, as shown in Eq. (6.15). By comparing the predicted recrystallization volume fraction (X_{DRX}) experimental value and the predicted value of the modified Yoda model, according to the error calculation method, the *R* value is 0.983, and the AARE value is 17.239%.

$$\begin{cases} X_{DRX} = 1 - \exp\left[-1.693\left(\frac{\varepsilon-\varepsilon_c}{\varepsilon_p}\right)^{1.813}\right] \\ \varepsilon_p = 0.0082 d_0^{0.133} Z^{0.087} \\ \varepsilon_c = 0.0072 d_0^{0.126} Z^{0.069} \end{cases} \tag{6.15}$$

Kopp [17] proposed a DRX dynamic model based on the Avrami function and introduced ε0.5-εc. Liu [21, 22] subsequently established an improved Kopp model, which can more accurately discuss the classical DRX kinetic model based on the Avrami form, and the modified DRX model can be expressed as Eq. (6.16). Comparing the experimental X_{DRX} with the X_{DRX} predicted by Liu's improved model, based on the above error calculation method, the *R* value and the AARE value are 0.988 and 12.406%, respectively. Table 6.3 shows the error comparison of different DRX models. Compared with other models, ε* is introduced in this section as the true strain value when the dynamic recrystallization rate is the fastest. The DRX kinetic prediction proposed as a parameter of the classical Avrami equation has the highest accuracy, and is more suitable for the prediction of dynamic recrystallization behavior in the continuous casting pressing process.

$$\begin{cases} X_{DRX} = \left[1 - 20.311^{(1-(\varepsilon-\varepsilon_c)/(\varepsilon_{0.5}-\varepsilon_c)}\right]^{-1} \\ \varepsilon_p = 0.0082 d_0^{0.133} Z^{0.087} \\ \varepsilon_c = 0.0072 d_0^{0.126} Z^{0.069} \end{cases} \tag{6.16}$$

On the basis of the previous discussion of the DRX dynamic model, the prediction map of the dynamic recrystallization grains of the slab under different deformation conditions is established. Upon the attainment of a steady state by DRX the dynamic recrystallization size can be calculated by Eq. (6.17) Z is the Zener-Hollomon parameter:

$$D_{\mathrm{DRX}} = 2.861 \times 10^4 Z^{-0.222} \tag{6.17}$$

As shown in Fig. 6.9, dynamic recrystallization is divided into three regions in the process of strand reduction: no dynamic recrystallization, partial dynamic recrystallization and complete dynamic recrystallization. When the strand is deformed at a given temperature and strain rate, the prerequisite for dynamic recrystallization is that the true strain exceeds the critical strain. When the true strain is less than the critical strain, the flow stress curve falls in the work hardening stage, and no dynamic recrystallization occurs.

When the strain exceeds the critical strain, the dynamic recrystallization volume fraction and the degree of austenite grain refinement increase with the increase of the deformation amount. Full occurrence of dynamic recrystallization requires the true strain to reach or exceed the steady-state strain.

Both the dynamic recrystallized volume fraction (X_{DRX}) and the recrystallized grain size (D_{DRX}) increased with the increase of the deformation temperature. As

Table 6.3 Comparison of errors of different models

Models	Modified model	LJ	Yoda	Liu
ARRE (%)	8.8407	24.821	17.239	12.406

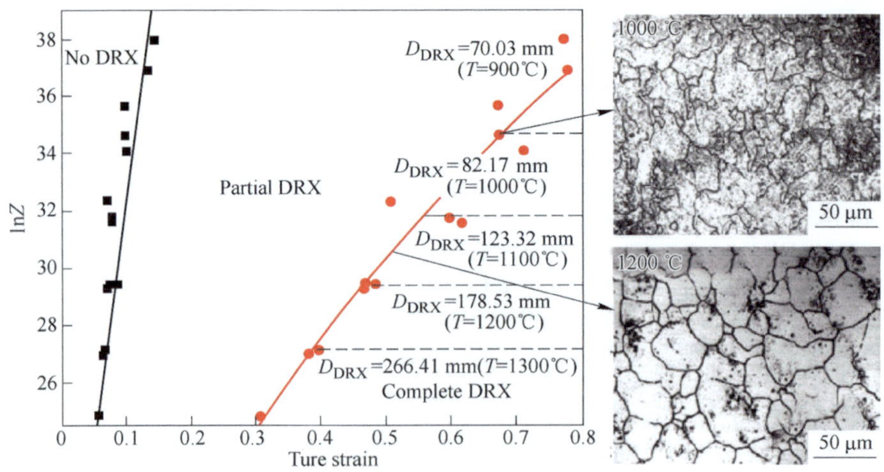

Fig. 6.9 DRX region map for different Z-parameters

shown in Fig. 6.9, when the deformation temperature is 1000 °C, 1100 °C and 1200 °C, the average grain size is 82.17 μm, 123.32 μm and 178.53 μm, respectively. Higher temperature can lead to increased deformation energy storage and increase the nucleation rate of DRX, but high temperature can promote the migration of austenite grain boundaries and accelerate the growth of austenite grains [23]. Consequently, the rational manipulation of the deformation temperature range constitutes a crucial determinant for the refinement of grains and the suppression of grain growth.

6.1.2 *Static Recrystallization Characteristics of Continuous Casting Strand for Reduction Process*

On the basis of the research on dynamic recrystallization, this section also takes microalloyed steel Q345 continuous casting slab as the research object. The material deformation and insulation process are separated. The static recrystallization behavior of the cast slab during the holding process was investigated using high-temperature confocal microscopy.

6.1.2.1 Multipass Compression Heat Simulation for Static Recrystallization

Static recrystallization occurs during the reduction pass interval. According to the thermodynamic law of wide and thick slabs in Chap. 3 (see Fig. 3.13), the temperature range for austenite static recrystallization is 900–1200 °C, and the strain rate and dynamic recrystallization strain rate are 0.001 to 0.1 s^{-1}, The interval between passes of the casting rolls is approximately 100 s. According to the discussion in the previous section, the necessary condition for the occurrence of static recrystallization is that the prestrain is lower than the critical strain for dynamic recrystallization ($\varepsilon c = 0.2$). Therefore, the strain range of the static recrystallization simulation test is designed to be 0.1–0.2, and the range of the pass interval is 0–100 s.

As shown in Fig. 6.10, the static recrystallization experiment process is divided into two parts. The first group is a single-pass thermal compression experiment, and the thermal history of the experimental process is shown in Fig. 6.10a. The single-pass compression experiment has three different deformation temperatures (1000 °C, 1100 °C, 1200 °C), three different strain rates (0.001 s^{-1}, 0.01 s^{-1}, 0.1 s^{-1}) and different prestrain (0.1, 0.15, 0.2). When the prestrain is below the critical strain for dynamic recrystallization to occur, no dynamic recrystallization occurs. After deformation, it was rapidly quenched to solidify the deformed microstructure, and then the sample was taken at the center position perpendicular to the pressing direction, and the size of the sample was 2 mm × 2 mm. For the purpose of investigating the microstructure evolution process during the static recrystallization procedure, the samples obtained by the single-pass compression experiment were placed in a high-temperature confocal laser scanning microscope device for in situ observation experiments. The samples were rapidly heated from the ambient temperature to 1000 °C, 1100 °C, and 1200 °C at a heating rate of 10 °C/s for 600 s. In situ observation of the evolution of static recrystallized structures. During the incubation, the samples were photographed every 0.1 s to record the evolution information of the microstructure.

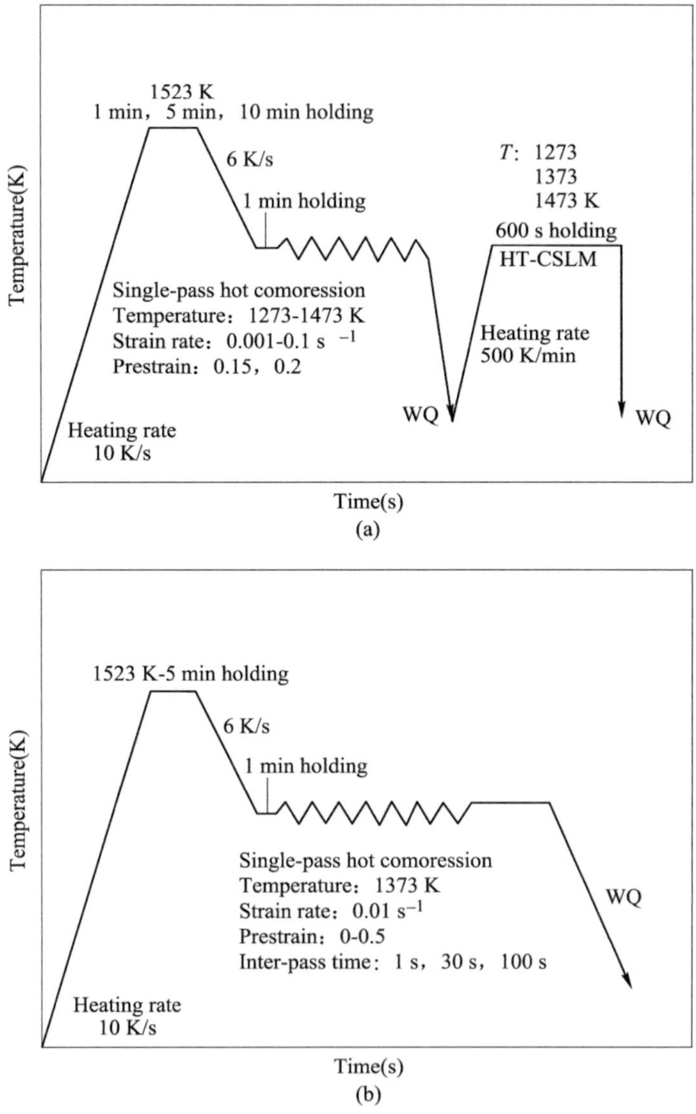

Fig. 6.10 **a** Experimental scheme for in situ experiments of the deformed microalloyed steel; **b** schematic illustration of the single-pass isothermal compression experiments

The first group of experiments used a high-temperature confocal microscope to observe the static recrystallization of the strand in the pass interval in situ, so as to study the evolution law of the microstructure. The thermal history of the second group of experiments is shown in Fig. 6.10b, and the isothermal compression experiment was carried out. The deformation temperatures are 1000 °C, 1100 °C, and 1200 °C. Strain rates 0.001 s^{-1}, 0.01 s^{-1} and 0.1 s^{-1}. The prestrain is 0.1, 0.1, and 0.5. They were quenched immediately after holding for 0 s, 30 s, and 100 s, respectively. The purpose of the second set of experiments was to quantitatively characterize the insulated tissue using electron backscattered diffraction scanning electron microscopy. In combination with the microstructure evolution process that is observed in situ, the microstructure evolution and laws of static recrystallization were analyzed.

6.1.2.2 Static Recrystallization Microstructure Evolution

In situ observation of static recrystallization (SRX) of microalloyed steel during heat preservation using a high-temperature confocal scanning microscope can accurately determine the onset time of static recrystallization. Figure 6.11 shows the evolution process of austenite static recrystallization during the process of rapid heating of the sample to 1200 °C for 600 s. The deformation temperature is 1200 °C, the initial austenite grain size is 85 μm, and the strain rate is 0.01 s^{-1}. When the temperature reaches 1200 °C, the initial structure of the sample is shown in Fig. 6.11a; no re-nucleated recrystallized grains were observed during the whole heating process, indicating that no static recrystallization occurred under this condition. At 1200 °C and the holding time was 43 s, some grain boundaries began to migrate to the center of grain boundary curvature, and static recovery began to occur at the grain boundaries, as shown in Fig. 6.11d. It shows that the deformation energy at the grain boundary is released after deformation during the heat preservation process, and the grain boundary gradually changes from a steady state to an unsteady state, and spontaneously migrates to the lower energy position. When the holding time reaches 92 s, the grain boundary reaches a stable state. And new recrystallized grain boundaries appeared due to strain-induced grain boundary migration. Therefore, 92 s was previously considered to be the latency of static recrystallization. As shown in Fig. 6.11f, as the holding time increases, more recrystallized grains appear in the field of view, and the static recrystallized volume fraction does not increase until 188 s. Virtually no alteration occurs in the austenite grains within the field of view during the period from 188 to 600 s. At this time, the internal deformation energy storage of the deformed grains is completely released, and the static recrystallization process ends.

Figure 6.11g is a schematic diagram of the evolution of the static recrystallization microstructure when the holding time is increased from 92 to 188 s. During the holding stage, there is no recrystallization nucleation. The energy reduction at grain boundaries is mainly achieved by strain-induced grain boundary migration. At the same time, with the increase of holding time, the common grain boundary of two

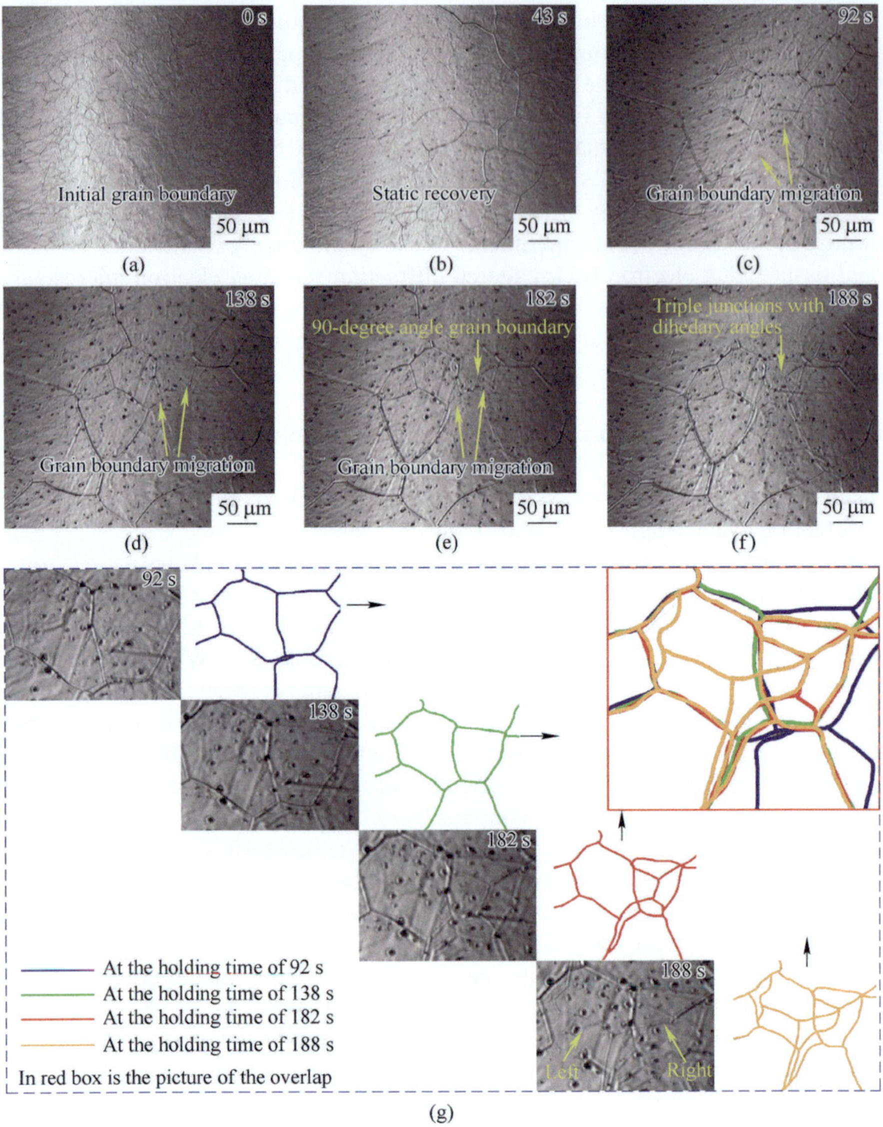

Fig. 6.11 Microstructure at strain rate of 0.01 s − 1, initial austenite grain size of 85 mm and insulation temperature of 1473 K: holding time of **a** 0 s, **b** 43 s, **c** 92 s, **d** 138 s, **e** 182 s, **f** 188 s, and **g** schematic of RX grain evolution with holding time from 92 to 188 s

adjacent grains begins to shorten and migrate to the center of grain boundary curvature with low energy.

The shape characteristics of the grain boundaries are mainly the mutually perpendicular interaction states and the diagonal trigeminal grain boundaries. It can be seen that static recrystallization is a process in which the microstructure reaches a stable state by consuming the internal defects of the microscopically deformed structure.

In Fig. 6.12, the deformation temperature is 1000–1200 °C, the strain rate is 0.1–0.001 s^{-1}, the initial austenite grain size is 67–99 μm, and the prestrain is 0.15 and 0.2.

Using Image Pro Plus software, the changes of recrystallized grain size and static recrystallized volume fraction under different holding time were statistically analyzed. Figure 6.12a, b, g describe the changes of recrystallized grain size and static recrystallized volume fraction of microalloyed steel under the same other deformation conditions at different holding temperatures. When the holding temperature was increased to 1200 °C, the microalloyed carbonitrides re-dissolved and the pinning effect on the recrystallized grains weakened. During the holding process, when the holding time was 300 s, the holding temperature increased from 1000 °C to 1200 °C, the recrystallized grain size increased from 30 μm to 150 μm, and the static recrystallization softening fraction increased from 10.6% to 13.6%. The 50% time to recrystallization ($t_{0.5}$) is much larger than the results obtained based on the true stress–strain curve, owing to the fact that the static recovery process is undergone prior to the onset of static recrystallization.

Figure 6.12a, c, and d reflects the effect of the strain rate range of 0.001–0.1 s^{-1} on the static recrystallization softening fraction. As the strain rate increases, the static recrystallization volume fraction increases. It means that the static recrystallization rate is increased. There exist two principal factors contributing to the aforementioned outcomes: on the one hand, the deformation energy storage is higher at high strain rates, and the static recovery is hindered; on the other hand, the dislocation density increases at high strain rates, and the driving force for recrystallization increases. The effect of prestrain on the static softening behavior is shown in Fig. 6.12e, h. The effect of prestrain on the static softening rate can be seen from the figure. Since the main mechanism between the two pass intervals under low strain conditions is work hardening or static recovery. Larger strains will lead to increased dislocation density and static recrystallization driving force, thus increasing the softening fraction. The effect of initial austenite grain size is shown in Fig. 6.12e, f, g. The softening fraction decreases with increasing grain size. The static recrystallization volume fraction increases the total area of grain boundaries when the initial austenite grain size is small, which promotes the increase of deformation energy storage and recrystallization nucleation sites.

Figure 6.13 shows the inverse pole figure and the distribution of misorientation under different strain conditions. Figure 6.13a and b shows that a large number of martensitic lath bundles are formed at the prior austenite grain boundaries after direct quenching. The grain boundary orientation is dominated by high-angle grain boundaries. However, the prior austenite grain boundaries can still be observed in the picture. Figure 6.13c and d reflects the critical strain that occurs when the strain

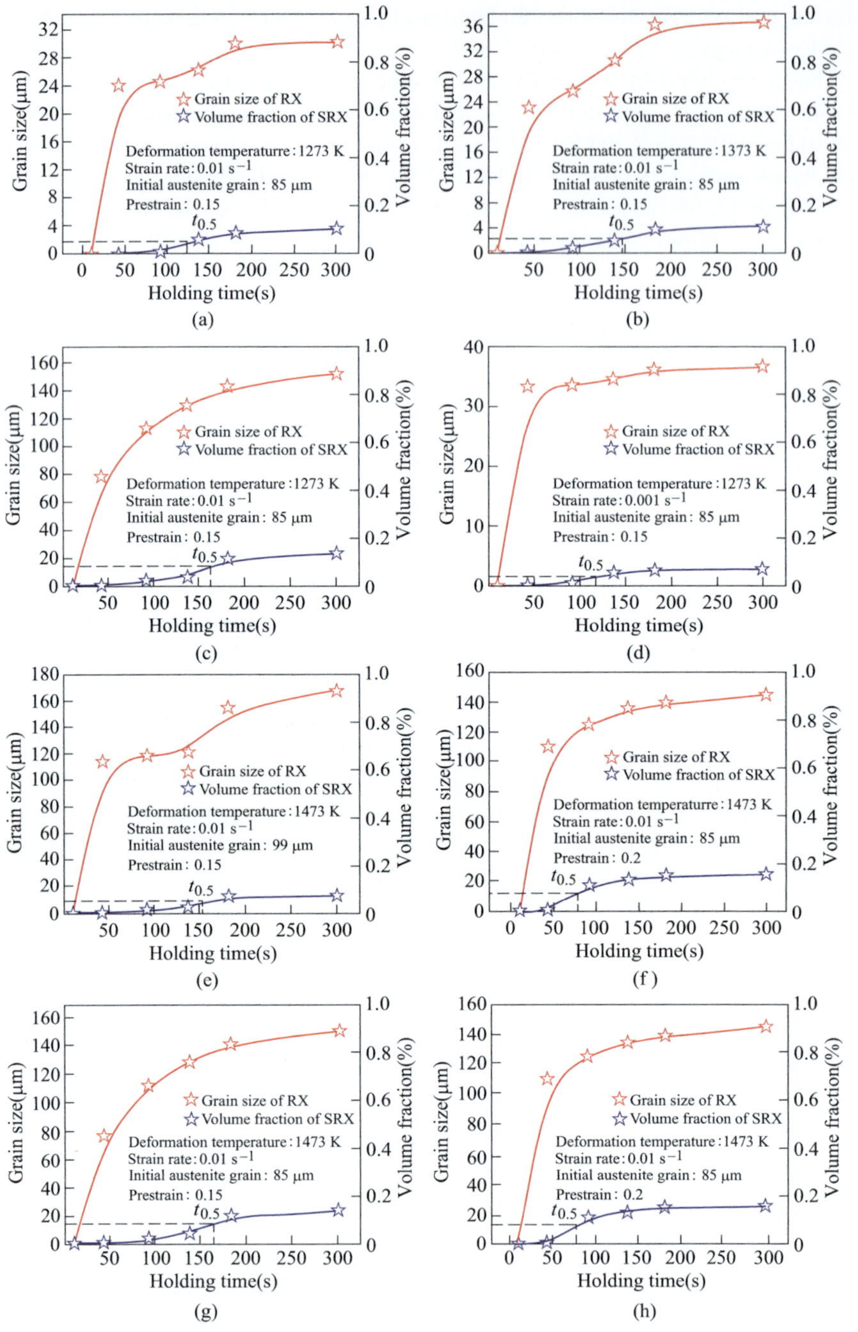

Fig. 6.12 Variations in the average RX grain size and SRX volume fraction with holding time under the different deformation conditions

is below the dynamic recrystallization. The microstructure evolution within the pass interval after deformation is mainly static recrystallization. And grain refinement, martensite nucleation particles increase, resulting in the increase of large-angle martensite lath bundles. The frequency of high-angle grain boundaries (HAGBs, $\theta > 15°$) increased from 14.7 to 21.4%. This is mainly due to the weakening of the restraining effect of the prior austenite on the formation of martensite due to grain refinement, which facilitates a larger angular deflection of the orientation of the surrounding austenite. As shown in Fig. 6.13e and f, as the amount of strain increases to 0.5, the small-angle grain boundaries less than 5° tend to decrease slightly, but still account for a relatively high proportion. The proportion of high-angle grain boundaries greater than 50° increased slightly, and the subgrain boundaries gradually became blurred. On the one hand, this is because the high strain promotes the occurrence of dynamic recrystallization, resulting in a decrease in the density of high-density dislocations. The martensitic lath is induced to recover, and the lath boundary gradually blurs and disappears, resulting in the reduction of small-angle lath grain boundaries less than 5°. On the other hand, dislocations form subgrain boundaries through sliding and climbing. Due to the short formation time, the intragranular orientation difference is small, so the subgrain boundaries are not clear and there are a large number of dislocations inside. The formation of low-angle subgrain boundaries will lead to an increase in the proportion of low-angle grain boundaries. Under the synergistic effect of the two, the small-angle grain boundaries did not disappear in large amounts, but only slightly decreased in proportion.

Figure 6.14 shows the inverse pole diagram and grain boundary misorientation distribution of the samples under different holding conditions with a prestrain of 0.15. A large number of ultra-low-angle grain boundaries (smaller-angle grain boundaries less than 5°) appeared after deformation without heat preservation. As shown in Fig. 6.14a and b, the proportion of high-angle grain boundaries is only 15%. As the holding time increased to 30 s, the ultra-low-angle grain boundaries decreased from 75.5 to 67.4%, as shown in Fig. 6.14c and d. As the strain-induced grain boundary migration process proceeds, subgrain boundaries are formed through dislocation slip and entanglement. Finally, recrystallized grain boundaries are formed, resulting in a continuous decrease in the density of other defects such as dislocations and vacancies in the original austenite grains. When the holding time increased to 100 s, the high-angle grain boundaries of 25–35° increased from 1.7 to 5.6%, and the frequency of ultra-low-angle grain boundaries decreased from 67.4 to 47.4%. Figure 6.14e and f show that strain-induced grain boundary migration is associated with the increase of high-angle grain boundaries.

Figure 6.15a shows the mass map of the original high-temperature austenite grains obtained by scanning electron microscopy (SEM). The degree of grain orientation is 25–35°. The martensitic laths initiate nucleation at the grain boundaries and elongate inwards along the austenite grain boundaries. The positions of the A1–A2 and B1–B2 lines are shown by the straight lines on the inverse pole figure in Fig. 6.15b. The change of orientation difference of the slats along different axes is shown in Fig. 6.15c and d. The dislocation angle between the laths is about 60°, indicating that the deformation energy storage of the martensitic laths is very strong. The cumulative

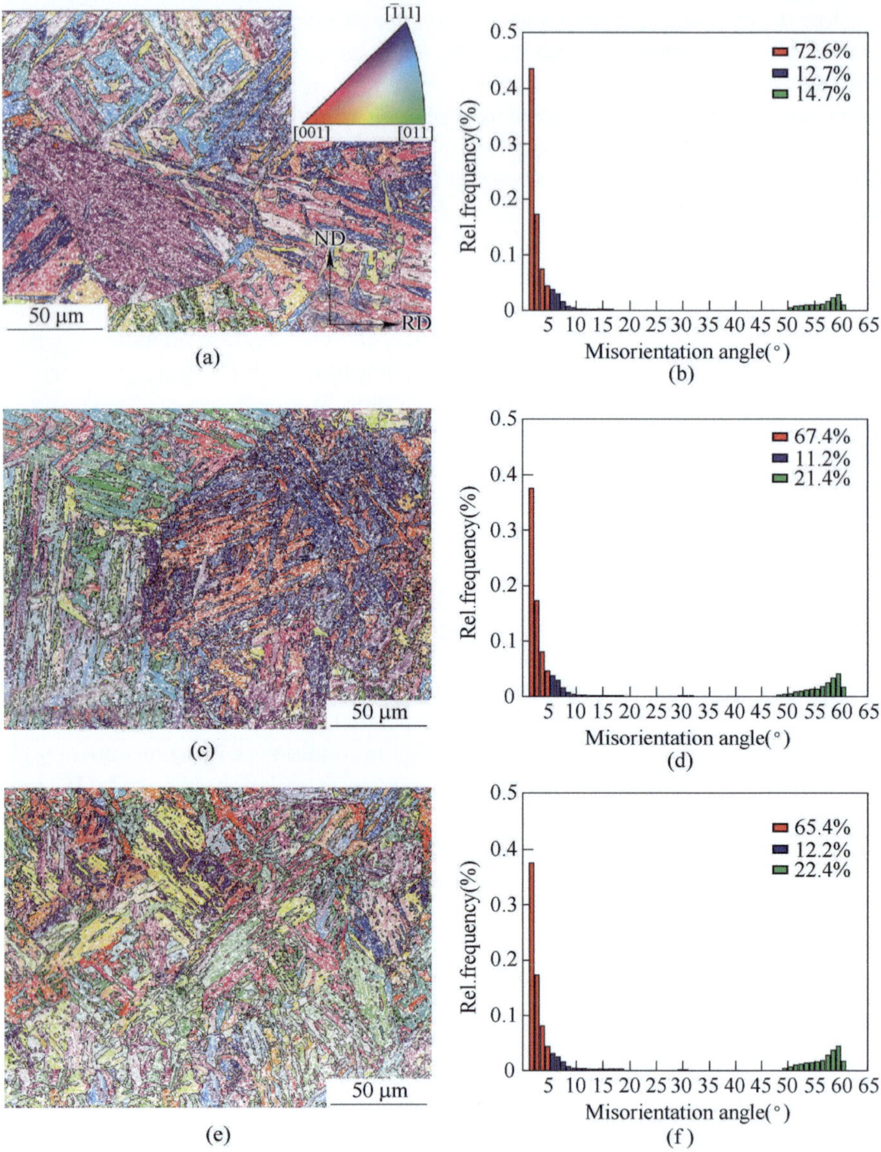

Fig. 6.13 Inverse pole figure maps and grain boundary misorientation distributions with different strains: **a**, **b** Direct quenching with no deformation; **c**, **d** 0.1; and **e**, **f** 0.5

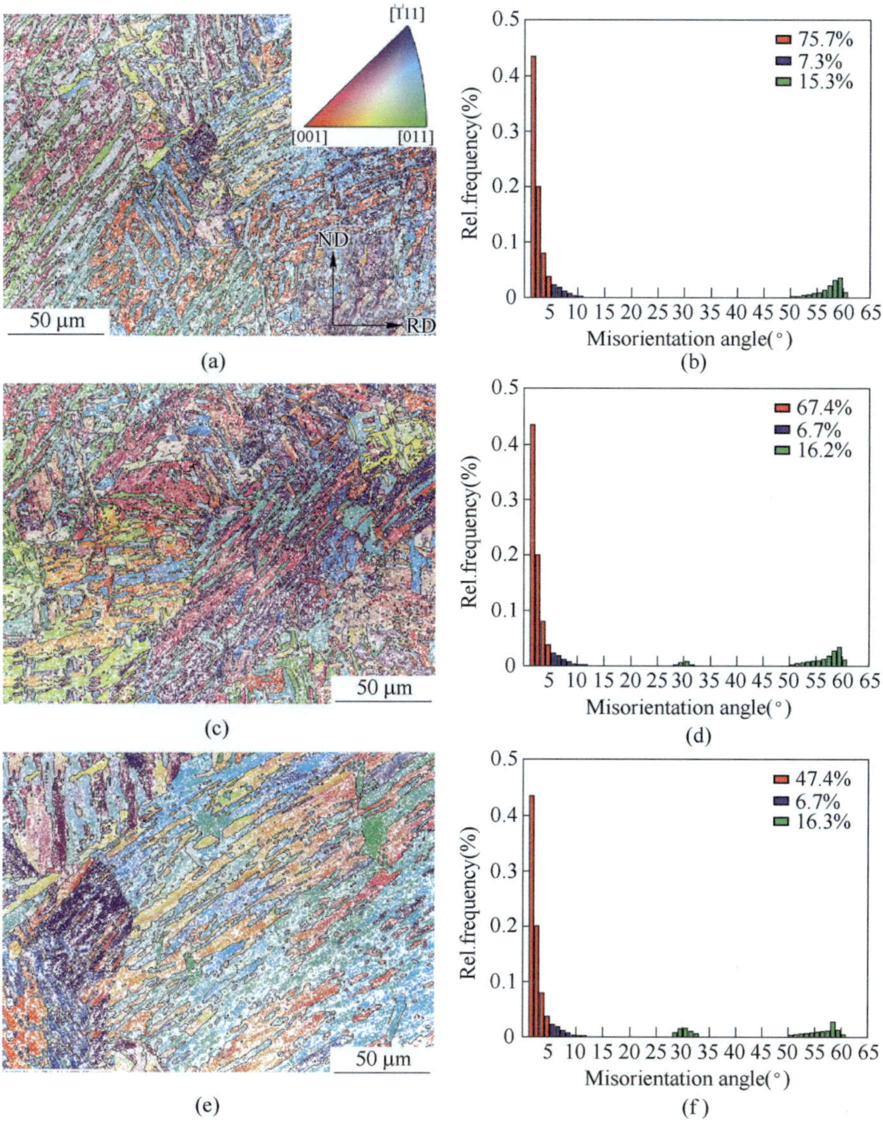

Fig. 6.14 Inverse pole figure maps and grain boundary misorientation distributions of different holding times (prestrain is 0.15): **a**, **b** Deformation quenching with no holding time; **c**, **d** 30 s, and; **e**, **f** 100 s

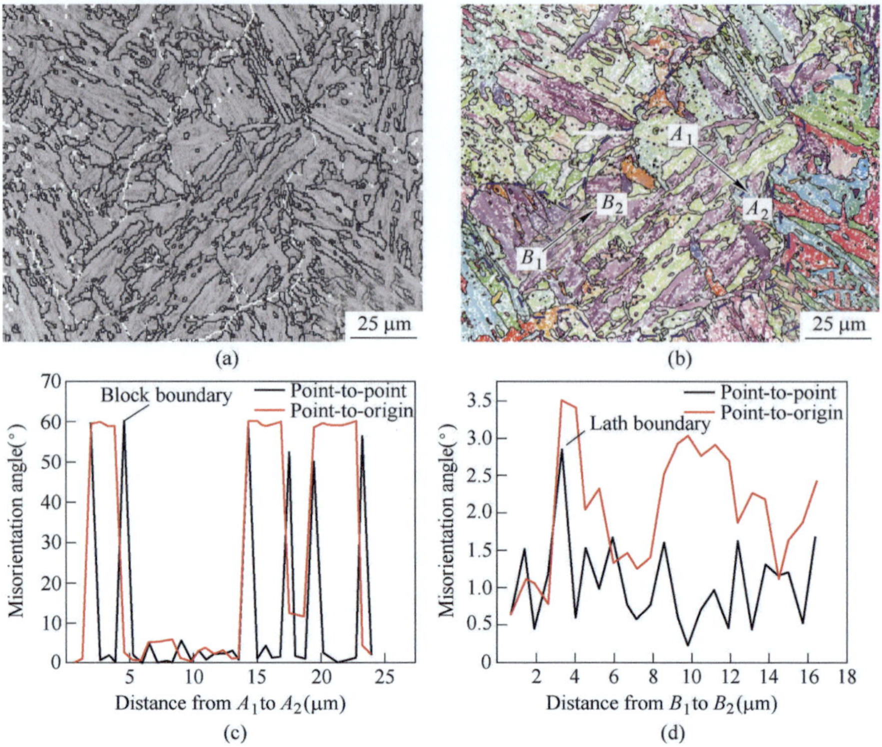

Fig. 6.15 **a** Image quality map with a prior austenite grain; **b** inverse pole figure; **c** misorientation profile along the transverse axis of a martensite block, lines A_1–A_2, and **d** misorientation profile along the transverse axis of a martensite lath, lines B_1–B_2

dislocation angle of the lath grain boundary is less than 3.5°, indicating that the dislocation density of the lath grain boundary is low.

6.1.2.3 Static Recrystallization Model Establishment

According to the results of in situ observations, a static recrystallization kinetic model was established to describe the microstructure evolution behavior of the microalloyed steel Q345E continuous casting slab during reduction. The model is based on the research of Huang et al. [24], and is revised in combination with the Avrami index to make the prediction results more accurate. The static recrystallization kinetic model equation is as follows [25]:

$$\begin{cases} X_s = X_s^{\infty}\left[1 - \exp(-0.693\left(\frac{t}{t_{0.5}}\right)^n)\right] \\ n = \lambda_1 \dot{\varepsilon}^{\lambda_2} \varepsilon^{\lambda_3} d_0^{\lambda_4} \exp(\frac{Q_n}{RT}) \\ X_s^{\infty} = \nu_1 \dot{\varepsilon}^{\nu_2} \varepsilon^{\nu_3} d_0^{\nu_4} Z_s^{\nu_5} \\ t_{0.5} = \omega_1 \dot{\varepsilon}^{\omega_2} \varepsilon^{\omega_3} d_0^{\omega_4} \left(\dot{\varepsilon}\exp(\frac{Q_s}{RT})\right)^{\omega_5} \end{cases} \quad ((6.18))$$

where $\lambda 1$–$\lambda 4$, Qn, $v1$–$v5$, and $\omega 1$–$\omega 5$ are material constants, which are also undetermined coefficients; XS is the static recrystallization volume fraction, %; n is the Avrami index; t is the holding time; XS∞ is the maximum static recrystallization volume fraction, %; $t_{0.5}$ is the recrystallization 50% time; $\dot{\varepsilon}$ is the strain rate, s^{-1}; d0 is the initial austenite grain size, μm; R is the gas constant, 8.314 J mol-1 K-1; T is the deformation temperature, K; QS is the appearance of static recrystallization activation energy, J/mol; ZS is the $\dot{\varepsilon}$ exp(QS/RT).

X^* is the variation of the static recrystallization volume fraction with the holding time, which is calculated from the Eq. (6.19):

$$X^* = X_S / X_S^{\infty} = 1 - \exp(-0.693\left(\frac{t}{t_{0.5}}\right)^n) \quad (6.19)$$

The model considers the effects of deformation temperature, strain rate, initial austenite grain size, and prestrain on the static recrystallized volume fraction. Taking the logarithm of the first expression of Eq. (6.19) yields:

$$\ln(\ln(1/(1 - X^*))) = \ln 0.693 + n\ln(t/t_{0.5}) \quad (6.20)$$

Linear fitting is performed on ln(ln(1/(1-X*))) and ln(t), and the results are shown in Fig. 6.16.

The Avrami index n corresponding to each deformation parameter is obtained. The mathematical expression for the value of n obtained by combining the empirical equation is as follows:

$$n = 0.26\dot{\varepsilon}^{-0.024} \varepsilon^{-0.426} d_0^{0.342} \exp\left(\frac{542.217}{RT}\right) \quad (6.21)$$

The logarithm of the static recrystallization volume fraction expression in Eq. (6.18) is selected to perform linear fitting of the maximum volume fraction of static recrystallization, deformation temperature, strain rate, initial austenite grain size and prestrain. The parameters $n1$, $n2$, $n3$, $n4$ and $n5$ are 6314.68, 0.149, 1.179, – 1.314, – 0.067, respectively. The comparison between the calculated results of X_s^{∞} and the experimental results is shown in Fig. 6.17a, and the correlation coefficient R is 0.913.

Take the logarithm of both sides of the fourth expression of Eq. (6.18), where Qs is 239027 J/mol. The values of parameters $\omega 1$, $\omega 2$, $\omega 3$, $\omega 4$, and $\omega 5$ were obtained by the method of multiple nonlinear regression. The comparison between the calculated

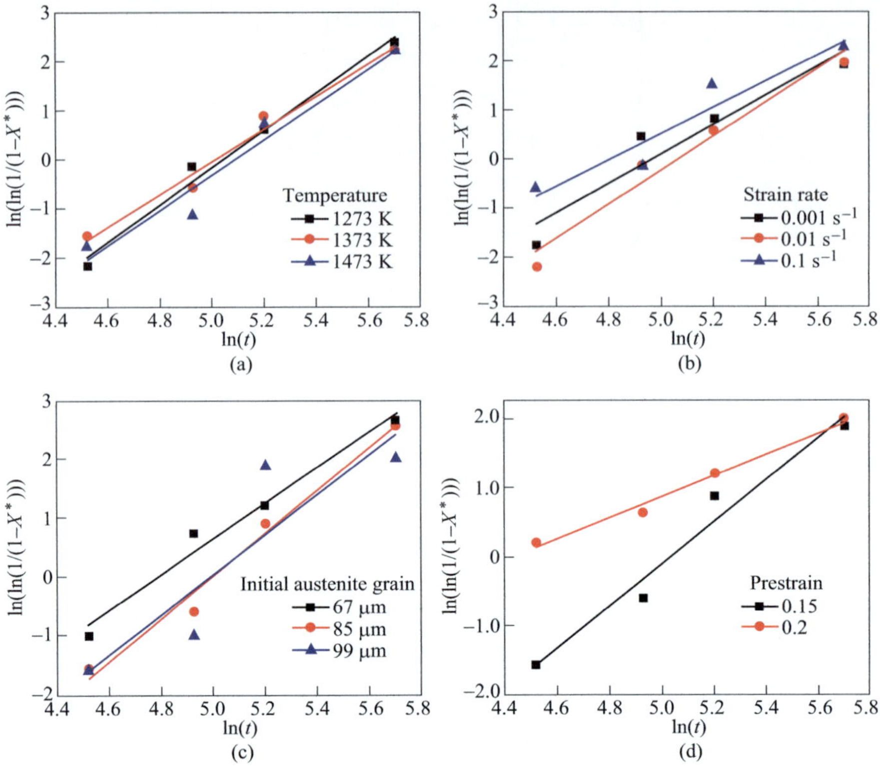

Fig. 6.16 Relationships between, ln(ln(1/(1-X*))) and ln(t) under different deformation conditions. **a** Deformation temperature; **b** strain rate; **c** initial austenite size, and **d** prestrain

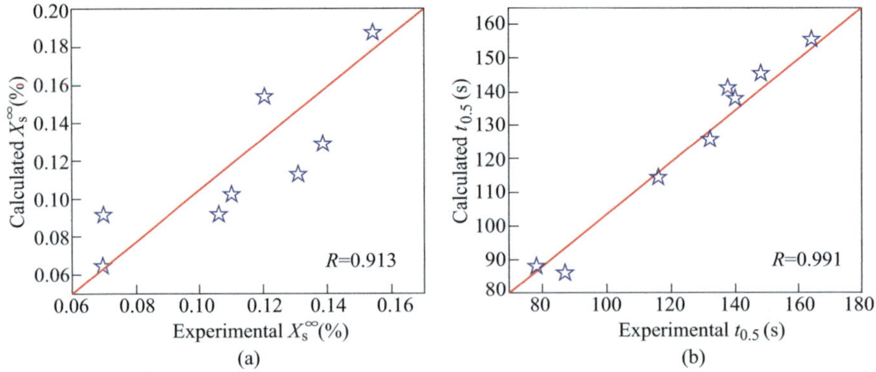

Fig. 6.17 **a** Comparison between the experimental and calculated X_S^∞; **b** comparison between the experimental and calculated $t_{0.5}$

results and the experimental value $t_{0.5}$ is shown in Fig. 6.17b, and the correlation coefficient R is 0.991.

Therefore, according to the above, the kinetic equation of static recrystallization is as follows [25]:

$$\begin{cases} X_s = X_s^{\infty}\left[1 - \exp(-0.693\left(\frac{t}{t_{0.5}})^n\right)\right] \\ n = 0.26\dot{\varepsilon}^{-0.024}\varepsilon^{-0.426}d_0^{0.342}\exp(\frac{542.217}{RT}) \\ X_s^{\infty} = 6314.68\dot{\varepsilon}^{0.149}\varepsilon^{1.178}d_0^{-1.314}Z_s^{-0.067} \\ t_{0.5} = 0.016\dot{\varepsilon}^{-0.041}\varepsilon^{-1.983}d_0^{1.35}(\dot{\varepsilon}\exp(\frac{239027}{RT}))^{-0.07} \end{cases} \tag{6.22}$$

6.1.3 *Metadynamic Recrystallization Characteristics of Continuous Casting Strand for Reduction Process*

Metadynamic recrystallization (M_{DRX}) occurs during the subsequent pass interval when the slab is strained beyond the critical strain for dynamic recrystallization during reduction. It cannot be captured by means of high-temperature confocal microscopy, so high-temperature compression experiments are used to study. It can be seen from the discussion in the previous section that the austenite temperature ranges from 900 °C to 1200 °C and the strain rate ranges from 0.001 s^{-1} to 0.1 s^{-1} during the interval between passes. The necessary condition for the occurrence of sub-dynamic recrystallization is that the prestrain is greater than the critical strain ($\varepsilon c = 0.2$) for dynamic recrystallization. Therefore, the design strain ranges from 0.25 to 0.35, and the pass interval ranges from 1 to 100 s.

6.1.3.1 Sub-Dynamic Recrystallization Multipass Heat of Compression Simulation

Single-pass and double-pass isothermal compression of Q345E was carried out using Thermecmaster-Z thermal simulation compression tester. Single-pass compression was used to study the microstructure of the slab after prestraining. The double-pass experiment is mainly based on the difference in stress or strain energy between the two passes caused by the softening of the austenite between the two passes. The metadynamic recrystallization volume fraction was calculated by the stress compensation method.

In this study, the 0.2% offset strain approach was employed to ascertain the M_{DRX} softening fraction within the two-pass compression procedure. The relative M_{DRX} softening fraction (fs) was calculated using Eq. (6.23) [26]:

$$f_s = \frac{\sigma_m - \sigma_2}{\sigma_m - \sigma_1} \tag{6.23}$$

where σ_m is the flow stress at the maximum strain in the first deformation step; σ_1 is the stress (offset strain of 0.2%) in the first deformation step; σ_2 is the stress (offset strain of 0.2%) in the second deformation step.

The M_{DRX} volume fraction (X_{MDRX}) was determined by the softening fraction (fs), as shown in Eq. (6.24):

$$X_{MDRX} = \frac{f_s - 0.2}{1 - 0.2} = \frac{f_s - 0.2}{0.8} \tag{6.24}$$

At the same time, the original microstructure of a single pass was compared to study the evolution law of microstructure under different deformation conditions. The sub-dynamic recrystallization experiment process is shown in Fig. 6.18, and two groups of experiments were carried out respectively. The first set of experiments is shown in Fig. 6.18a Condition I. First, it was heated to 1250 °C at 10 °C/s, held for 1 min, 5 min, and 10 min, and then directly quenched. The purpose was to test the influence of different initial austenite grain sizes on the behavior of sub-dynamic recrystallization. With the aim of investigating the impact of diverse deformation parameters on sub-dynamic recrystallization under varying initial austenite grain dimensions a single-pass compression experiment was performed. Condition II is shown in Fig. 6.18a. Heat to 1250 °C at 10 °C/s and hold for 1, 5, and 10 min respectively. Then, the cooling rate was 6 °C/s to the deformation temperature, and the temperature gradient was eliminated by holding for 1 min. The deformation temperatures are 900 °C, 1000 °C, and 1100 °C, the prestrains are 0.25, 0.3, and 0.35, and the strain rates are 0.001 s^{-1}, 0.01 s^{-1}, and 0.1 s^{-1}. After holding for 1–100 and then directly quenched. The purpose is to study the effect of different deformation parameters on the evolution of the sub-dynamic recrystallization microstructure. The second group of experiments is shown in Fig. 6.18b, and the experimental steps are the same before the second charging. A second pass of compression followed by direct quenching was performed to determine the effect of different deformation parameters on the sub-dynamic recrystallization softening fraction. Cut along the pressing direction, and use picric acid reagent to add corrosion inhibitor in a constant temperature water bath at 80 °C for 30 min. Austenite grains in different regions were observed using a Zeiss optical microscope (OM). The average austenite grain size was calculated by the line intercept method.

6.1.3.2 Sub-Dynamic Recrystallization Microstructure Evolution

As shown in Fig. 6.19a, the yield stress of the second pass was less than that of the first pass under the same deformation conditions, which is attributed to the softening effect at the first deformation. Considering the experimental error and microstructural heterogeneity, the flow curves of the first pass and the second pass at an inter-pass time of 10 s were significantly higher than those under other conditions. With an increase in the inter-pass time from 30 to 50 s, the flow curves of the second pass decreased rapidly, indicating that the increases in the M_{DRX} softening fraction, residual stress,

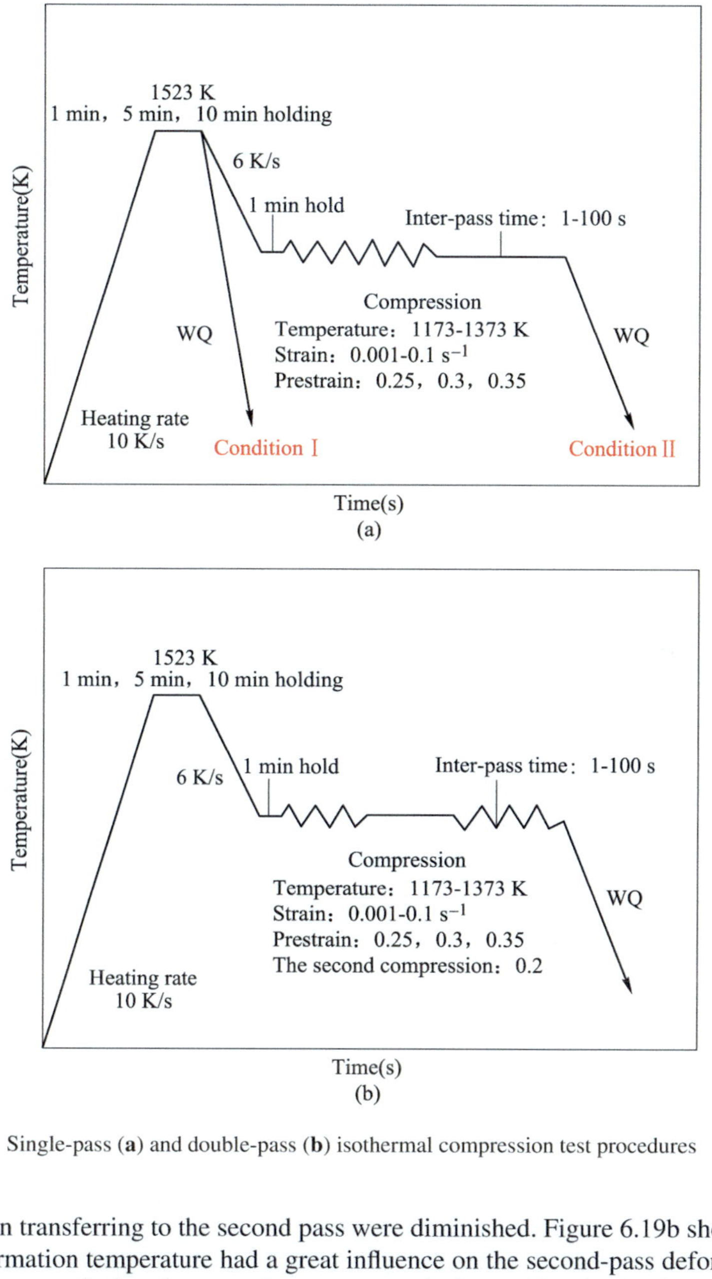

Fig. 6.18 Single-pass (**a**) and double-pass (**b**) isothermal compression test procedures

and strain transferring to the second pass were diminished. Figure 6.19b shows that the deformation temperature had a great influence on the second-pass deformation. The true stress during the second pass was equivalent to that in the first pass at a temperature of 1173 K, indicating that the softening rate was very small. However, at deformation temperatures of 1273 K and 1373 K, the true stress in the second pass decreased significantly, mainly because the amount of stored energy decreased and the DRX volume fraction increased when the deformation temperature increased.

The influence of the strain rate is presented in Fig. 6.19c; the higher the strain rates, the more softening that occurs at 1273 K. At a strain rate of 0.001 s^{-1}, the softening phenomenon was hardly visible. Figure 6.19d shows the true stress–strain curves for different initial austenite grain sizes. With the augmentation of the austenite grain size, the true stress–strain curve did not significantly increase. As shown in Fig. 6.19e, the M_{DRX} softening rate increased in the first deformation step as the maximum strain increased because the amount of stored energy increased.

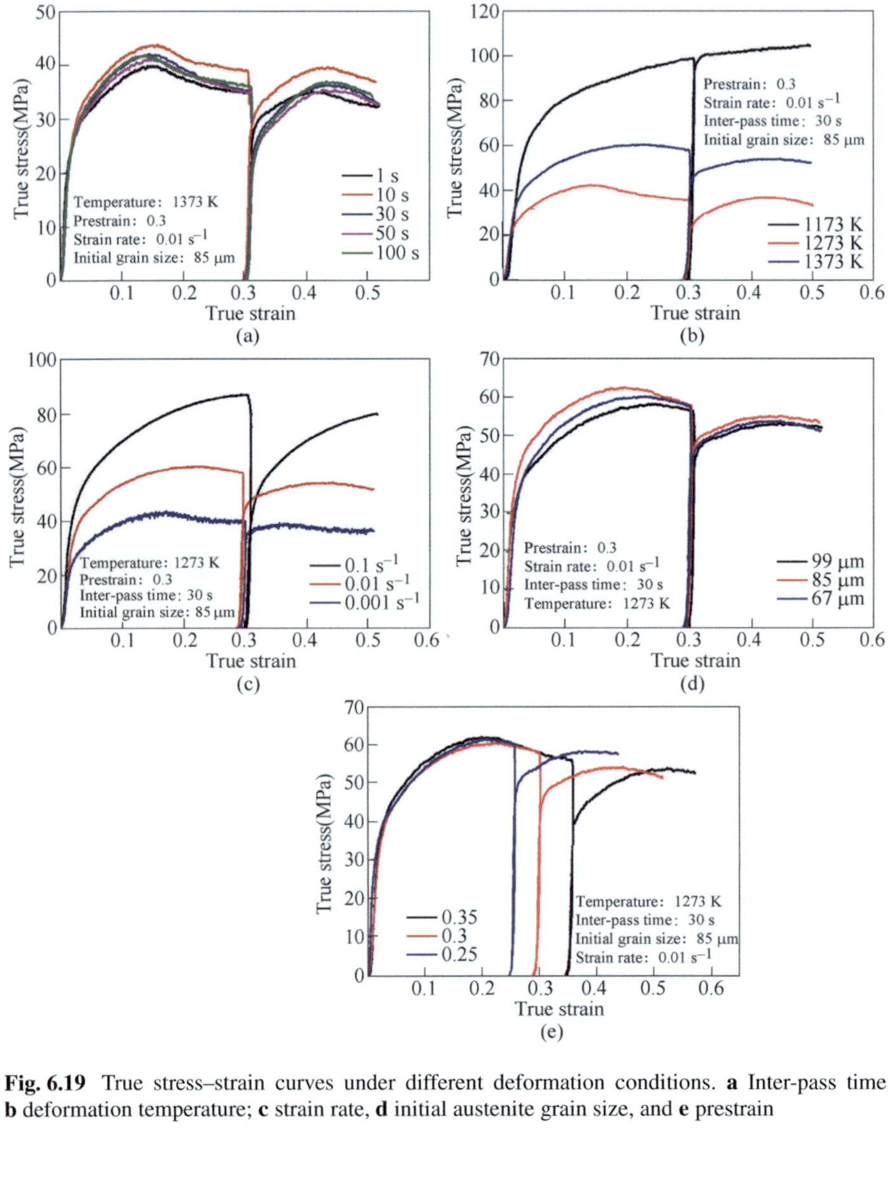

Fig. 6.19 True stress–strain curves under different deformation conditions. **a** Inter-pass time; **b** deformation temperature; **c** strain rate, **d** initial austenite grain size, and **e** prestrain

The M_{DRX} volume fraction as a function of the inter-pass time is shown in Fig. 6.20a for a prestrain of 0.3, a strain rate of 0.01 s^{-1} and an initial austenite grain size of 85 μm. The volume fraction of M_{DRX} increased from 2.4 to 27% and 18.6% from 53.3% when the deformation temperature was raised from 1173 to 1373 K and the inter-pass time was extended from 1 to 100 s, respectively. These increases are explained mainly by the substantial increase in the high-angle grain boundary (HAGB) mobility with increasing temperature. The sub-dynamic recrystallization softening curve of the deformation temperature conforms to the Avrami equation. Since the sub-dynamic recrystallization is a process of austenite grain boundary migration during the interval between passes after the dynamic recrystallization nucleation occurs, the internal deformation energy stored within the high-temperature deformed sample is considerable, and the grain boundary migration rate is fast, so the sub-dynamic recrystallization can be completed in a short time, resulting in a gentle curve after the interval of 30 s at 1100 °C.

Figure 6.20b, c and d show the microstructure of steel under the same conditions (an inter-pass time of 30 s at deformation temperatures ranging from 1173 to 1373 K. The corresponding austenite grain sizes were 80 μm, 60 μm, and 56 μm. As shown in Fig. 6.20b, the prior austenite grains were flattened and elongated. Fine recrystallized

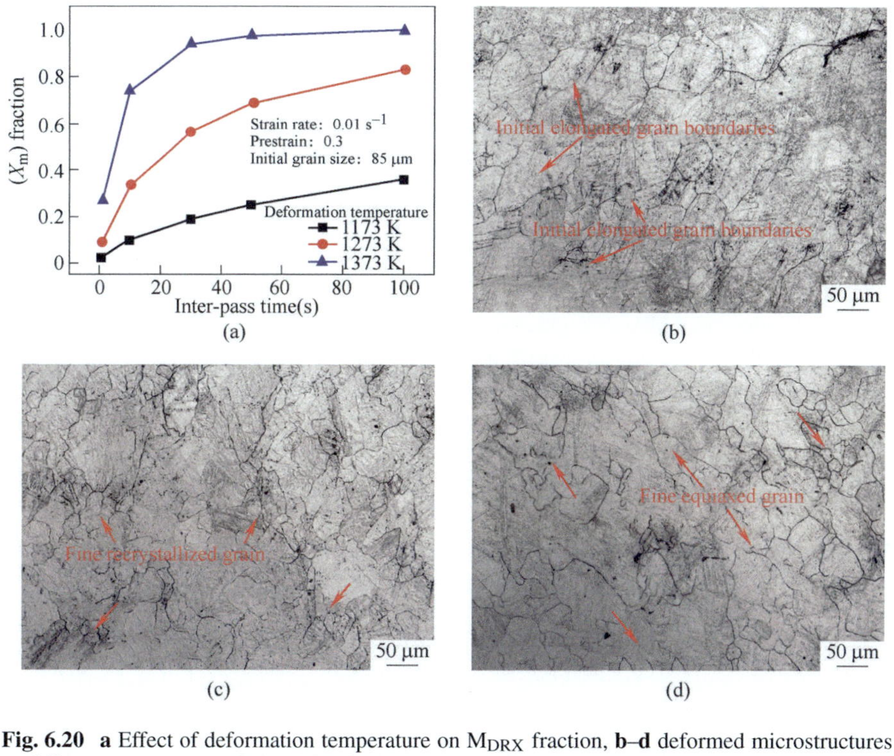

Fig. 6.20 **a** Effect of deformation temperature on M_{DRX} fraction, **b–d** deformed microstructures at different temperatures; **b** 1173 K—inter-pass time of 30 s; **c** 1273 K—inter-pass time of 30 s, and **d** 1373 K—inter-pass time of 30 s

grains were distributed at the boundaries of the prior austenite grains and formed a necklace or lump microstructure, which implies that a partial M_{DRX} phenomenon took place within the specimen. As shown in Fig. 6.20c, the size of the fine recrystallized grains increased significantly, and the coarse austenite grains inside the sample were replaced by fine grains at a deformation temperature of 1273 K. As the deformation temperature increased to 1373 K, as shown in Fig. 6.20d, the inside of the sample contained small equiaxed M_{DRX} grains. The effect of strain on the recrystallization behavior was divided into three main regions. When the deformation temperature increased from 1173 to 1273 K, M_{DRX} occurred during the inter-pass time and was accompanied by SRX, the occurrence of which required an incubation period to eliminate internal defects in the deformed microstructure. At a deformation temperature of 1373 K, pure M_{DRX} occurred during the inter-pass time, consequently, the metallographic structure exhibited a fine and uniform characteristic.

The effects of the strain rate and initial austenite grain size on the M_{DRX} volume fraction and recrystallized grains are shown in Fig. 6.21, which shows a similar trend as the results of previous studies. Since the DRX volume fraction is not sufficient for pure M_{DRX} to occur within the inter-pass time, the influence of prestrain on recrystallization cannot be ignored. As the prestrain increased from 0.25 to 0.35, the M_{DRX} volume fractions with an inter-pass time of 30 s were 46.6, 56.4, and 63.2%. When the austenite grains were subjected to a deformation temperature of 1273 K, the austenite grain sizes were 62 μm, 60 μm, and 53 μm with prestrains of 0.25, 0.3, and 0.35, respectively.

Figure 6.22 shows the Inverse Pole Figure (IPF) maps and misorientation distribution of the samples deformed to a strain of 0.3 at 1273 K with a strain rate of 0.01 s^{-1} followed by different inter-pass times. Figure 6.22a shows that the grains were mainly sub-block and lath boundaries after an inter-pass time of 5 s. As the inter-pass time increased to 60 s, packet or block boundaries formed. The grain boundary misorientation distributions are shown in Figs. 6.22b, d and f. The Rel. frequency of very low-angle grain boundaries (VLAGBs) decreased from 58.8 to 52.1%, which the Rel. frequency of HAGBs increased from 28.5 to 38.9%, but VLAGBs were still dominated, indicating that recovery constituted the predominant softening mechanism. The total grain boundary length (TGBL) decreased from 2.247 × 104 mm to 7.007 × 103 mm, suggesting that subgrain growth occurred with the inter-pass time increasing from 5 to 60 s.

6.1.3.3 Modeling Kinetics of M_{DRX}

To accurately understand the microstructure evolution of the continuously cast slab during the two-pass reduction process, it was necessary to establish an M_{DRX} kinetic model, which can be expressed by the JMAK equation, as shown in Eq. (6.25):

$$X_{MDRX} = 1 - \exp\left[-0.693\left(\frac{t}{t_{0.5}}\right)^{n_{MDRX}}\right] \quad (6.25)$$

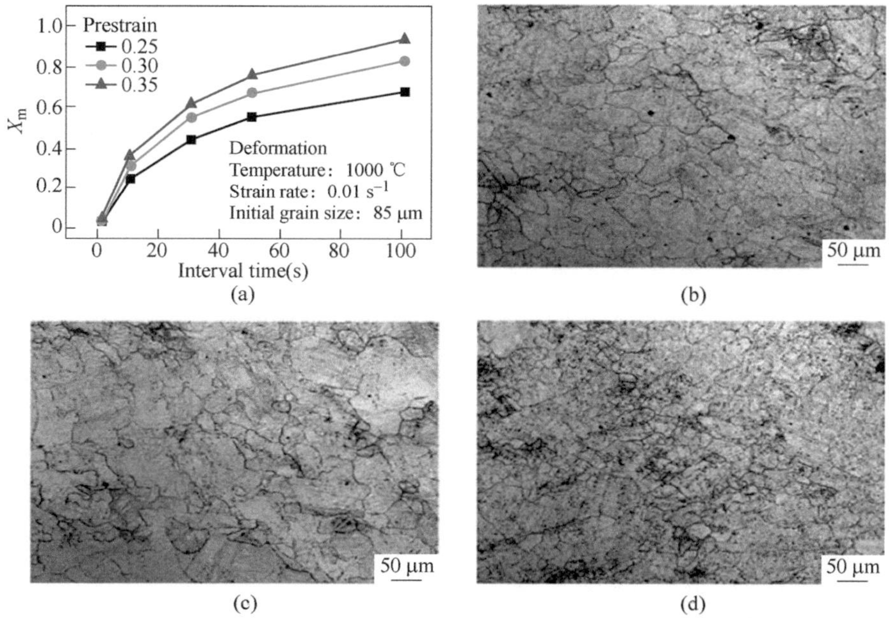

Fig. 6.21 M_{DRX} volume fraction **a** Under different prestrain conditions and austenite metallographic structure under different prestrain conditions; **b** 0.25, **c** 0.3, **d** 0.35

X_{MDRX} is related to factors including the deformation temperature, strain rate, prestrain, and initial austenite grain size.

For M_{DRX},

$$t_{0.5} = A_{MDRX}\dot{\varepsilon}^p \varepsilon^q d_0^s \exp\left(\frac{Q_{MDRX}}{RT}\right) \quad (6.26)$$

where $t_{0.5}$ is the time for 50% recrystallization, s; nM_{DRX}, A_{MDRX}, p, q, and s are material dependent constants; $\dot{\varepsilon}$ is the strain rate, s^{-1}; ε is the prestrain; d0 is the initial austenite grain size, μm; R is the gas constant, $R = 8.314$ J/(mol K); T is the deformation temperature, K; and Q_{MDRX} is the apparent activation energy of M_{DRX}, J/mol.

Equation (6.27) is obtained by taking the logarithm of both sides of Eq. (6.26):

$$\ln\left(\ln\left(\frac{1}{1 - X_{MDRX}}\right)\right) = \ln 0.693 + n_{MDRX} \ln t - n_{MDRX} \ln t_{0.5} \quad (6.27)$$

The Avrami exponent n_{MDRX} was obtained under different deformation temperatures, strain rates, initial austenite grain sizes, and prestrains. As shown in Fig. 6.23, the Avrami exponent n_{MDRX} was 0.653 according to a linear regression.

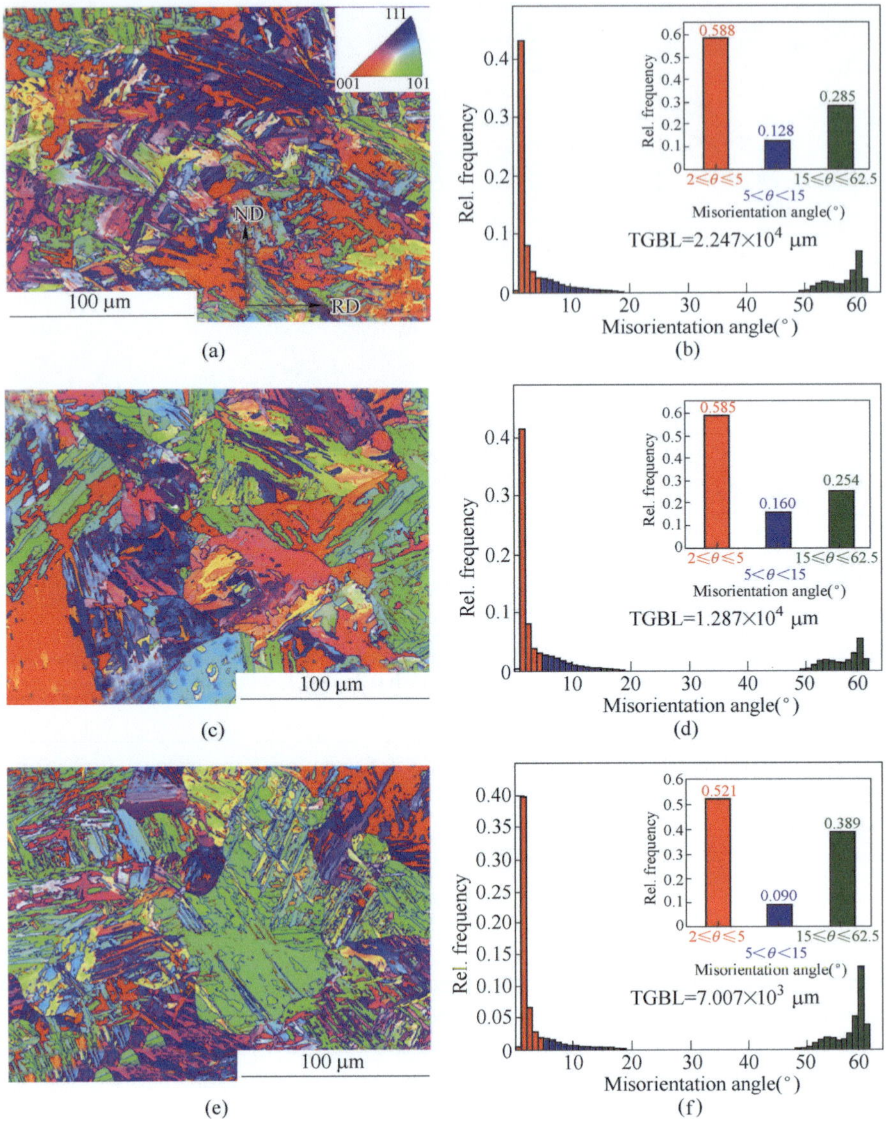

Fig. 6.22 IPF maps and grain boundary misorientation distributions of tested slab with different inter-pass times. **a**, **b** Inter-pass time of 5 s; **c**, **d** inter-pass time of 30 s, and **e**, **f** inter-pass time of 60 s

To verify the M_{DRX} kinetic model obtained above, the values of n_{MDRX} was compared with some published research data. In addition, the apparent activation energy is an important material constant, and the activation energies of different types of steel from previous studies, indicate that the values of n_{MDRX} and Q_{MDRX} in this article are within reasonable ranges.

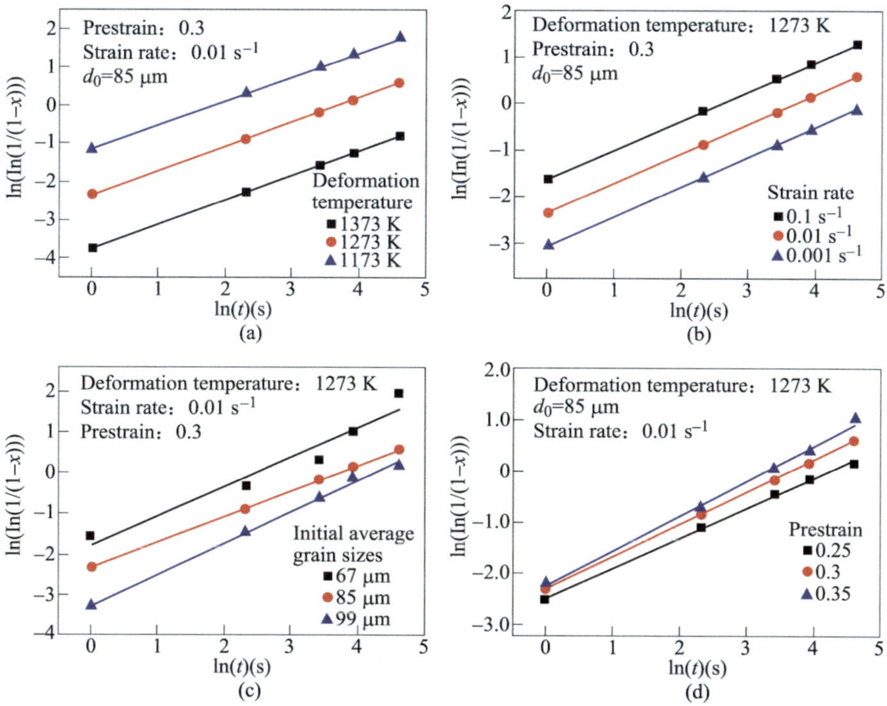

Fig. 6.23 Relationships between ln(ln(1/(1-x))) and ln(t) under different deformation conditions. **a** Deformation temperature; **b** strain rate; **c** initial austenite grain size, and **d** prestrain

Taking the logarithm of both sides of Eq. (6.27) yields the following equation:

$$\ln t_{0.5} = \ln A_{\mathrm{MDRX}} + \ln \dot{\varepsilon}^{P} + \ln \varepsilon^{q} + \ln d_0^{s} + \left(\frac{Q_{\mathrm{MDRX}}}{\mathrm{RT}}\right) \tag{6.28}$$

Through the corresponding relationship between the sub-dynamic recrystallization volume fraction and the pass interval time, the index of the 50% recrystallization time ($t_{0.5}$) and different deformation conditions is obtained, as shown in Fig. 6.24.

In short, the M_{DRX} kinetic equation can be reasonably expressed as follows:

$$\begin{aligned} X_{\mathrm{MDRX}} &= 1 - \exp\left[-0.693\left(\frac{t}{t_{0.5}}\right)^{0.653}\right] \\ t_{0.5} &= 2.143 \times 10^{-17}\dot{\varepsilon}^{-0.437}\varepsilon^{-1.76}d_0^{3.049}\exp\left(\frac{251747.9}{\mathrm{RT}}\right) \end{aligned} \tag{6.29}$$

Based on the above sub-dynamic recrystallization kinetic model, the comparison between the experimental value and the predicted value at 50% volume fraction of recrystallization ($t_{0.5}$) was carried out, and the results are shown in Fig. 6.25.

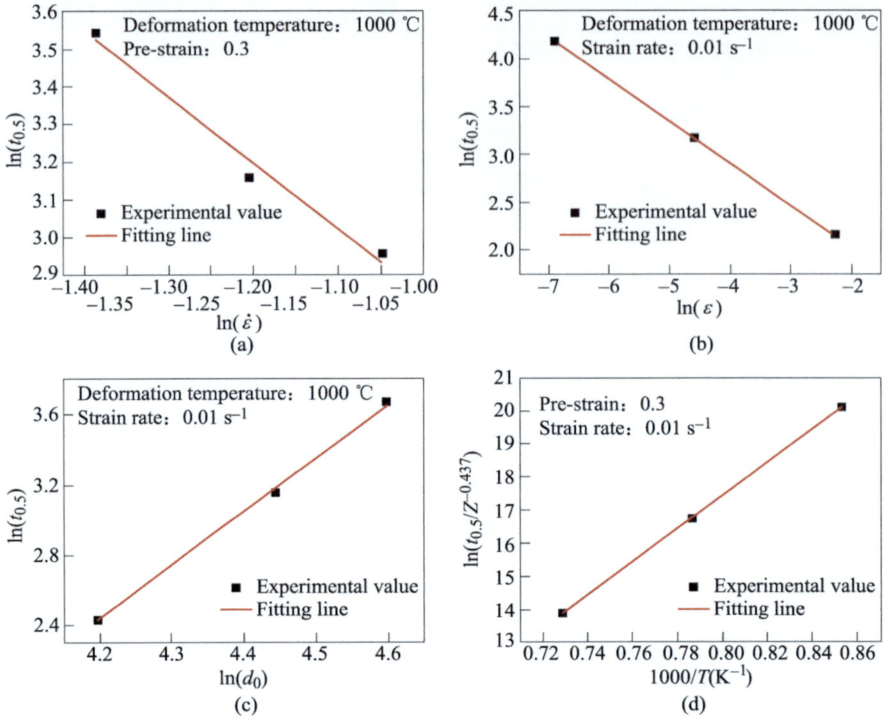

Fig. 6.24 Relationship between ln($t_{0.5}$) and different deformation parameters. **a** ln($\dot{\varepsilon}$); **b** ln(ε); **c** ln(d0); **d** 1000/T

Fig. 6.25 Experimental and predicted values of 50% recrystallization time

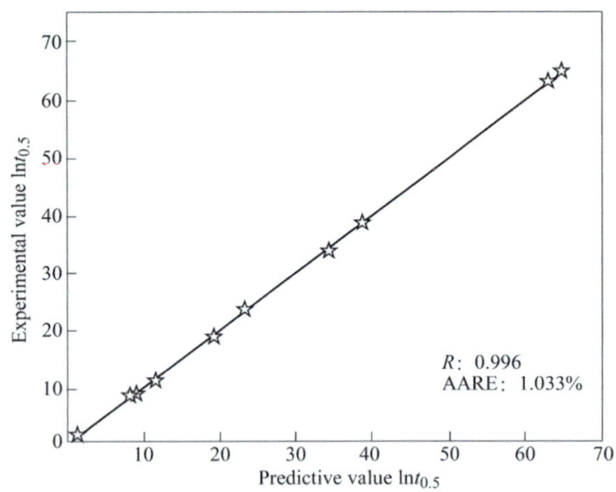

In accordance with the experimental outcomes and the analysis of M_{DRX}, the deformation temperature, strain rate, initial austenite grain size and prestrain greatly influence the post-M_{DRX} grains during HR. Based on the above analysis, the d_{MDRX} prediction model was expressed as Eq. (6.30):

$$d_{\text{MDRX}} = Cd_0^{k_1}\varepsilon^{k_2}\dot{\varepsilon}^{k_3}\exp(\frac{Q_{\text{MDRX}}}{\text{RT}})^{k_4} \tag{6.30}$$

where d0 is the initial austenite grain size, μm; C, k_1–k_4 are material constant, which is also an undetermined coefficient; E is the strain rate, s-1; R is the gas constant, J/(mol K); T is the deformation temperature, K.

Taking the logarithm of both sides of Eq. (6.30) results in Eq. (6.31).

$$\ln d_{\text{MDRX}} = \ln C + k_1 \ln d_0 + k_2 \ln \varepsilon + k_3 \ln \dot{\varepsilon} + k_4 \frac{Q_{\text{MDRX}}}{\text{RT}} \tag{6.31}$$

As shown in Fig. 6.26, the linear relationship between ln(d_{MDRX}) and ln(d0) was obtained, and the k_1 value obtained by fitting was 1.247. Similarly, as shown in Fig. 6.26 b–d, the values of k_2, k_3 and k_4 were determined to be − 0.404, − 0.061, and 0.096, respectively. Based on the material constants k_1, k_2, k_3, and k_4, Eq. (6.27) was used to determine a value of 0.0115 for the constant C.

According to the above calculations, the M_{DRX} grain prediction models under different conditions were obtained, as shown in Eq. (6.32):

$$d_{\text{MDRX}} = 0.0115 d_0^{1.247}\varepsilon^{-0.404}\dot{\varepsilon}^{-0.061}\exp(\frac{251747.9}{\text{RT}})^{0.096} \tag{6.32}$$

A comparison of the predicted and experimental values of the austenite grain size under different deformation conditions is presented in Table 6.4.

6.2 The Law of Tissue Evolution in the Heating Process

The above research proves that the reduction at the end of solidification of slab solidification induces austenite recrystallization and significantly refines the austenite grains. Whether the refined austenite grains will be further retained during the subsequent heating of the charging process remains to be further investigated. Therefore, this section will focus on the effect of the prereduction of the continuous casting slab on the microstructure evolution under the actual charging process. The root causes of the evolution of the slab structure are also analyzed mechanistically. The effects of different reductions and charging processes on the microstructure evolution before slab charging are clarified. It provides important theoretical support for the later formulation and optimization of the continuous casting slab hot charging and hot delivery process.

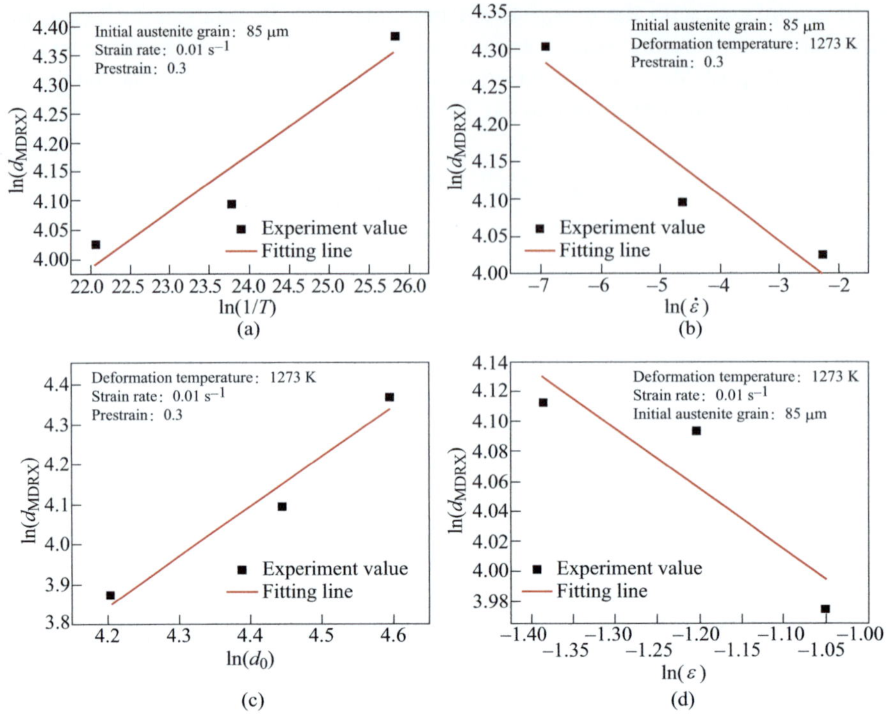

Fig. 6.26 **a** Relationship between ln(d_{MDRX}) and ln(1/T); **b** relationship between ln(d_{MDRX}) and ln($\dot{\varepsilon}$); **c** relationship between ln(d_{MDRX}) and ln(d0), and **d** relationship between ln(d_{MDRX}) and ln(ε)

Table 6.4 Comparison of predicted and experimental values of austenite grain size under different deformation conditions

Temperature (K)	Strain rate (s-1)	Prestrain	d_0 (μm)	Experimental d_{MDRX} (μm)	Predicted d_{MDRX} (μm)	Error value (%)
1173	0.01	0.3	85	80	75	0.063
1273	0.01	0.3	85	60	61	0.017
1273	0.001	0.3	85	74	71	0.041
1273	0.1	0.3	85	56	54	0.035
1273	0.01	0.3	67	48	46	0.042
1273	0.01	0.3	99	79	75	0.051
1373	0.01	0.3	85	56	52	0.072
1273	0.01	0.25	85	62	65	0.048
1273	0.01	0.35	85	53	57	0.076

6.2.1 The Microstructure Evolution of the Wide and Thick Slab for Delivery and Charging Process

High-strength low-alloy steel slabs are finding growing applications across various fields, and their initial microstructure before hot rolling exerts a pivotal influence on the microstructure and properties of the final rolled products. Therefore, the microstructure evolution of the reduced slab during the heating process is demonstrated. It is an important theoretical basis for realizing the optimization of the production process of the thick plate and the improvement of the structure and performance.

6.2.1.1 Comparison of Original Microstructures at Different Positions of Wide and Thick Slabs with Different Reductions

The Q345E slab samples at different positions on the surface and the center with a reduction of 9 mm, 15 mm, and 25 mm were taken, and the microstructure was observed for comparison.

(A) **Structure Comparison of Slab Core**

Figure 6.27 shows the enlarged comparison of the core and local tissue of slabs with different reductions.

It can be seen from the comparison of metallographic results. With the increase of slab rolling reduction, the original austenite grains of the slab gradually decrease, and the change of ferrite grain morphology also has a decreasing trend. Among them, when the reduction is 25 mm, some ferrite has been transformed into acicular ferrite in the structure. The average grain size of prior austenite with a reduction of 25 mm is 810 μm. Compared with the average size of prior austenite grains of 1250 μm with a light reduction of 9 mm, the grain size is reduced by 65%. Due to the reduction of austenite grains, the thermoplasticity of the slab surface can be effectively improved. As a result, the probability of crack generation and extension is indirectly reduced, and the microstructure and properties of the slab are improved.

(B) **Structure Comparison of the Center of the Surface Layer of the Slab**

The center position of the slab surface and its local structure are enlarged and compared with different reductions in Fig. 6.28. It can be seen that the as-cast microstructure of the continuous casting slab at the center of the surface layer changes significantly, and the austenite grains gradually become smaller. The grain size of ferrite gradually decreases with the increase of rolling reduction. In this case, the slaty ferrite has structurally evolved into acicular ferrite at a reduction of 25 mm. The average grain size of austenite with a reduction of 25 mm is 795 μm. Compared with the average grain size of austenite with a reduction of 9 mm of 1200 μm, it is refined by 66.25%.

The austenite is deformed during the reduction process. It promotes the creation of a large number of deformation bands, twins, dislocations and other defects in the

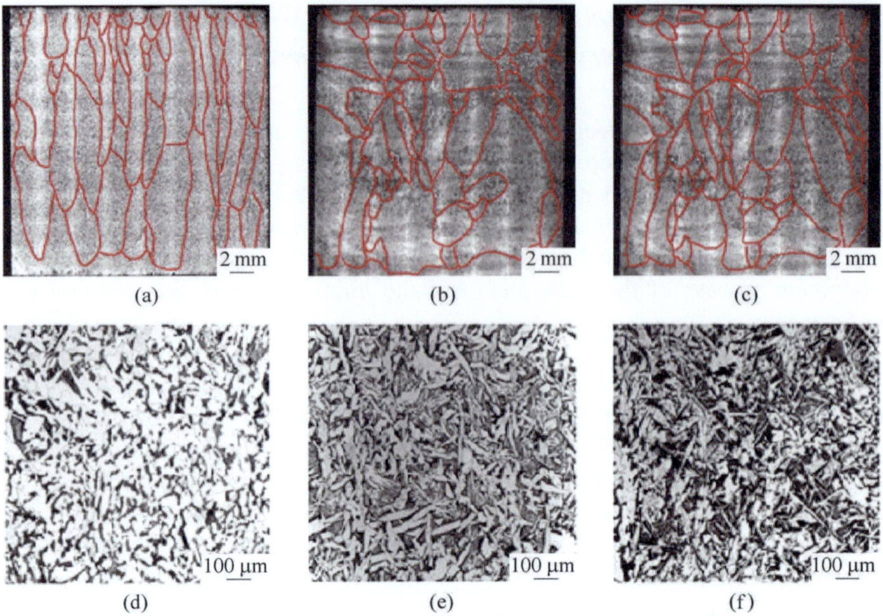

Fig. 6.27 Enlarged comparison of the core and local structures of slabs with different reductions the microstructure of the core. **a** The reduction is 9 mm; **b** the reduction is 15 mm; **c** the reduction is 25 mm; partially enlarged core microstructure; **d** reduction of 9 mm; **e** reduction of 15 mm; **f** reduction of 25 mm

grain boundaries. It provides more nucleation particles for the phase transformation of austenite to ferrite, thereby achieving the purpose of refining the ferrite structure. In parallel, the refinement of ferrite grains can promote the nucleation of austenite during the charging and heating process, thereby achieving the effect of refining austenite grains.

6.2.1.2 Comparison of Charging Effects of Wide and Thick Slabs with Different Reductions in Different Positions

Because the as-cast structure sampled on site is a cold structure. This section simulates the on-site cold-charging experiment and formulates the thermal history curve of the cold-charging process. The purpose is to compare the microstructure evolution of the experimental steel under the same charging conditions with different reduction processes.

(A) The Position of the Center of the Cross-section of the Slab

According to the calculation results of on-site slab temperature field in Chap. 2, a reasonable heating scheme for different positions of the slab is formulated. The

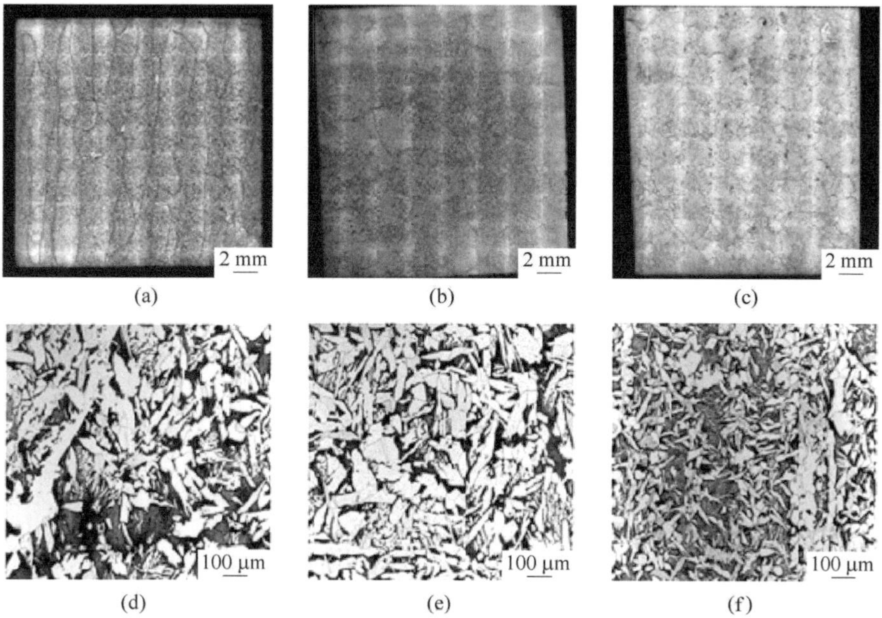

Fig. 6.28 Center position of the slab surface and its local microstructure magnification comparison with different reductions: The central position of the surface organization. **a** The amount of depression is 9 mm; **b** the amount of depression is 15 mm; **c** the amount of depression is 25 mm; partial enlargement of the tissue at the center of the surface layer; **d** the amount of depression is 9 mm; **e** the amount of depression is 15 mm; **f** the amount of depression is 25 mm

thermal history curve is shown in Fig. 6.29a, and the sampling location is shown in Fig. 6.29b.

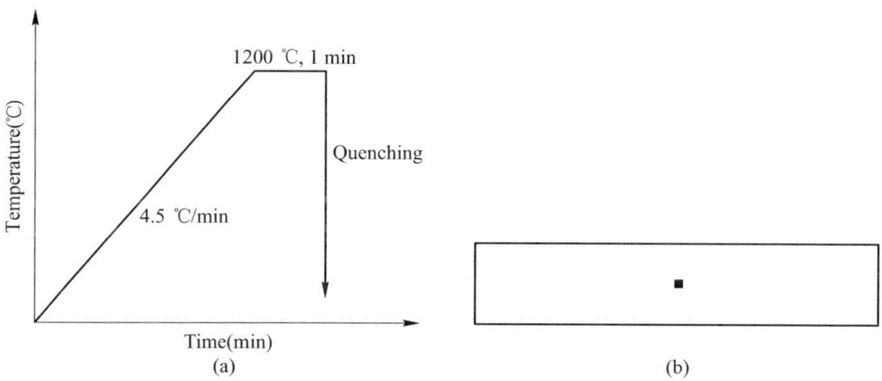

Fig. 6.29 Loading system and sampling scheme of casting billet. **a** Thermal resume curve of the casting core; **b** experimental sampling location

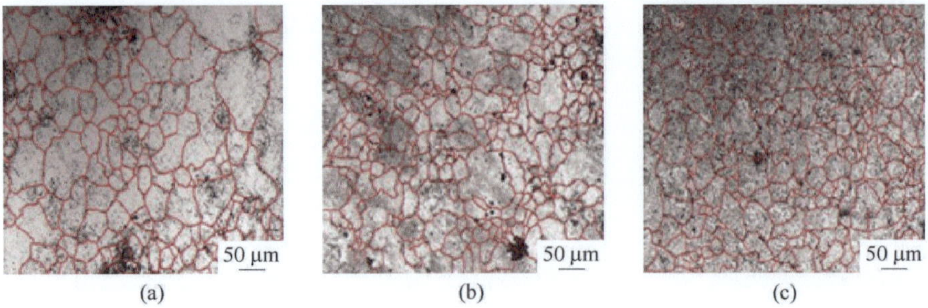

Fig. 6.30 Comparison of the central microstructure of the billet after cold charging with different reductions. **a** Reduction 9 mm; **b** reduction 15 mm; **c** reduction 25 mm

The sample was heated to 1200 °C at 4.5 °C/min and then kept for 60 s. Etching with saturated picric acid solution after quenching revealed the microstructure of the cast slab. The experimental results are shown in Fig. 6.30.

Figure 6.30 is a comparison of the central microstructure of the slab after cold charging with different reductions. It can be seen from the figure that the austenite grain size gradually decreases with the increase of the rolling reduction. When the rolling reduction is 9 mm, the maximum austenite grain size is 105 μm, the minimum austenite grain size is 31 μm, and the average grain size is 53 μm. When the reduction is 15 mm, the maximum grain size is 95 μm, the minimum is 23 μm, and the average grain size is 34 μm. When the reduction is 25 μm, the maximum grain size is 60 μm, the minimum is 18 μm, and the average grain size is 31 μm.

It is clear from the statistical results. The larger the reduction, the smaller the grain size of the slab, and the better the isotropy of the grains. During the reduction process, with the increase of the deformation amount, the recrystallization rate of the original austenite increases, the austenite grains are obviously refined, and the total area of the grain boundaries increases. The ferrite transformation process provides more nucleating particles, which promotes the refinement of the ferrite structure.

(B) **The Center Position of the Surface Layer of the Cross-section of the Slab**

Formulate a reasonable heating scheme for the surface center position of the billet's cross-section based on the calculation results of the on-site billet temperature field in Chap. 2. The thermal history curve is shown in Fig. 6.31a, and the sampling location is shown in Fig. 6.31b.

During the experiment, the samples with reductions of 9 mm, 15 mm, and 25 mm were heated to 900 °C at a rate of 6.5 °C/min. The samples were then heated to 1200 °C at a heating rate of 4.5 °C/min in the temperature range of 900–1200 °C for 60 s, and then quenched to observe the metallographic structure. The experimental results are shown in Fig. 6.32.

Figure 6.32 shows the microstructure comparison of the center position of the cold-loaded surface layer of the slab with different reductions. It can be seen from the figure that the austenite grain size gradually decreases with the increase of the rolling reduction. The grains with a reduction of 9 mm have a maximum size of

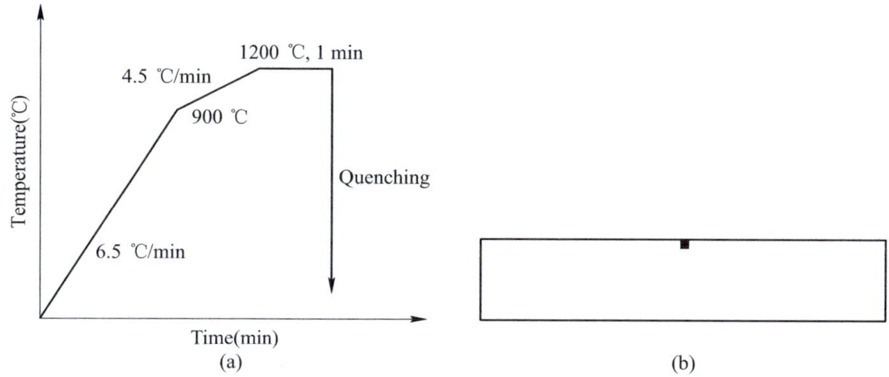

Fig. 6.31 Loading system and sampling scheme of casting billet. **a** The central thermal resume curve of the slab surface; **b** experimental sampling location

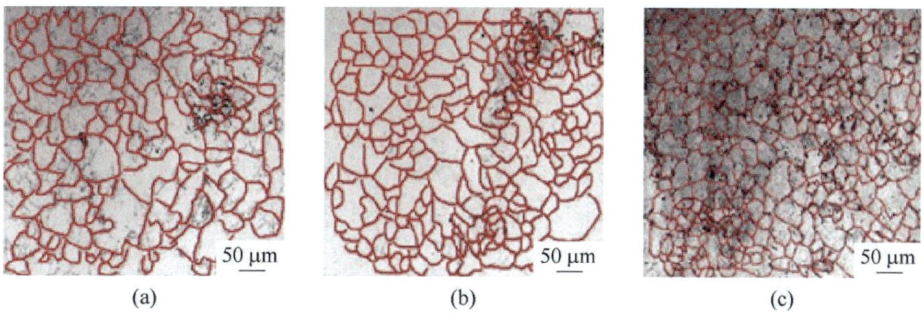

Fig. 6.32 Microstructure comparison of the center position of the cold-loaded surface of the billet with different reductions. **a** Reduction 9 mm; **b** reduction 15 mm; **c** reduction 25 mm

95 μm, a minimum size of 26 μm, and an average grain size of 52 μm. The maximum grain size for a reduction of 15 mm is 86 μm, the minimum is 20 μm, and the average grain size is 31 μm. The maximum grain size is 59 μm and the minimum is 18 μm with a reduction of 25 mm, and the average grain size is 25 μm. When the rolling reduction is 25 mm, the grain refinement of the slab is remarkable and the isotropy of the grains is better. It can be seen from Figs. 6.30 and 6.32 that the surface structure of the slab is smaller than the core structure. Because of the faster cooling rate, finer ferrite grains form on the slab surface, resulting in a finer austenite grain size after reheating.

The reduction induces recrystallization of austenite and increases the deformation energy storage inside the crystal. And provide more energy and nucleation particles for the phase transition of the organization. Therefore, as the reduction amount increases, the deformation energy storage will become larger and larger. The grain-refining impact of austenite becomes more pronounced with increasing reduction.

(C) Slab Cross-section 1/4 Center Position

According to the simulation data, a reasonable heating plan for the position of the core of 1/4 of the cross-section of the slab is formulated to make it closer to the on-site heating conditions. The thermal history curve is shown in Fig. 6.33a, and the sampling location is shown in Fig. 6.33b.

Figure 6.34 shows the comparison of the core structure of the cold-packed 1/4 slab with different reductions. It can be seen from the figure that the austenite grain size gradually decreases with the increase of the rolling reduction. The grains with a reduction of 9 mm have a maximum size of 106 μm, a minimum size of 35 μm, and an average grain size of 57 μm. The grain size with a reduction of 15 mm is 93 μm at the maximum and 21 μm at the minimum, and the average grain size is 41 μm. The maximum grain size for a reduction of 25 mm is 63 μm, the minimum is 23 μm, and the average grain size is 33 μm. When the reduction is 25 mm, the reheated austenite grains are significantly refined compared with 9 mm.

The main reason for this is roughly the same as the mechanism of grain refinement at the surface location. The structure of 1/4 of the core of the slab is almost the same

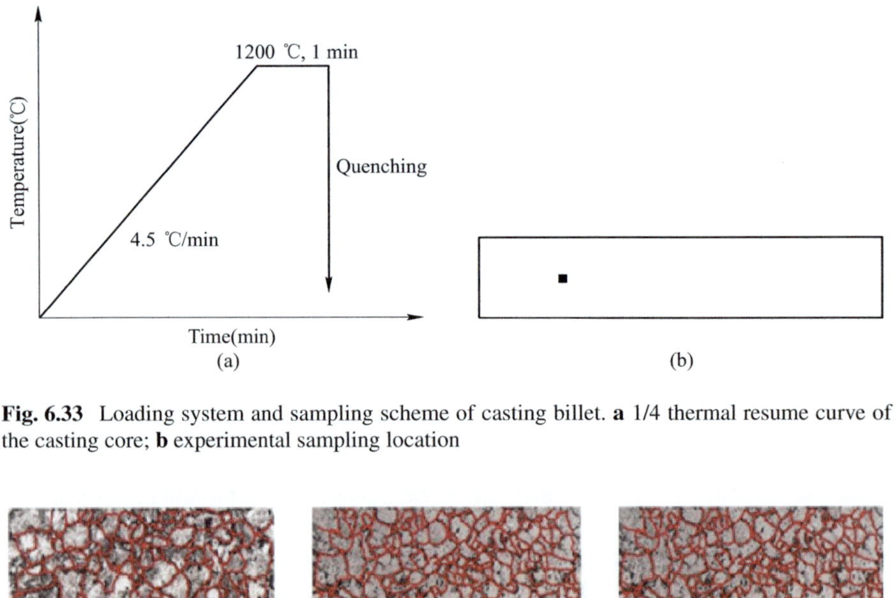

Fig. 6.33 Loading system and sampling scheme of casting billet. **a** 1/4 thermal resume curve of the casting core; **b** experimental sampling location

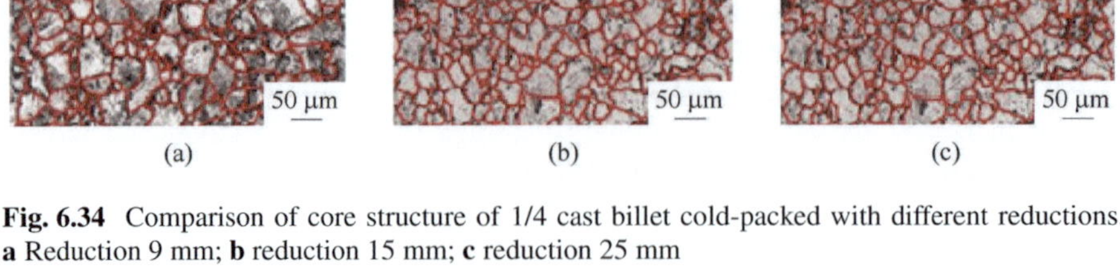

Fig. 6.34 Comparison of core structure of 1/4 cast billet cold-packed with different reductions. **a** Reduction 9 mm; **b** reduction 15 mm; **c** reduction 25 mm

as the grain size of the core of the slab. The reason of this is that the thermal history of the charging process is the same in both sites. The differences in the causes of organizational variability are relatively small.

(D) The Position of the Corner of the Cross-section of the Slab

According to the calculation results of the slab temperature field in Chap. 2, a heating scheme is formulated. The thermal history curve is shown in Fig. 6.35a, and the sampling location is shown in Fig. 6.35b.

During the experiment, samples with different reductions were heated to 900 °C at 14 °C/min. The sample was then heated to 1200 °C at a heating rate of 5 °C/min in the temperature range of 900–1200 °C for 60 s. After the final quenching, the tissue was corroded with saturated picric acid solution. The experimental results are shown in Fig. 6.36.

Figure 6.36 shows the comparison of cold-packed microstructures at the corners of slabs with different reductions. According to the statistics of austenite grain size, the maximum grain size of the slab with a rolling reduction of 9 mm is 80 μm, the minimum size is 24 μm, and the average grain size is 36 μm. The grain size with

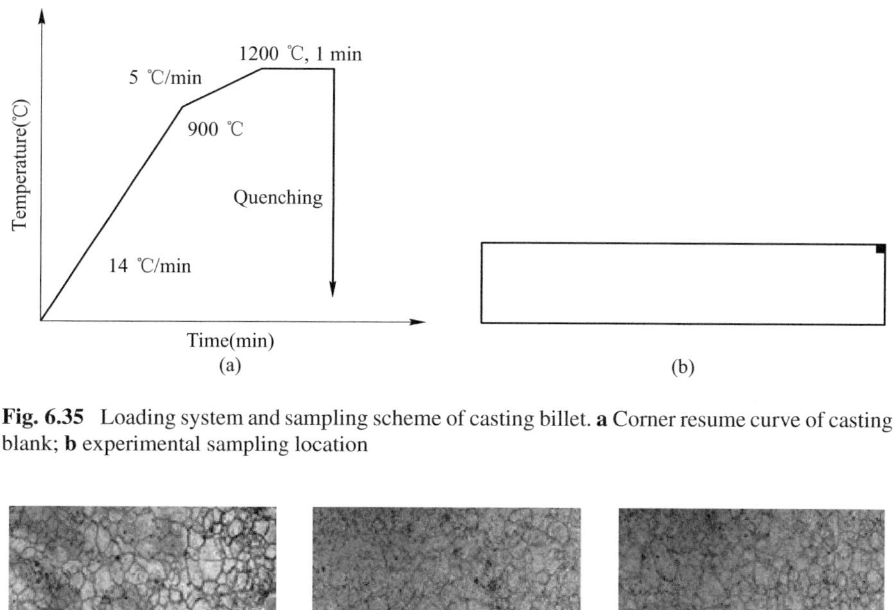

Fig. 6.35 Loading system and sampling scheme of casting billet. **a** Corner resume curve of casting blank; **b** experimental sampling location

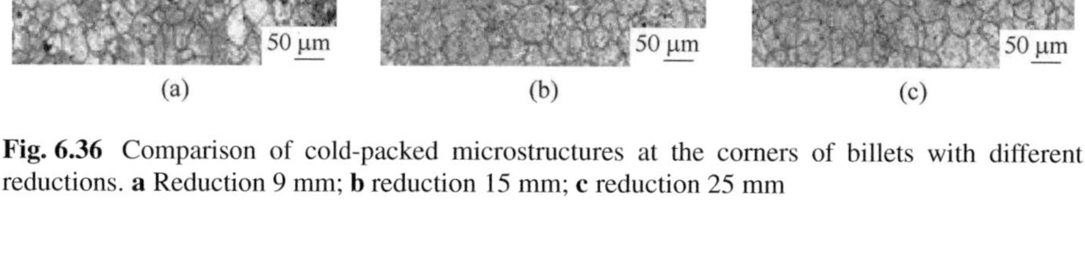

Fig. 6.36 Comparison of cold-packed microstructures at the corners of billets with different reductions. **a** Reduction 9 mm; **b** reduction 15 mm; **c** reduction 25 mm

a reduction of 15 mm is 76 μm at the maximum, 20 μm at the minimum, and the average grain size is 35 μm. The grain size with a reduction of 25 mm is 73 μm at the maximum, 18 μm at the minimum, and the average grain size is 33 μm. It can be seen from the statistical results there is little difference in austenite grain size.

Therefore, the structure of the slab corners does not change significantly with the increase of the rolling reduction. The reason for this is that the cooling rate of the corners is much higher than that of other positions during the continuous casting process. Mechanical stress has little effect on the nucleation of the organization. Therefore, increasing the rolling reduction cannot refine the corner structure of the slab.

(E) **The Center Position of the Narrow Face of the Slab Cross-section**

According to the calculation results of the actual slab temperature field in Chap. 2, a reasonable heating plan for the center position of the narrow surface of the slab cross-section is formulated. Figure 6.37a is the heating history curve of the center of the narrow face of the slab, and Fig. 6.37b is the sampling position of the slab.

The experimental procedure is as follows: the samples with different reductions are heated to 900 °C at 10 °C/min. Then heated to 1200 °C at 5 °C/min for 60 s and then quenched. The microstructure and evolution of the slab samples with different reductions were observed by metallographic microscope, as shown in Fig. 6.38.

Figure 6.38 shows the comparison of the cold-packed structure of the narrow face center of the slab with different reductions. As can be seen from the figure, there is almost no significant change in austenite grain size with increasing reduction.

The statistical results of austenite grain size show that the largest grain size is 90 μm, the minimum size is 38 μm, and the average size is 47 μm when the reduction is 9 mm. The grain size with a reduction of 15 mm is 73 μm at the maximum, 26 μm at the minimum, and 39 μm in average. The grain size with a reduction of 25 mm is 77 μm at the maximum, 18 μm at the minimum, and the average size is 35 μm.

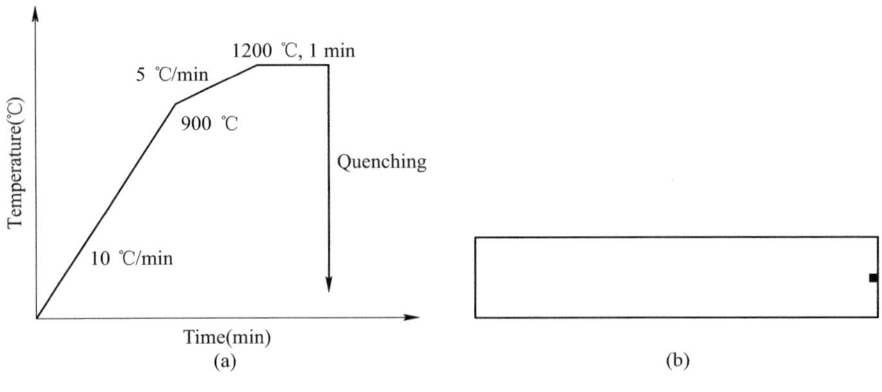

Fig. 6.37 Loading system and sampling scheme of casting billet. **a** The central resume curve of the narrow face of the casting blank; **b** experimental sampling location

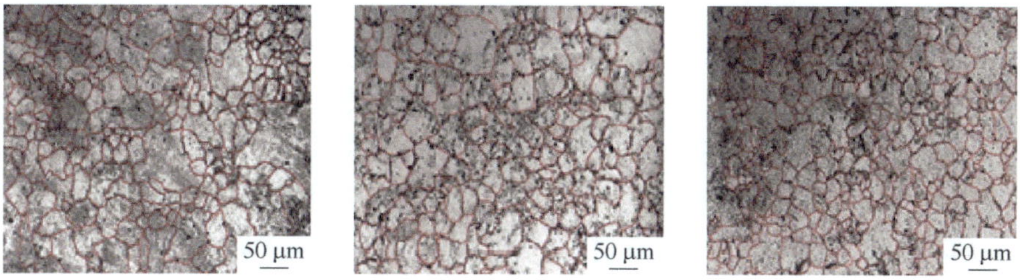

(a) (b) (c)

Fig. 6.38 Comparison of cold-packed microstructure of the narrow face center of the slab with different reductions. **a** Reduction 9 mm; **b** reduction 15 mm; **c** reduction 25 mm

The difference in the austenite grain size of the cast slab decreases with the increase of the reduction. It can be obtained from by observing the metallographic structure. The degree of mixed crystallinity of austenite also decreases with the increase of reduction. However, due to the structure in the center of the narrow face, the cooling rate before charging is larger. Therefore, the ferrite grains are also relatively fine, and the austenite grain size will be reduced accordingly after charging and reheating.

Figurc 6.39a shows the average size of prior austenite grains at different positions of the slab before charging as a function of reduction. It can be seen from the figure that with the increase of the reduction, the grain size of the original austenite gradually decreases.

The average size of prior austenite grains with a reduction of 9 mm is 1150 μm. The average size of prior austenite grains with a reduction of 25 mm is 700 μm, and the grain refinement is 60%. Figure 6.39b shows that the average grain size of the lightly reduced austenite after charging is 50 μm. The average austenite grain size under heavy pressure after charging was 28 μm, and the grain size decreased by 56%.

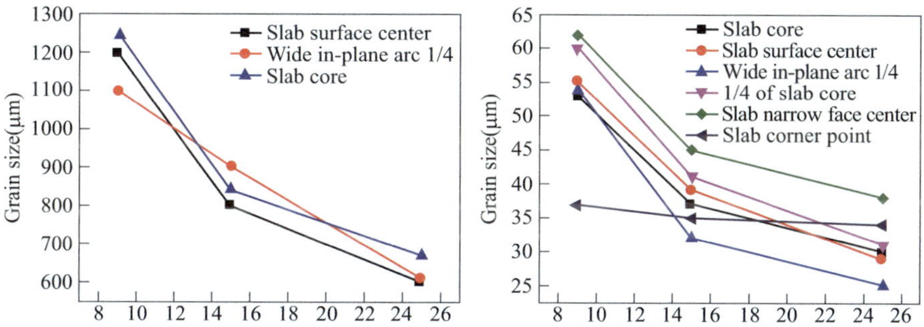

(a) (b)

Fig. 6.39 Relationship between austenite grain boundary size and reduction before and after charging. **a** The relationship between the size of the original austenite grain boundary and the reduction; **b** the relationship between the size of the austenite grain boundary and the reduction after charging

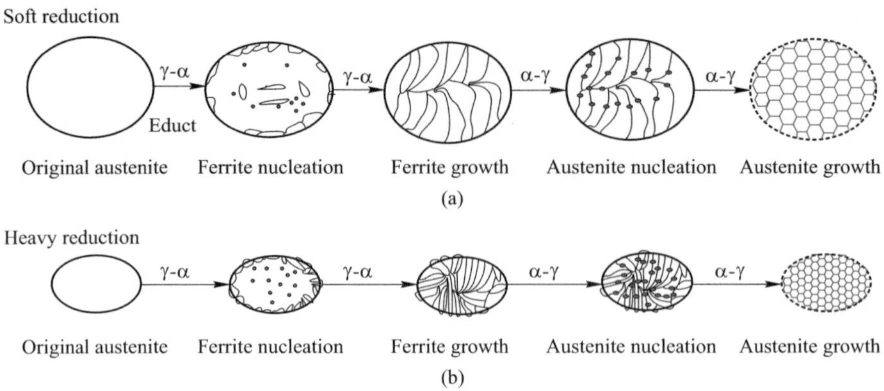

Fig. 6.40 Microscopic schematic diagram of microstructure evolution from continuous casting to charging. **a** Press lightly; **b** press hard

(F) **Conclusion**

In summary, the microstructure evolution of different reduction modes during continuous casting and charging is shown in Fig. 6.40. Figure 6.40a is a diagram of the evolution of the organization during the light reduction process. It can be seen that the prior austenite grains of the slab at light reduction are relatively large. Ferrite nucleates preferentially at the previous austenite grain boundaries during γ-α transformation during cooling. And ferrite will grow inside the grain boundaries during the growth process. When the temperature is lowered, the interior of the grain is completely transformed into ferrite. Due to the larger original austenite grains in the slab, the ferrite within the grains has sufficient growth space. After the phase transformation, the ferrite grain size is larger.

Figure 6.40b shows the microstructure evolution under the process conditions under heavy pressure. The reasons for the refinement of the reheated structure are mainly as follows: (1) The increase of the rolling reduction makes the original austenite grains of the cast slab deformed. At this time, many defects such as deformation bands, dislocations, and twins will be generated in the grain boundaries of the prior austenite. These positions will provide more nucleation sites and energy for the phase transformation, and contribute to the ferrite transformation. The refinement of ferrite promotes the increase of its volume fraction, which leads to the fact that the ferrite in the slab under heavy pressing mainly uses grain boundaries as nucleation particles. When the ferrite grains nucleate and grow at the austenite grain boundaries, it will have a significant division effect on the prior austenite. The ferrite film is formed to a certain thickness. (2) When the ferrite transformation occurs, most of the ferrite grains grow inside the original austenite grain boundaries. Due to the recrystallization of the slab under heavy pressure, the grain size of the original austenite is smaller. Therefore, less space is reserved for ferrite growth. This limits the growth of ferrite to a certain extent and plays a role in microstructure refinement. (3) Under normal circumstances, with the increase of slab reduction, the deformation of austenite is larger. And the dislocation density inside the austenite grains is larger. Under the condition of heavy pressure, the more carbonitride precipitation nucleation particles, the finer and dispersed the precipitation. This has a more obvious effect on the pinning effect of the subsequent heating process on the growth of austenite, and the size of the austenite grains is relatively small.

6.2.1.3 Comparison of Microstructure and Properties of Rolled Products of Wide and Thick Slabs with Different Reductions

The microstructure evolution of the above-mentioned slabs (2300 mm × 280 mm) with different reductions after being rolled according to the same rolling process (rolling reduction ratio of 2:1) was studied. Figure 6.41 shows the comparison of the surface layer and the core structure of the rolled products with different reductions. It can be seen from the figure that the austenite grain size gradually decreases with the increase of the rolling reduction. The statistical results of grain size show that the average grain size of the surface layer of the slab with a rolling reduction of 9 mm is 46 μm. The average grain size at a reduction of 25 mm is 27 μm, which is 41% smaller than the average grain size of a reduction of 9 mm. The rolling reduction is 9 mm, and the average grain size of the slab surface is 58 μm. The average grain size is 36 μm at a reduction of 25 mm, which is 37% smaller than that of a reduction of 9 mm.

A universal material testing machine is used to test the mechanical properties of the finished rolled products. A study of the effect of microstructure of microalloyed steel continuous casting slabs on the mechanical properties of rolled products. Figure 6.42 shows the position of the surface and core of the billet at different reductions. Ten specimens were taken from each position for tensile and impact tests. The test results show that the yield strength of the 25 mm rolled material under heavy pressure is slightly higher than that of the conventional continuous casting. The impact energy of 25 mm rolled material under heavy pressure is much higher than that of conventional rolled material. According to the Hall–Petch equation, grain boundaries can block the movement of dislocations and increase the total area of grain boundaries, which can effectively improve the mechanical properties of materials. Therefore, the grain refinement of the center and surface under heavy pressure is beneficial to improve the strength. When the rolling reduction ratio of conventional continuous casting Q345 steel is 2:1, the yield strength cannot reach the national standard. It is further proved that the implementation of heavy reduction technology is an effective way to achieve low rolling reduction ratio to prepare high-quality rolled products.

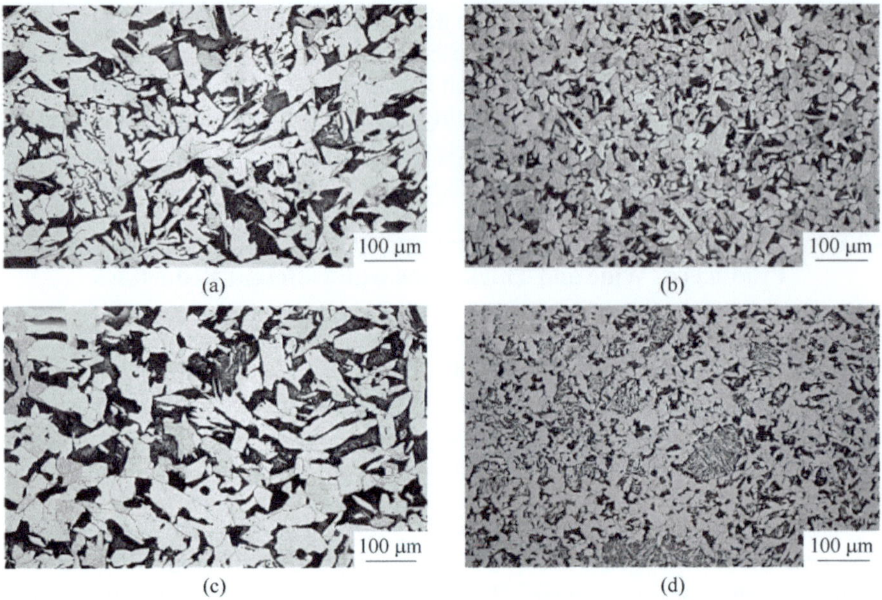

Fig. 6.41 Comparison between the broken surface of the wide and thick plates with different reduction amounts and the original austenite particle size of the core. **a** Surface pressure 9 mm; **b** surface reduction of 25 mm; **c** heart pressure 9 mm; **d** heart pressure 25 mm

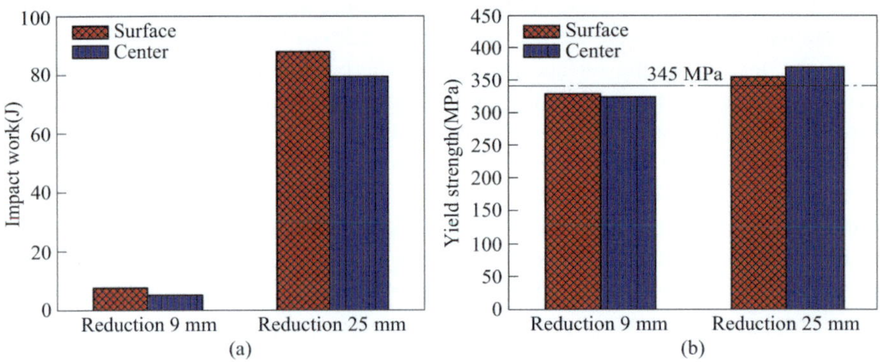

Fig. 6.42 Mechanical properties of different reductions. **a** Impact energy; **b** yield strength

6.2.2 *The Microstructure Evolution of the Bloom for Delivery and Charging Process*

Compared with microalloyed steel blooms, blooms can be produced in a wider range of steel grades and with greater differences in carbon content. During the reduction process, the structure is complex and changeable, which is quite different from that of the microalloyed blooms. In view of this, this section takes GCr15 bearing steel as an example to study the microstructure evolution law of the reduced casting billet during the heating process.

6.2.2.1 Grain Evolution of Prior Austenite with Different Reductions at Surface and Core Positions of Bloom

Take 10 mm × 10 mm × 10 mm small pieces on the surface and core of blooms with three reductions of 18 mm, 24 mm and 34 mm. After grinding and polishing, use 4% nitric acid alcohol solution to corrode and observe the tissue morphology. The structure of the tissue is shown in Fig. 6.43.

Figure 6.43 shows the comparison between the surface layer and the local structure of the core of the bloom with different reductions. As the reduction increases, the prior austenite grains gradually become smaller. On the surface of the bloom, the prior austenite grain size is 1358 μm when reduced by 18 mm. When the reduction was increased to 34 mm, the prior austenite grain size was 688 μm. The grains were refined by 49.3% after the reduction was increased. At the core of the bloom, the prior austenite grain size is 1675 μm when the reduction is 18 mm. When the reduction

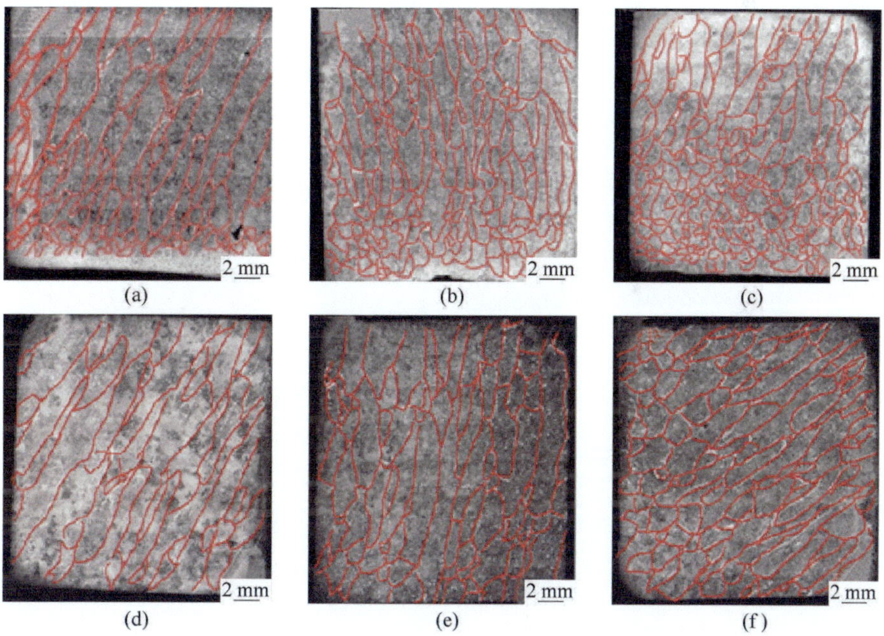

Fig. 6.43 Comparison of the grain size of the original austenite on the surface and the core of the bloom with different reductions. **a** Surface pressure 18 mm; **b** surface reduction 24 mm; **c** surface reduction 34 mm; **d** core reduction 18 mm; **e** depression of the heart is 24 mm; **f** depression of the heart is 34 mm

amount reaches 34 mm, the prior austenite grain size is 943 μm. The grains were refined by 43.7% with increasing reduction.

6.2.2.2 Evolution of Austenite Grains at Different Reductions During the Charging Process at the Surface and Core Positions of Blooms

Because the as-cast structure sampled on site is a cold structure. Therefore, we used finite element simulation to calculate the thermal history of the bouquet loading and heating process. Figure 6.44 shows the cloud map of the simulated temperature distribution of the bloom-heating furnace. The heating and heating curves at different positions are shown in Fig. 6.45. The curves are based on finite element simulation to provide the thermal history of bloom charging for laboratory simulations. The sample is heated according to the charging process simulated by the finite element. After quenching, the microstructure of the cast blooms is corroded with saturated picric acid solution.

Figure 6.46a–c is the comparison of the surface structure of the blooms after cold charging with different reductions. It can be seen from the figure that the austenite grain size decreases with the increase of the rolling reduction. The average austenite grain size was 158 μm at a reduction of 18 mm. The reduction was 24 mm and the

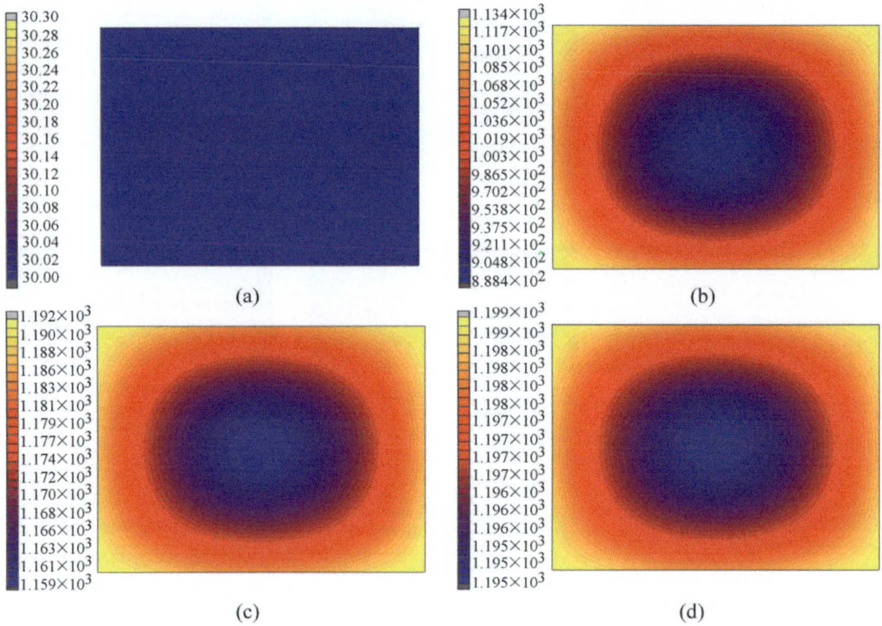

Fig. 6.44 Cloud map of temperature distribution during bloom heating. **a** Start heating; **b** heating for 60 min; **c** heating for 120 min; **d** heating for 180 min

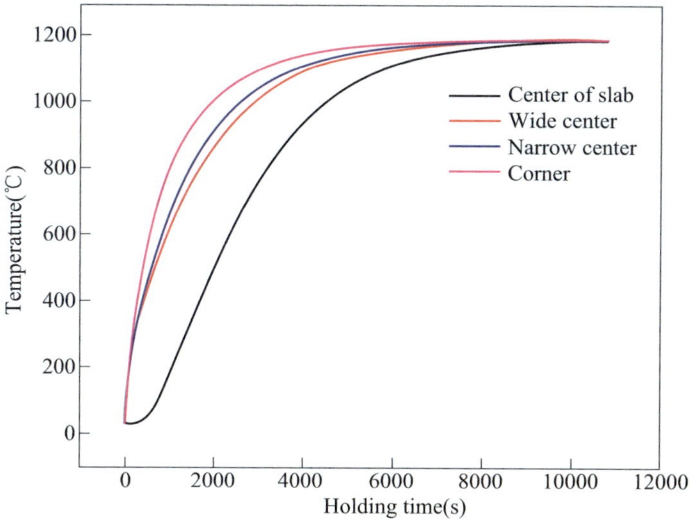

Fig. 6.45 Thermal history curves of blooms at different positions

average grain size was 146 μm. The average grain size was 125 μm at a reduction of 34 mm. The average grain size was refined by 20% with increasing reduction. Figure 6.47d–f is the comparison of the core structure of the cast blooms after cold charging with different reductions. It can be seen from the figure that the austenite grain size gradually decreases with the increase of the rolling reduction. From the statistical results of grain size, it can be known that the average austenite grain core size is 195 μm when the reduction is 18 mm. The average grain size was 182 μm at a reduction of 24 mm. The average grain size was 177 μm at a reduction of 34 mm. The grain refinement was 9.23% with increasing reduction.

The relationship between the cold blooms with different reductions and the microstructure after charging is analyzed. The effect of reduction on the evolution of austenite structure is explained mechanistically. Similar to the microalloyed steel, the ferrite conversion rate of the blooms under heavy pressing is higher than that of the blooms under light pressing per unit area. The higher the conversion of ferrite, the more nucleation particles are provided for the nucleation of austenite during charging, thereby refining the austenite grain. In addition, the fine grain size of ferrite is conducive to the nucleation of austenite during heating.

6.2.2.3 Evolution of Austenite Grains on the Surface and Core of Blooms Rolled with Different Reductions

The blooming products were sampled, and the treatment method was the same as that in the previous section. The experimental results are shown in Fig. 6.46.

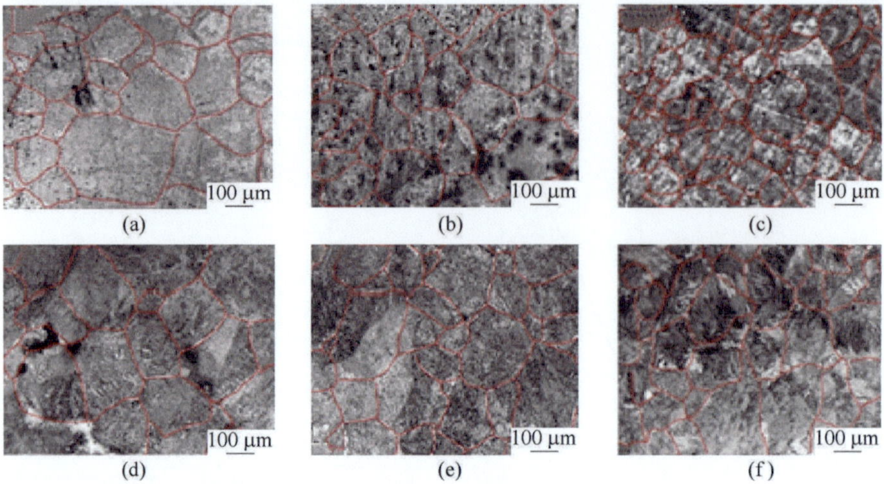

Fig. 6.46 Comparison of grain size between surface and core prior austenite of blooms with different reductions. **a** Surface reduction is 18 mm; **b** surface reduction is 24 mm; **c** surface reduction is 34 mm; **d** depression of the heart is 18 mm; **e** depression of the heart is 24 mm; **f** depression of the heart is 34 mm

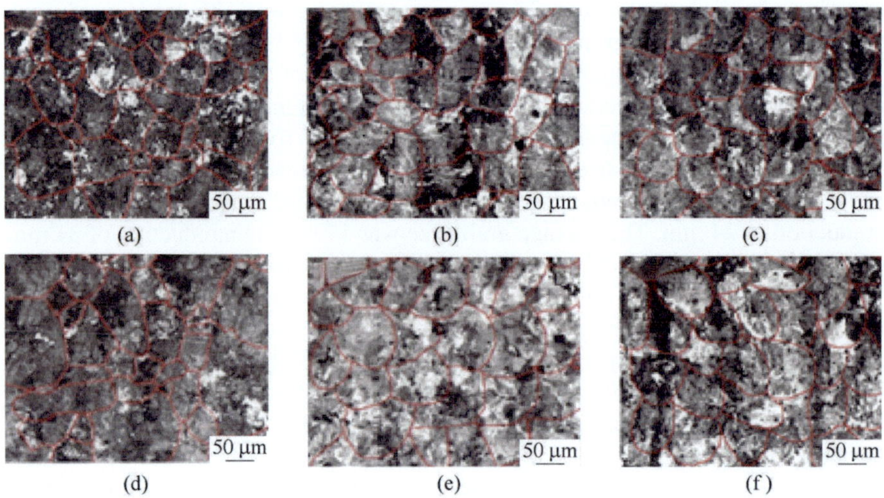

Fig. 6.47 Comparison of austenite grain size between surface and core of rolled products with different reductions. **a** Surface reduction is 18 mm; **b** surface reduction is 24 mm; **c** surface reduction is 34 mm; **d** depression of the heart is 18 mm; **e** depression of the heart is 24 mm; **f** depression of the heart is 34 mm

Figure 6.47a–c is the comparison of the surface structure of the rolled products with different reductions. It can be seen from the figure that the austenite grain size gradually decreases with the increase of the rolling reduction. The grain size statistics show that the average austenite grain size is 78 μm when the reduction is 18 mm, the average grain size is 71 μm when the reduction is 24 mm, and the average grain size is 65 μm when the reduction is 34 mm. Figure 6.47d–f is the comparison charts of the core structure of the cast blooms after cold charging with different reductions. It can be seen from the figure that the austenite grain size gradually decreases with the increase of the rolling reduction. The average austenite grain size is 81 μm when the reduction is 18 mm, the average grain size is 71 μm when the reduction is 24 mm, and the average grain size is 68 μm when the reduction is 34 mm.

Comparison of the austenite grains of the cold blooms, hot charging and rolled products by different reductions. It can be seen that as the reduction amount increases, the structure is gradually refined. Compared with the cast blooms with a large reduction, the structure of the rolled product is relatively fine. It can be seen that the fineness of the casting blank structure will still be "inherited" to the rolled product after charging and rolling.

References

1. Gong MN, Li HJ, Wang B et al (2020) Research on dynamic recrystallization behavior of Nb-Ti continuous casting slab during hot core rolling under large reduction. Steel Roll 233(01):16–21
2. Zhao XK, Zhang JM, Lei SW et al (2013) Dynamic recrystallization (DRX) analysis of heavy reduction process with extra-thickness slabs. Steel Res Int 85(5):811–823
3. Yang Q, Ji C, Zhu MY (2019) Modeling of the dynamic recrystallization kinetics of a continuous casting slab under heavy reduction. Metall Mater Trans A 50(1):357–376
4. Wei ZJ, Ji C, Chen TC, Zhu MY, Wu CH (2021) Metadynamic recrystallization and microstructure evolution behavior of a continuously cast Ti-microalloyed steel slab in a heavy reduction process. J Iron Steel Res Int
5. Wei ZJ, Ji C, Chen TC, Zhu MY (2022) In situ observations and microstructure evolution behavior of static recrystallization of microalloyed continuous casting slabs in a solidification end reduction process. Steel Res Int 93(1):2100348
6. Chen T et al. (2024) Microstructure evolution and its influence on thermoplasticity of wide and thick continuous casting slab with heavy reduction. Steel Res Int 31(9):2196–2206
7. Sellars CM, Mctegart WJ (1966) On the mechanism of hot deformation. Acta Mater 14(9):1136–1138
8. Youssef KM, Wang YB, Liao XZ et al (2011) High hardness in a nanocrystalline Mg97Y2Zn1 alloy. Mater Sci Eng A 528(25–26):7494–7499
9. Mcqueen HJ, Yue S, Ryan ND et al (1995) Hot working characteristics of steels in austenitic state. J Mater Process Technol 53(1–2):293–310
10. Han Y, Wu H, Zhang W et al (2015) Constitutive equation and dynamic recrystallization behavior of as-cast 254SMO super-austenitic stainless steel. Mater Des 69:230–240
11. Zener C, Hollomon JH (1944) Effect of strain rate upon plastic flow of steel. J Appl Phys 15(1):0–22
12. Poliak EI, Jonas JJ (1996) A one-parmenter approach to determining the critical conditions for the initiation of dynamic recrystallization. Acta Mater 44(1):127–136
13. Najafizadeh A, Jonas JJ, Stewart GR et al (2006) The strain dependence of postdynamic recrystallization in 304 H stainless steel. Metall Mater Trans A 37(6):1899–1906

14. Wahabi ME, Cabrera JM, Prado JM (2003) Hot working of two AISI 304 steels: a comparative study. Mater Sci Eng A 343(1–2):116–125
15. Mecking H, Kocks UF (1981) Kinetics of flow and strain-hardening. Acta Metall 29(11):1865–1875
16. Estrin Y, Mecking H (1984) A unified phenomenological description of work hardening and creep based on one-parameter models. Acta Mater 32(1):57–70
17. Kopp R, Luce R, Leisten B et al (2001) Flow stress measuring by use of cylindrical compression test and special application to metal forming processes. Steel Res Int 72(10):394–401
18. Laasraoui A, Jonas JJ (1991) Recrystallization of austenite after deformation at high temperatures and strain rates—analysis and modeling. Metall Trans A 22(1):151–160
19. Serajzadeh S, Taheri AK (2003) Prediction of flow stress at hot working condition. Mech Res Commun 30(1):87–93
20. Senuma T, Yada H, Shimizu R et al (1990) Textures of low carbon and titanium bearing extra low carbon steel sheets hot rolled below their AR3 temperatures. Acta Mater 38(12):2673–2681
21. Liu J, Cui Z, Ruan L (2011) A new kinetics model of dynamic recrystallization for magnesium alloy AZ31B. Mater Sci Eng A 529:300–310
22. Andrade HL, Akben MG, Jonas JJ (1983) Effect of molybdenum, niobium, and vanadium on static recovery and recrystallization and on solute strengthening in microalloyed steels. Metall Trans A 14(10):1967–1977
23. Lin YC, Fang X, Wang YP (2008) Prediction of metadynamic softening in a multi-pass hot deformed low alloy steel using artificial neural network. J Mater Sci 43(16):5508–5515
24. Zhao MJ, Huang L, Zeng R et al (2019) In-situ observations and modeling of static recrystallization in 300 M steel. Mater Sci Eng A 765:138–300
25. Xu Y, Zhao Y, Liu J (2020) A modified kinetics model and softening behavior for static recrystallization of 12Cr ultra-super-critical rotor steel. Mater Res Express 7(5):056–507
26. Bo MA, Peng Y, Jia B et al (2010) Static recrystallization kinetics model after hot deformation of low-alloy steel Q345B. J Iron Steel Res 17(8):6

Chapter 7
Control Equipment and Its Application of the Solidification End Reduction Technology for Continuous Casting Process

Along with the adequate theoretical basis, the corresponding control technology is also needed to ensure the effective implementation of the mechanical reduction. In the first place, the solidified end of the slab should be accurately located to determine the reduction region. After that, the transfer efficiency of deformation to the center should be increased to improve the effect of reduction process. Meanwhile, cracks and other slab quality defects cannot occur. In addition to the control technology, the equipment shall also have ample mechanical reduction capability.

This chapter mainly introduces the key technologies and equipment developed by our team in the field of the solidification end reduction technology, including the real-time heat-tracking technology, the online control technology, the design method of enhanced compact segment (ECS) for wide-thick slabs, and curved surface convex roll (CSC-Roll) for bloom. And the practical application effect is also recommended.

7.1 Solidification End Reduction Technology

In view of the individual characteristics of wide-thick slabs and blooms, high-precision online calibration technology for solidification end position and shape based on non-uniform distribution of solutes and real-time feedback of pressure drop, a two-stage continuous heavy reduction process, precise control of driving torque to enhance the deformation of the billet heart of efficient extrusion control technology, innovation into the "accurate, efficient, stable" reduction technology have been developed. The details are shown in Table 7.1.

C. Ji et al., *Solidification End Reduction of Continuous Casting Process*, Key Steel Manufacturing Technologies by Collaborative Innovation,
https://doi.org/10.1007/978-981-96-8544-8_7

Table 7.1 Integrated technology of heavy reduction

Technical characteristics	Technical contents	Key technologies
Accurate	Accurate calculation of solidification end position and morphology based on real-time data of pressure and reduction amount of hydraulic cylinder combined with heat-tracking calculations that take into account the non-uniform distribution of solutes	Pressure feedback-based solidification end detection technique [1, 2]
	In order to solve the serious problem of center segregation on both sides in wide direction, the pressing process at the front of non-uniform solidification of the slab and the method of establishing the base roll gap system of the casting machine based on weighted average method were developed	Non-uniform solidification front reduction technique [3, 4]
	Based on the results of solidification heat transfer calculations, combined with the billet thickness as a ruler to detect the roller gap deviation, the bloom continuous casting straightening machine roller gap online calibration technology was developed	Bloom continuous casting straightening machine roller gap online calibration technology [5]
Efficient	Through the unbalanced torque control, the extension dissipation of the reduction amount is reduced, so that the billet is in the state of extrusion and deformation, and the transfer efficiency of the reduction amount to the center of the billet is significantly improved	Efficient extrusion control technology [6]
	Through the implementation of two stages of continuous heavy reduction before and after the coagulation endpoint, the improvement of central segregation was achieved	Two-stage continuous heavy reduction process [7]
	Single-point reduction is used to increase the strain rate in the billet center to weld shrinkage; continuous reduction is used to compensate for porosity caused by the difference in shrinkage between the inner and outer billets	"Single point + continuous" heavy reduction [8]
Stable	Avoid reduction leading to crack initiation and expansion, and formulate reasonable and efficient cooling process and roll arrangement according to the distribution of crack initiation risk in billet casting	Crack risk prediction and control technology [9–11]
	Ensures stable and efficient implementation of reduction, with process optimization functions based on rolled material retrospective quality data, which significantly reduces equipment losses and improves the yield rate	Reduction process control system [12]

7.1.1 Two-Stage Continuous Reduction Process

Because of the characteristics of metal solidification segregation crystallization, solute segregation elements first form at the solid–liquid interface and gather in the center of the slab. Upon entering the two-phase region, they are drawn into the center of the slab due to solidification shrinkage and then form a central deviation. As the solidification process progresses, the liquid steel cannot flow freely, and at this time, the solidification shrinkage cannot be compensated by the liquid steel, forming central porosity and shrinkage defects. Therefore, the center of segregation is in the two-phase area of the formation of the latter part (as shown in Fig. 7.1), when the liquid steel still has a certain degree of mobility; and the center of porosity is formed in the two-phase area at the end of the formation of the casting billet, even in the casting billet completely solidified, due to the high-temperature casting billet core shrinkage to be much larger than the outer layer has been gradually cooled the amount of shrinkage of the billet, and in the outer layer of the billet casting of support role, will continue to form porosity defects.

On the basis of accurate description of solidification heat transfer and billet shell deformation behavior in the heavy reduction process of wide slab and bloom continuous casting, a two-stage continuous heavy reduction process as shown in Fig. 7.1 has been developed based on the principle of solidification and complementary reduction, and the specific implementation scheme is shown in Fig. 7.2.

At the end of solidification, under the action of billet contraction, the liquid steel rich in solute segregation elements before the end of solidification converges to the heart of the billet, i.e., the solute elements accelerate the enrichment, which is the main reason for the formation of segregation in the center of the billet. The first stage of the two-stage continuous heavy reduction process is aimed at the poor fluidity and easy segregation of the paste area at the end of solidification, and the solidification end reduction is used to inhibit the accelerated enrichment of solute elements and enhance the homogeneity of the billet; the second stage of the reduction process improves the porosity defects of the billet further through continuous reduction. The

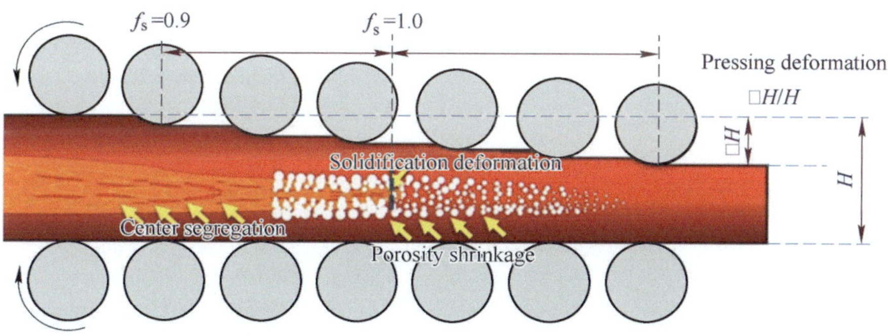

Fig. 7.1 Mechanism of center segregation and looseness formation in continuous casting slab and two-stage continuous heavy reduction process

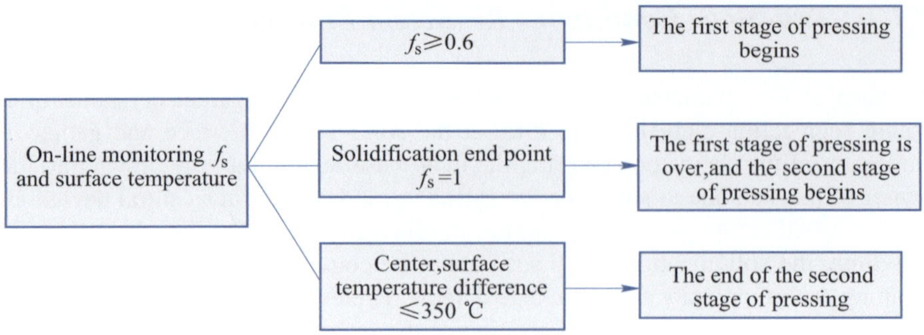

Fig. 7.2 Two-stage continuous heavy reduction process implementation flow

technology is universal, i.e., the principle of the process is applicable to different steel grades and sections, but there are differences in critical reduction intervals and values of reduction corresponding to different steel grades and billet cross-sections. The implementation characteristics and process parameters of the two-stage continuous heavy reduction process are shown in Table 7.2.

The solidification end reduction technology is applicable to different steel grades and sections, but there are differences in the values of critical reduction intervals and reduction rate (reduction amount) corresponding to different steel grades and sections of billet casting. As shown in Table 7.3, the billet section is the most important factor affecting the critical reduction rate (reduction amount), with the increase of section size, the heavy reduction process deformation penetration efficiency to the center of

Table 7.2 Two-stage continuous reduction process

Item	Stage I $fs = 0.6$–0.9	Stage II $fs > 0.9$
Solidification characteristics	Poor liquidity and segregation formation	Affected by solidification shrinkage and thermal shrinkage, porosity begins to occur
Purpose of implementation	Improves segregation and homogeneity	Closure of shrinkage by using internal and external temperature difference (350 °C)
Reduction process	Wide-thick slab:1–2 segments, reduction rate 2–5 mm/m Bloom: 2–4 straightening machines, total reduction amount $\leq$ 10 mm	Wide-thick slab: $\geq$ 2 segments, reduction rate 2–10 mm/m Bloom: 5–7 straightening machines, total reduction amount $\geq$ 20 mm
Implementation equipment	Wide-thick slab: conventional segment Bloom: mainly reduction by flat roll	Wide-thick slab: ECS segment Bloom: mainly reduction by CSC-Roll

Table 7.3 Influence of continuous casting section and steel grade on reduction

Section	Steel grade	Reduction zone	Reduction rate/amount
280 mm thick	No.45 (w(C) = 0.45%)	$fs \geq 0.907$	$\geq$ 3.1 mm/m
	Q345B (w(C) = 0.22%)	$fs \geq 0.902$	$\geq$ 3.0 mm/m
300 mm thick		$fs \geq 0.903$	$\geq$ 3.3 mm/m
320 mm × 410 mm	U78CrV (w(C) = 0.78%)	$fs \geq 0.899$	$\geq$ 11 mm
360 mm × 450 mm		$fs \geq 0.901$	$\geq$ 15 mm
	No.45 (w(C) = 0.45%)	$fs \geq 0.895$	$\geq$ 10 mm

the billet decreases dramatically, and at the same time, the carbon content increases, the viscosity of the steel increases, and the required reduction amount of deformation increases accordingly.

7.1.2 *Efficient Extrusion Control Technology*

Taking advantage of the large temperature difference between the inside and outside of the billet during the continuous casting process, solidification end reduction technology can efficiently transfer the deformation to the center of the billet to improve internal defects. However, with the increase in the amount of reduction, it will be more by the casting billet broadening and ductile deformation dissipation, the casting billet extrusion effect is poor, and even stretching phenomenon (Fig. 4.48), can be through the control of its ductile strain, so that the casting billet along the casting direction and compact direction are in the state of extrusion deformation, and significantly and effectively improve the reduction amount of the transfer efficiency, so as to achieve high efficiency to improve the porosity, closured shrinkage holes, and enhance the densification process.

In view of this, high efficiency extrusion deformation control technology has been developed. By controlling the unbalanced torque of each drive roll (see Fig. 7.3), the tension in the billet traveling process is effectively eliminated, so that the billet can really be milled thin rather than stretched thin, and the efficiency of the reduction amount to the billet heart can be effectively improved.

For example, by adjusting the roll speed of the non-heavy reduction straightening machine (segment) before the heavy reduction straightening machine (segment), a roll speed difference is formed between the casting rolls of the former straightening machine and the casting rolls of the heavy reduction straightening machine. In the roll speed difference of 0.8–1.2 range, the billet density with the roll speed difference increases, the center of the shrinkage along the casting direction and the width direction of the extension has been effectively suppressed, shrinkage closured better, as shown in Fig. 4.62. Meanwhile, the continuous casting billet can be squeezed before the increase in roll speed, increasing the reduction of the process of porosity on the center of the reduction and the improvement of the effect. However, after the roll

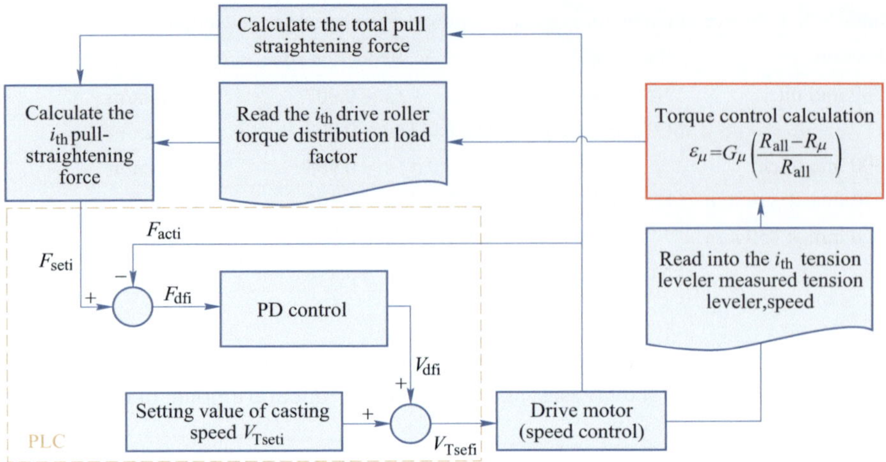

Fig. 7.3 Unequalized torque control process

speed difference exceeds 1.05, the improvement of densification fails as the rolling friction turns into sliding friction, so the roll speed difference cannot be too large. The speed control flow of each drive roll is shown in Fig. 7.4.

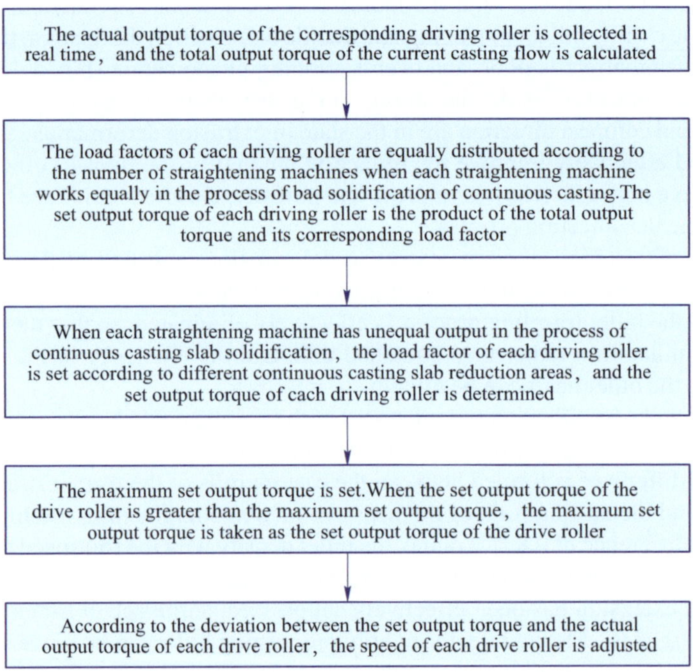

Fig. 7.4 Flowchart of torque control for tensioning machine with reduction effect

Specific implementation methods are as follows:

(1) In the continuous casting process, the straightening machine system of each frequency converter real-time acquisition of the corresponding drive rollers of the actual output torque T_{act}^1, , T_{act}^2, ..., T_{act}^n, T_{act}^i is the actual output torque of the ith drive roller, n is the number of drive rollers involved in the load distribution of this casting stream, and the current total output torque of this casting stream is calculated as $T = T_{act}^1 + T_{act}^2 + \ldots + T_{act}^n$.
(2) According to the current total output torque of the casting flow and its corresponding load factor to determine the set output torque of each driving roll T_{set}^1, T_{set}^2,...,T_{set}^n; when the equal output of each straightening machine pulling casting, the load factor of each driving roll according to the number of straightening machines equally distributed, set the output torque for the total output torque and its corresponding load factor product; in the case of the non-equal output of each straightening machine pulling casting, set the load factor for different areas of the continuous casting billet: set the continuous casting billet before reduction of load factor of 50–150% of the total load, set the load factor of the continuous casting billet in the region of reduction of – 50 to 100%, set the load factor of the total load of – 50 to 100%, set the load factor of the continuous casting billet in the region of reduction of – 100%. The load factor before the reduction of the continuous casting billet is 50–150% of the total load, the load factor within the reduction area of the continuous casting billet is set at – 50 to 100% of the total load, the load factor after the reduction area of the continuous casting billet is set at – 50 to 50% of the total load, and the load factor of each area of the continuous casting billet sums up to 100%.
(3) Calculate the set output torque of each straightening machine based on the reduction amount of each drive roller in the reduction zone.

$$\varepsilon_\mu = G_\mu \left(\frac{R_{all} - R_\mu}{R_{all}} \right) \tag{7.1}$$

where μ is a constant, 1, 2,..., m; m is the number of driving rolls in the reduction zone; ε_μ is the set output torque of the μth straightener in the reduction area; G_μ is the control gain of the output torque of the μth straightening machine; R_μ is the reduction amount of the μth straightening machine or segment; R_{all} is the total amount of reduction.

(4) Before and after the reduction zone, the set output torque of each drive roll is the product of the total output torque and the load factor; the maximum set output torque T_{max}, when the set output torque T_{set}^i of the drive roll is greater than the maximum set output torque T_{max}, the maximum set output torque T_{max} is taken as the set output torque of the drive roll.
(5) Adjust the speed according to the deviation a between the setting of each drive roller and the actual output torque; the deviation value b was calculated, and the speed of each drive roller was gradually adjusted according to the deviation to set the fine-tuning speed of – 0.20 to 0.20 m/min. If the deviation of the

adjusted driving roller speed is more than ± 20% of the initial set pulling speed, the driving roller speed will not be adjusted, if not, the next judgment will be carried out; if the output torque of the drive roller reaches the set output torque of the drive roller, the speed will not be adjusted. Otherwise, the speed will be fine-tuned.

7.1.3 Non-uniform Solidification Front Reduction Process for Wide-Thick Slab

In the process of continuous casting billet of wide-thick slab, the problem of uneven lateral cooling is prominent. The uneven cooling will inevitably produce uneven thermal shrinkage deformation. At the same time, the continuous casting process adjacent to the two support rolls between the solidified billet shell in the steel static pressure under the action of the bulging deformation, the two together to promote the dendritic crystal between the high solute concentration of liquid steel to the center of the casting billet flow, which further exacerbates the wide to the center of the 1/4–1/8 region.

As shown in Fig. 7.5, for Q235 steel with the section of 1800 mm × 280 mm and the casting speed of 0.9 m/min, the lengths of soft reduction intervals determined by the billet transverse 1/2, 1/8, and the whole width direction increased by 0.24 m, 0.23 m, and 0.51 m, respectively, for every increase of the casting speed of 0.1 m/min. Therefore, in order to improve the center segregation within the whole cross-section of the billet, full consideration should be given to the design parameters of liquid core extension at the transverse 1/8, i.e., the endpoint of light reduction at 1/8 position in the width direction should be adopted as the endpoint of soft reduction, and the reduction zone should be expanded from the original homogeneous solidification of 1.6–2.1 m to the non-homogeneous solidification of 3.6–4.6 m.

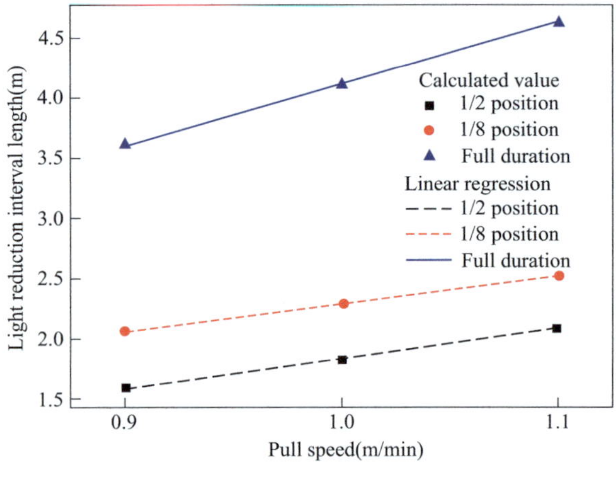

Fig. 7.5 1800 mm × 280 mm continuous casting billet reduction zone

In the process of continuous casting of wide slabs, in addition to the solidification end reduction zone, in addition to the solidification end reduction zone, the setting of the roll gap also needs to consider the characteristics of its uneven solidification. As mentioned in Sect. 1.1.1, when the continuous casting billet passes through the support rolls, the solidified billet shell undergoes periodic bulging deformation, which promotes the relative movement of the solid and liquid phases in the two-phase area, thus aggravating the central segregation. Excessive roll gaps can exacerbate bulging and thus greatly worsen segregation; at the same time, it will significantly increase the solidification front stress, strain, resulting in a significant increase in the risk of internal cracking. If the roll gap is set too small, it will aggravate the wear of the casting rolls, reduce the life of the equipment, and even lead to the risk of stalling the billet. In view of this, the development of a weighted thermal shrinkage holes can be fully considered non-uniform distribution of the base roll gap process, the process shown in Fig. 7.6. First of all, according to the actual conditions of wide slabs, FEM simulation is used to get the cast billet in the casting flow at different locations of thermal shrinkage and bulging deformation law and then according to the different locations along the casting of the billet width to the contraction characteristics of the actual roll gap value. Specific weighted thermal shrinkage base roll gap calculation method is shown in Fig. 7.7; the calculation equation is shown in Eq. (7.2).

$$\overline{C} = 2 \times \left(\sum_{i=S_b}^{S_e} C_i \eta_i \gamma_i + \sum_{j=M_b}^{M_e} C_i \eta_i \gamma_i \right) \tag{7.2}$$

where $\overline{C}$ is the weighted thermal shrinkage, mm; S_b/S_e and M_b/M_e are the start/end position nodes in the solidified and unsolidified regions, respectively; C is the node thermal shrinkage value, mm; η and γ are the node and region weighting coefficients, respectively.

Combined with the actual site, it can be determined in the casting flow solidification before the end of the early casting stage, γi reasonable value of 0.6–0.7, γj reasonable value of 0.3–0.4. After the end of the solidification, casting billet has been completely solidified; there is no cross-section of the liquid-phase or two-phase area of the liquid phase does not solidify, the stage of $\gamma i = 0$ and $\gamma j = 1$.

7.1.4 Online Calibration Technology for Roll Gap of Bloom Straightening Machine

In the process of solidification end reduction, the blank pressing force of the bloom straightening machine is 3–10 times that of the normal production, and the hydraulic cylinder of the frame moves continuously with the online adjustment of the reduction amount, which is easy to form the roll gap detection deviation, resulting in the actual reduction amount in the production process is too large or too small, which not

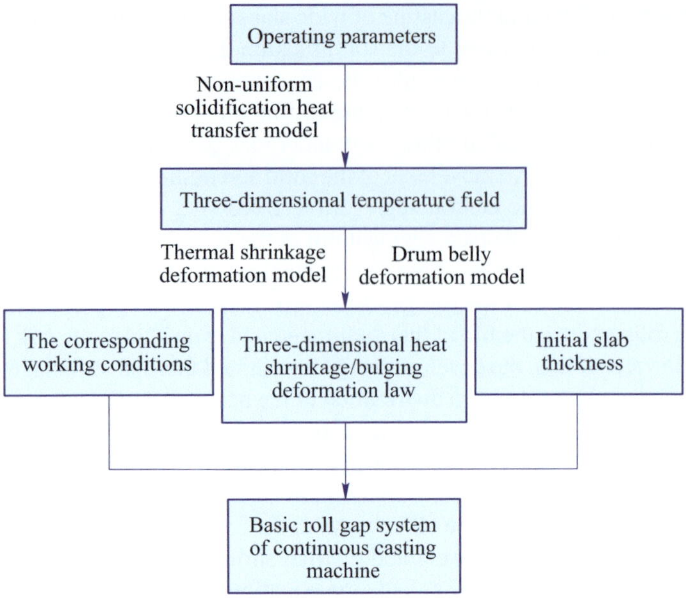

Fig. 7.6 Basic roll seam system development process

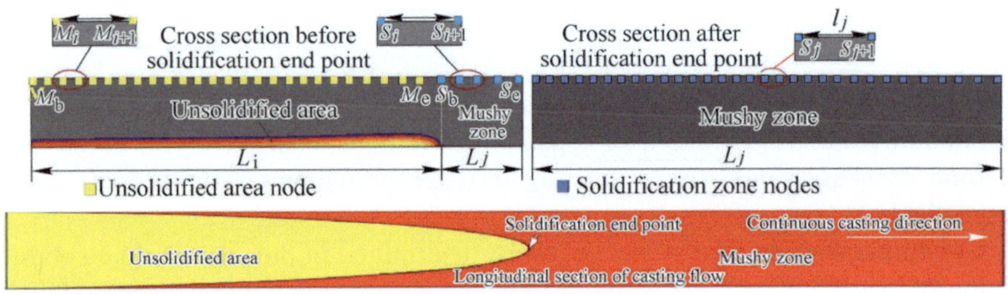

Fig. 7.7 Schematic diagram of the weighted thermal shrinkage calculation method

only cannot play the effect of the reduction process, but also can cause quality problems such as bulging and central crack. Therefore, developed a continuous casting billet straightening machine roll gap online calibration method, which can realize the online calibration of roll gap of continuous casting bloom and ensure the accurate implementation of reduction.

There are usually 7–9 straightening frames in the air-cooling zone of bloom continuous casting machine, and the first frame is the detection frame. If the detection error of each tension leveler sensor can be accurately detected, the real value of the roll gap in the production process can be accurately calculated through error correction in the production process. Therefore, the deviation of the roll gap value of each roll is calibrated by referring to the detection value of the No.1 detection roll that does not participate in the soft reduction, as follows.

(1) Determine whether the current casting speed is stable and whether it can be calibrated.

The natural thermal shrinkage of bloom is different under different casting speeds, so it is necessary to select the hot slab under stable casting speed as the thickness scale to measure the roll gap deviation of each roll.

The bloom is divided into several "tracking units" that are constantly "born" and move with the casting speed according to the fixed period (5 *s*) along the casting direction. The properties of these units include: casting speed, casting length, furnace number, billet age, etc. Tracking device average casting speed through the No.1 straightening frame is defined as:

$$\overline{v} = \sum_{1}^{n_i} \frac{v_{n_i}}{n_i} \tag{7.3}$$

where v_{n_i} is the instantaneous casting speed of the slice unit recorded every 5 *s*; and n_i is the number of instantaneous casting speeds counted from the birth of the crystallizer. The average speed variance of the tracking unit through the No.1 straightening frame is defined as:

$$\delta = \sqrt{\frac{\sum_{1}^{n_i} (v_{n_i} - \overline{v})^2}{n_i}} \tag{7.4}$$

When $v_0 - 0.01 > v_{\text{mean}} > v_0 + 0.01$ ($v0$ is instantaneous casting speed), and the average casting speed variance $\delta < 0.008$, it is considered that the casting speed is stable at this time, and the roll gap deviation can be measured (Fig. 7.8).

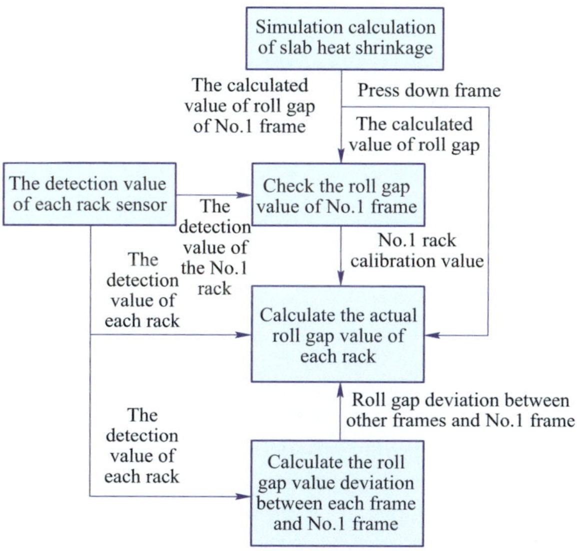

Fig. 7.8 Roll gap online calculation process diagram

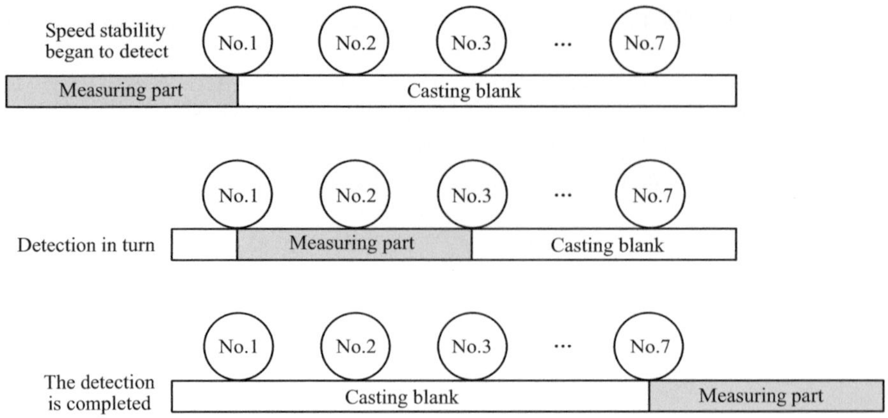

Fig. 7.9 Online detection diagram of roll gap

(2) Based on the roll gap of No.1 straightening machine, the roll gap deviation of other straightening machines is determined.

As shown in Fig. 7.9, when the casting speed is stable, a 3 m long slab (shown in the shaded part of the figure) is selected to start online calibration.

When the detection part passes through the *i*th straightening rack, the *j*th real-time roll gap value detection value Gap_i^j of the *i*th straightening rack is recorded every 5 *s*, and the total number of n_i times is recorded. At the end of the inspection, the average roll gap inspection Gap_i^* value of the ith straightening frame is:

$$\mathrm{Gap}_i^* = \sum_1^{n_i} \mathrm{Gap}_i^j / n_i \tag{7.5}$$

The difference between the roll gap detection value of the *i*th straightening frame and the No.1 tension straightening frame is D_i^*:

$$D_i^* = \mathrm{Gap}_i^* - \mathrm{Gap}_1^* \tag{7.6}$$

Under the condition of stable casting speed, the deviation between the calculated value and the set value of the roll gap of No.1 straightening frame is $\Delta\mathrm{Gap}_1$:

$$\Delta\mathrm{Gap}_1 = \mathrm{Gap}_1^{\mathrm{cal}} - \mathrm{Gap}_1^* \tag{7.7}$$

where $\mathrm{Gap}_1^{\mathrm{cal}}$ is the roll gap value of the No.1 straightening frame at the casting speed obtained according to the calculation results of solidification thermal shrinkage.

During the casting production process, the relative error of the roll gap detection sensors for each straightening frame can be considered to be stable, that is, the D_i^* and $\Delta\mathrm{Gap}_1$ obtained after the detection remain unchanged.

When a tracking unit P_0 to the No.1 tension straightening frame, record the current roll gap detection value c and write it into the slice unit attribute array. At this time, the actual value of the No.1 tension straightening frame roll gap Gap1 is:

$$\mathrm{Gap}_1 = \mathrm{Gap}_1^m + \Delta\mathrm{Gap}_1 \tag{7.8}$$

(3) The current actual roll gap is calculated according to the roll gap deviation of each straightening machine.

When the tracking unit P0 is moved to the ith straightening frame, the roll gap value Gapi of the frame is:

$$\mathrm{Gap}_i = \mathrm{Gap}_1 + D_i^* \tag{7.9}$$

According to Eq. (7.8) and Eq. (7.9), with the continuous generation and movement of the tracking unit, the roll gap value of each frame will be calculated once in each cycle, which provides a guarantee for the accurate implementation of the reduction.

7.1.5 Online Positioning Technology of Solidification End Considering Real-Time Verification of Solute Distribution and Shell Thickness

The accurate prediction of the position of the solidification end in the continuous casting process is the prerequisite for the effective implementation of the reduction. In the actual situation, as the solidification process progresses, macrosegregation will inevitably occur in the billet, resulting in differences in thermal properties. However, the previous heat transfer models only use the constant element content to calculate the thermophysical parameters and cannot consider the difference in thermophysical parameters caused by solute segregation at different positions along the thickness direction of the slab.

Therefore, a "drift" model of solidification end based on solute distribution was developed. Evolution of solute distribution during solidification of slab sections based on multiphase solidification theory calculations, the distribution law of corresponding thermophysical parameters is obtained, and the "drift" model of solidification end based on solute segregation distribution is established by coupling with two-dimensional heat transfer model. Figure 7.10a is the relationship between "pressure-reduction-shell thickness" obtained by the actual and thermodynamic simulation of a bloom caster in a steel plant. Under the steady-state or unsteady-state casting conditions in actual production, the steel grade-pressure-reduction corresponding relationship data of each tension leveler during periodic reduction is measured online, and the shell thickness at the current position is calculated according to the Eq. (7.10).

$$\delta = 0.075 \times \left(0.039 - 1.879 \times 10^{-5}p\right) \times R^{0.379} \tag{7.10}$$

where δ is shell thickness; p is pressure; R is reduction amount.

Furthermore, the grid mapping method (Fig. 7.10b) is used to correct the "drift" model of the solidification end point, and the accurate prediction of the solidification end point under the factors such as reduction deformation and solute migration is realized (Fig. 7.10c).

The specific implementation logic is shown in Fig. 7.11, including offline data acquisition and online calculation.

(1) Offline data acquisition

(1) Offline calculation of thermophysical parameters of steel: The phase fraction of steel under different element segregation degree was calculated by microsolidification model, and the thermophysical parameters of steel were calculated based on the phase fraction.

(2) Offline obtain the growth law of solidified shell at the position of 1/2 width of the slab: the macrosegregation detection is used to obtain the segregation

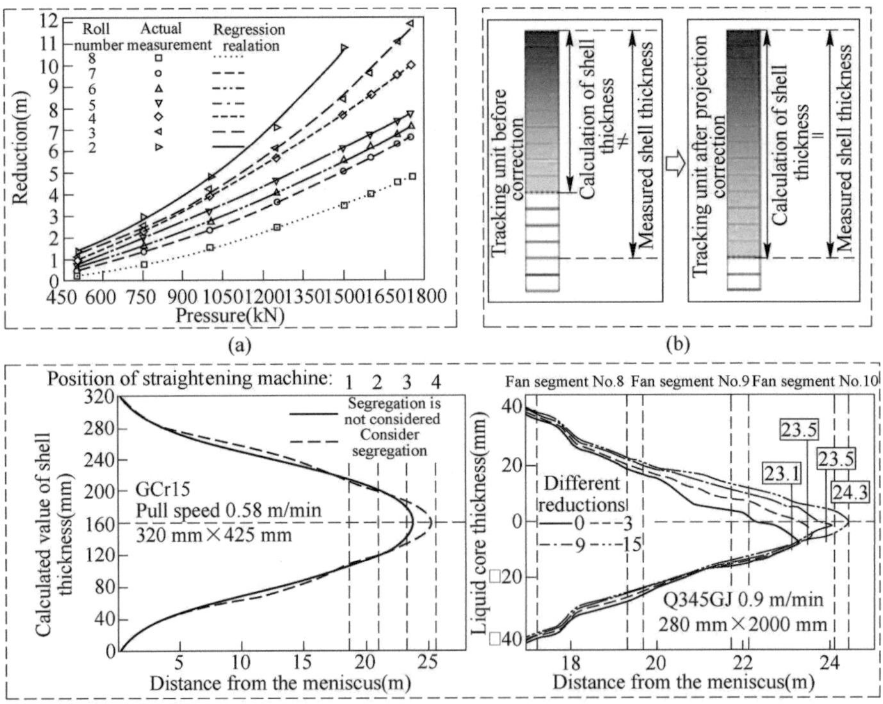

(c)

Fig. 7.10 High-precision solidification end prediction technology. **a** Pressure-reduction-shell thickness relationship; **b** compute grid mapping correction; **c** solidification end point drift

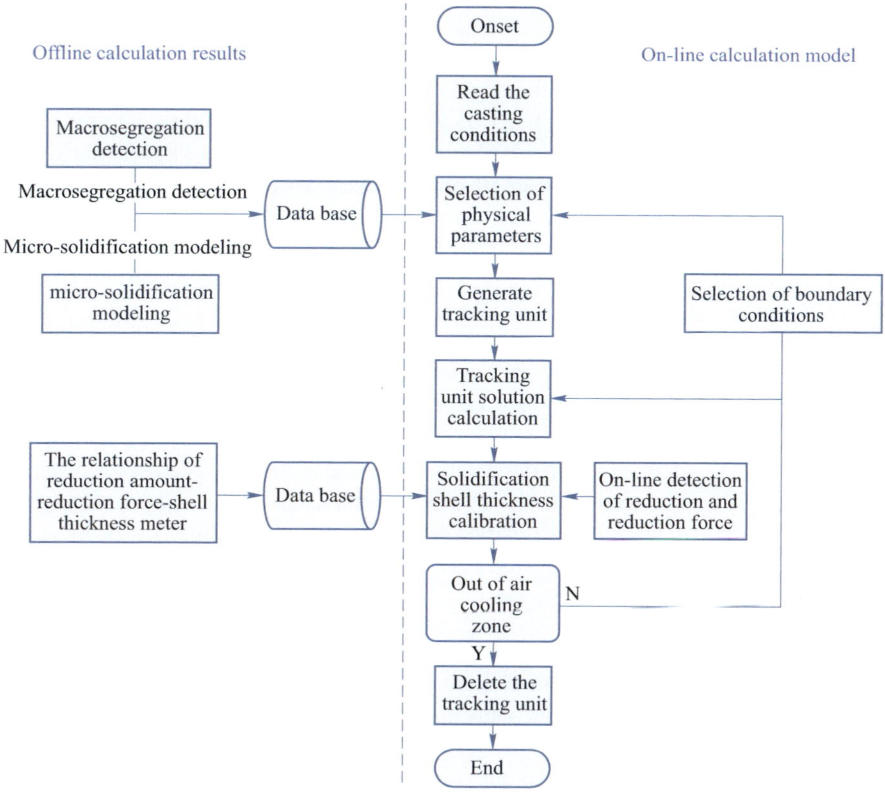

Fig. 7.11 Pressure-reduction control model based on solute distribution

distribution law of the slab along the thickness direction at the position of 1/2 width of the slab;

(3) The growth law of solidified shell at different positions in the wide direction of the slab is obtained offline. According to the two-dimensional macroscopic solidification heat transfer and microscopic solute segregation fully coupled calculation model, the growth law of solidified shell at 1/8 position, 1/4 position, and 1/2 position in the wide direction of the slab is obtained.

(2) Online thermal tracking calculation model

(1) Obtain and read the initial casting conditions and casting process information of the billet.

(2) Generation of tracking unit: In a cycle, a new tracking unit is generated at the meniscus position of the crystallizer, and the temperature and position of the tracking unit are initialized.

(3) Select the thermophysical parameters: In a tracking unit, the initial casting conditions and casting process information of the slab are called to determine the thickness position of the slab. The degree of element segregation

corresponding to the thickness position is selected in the database, and calls the thermophysical parameters corresponding to the position in the database according to the degree of elemental segregation.

(4) Tracking unit calculation: In a cycle, call the thermal physical parameters, starting from the mold level of billet width to 1/2 position of the whole continuous casting flow length of all tracking unit to complete a temperature field calculation.
(5) Calculation of non-uniform solidification front: Through the obtained growth law of solidified shell at different positions of the slab, the growth law of solidified shell at 1/4 and 1/8 positions in the width direction of the billet is obtained.
(6) Determination of tracking device position: According to the total casting length recorded during the generation of the tracking unit minus the total casting length at the current time, the position of the tracking unit in this cycle is calculated, and the position is saved to the attributes of the tracking unit. The boundary conditions are selected according to the position of the tracking unit, and the position of the last tracking unit is judged at the same time. If it exceeds the length of the casting machine, the next stage of calculation is performed.

Table 7.4 shows the different values of the solidification end point calculated by using the constant carbon content and the carbon content detected at 10 °C, 20 °C, and 30 °C. The solidification endpoint difference at 10 °C, 20 °C and 30 °C was 0.597 m, 0.608 m, and 0.623 m, respectively.

Table 7.5 is the influence of different casting speeds on the position of solidification end when casting No.45 steel with 2000 mm × 290 mm section at 1516 °C. It can be seen that with the increase of casting speed, the "drift" phenomenon of the final solidification temperature to the casting machine outlet becomes more and more prominent.

The main advantages of this technique over traditional methods are as follows:

(1) Fully consider the difference of physical parameters in different regions caused by solute segregation in the cross-section of the slab, and select the thermal physical parameters at the corresponding position according to the segregation distribution of the slab, so as to realize the accurate prediction of the "drift" phenomenon at the solidification end.

Table 7.4 Different methods were used to calculate the solidification end point results at different superheats

Superheat (°C)	20	30	40
Solidification end point of fixed carbon content (m)	23.761	24.044	24.409
Real-time solidification end point of carbon content (m)	24.358	24.652	25.032
Offset (m)	0.597	0.608	0.623

Table 7.5 Different methods were used to calculate the solidification end point results at different casting speed

Casting speed (m/min)	0.75	0.8	0.85
Solidification end point of fixed carbon content (m)	20.506	22.133	23.761
Real-time solidification end point of carbon content (m)	20.993	22.657	24.358
Offset (m)	0.487	0.524	0.597

(2) Based on the calculation of the solidification process of the central thickness of the slab, according to the relationship between the growth law of the solidified shell at different positions of the slab width obtained offline, the synchronous prediction of the solidification process of the typical position of the slab width is realized.

(3) The prediction accuracy of the model is further improved. The "pressure-reduction amount" online measurement data is used to verify the position of the solidification end, that is, the "pressure-reduction amount" of the clamping cylinder at the inlet and outlet of the fan-shaped section of the slab continuous casting machine, the "pressure-reduction" of the intermediate drive roll reduction cylinder, or the data measured in real time by the bloom continuous casting straightening machine to verify the predicted thickness of the shell online. In addition, according to the "shell thickness-reduction-pressure" relationship established by the offline thermal–mechanical coupling simulation results and the nail shooting test results, the shell thickness at each reduction position is derived, so as to provide online correction support for the calculation results of the solidification end.

7.1.6 "Single Point + Continuous" Heavy Reduction Process

In the process of soft reduction of conventional slab, the roll gap between the two adjacent segments is mostly consistent, so as to avoid the bulging of slab surface caused by improper connection. Compared with the continuous reduction, single point deformation is more conducive to the penetration of the reduction to the center of the slab. However, the use of roll reduction similar to Sumitomo Metals in Japan cannot suppress the macrosegregation defects caused by the inability to improve the long-distance transport of solutes.

The pressure capacity of the hydraulic cylinder at the entrance of the section is fully utilized, taking into account the characteristics of the ECS section under heavy load deceleration, as shown in Fig. 7.12. The specific implementation method of "single point + continuous" is:

(1) The first inner arc roller at the entrance of the ECS segment is used to implement a large reduction. The first casting roller at the entrance can continuously increase the reduction of the first roller (the proportion of the reduction of the first roller

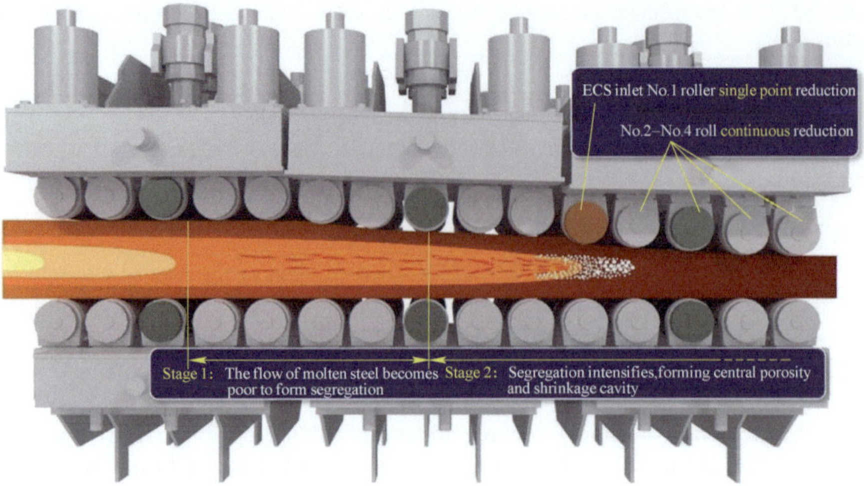

Fig. 7.12 Single point + continuous reduction process diagram

to the total reduction) when the equipment conditions allow and does not crack. The large reduction of the first roller can effectively use the internal and external temperature difference to improve the efficiency of the shell deformation to the center, which is conducive to the improvement of the density of the center of the slab (as shown in Fig. 4.24). In the first sector after the solid fraction $fs \geq 0.9$ position, the reduction process is applied, in which the first support roller implements a single point reduction of 3–20 mm to achieve efficient penetration of the reduction.

(2) At the same time, the subsequent rollers continue to reduction, and each roller adopts a rate of 1.0–5.0 mm/m to continuously press the billet to ensure that the billet does not rebound after the reduction, and at the same time, the billet shell is forced to continue to compress, so as to fully guarantee the improvement of the synchronous shrinkage cavity inside and outside the billet and avoid the porosity caused by the inconsistent shrinkage rate inside and outside the billet.

The flow chart of the implementation of the single point + continuous reduction process is shown in Fig. 7.13.

7.1.7 Crack Risk Predictive Control Technology

The pressing process is prone to two types of cracks and defects: intermediate crack initiation and surface crack propagation. The intermediate crack is generated in the brittle solidification zone. In the range of zero feeding temperature (LIT) to zero plastic temperature (ZDT), the generation of the intermediate crack mainly depends

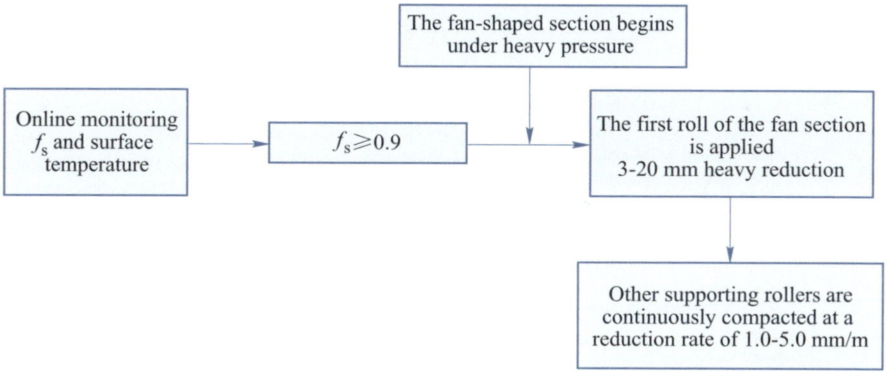

Fig. 7.13 Single point + continuous reduction process flow chart

on the stress and strain that the solidification front edge can withstand. In this temperature range, when the strain at the solidification front exceeds the critical strain, the crack is generated. The surface crack is mainly caused by vibration in the crystallizer and expands in the subsequent straightening and pressing process.

Critical strain measurements for intermediate crack initiation are shown in Fig. 7.14: (1) Sampling from the slab thickness, width 1/4 line and the distance from the surface of 10 mm. (2) The high-temperature tensile test was carried out at about 30 °C below the theoretical solidus, and the true stress–strain curve of the tensile specimen was obtained. (3) The actual tensile temperature (viscosity temperature LIT) was determined according to the fracture morphology and liquid fraction of the actual tensile specimen. (4) The critical strain range of intermediate crack initiation was determined by the stress–strain curve of the tensile specimen at the actual viscosity temperature, and the high-temperature tensile tests with different strains were carried out. (5) Microscopic observation was carried out on the samples corresponding to different strains. The critical strain is determined based on the presence or absence of microcracks, and the strain that just produced cracks was used as the critical strain for the initiation of intermediate cracks. (6) The crack can be avoided when the strain in the reduction process is less than the critical strain. The specific implementation case is shown in Sect. 7.3.2.1.

The surface crack critical strain measurement method is similar to the intermediate crack measurement method, as shown in Fig. 7.15. The main steps include: (1) Using the three-dimensional thermal/mechanical coupling model to determine the strain concentration range of the slab during the reduction process and the corresponding test temperature. (2) The critical strain of surface crack propagation at different temperatures was tested by tensile specimens. (3) The microstructure of the tensile specimen was observed to obtain the critical strain of the surface crack propagation of the continuous casting billet within the test temperature range. (4) Based on the thermal/mechanical simulation analysis under different deformation conditions (including solidification end reduction), the crack propagation risk under

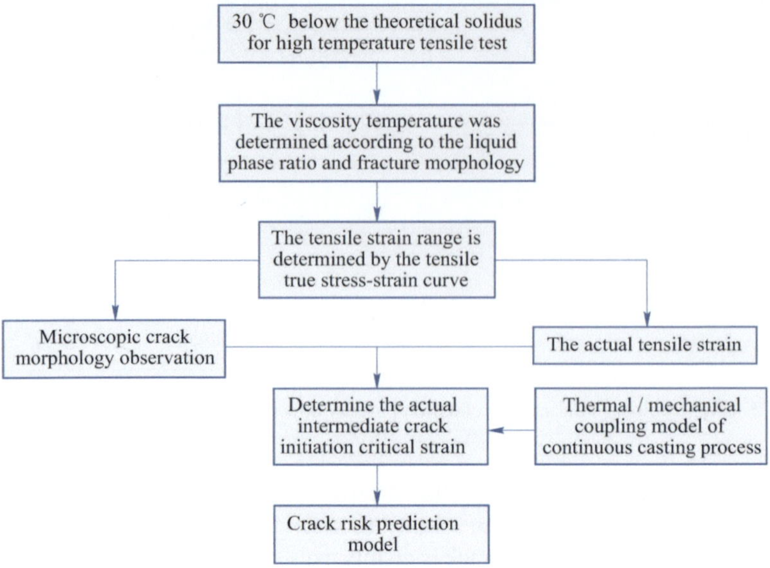

Fig. 7.14 Determination process of intermediate crack critical strain of continuous casting billet

different solidification end reduction conditions was analyzed. The crack propagation risk obtained by this method can determine the maximum reduction of the crack risk at the solidification end and avoid the generation of cracks during the reduction process. The specific implementation case is shown in Sect. 7.3.2.2.

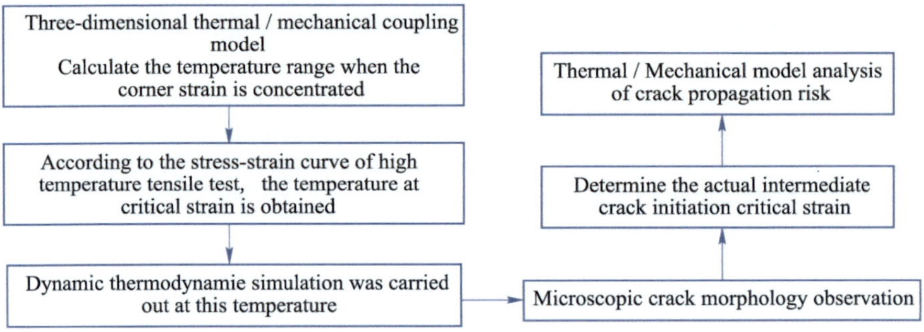

Fig. 7.15 Continuous casting slab surface crack critical strain measurement and control technology

7.1.8 *Intelligent Control System of Solidification End Pressing Down*

Equipped with control system which conforms to the process characters tics of the mechanical reduction at solidification end is a necessary condition for stable, safe, and efficient implementation of this technology. The basic functions of data acquisition, data receiving, and data transmission should be included. This section describes the functions of the intelligent control system developed by our team, including system architecture, program flow, sub-module, and three-dimensional model.

7.1.8.1 System Architecture

As shown in Fig. 7.16, the structure diagram of the reduction control system consists of the following main parts: (1) System security, ensuring data security of the system, for data backup and encryption; (2) display system, in which the data display, data processing and reduction condition analysis; (3) data model, which is the core of the reduction process control to ensure the stability and accuracy; (4) data access interface to connect display subsystem and storage subsystem; (5) communication system transmits data model and display system commands to data points of primary PLC system to realize precise control of reduction equipment. This system ensures that the operator can realize intelligent and stable control of the pressing process only by inputting data at the front of the control interface.

7.1.8.2 System Functional Structure

The functions of the mechanical reduction process control system mainly include switching steel module, dynamic reduction module, real-time temperature calculation module, basic roll gap module, parameter maintenance module, and usage help, as shown in Fig. 7.17.

(A) Dynamic reduction module

The dynamic reduction module is divided into four parts: roll gap cone diagram, secondary cooling water, dynamic reduction, and reduction setting, and the interface is shown in Fig. 7.18.

The roll gap cone diagram shows the roll gap reduction position and the solid, liquid, and two-phase region positions of each segment. The secondary cooling water interface reveals the corresponding parameter data of each region. Dynamic reduction interface can be used to observe data of entry and exit parameters of roll gaps and sector segments. And in the reduction setting interface, users can switch reduction parameters.

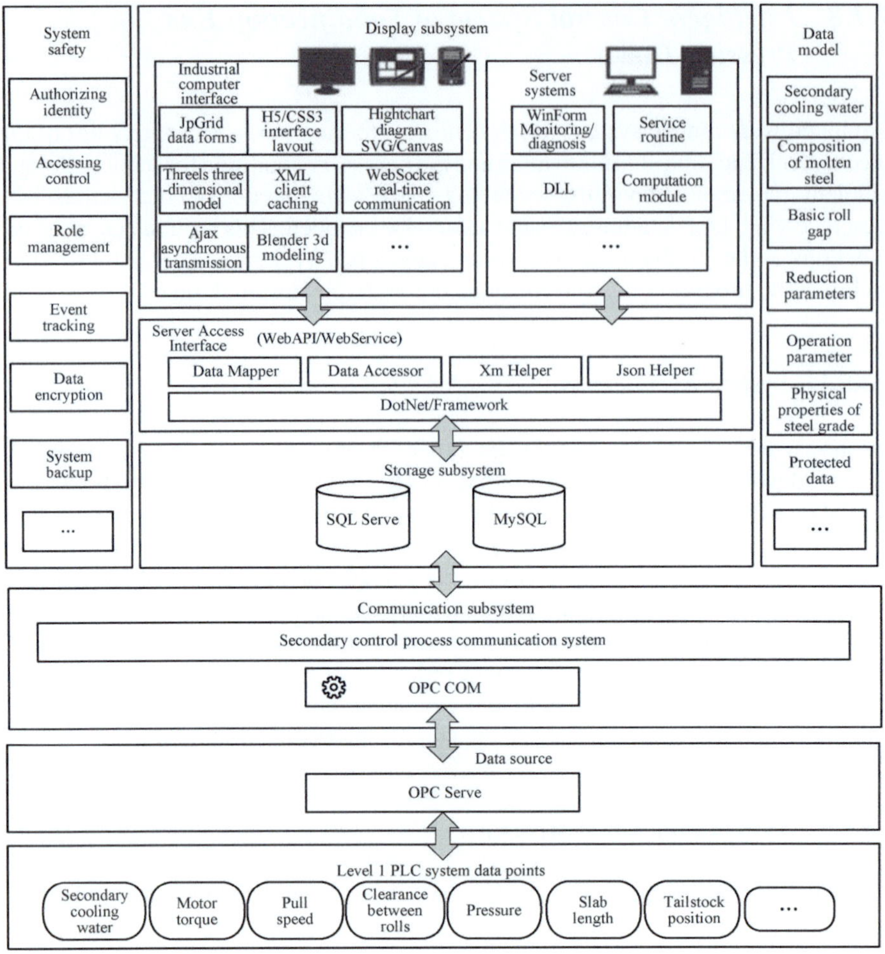

Fig. 7.16 System architecture diagram

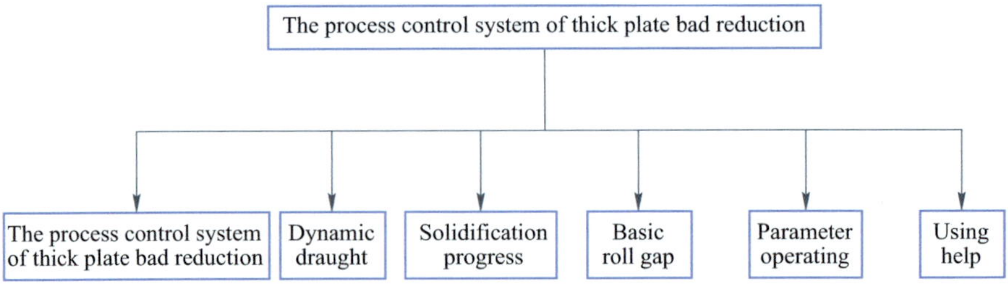

Fig. 7.17 System functional structure

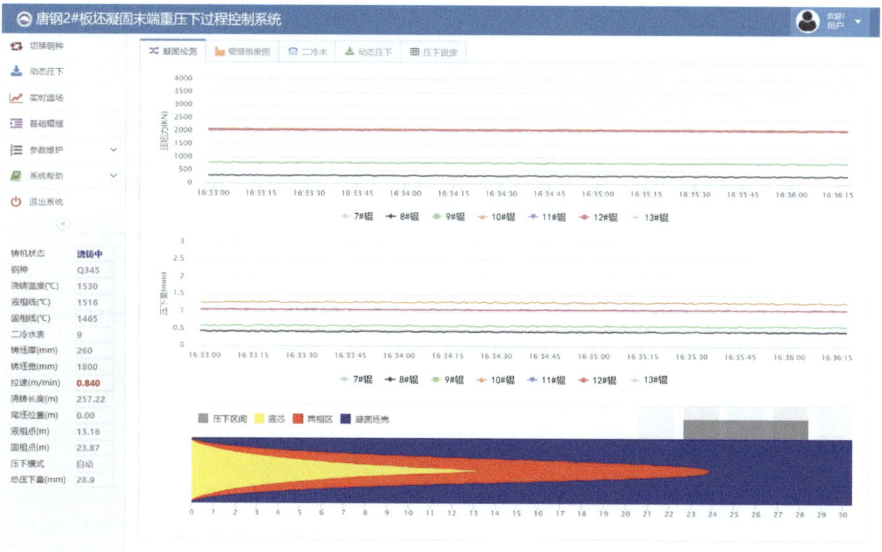

Fig. 7.18 Interface of dynamic reduction module

(B) Real-time calculation module for solidification process

This interface can show the real-time temperature curve of the whole continuous casting process, and the relevant information at the mouse position can be obtained in real time, like center temperature, surface temperature, and shell thickness.

(C) Basic roll gap module

The interface is able to modify and edit the data of the base roll gap and the entrance and exit of each segment section with different slab thickness, add new slab thickness parameters, and search the corresponding data table according to the casting conditions.

(D) Parameter maintenance module

This module is divided into six interfaces with functions of: (1) The reduction parameter setting interface can modify the parameter data for different steel grades and different slab thickness. (2)The protection setting interface at low casting speed can set the protection setting for entry and exit pressure of each segment after segment 7 at low casting speed, and the parameters can be modified and added. (3) Component setting interface for steel can be used to set the content of different steel components. (4) Parameter setting interface for steel can set corresponding parameters for different steel grades. (5) On the basic water meter management interface, users can manage water meters with different numbers. (6) The monitoring interface of simulated state can monitor the parameters of current casting state and display the data with tree

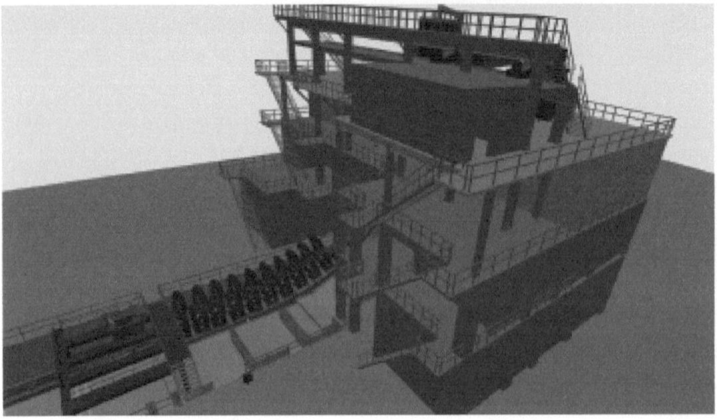

Fig. 7.19 3D simulation model

structure. When the status of monitoring point is red, the next level branch data still exists, click for further observation.

(E) 3D simulation model

According to the actual operation mechanism and conditions of the reduction process and combined with the actual field conditions, a three-dimensional model of the caster is built to simulate the pressing process, as shown in Fig. 7.19.

7.2 Design of the Key Components of the Enhanced Compact Segment of the Wide-Thick Slab

Segment is an important part of slab continuous casting machine, which plays the role of slab guidance, cooling, straightening, and reduction. Its structure has complex and precise characteristics [13]. Especially in the process of heavy reduction and large deformation, the pressing force of the segment is more than 1500 *t*, and it is necessary to ensure that the roll gap accuracy error is not more than 0.1 mm. Therefore, the rationality and reliability of the design of the key components of the segment under heavy reduction is the fundamental guarantee for the stable operation of the heavy reduction process. This section will explain the design of the key components of the enhanced compact segment (ECS) and its key components-cast roller structure design, such as bearing group coordination, drive system configuration, and optimization analysis and design of the key support structure of the segment. The enhancement of the enhanced compact segment refers to that the compact force is 4 times higher than that of the ordinary segment. Compact means that its length is the same as that of the ordinary segment, which creates conditions for the interchange between the heavy reduction section and the ordinary section. The physical picture is shown in Fig. 7.20.

Fig. 7.20 Segment for slab

7.2.1 Design and Check of Compact Segment Casting Roller Structure

7.2.1.1 Compact Segment Casting Roller Structure Design

As a directly accessible core component in the compact car segment, the casting roller works in the harsh environment of high temperature, water vapor, thermal stress, and mechanical load for a long time, and the surface of the casting roller is also affected by the oxide scale and the protective slag scale. Therefore, it is necessary to reasonably design and check the structure of the compact segment continuous casting roller [14–16]. According to the structure and support mode of the casting roller, it can be divided into three categories: integral roller, segmented roller, and combined roller. The integral roller can be divided into: integral type and mandrel type; the sectional roller can be divided into: integral sectional roller and mandrel sectional roller; the combined roller can be divided into: parallel combined segmented roller and series combined segmented roller. The characteristics, advantages, and disadvantages of each type of segment casting roller are shown in Table 7.6. Combined with the characteristics of the solidification end reduction technology, the reliability of the mandrel segmented roller meets the strength and stiffness requirements of the lower casting roller with large reduction.

Table 7.6 Characteristics of segment casting roller

Item	Integral roll		Integral roll		Combined roll	
	Integral type	Mandrel type	Integral type	Mandrel type	Parallel combined	Series combined
Characteristic	The simplest casting roller is made of the whole bar, and the bearings are installed at both ends of the casting roller	The roller sleeve is connected with the mandrel by keys, and the bearing is installed on the mandrel and set at both ends of the casting roller	The whole roll is divided into multiple sections to increase the number of bearing supports and improve the overall stiffness of the casting roll	The segmented roller sleeve is fixed on the mandrel by key, and the bearing seat is set between the roller sleeves	Multiple segmented rollers are combined in parallel, and each segmented roller has independent bearing support at both ends	The segmented roller is connected by a matching type, and the structural strength is large

(continued)

Table 7.6 (continued)

Item	Integral roll		Integral roll		Combined roll	
	Integral type	Mandrel type	Integral type	Mandrel type	Parallel combined	Series combined
Advantage	Simple structure, easy processing and assembly, low cost	The roller sleeve can be replaced, and the surface surfacing layer can be used repeatedly, which has a certain control effect on the deflection of the casting roller	The coaxiality is good, the elevation of the segmented roller is consistent, the assembly workload without matching gap is small, there is no water jacket, and the production reliability is high	There are many supporting points, good stress condition, small deflection of the mandrel, no water jacket connection, reducing the risk of cooling water leakage, and low use cost	The structure of each segment roller is the same and can be used independently. The parts are interchangeable, the equipment management is convenient, and the production cost is reduced	The series type has all the advantages of the parallel type and the number of bearings is small, the use cost is low, and the maintenance amount is small
Disadvantage	When the force is applied, a large deflection will be produced, and the bulging of the billet is easy to affect its size and quality	The excessive deflection in the middle of the mandrel affects the service life, and has little advantage over other types of mandrel rollers	The split bearing used is expensive and has a high failure rate. Once a failure occurs, it will affect the quality of the slab and subsequent production	The installation between the mandrel length and the roller sleeve is not easy, and the disassembly and maintenance need special drawing equipment	There is a risk of cooling water leakage in the inner water jacket connected between the segmented rollers, and there are more bearings, which increases the workload of installation and maintenance	Because the shaft and hole at the connection between the segmented rollers are special-shaped, the processing requirements are high, and the manufacturing process needs to be strictly controlled

7.2.1.2 Design Check of Segment Casting Roller

The deflection and stiffness of the core shaft four-section roller structure used for the segment casting roller under heavy reduction load are checked. The schematic diagram of the core shaft section roller structure is shown in Fig. 7.21. In the diagram, each segment roller continuously crosses multiple intermediate bearing supports and forms a multispan beam through a mandrel in series. When establishing the mathematical model of the casting roller, it can be simplified into a computable continuous beam model. According to the number of intermediate supports, it can be determined that the mandrel four-section roller is a three-time statically indeterminate structure. By establishing a mathematical model of three bending moments of continuous beam, the finite difference numerical calculation method is used to solve the overall deflection distribution and stress value of the casting roller in the sector section and check whether the casting roller meets the conditions of heavy reduction process.

In order to make the mathematical model more concise and the deduction logic clear, the following assumptions are made for the mathematical model [17–19]:

(1) It is considered that the radial load acting on the casting roller is uniform load.
(2) The deformation range of the casting roller is in the elastic range.
(3) The shear stress is in the same direction and uniformly distributed in the cross-section.
(4) Without considering the deformation of the bearing, the bearing support on one side of the simplified segmented roller bearing is set as a fixed hinge support, and the others are all movable hinge supports.

According to the radial load of the four-section mandrel roller, the force analysis of the four-section roller is shown in Fig. 7.22. Where a is the length of the unstressed area on one side of the casting roller, and the value is determined according to the length of the slab.

According to the force analysis, the basic mechanical equilibrium equations of the four-section roller are listed as follows:

$$\begin{cases} F_{\mathrm{RA}} + F_{\mathrm{RB}} + F_{\mathrm{RC}} + F_{\mathrm{RD}} + F_{\mathrm{RE}} = qb \\ F_{\mathrm{RB}} l_1 + F_{\mathrm{RC}}(l_1 + l_2) + F_{\mathrm{RD}} \sum_{i=1}^{3} l_i + F_{\mathrm{RE}} l = \frac{qbl}{2} \end{cases} \tag{7.11}$$

where F_{RA}, F_{RB}, F_{RC}, F_{RD}, F_{RE} are support reaction force of each support, N;q is the casting roller is subjected to uniform load, N/m;b is the total action length of load, m;

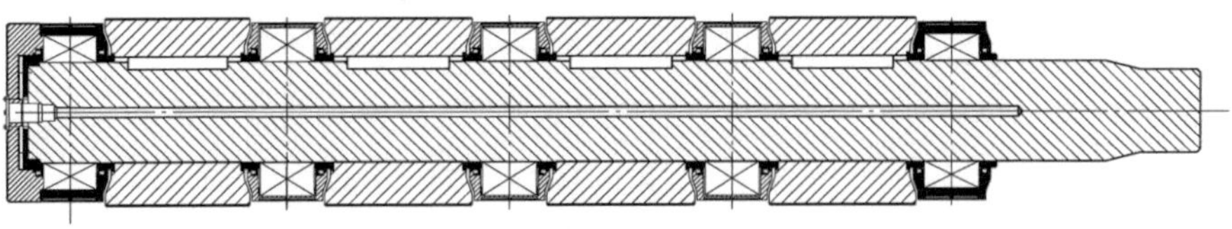

Fig. 7.21 Mandrel four-section roller schematic diagram

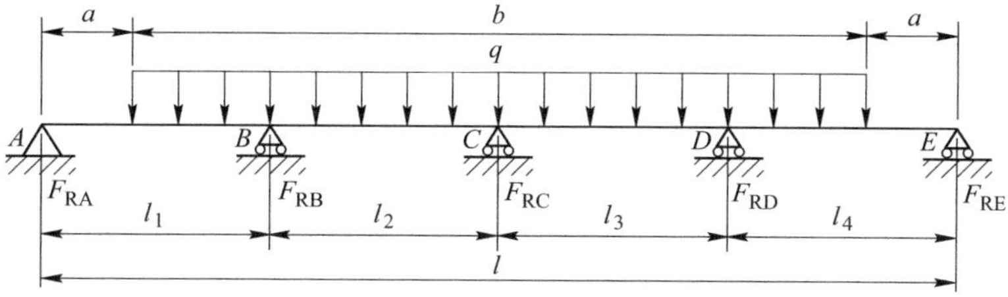

Fig. 7.22 Force diagram of the mandrel four-section roller

l_1, l_2, l_3, l_4 are length of each segment roller, m;l is the total length of casting roller, m.

7.2.1.3 Casting Roller Stress Calculation and Checking

In order to obtain the moment of each point, the stress value at any point of the mandrel is calculated by Eq. (7.12) and whether this point meets the allowable stress value of the casting roller material is checked.

$$\sigma_{(x)} = \frac{M_{(x)}}{\frac{\pi d_r^3}{32}} \leq [\sigma] \tag{7.12}$$

where $\sigma_{(x)}$ is the stress value of any point of casting roller, Pa; $M_{(x)}$ is the torque at any point of the casting roller, $N{\cdot}m$; d_r^3 is the cast roller mandrel diameter, mm; $[\sigma]$ is the allowable stress of casting roller material, Pa.

Taking the compact segment casting roller of wide-thick slab as the specific research object, the stress value and deflection distribution calculation model of the four-segment mandrel of the slab segment are programmed. In order to make the calculation program clearer, the design indexes are listed in Table 7.7.

Through the above theoretical derivation and combined with the actual data, the overall stress distribution of the casting roller mandrel and whether the stress value of each equidistant node meets the allowable stress value of the material used can be explored.

From the calculation results of the stress distribution of each equidistant node in Fig. 7.23, it can be clearly obtained that the maximum absolute value of stress is 43.075 MPa at 260 mm roll diameter. In order to ensure the strength of the mandrel, the material used for the casting roller mandrel under single point heavy reduction described in this paper is 42CrMo. The allowable stress value of the material is 550 MPa, which is much larger than the maximum stress value of the casting roller. As a result, cast roller mandrels at each roll diameter meet the strength requirements.

Table 7.7 Design index of casting roller

Parameter	Parameter meaning	Index
q	Uniform load	2.0×10^6 N/m
l	Length of each segment roller	0.6 m
L	Total length of casting roller	2.4 m
a	Length of unilateral unstressed area	0.1 m
b	Slab width	2.2 m
10	Starting point of coordinates	0 m
h	Step length	0.01 m
D	Cast roller mandrel diameter	0.26/0.28/0.3 m
d	Core shaft water pipe diameter	0.02 m
E	Elastic modulus of mandrel material	2.06×10^{11} Pa
B	Constant term on the right side of equal sign	

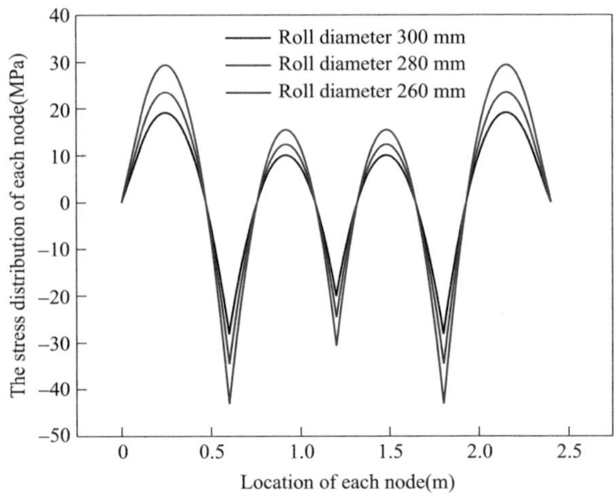

Fig. 7.23 Stress distribution of each equidistant node

7.2.1.4 Casting Roll Deflection Calculation and Checking

The numerical method used to calculate the deflection of the casting roller is mainly the finite difference method. Taking the bearing support on the left side of the casting roller as the coordinate origin, the length direction of the casting roller is the positive direction of the x-axis, and the deflection curve equation $\omega = f(x)$ of the four-segment roller is established. The casting roller is divided into n parts by using the discrete idea, and the equal nodes are marked as$x_0, x_1, \cdots, x_i, \cdots, x_n$, and the pitch of the two adjacent nodes is $h = x_i - x_{i-1}$. The nodal finite difference equations obtained along the positive direction of the transverse axis of the beam are solved algebraically. The solution obtained is the approximate value of the deflection of each node on the casting roller, and the connection of each approximate value point is the deflection curve of the casting roller under force.

The first derivative $\frac{d\omega}{dx}$ of the deflection ω at Point A on the x-axis i point is approximately equal to the slope at Point A. According to the differential mean value theorem, it is approximately equal to the slope of the chords of Point B and Point C. It can be concluded that:

$$\left(\frac{d\omega}{dx}\right)_i \approx \frac{\omega_{i+1} - \omega_{i-1}}{2h} \tag{7.13}$$

The second derivative $\frac{d^2y}{dx^2}$ of the Eq. (7.13) represents the change rate of the first derivative. Therefore, at the i point, the slope on the right side of the i point can be subtracted from the slope on the left side of the i point, and then the value obtained by removing the interval length is approximately obtained. The expression is:

$$\left(\frac{d^2y}{dx^2}\right)_i \approx \frac{\frac{y_{i+1}-y_i}{a} - \frac{y_i-y_{i-1}}{a}}{a} = \frac{y_{i+1} - 2y_i + y_{i-1}}{a^2} \tag{7.14}$$

The finite difference representation obtained by Eq. (7.13) and Eq. (7.14) is the central difference method. The higher-order derivatives above the second order can also be expressed in the form of difference, but it is not necessary to solve the deflection of the mandrel. The following is the differential expression of the second derivative and the differential equation of the deflection curve:

$$\frac{d^2\omega}{dx^2} = \frac{M}{EI} \tag{7.15}$$

The finite difference equation of multi-span continuous beam can be obtained:

$$\omega_{i+1} - 2\omega_i + \omega_{i-1} = h^2\frac{M_i}{(EI)_i} \quad i = 1, 2 \cdots n - 1 \tag{7.16}$$

In Eq. (7.16), a series of equidistant nodes are selected along the positive direction of the x-axis. The finite difference equation of the corresponding node can be proposed for the section of each node. Then, the algebraic equations composed of nodes are formed by the boundary strips with zero deflection of each hinge support point. Solving the equations, the deflection values of the corresponding sections of each equidistant node on the casting roll can be obtained. The deflection results obtained should meet the allowable deflection requirements of Eq. (7.17)

$$\omega_{(x)} < [\omega] < 1\text{mm} \tag{7.17}$$

Since the four-section cast roller mandrel is a symmetrical structure, the deflection results on the left and right sides along the centerline of the casting roller mandrel are the same. Therefore, in order to cooperate with the finite difference numerical calculation method, the deflection analysis and solution are carried out on the

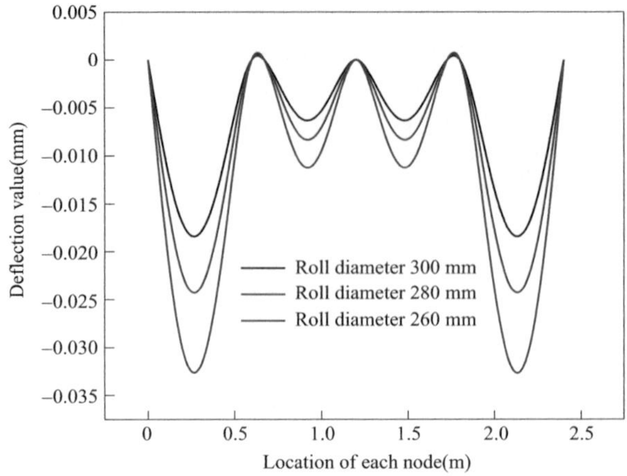

Fig. 7.24 Deflection distribution of each equidistant node

1/2 section of the side of the casting roller mandrel, which can greatly reduce the calculation amount and improve the calculation speed. Let Eq. (7.16) be simplified, and let the constant term on the right side be set to *B*, that is:

$$B_{(i)} = h^2 \frac{M_i}{(EI)_i} \quad i = 1, 2 \cdots n - 1 \tag{7.18}$$

The overall deflection distribution of the casting roller under the shaft diameter is shown in Fig. 7.24.

It can be seen from Fig. 7.24 that the absolute value of the deflection of the casting roller at each equidistant joint section under each roller diameter is less than 1 mm, indicating that the casting roller mandrel meets the stiffness requirements. Combined with the calculation results of the overall stress value of the casting roll, it shows that the structure of the casting roll under each roll diameter can meet the requirements of the reduction process. The roll diameter and bearing model can be selected by considering the investment cost, frame size and spray layout.

7.2.2 *Matching Design of Compact Segment Casting Roller Bearing Group*

There are two main factors involved in the selection of segmented cast roller bearings. One is the selection of bearing type according to the application environment and conditions. The second is the static strength check of the selected bearing. Due to the long-term use of the casting roller under high temperature and heavy load conditions, the bending deformation of the mandrel is an inevitable problem. According to the force analysis, the casting roller mainly bears the radial load caused by the superposition of the pressure force transmitted from the hydraulic device to the casting roller and the static pressure vector of the molten steel inside the casting billet, and the bearing speed is low during the drawing process. Therefore, the type of support bearing at both ends should select the standard series of self-aligning roller bearings. In addition, because the heat in the middle of the segmented roller is higher than that at both ends, the casting roller is more susceptible to heat expansion along the axial and radial directions. In order to absorb the expansion amount while ensuring the reduction, the intermediate bearing should select the CARB ring roller bearing that can move axially, and the clearance fit between the inner ring of this type of bearing and the mandrel should be selected to ensure that the inner ring of the bearing does not fail.

For the low-speed (speed $\leq 10\ r$/min) segment roller bearing, the main failure mode is excessive plastic deformation, so it is necessary to check and calculate the bearing according to the static strength. The calculation equation of rated static load of rolling bearing is shown in Eq. (7.19) [20].

$$C_0 = S_0 P_0 < C_{0r} \tag{7.19}$$

where C_0 is the basic rated static load calculation value, N; P_0 is the equivalent load, casting roller mainly bear radial load, so $P_0 = F_r$,F_r is the radial load, N;S_0 is the safety factor, for the normal use of roller bearing safety factor should be taken: $1 \leq S_0 \leq 3.5$; C_{0r} is the radial basic rated static load, N.

Selection of bearings suitable for the reduction process according to the diameter of the mandrel and the force conditions: the bearing type used in the two sides of the edge position, namely A and E, is the self-aligning roller bearing, which is selected according to the different shaft diameters. The basic rated static load C_0 under the bearing model is 2770 kN (260 mm), 3000 kN (280 mm), 3690 kN (300 mm); the bearing type at the middle position, namely Point B, Point C, and Point D, is CARB ring roller bearing. The basic rated static loads under the bearing model are 2850 kN (260 mm), 3100 kN (280 mm), and 3750 kN (300 mm), respectively. The bearing reaction force output results of each position are FRA = 292.79 kN, FRB = 1343 kN, FRC = 1128.4 kN, FRD = 1343 kN, FRE = 292.79 kN, and the static strength safety factor of each bearing position under different shaft diameters is calculated as shown in Fig. 7.25.

The safety factor results of each bearing position under different shaft diameters are evaluated. The minimum value of static strength safety factor is 2.12 at Point B and Point D under 260 mm shaft diameter. The static strength safety factor $S \geq 2$ indicates that the selected bearing models meet the requirements.

7.2.3 Configuration Design of Compact Segment Casting Roller Drive System

The driving system configuration of the segment mainly includes three parts: the coupling connected to the casting roller, the reducer providing the specified speed,

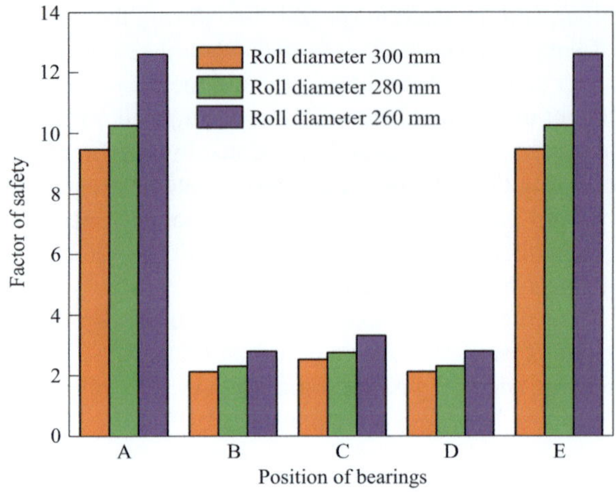

Fig. 7.25 Bearing safety factor

and the driving motor providing the power output. In order to meet the requirements of the change of the opening speed of the casting roll and the rework of the driving roll, the cross universal coupling should be selected between the driving roll and the reducer. Reducer selection of multistage planetary reducer in the hardened planetary reducer. Under the condition that the weight is close to the ordinary planetary reducer, the bearing capacity of this type of reducer is doubled. The selection of the drive motor should meet the requirements of strong adaptability and not affected by the ambient temperature. In addition, it should be convenient to control and can output the corresponding torque according to the change of the pulling speed. Therefore, a six-pole AC servo motor is selected. The six-pole AC servo motor has excellent reliability and overload capacity. The segment drive system is shown in Fig. 7.26.

Fig. 7.26 Segment drive system

The design and parameter selection of the segment drive system mainly include two aspects: the reduction ratio of the reducer and the power of the drive motor. These two parameters are directly related to the casting speed and the casting capacity. The calculation of the drawing force of the drive system is shown in Eq. (7.20) [21, 22].

$$F_d = \frac{\frac{9550 \times P}{n} \times i \times \eta}{r} \tag{7.20}$$

where F_d is the punching force of driving system, N;P is the drive motor speed, kW;n is the drive motor speed, $r/\min$;i is the reduction ratio of reducer; η is the transmission efficiency; r is the casting roller radius, m.

When the casting force of the driving system is greater than the casting force required by the process, and the surplus is controlled within a certain range to ensure that the casting roller and the casting billet do not slip, it is to meet the use conditions. According to this, the power of the motor and the reduction ratio of the reducer can be selected. In addition, the frequency of the motor should be calculated when it is working. Because the casting is a continuous and uninterrupted production, it is necessary to avoid the low power or high power operation of the motor for a long time, so as to avoid the overheating or frequency overload of the motor. The motor frequency calculation is shown in Eq. (7.21).

$$f_{\text{operating}} = \frac{v \times i}{\pi \times d} \times \frac{f_{\text{rated}}}{n_{\text{rated}}} \tag{7.21}$$

where $f_{\text{operating}}$ is the motor operating frequency, Hz; v is the casting speed, m/min; i is the reduction ratio of reducer; d is the casting roller diameter, m; f_{rated} is the motor rated frequency, Hz, usually take 50 Hz; n_{rated} is the rated speed of the motor, r/min, the six-pole AC motor usually takes 960–1000 r/min.

7.2.4 Optimization Analysis of Key Supporting Components of Compact Segment

The design principle of segment structure is to find the optimal solution of the objective function in the mathematical model by controlling the variables on the basis of meeting the structural design requirements (such as strength, stiffness, mass, etc.). The optimization design method first appeared in the problem of finding the optimal solution in mathematics. However, with the development of computer technology, the optimization design problem of structure has also developed rapidly with the help of computer, especially in the fields of machinery, architecture and aerospace. The optimization design of compact sector research mainly includes structural topology optimization [23], that is, structural shape optimization and structural parameter optimization, that is, two-level optimization design that controls the specific parameterization of design variables.

7.2.4.1 Extension Optimization Analysis

In this section, the key support components of the segment frame refer to the hydraulic rod support structure. Through the optimization design of the appearance shape of the support structure under the constraints of the hydraulic system load and the frame displacement boundary, the hydraulic rod support structure is guaranteed to not lose the basic function and ensure the best material distribution under the premise of satisfying the stiffness and strength design conditions. From the deformation law of the hydraulic rod after increasing the reduction of the conventional segment and the static analysis results of the overall frame, it is found that the hydraulic rod needs to be supported to reduce its deformation and ensure the stability and long service life of the segment.

The steps of topology optimization design of hydraulic rod supporting structure are as follows:

(1) Import the three-dimensional geometric model.
(2) Give material properties.
(3) Meshing: The automatic meshing method is used to divide the mesh. The divided mesh is tetrahedral mesh and then the mesh size is further refined. The mesh division is shown in Fig. 7.27.
(4) Applying constraints and loads: According to the previous calculation results, six-degree-of-freedom constraints are applied to the bottom plate and corresponding loads are applied to the upper part of the left and right bucket supports.
(5) Static analysis and calculation: The static analysis results of the simulated hydraulic rod support structure are shown in Fig. 7.28.
(6) Topology optimization calculation, the optimization results are as follows.

From the results of topology optimization analysis in Fig. 7.29, the removal part (orange area) is mainly concentrated on the lower plate of hydraulic rod support, and the boundary part (brown area) is also concentrated on the lower plate. The

Fig. 7.27 Tetrahedral mesh division of hydraulic rod support structure

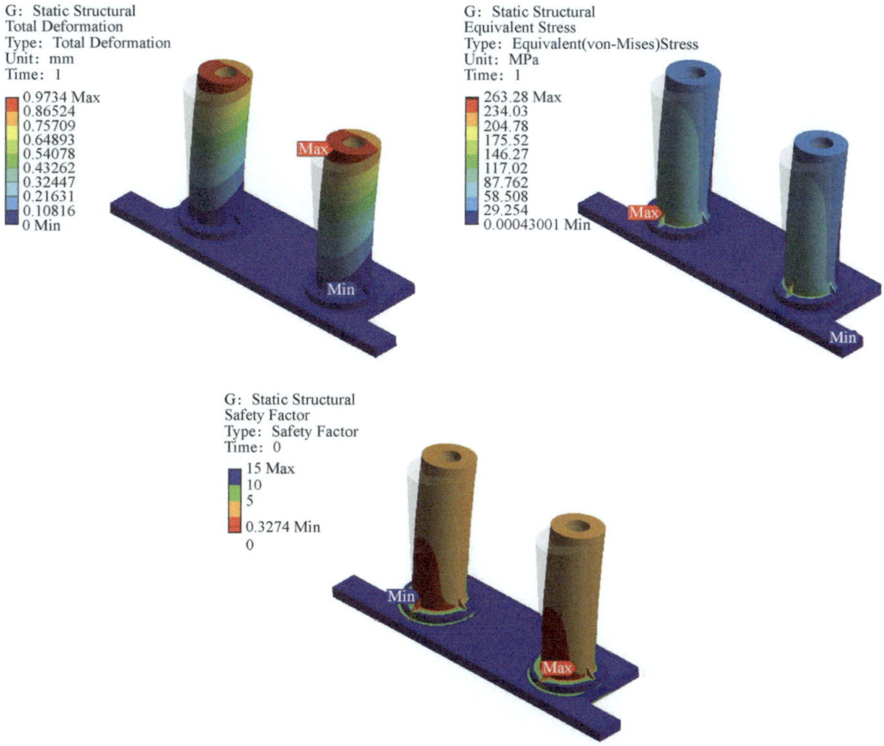

Fig. 7.28 Analysis result of external fixing sleeve of hydraulic rod. **a** Deformation nephogram; **b** stress nephogram; **c** safety factor nephogram

thickness of the lower plate can be optimized to a certain extent. The upper half of the reserved part (gray area) has no removal part except for two points, indicating that the material that can be removed from the thickness of the lower plate can be supplemented to the reinforced rib plate to further reduce the overall deformation and stress concentration at the rib plate. In order to further optimize the external fixing sleeve of the hydraulic rod, the thickness of the reinforcing rib plate is taken as the input variable, and the parameter optimization design of the reinforcing rib plate is carried out in combination with the structural deformation, stress concentration area and density.

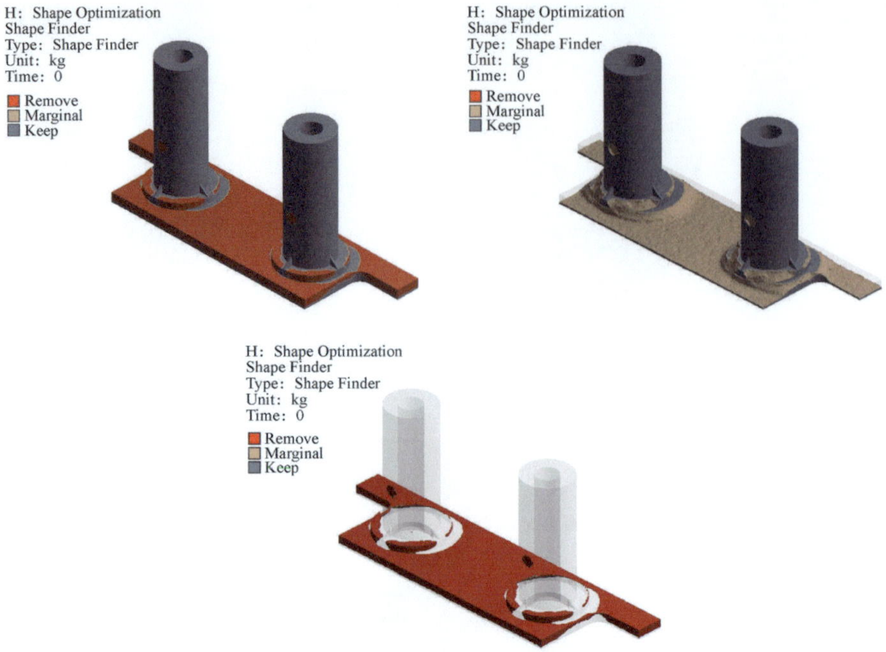

Fig. 7.29 topology optimization analysis results. **a** Overall results of topology optimization; **b** material retention and boundary part; **c** material removal part

7.2.4.2 Parameter Optimization Analysis

The parameter optimization method of hydraulic rod supporting structure is as follows:

(1) Establish a parametric model.
(2) Set the objective function and constraint conditions: The rib thickness is defined as the optimization objective. There are two main constraints in the parameter optimization design. One is that the stress of the rib plate should not exceed the allowable yield stress of the selected material, and the other is that the maximum deformation of the outer fixing sleeve of the whole hydraulic rod should not exceed 2 mm.
(3) Response surface: When the input and output parameters and function settings are completed, the classical reaction type is selected, and the value range of 8–16 mm continuous variables is set for the design point of rib thickness, and a series of design points are generated. The calculation results of the five best design point candidate points are shown in Table 7.8.
(4) View the response surface calculation results. The optimization calculation results of the surface response parameters will eventually fit the input and output parameters into a function curve image, including the relationship between the

Table 7.8 Design point solution results

A	B	C	D	E
Variable	Rib thickness	Maximum stress (MPa)	Total deformation (mm)	Mass (kg)
1	12	239.44	0.97136	1578.4
2	8	245.54	0.975	1577.9
3	16	279.56	0.9683	1579
4	10	237.56	0.97313	1578.2
5	14	241.32	0.96987	1578.7

design point and the mass, the maximum deformation and the equivalent stress. The calculation results are shown in Figs. 7.30, 7.31, 7.32.

The relationship between the design points and the quality of the supporting structure is shown in Fig. 7.31. As the thickness of the ribbed plate increases, the quality of the hydraulic rod supporting structure is also increasing, which is in line with the actual logic. Figure 7.31 reflects the relationship between the thickness of the rib and the maximum deformation of the structure. It can be concluded that for a given range of rib thicknesses, the maximum deformation of the structure decreases linearly with increasing rib thickness. Figure 7.32 reflects the relationship between the thickness of the rib plate and the maximum equivalent stress. In the given thickness range of the rib plate, the equivalent stress shows an increasing trend as a whole. Specifically, the maximum equivalent stress in the figure decreases first, reaches the minimum value at 10.3 mm and then increases, and decreases again at 12 mm. After

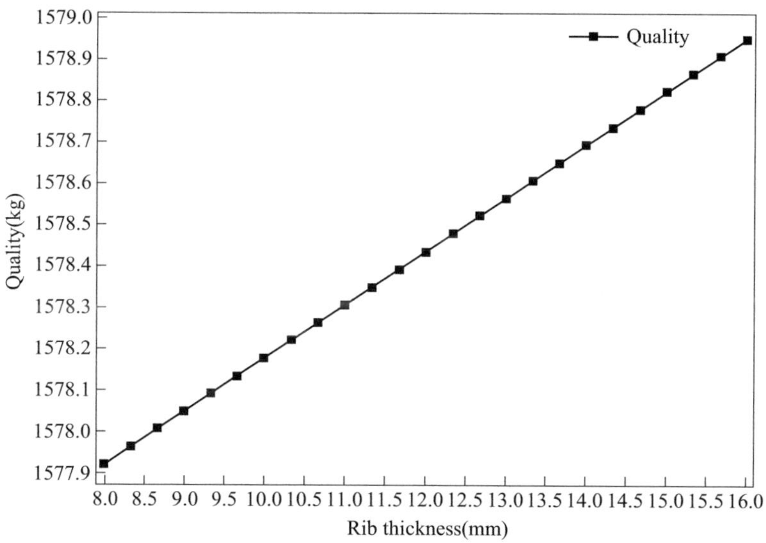

Fig. 7.30 Functional relationship between design point and mass

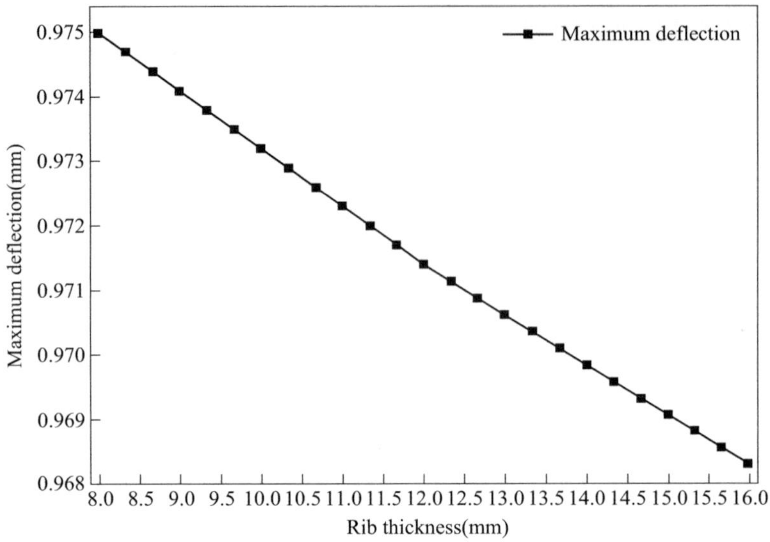

Fig. 7.31 Function relations between design point and maximum deformation

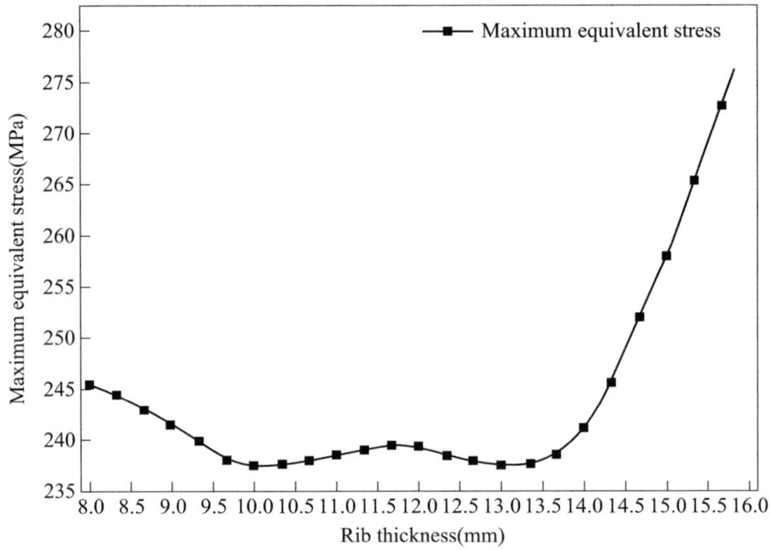

Fig. 7.32 Functional relationship between the design point and the maximum equivalent stress

the thickness is reduced to 13.2 mm, the equivalent stress has been in a stable state of increase, and the minimum value of the maximum equivalent stress is 279.56 MPa.

Figure 7.33 is the sensitivity analysis diagram, which can visually display the thickness of the rib plate represented by the design point, as the input design variables on the quality of the hydraulic rod support structure, the maximum deformation, and the maximum equivalent stress output variables. The height of the bar chart reflects the influence of the input variable on the target output variable. The higher the height, the greater the influence of the input design variable on the target output variable. It can also be seen from the pie chart that the closer the input variable is to the center of the circle, the greater the impact on the target output variable. It can be seen from the synthesis of the two figures that the quality factor has little effect on the thickness of the whole rib, and the maximum equivalent stress is the main factor determining the thickness of the rib. Figure 7.34 reflects the cobweb diagram of the balance relationship between the quality of the hydraulic rod support structure, the maximum equivalent stress, and the maximum deformation.

The design points generated by the optimization design algorithm can obtain the output parameter values covering the design points of the whole design requirements through iterative calculation. Finally, the limited design objectives are given by comprehensively analyzing the relationship between the design point and the maximum equivalent stress, the maximum deformation, and the quality and the design point to obtain the optimal design scheme. The maximum stress of the scheme should be less than 345 MPa, and the maximum deformation should not exceed 2 mm.

In the multi-objective optimization design, because each input parameter is in a balanced state, it is not possible to have every input target optimally designed, because the change of one parameter often causes the linkage change of other parameters. Therefore, in the optimization design, it is necessary to solve the optimal solution according to the working requirements of the hydraulic rod support structure. The

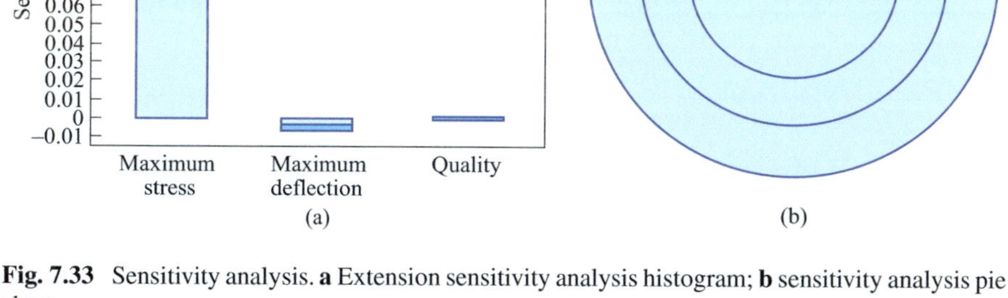

Fig. 7.33 Sensitivity analysis. **a** Extension sensitivity analysis histogram; **b** sensitivity analysis pie chart

Fig. 7.34 Hydraulic rod support quality, the maximum equivalent stress and the maximum deformation under the interaction of the balance relationship cobweb diagram

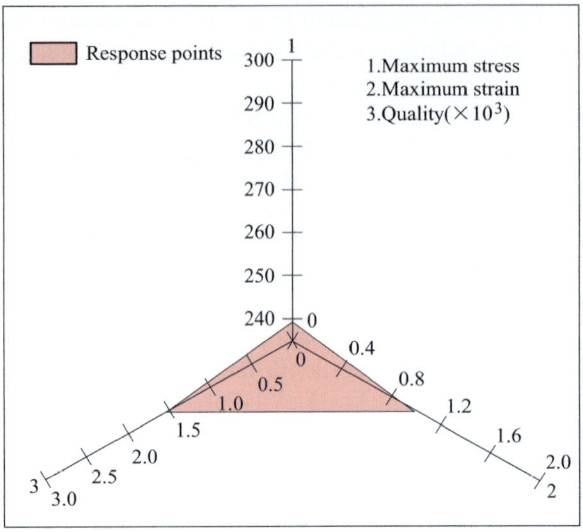

Reference	Name	P5 - DS_D1@Extrude6@s3.Part	P2 - Equivalent Stress Maximum (MPa)		P3 - Total Deformation Maximum (mm)		P4 - Geometry Mass (kg)	
			Parameter Value	Variation from Reference	Parameter Value	Variation from Reference	Parameter Value	Variation from Reference
◉	Candidate Point 1	10.324	237.46	0.00%	0.97284	0.00%	1578.2	0.00%
○	Candidate Point 2	13.084	237.76	0.13%	0.97054	-0.24%	1578.6	0.02%
○	Candidate Point 3	10.892	238.28	0.35%	0.97232	-0.05%	1578.3	0.00%

Fig. 7.35 Alternate optimization scheme diagram

optimization described in this section is to ensure that the maximum deformation and the maximum equivalent stress minimize the structural mass while meeting the limited requirements. The alternate optimization scheme given from 1000 calculation points within a given optimization range is shown in Fig. 7.35.

Through the above analysis, the Candidate Point 1 scheme can be determined as the final design scheme. Comparing the effects of the three optimization schemes with Candidate Point 1 as the reference point, it is found that the maximum equivalent stress of the other two schemes is greater than 0.13% and 0.35% than that of scheme 1. Although the maximum deformation of scheme 2 and scheme 3 is 0.24% and 0.05% better than that of scheme 1, from the reference of each design index, the rib thickness of scheme 1 is 10.324 mm as the optimal solution.

7.3 Design of Gradient Curvature Convex Roller for Bloom

The side shell of the bloom is solidified completely on the solidification end of strand, which provides the main resistance to HR deformation. The use of convex roll could avoid the bloom side part and yield a high reduction amount. At the same time, in order to increase the amount of reduction, the equipment capacity needs to be improved and the cost is higher. According to the solidification characteristics of the slab, the liquid core is conical, while the deformation resistance during reduction is mainly concentrated in the solidified shell on both sides of the slab. Therefore, the convex roll reduction technology capable of avoiding slab corners has been proposed.

Since 1990s, Disk-Roll reduction was firstly proposed by Nippon Steel [24] (Fig. 7.36a), which features a large-diameter part that projects radially outward in the axial direction from the center region, in order to reduce the contact resistance of the solidified shell on both sides of the slab to the roll during the reduction process. And then Nippon Steel and Sumitomo Metal improved the transition between the convex part and the unconvex part to a linear transition [25] (Fig. 7.36b). The field application shows that slab segregation can be significantly improved by reduction using the Disk-Roll at the solidification end. Ogibayashi et al. investigated that the shape of the roll affects the size of the two-phase zone and the convex roll is more effective in reducing the center segregation [26]. Chang et al. proposed a convex roll a convex roller with two arc shoulders for soft reduction (POSCO-Roll) as shown in Fig. 7.36c. The centerline segregation grade was improved by 87.5% compared with the conventional roll [27]. Chang et al. have made a further improvement in the center segregation of the slab and have taken into consideration the use of arc transitions for reducing stress concentration. If the shape of the gradient zone is optimized, the center quality of the slab can be improved while the surface defects can be reduced. The author argues that the double 1/4 arc transition structure between the convex surface and the plane of the POSCO-Roll, which will cause the concentration of stress–strain, thus making the slab more prone to cracking. A curved surface convex roll is proposed as shown in Fig. 7.36d to solve this problem. The convex roll was connected with an arc line with slowly changing curvature between the convex part and the plane, so that the stress–strain concentration was greatly reduced during the reduction process [28].

7.3.1 Design Principle of CSC-Roll

The shape and size of the convex roll significantly affect the reduction effect. However, previous study only considered the effect of different roll shapes on the quality of slab, and did not further explore the influence of the convex roll size on the deformation of slab. The surface transition zone structure of the convex roll affects its service life and is more prone to cracks and other defects during long contact with the slab. Therefore, it is essential to establish a deformable body model for analyzing the temperature and deformation of the convex roll during reduction process, so as to obtain a more appropriate structure of the transition zone.

The convex roll consists of edge zone, transition zone, and convex platform region, as shown in Fig. 7.37.

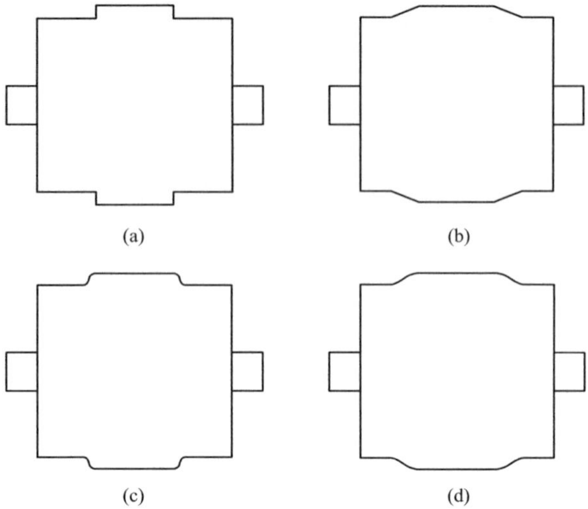

Fig. 7.36 Different convex roller diagram. **a** Rectangular convex roller; **b** beveled convex roller; **c** arc convex roller; **d** gradient curvature convex roller

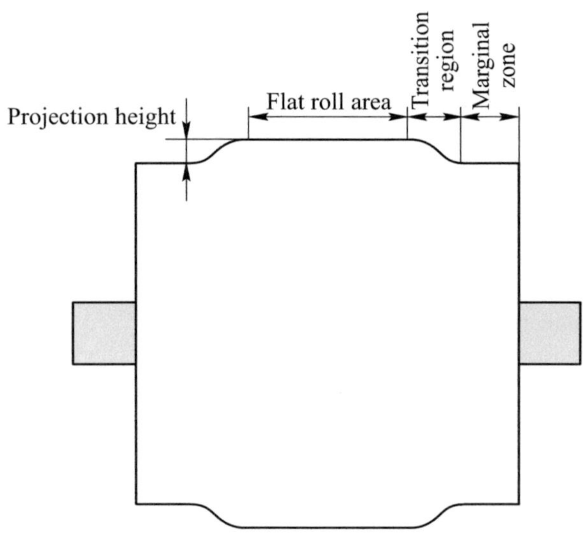

Fig. 7.37 Schematic diagram of CSC-Roll

The design of the convex roll should not only achieve the purpose of improving the internal quality of the slab, but also prolong the service life of the convex roll. Specific design principles are as follows:

(1) The maximum stress on the slab surface should be as small as possible. Stress concentration emerges in the slab where it is in contact with the edge of the convex platform. The greater contact stress will accelerate the wear of the roll sleeve and reduce the service life of the equipment. Stress concentration can be

mitigated by reducing the height of the convex platform or increasing the length of the gradient zone.

(2) The deformation of the slab liquid core should be as large as possible. There is a certain length of liquid core at the solidification end of the slab. To achieve a better segregation improvement effect, sufficient deformation of the liquid core is required. The deformation can be controlled by adjusting the convex length.

(3) The reduction force does not exceed the maximum output pressure of the equipment. The output pressure of the sector hydraulic cylinder is limited. It is necessary to control the length of the convex platform, the height of the convex platform, and the length of the gradient zone so that the reduction force under the target reduction is less than the maximum output pressure of the equipment.

7.3.2 *Design of CSC-Roll for Bloom*

The design and application effect of a curving surface convex roll for a 360 mm × 450 mm No.45 steel continuous casting bloom was introduced in this section.

7.3.2.1 Finite Element Model

A 3D thermal–mechanical direct coupling model was developed as shown in Fig. 7.38. XO is the thickness direction, YO is the casting direction, ZO is the width direction, and XOY is the width direction symmetrical surface, and the casting speed is 0.55 m/min. In order to describe the deformation and heat transfer behavior of the bloom, the measuring points are selected as shown in Fig. 7.38, where Point A–Point E are the center of the bloom, 1/4 inner surface node, inner surface center node, 1/4 corner node and outer surface center node, respectively. Before the reduction process is implemented, the measured porosity zone does not exceed 80 mm × 120 mm in general, as shown in Fig. 7.39. The porosity zone of the model is shown in Fig. 7.39b and the node on the elliptical half-shaft is selected as the analysis object to describe the transfer of the reduction amount to the center of the bloom.

7.3.2.2 Model Validation

The accuracy of the model is evaluated through comparing the data computed by the model with the field measured data. As shown in Fig. 7.40, at the casting speed of 0.55 m/min and the specific water flowrate in the secondary cooling zone is 0.26 L/kg, the average reaction force calculated with 10 mm reduction amount on the No.6 withdrawal machine is 1686.368 kN, and the force measured by the pressure sensor online is 1626.78 kN. The relative error between the calculated result and the actual measured value is 3.636%. Figure 7.41 is the comparison of the simulated and actual

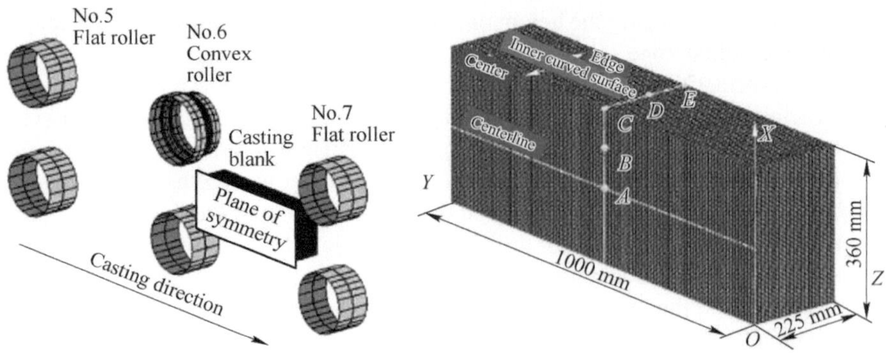

Fig. 7.38 Finite element model and position of measuring points

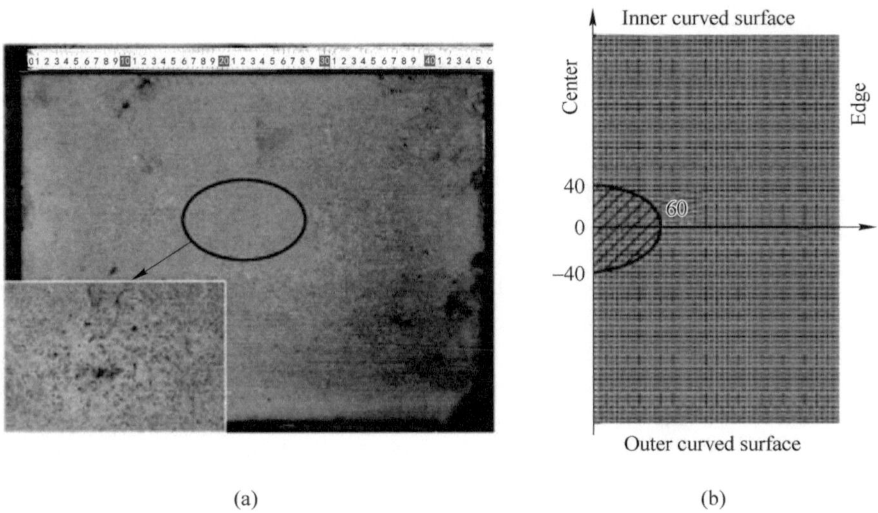

Fig. 7.39 Schematic diagram of porosity zone. **a** Measured; **b** calculated

bloom after reduction. It shows that the simulated results are in good agreement with the real measured, which also proves the accuracy of the model.

7.3.2.3 Design of CSC-Roll Structure

(A) The shape of the transition zone

As shown in Fig. 7.42, the deformation distributions in the thickness and width directions of the porosity zone after reduction with four rolls of different transition zone shapes are compared using the finite element model. The length and height of

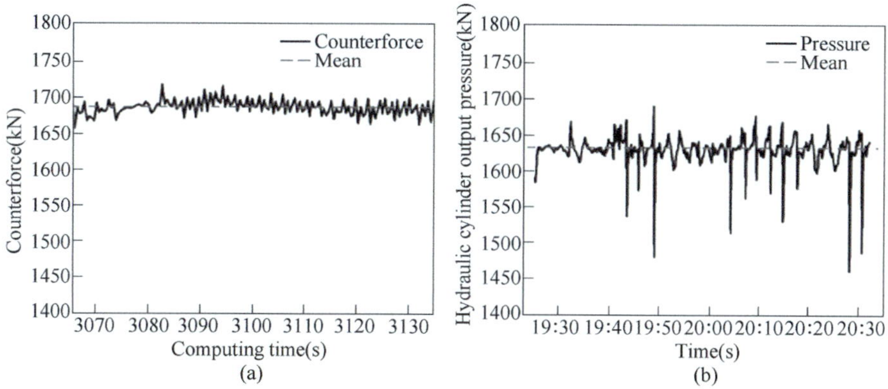

Fig. 7.40 **a** Calculated counterforce; **b** measured pressure

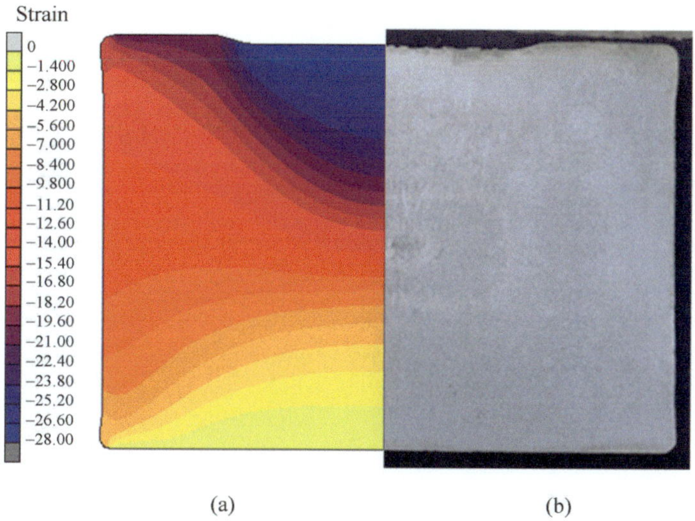

Fig. 7.41 **a** Simulation results of strain distribution of bloom; **b** real deformation shape

the convex platform region for each roll are the same, and the amount of reduction is 10 mm.

As shown in Fig. 7.42a, the deformation in the center area of the bloom gradually decreases from the inner surface to the outer surface in the thickness direction, and the deformation of the CSC-Roll is larger than that of other rolls. As shown in Fig. 7.42b, the amount of deformation transferred to the bloom center by the CSC-Roll is larger than that by other rolls, which means that the CSC-Roll is more beneficial to improve the transfer efficiency of the pressure to the center of the bloom, and the improvement of the internal quality of the bloom is better at the same reduction amount.

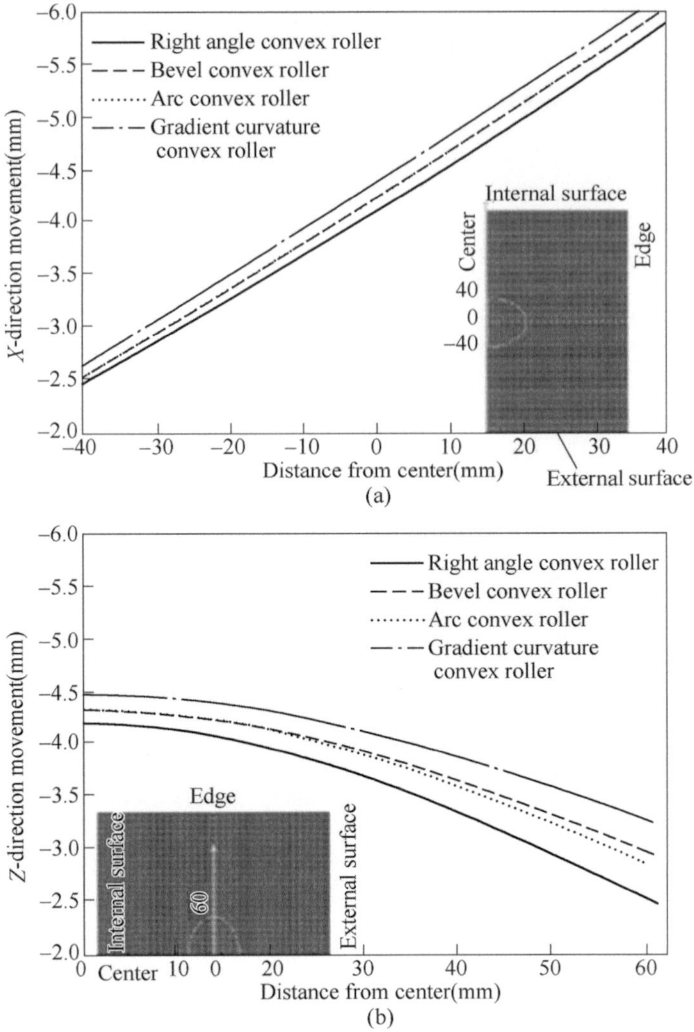

Fig. 7.42 Deformation distributions of the porosity zone. **a** In the thickness direction; **b** in the width direction

It can be seen in Fig. 7.43 that the equivalent strain of the inner surface increases rapidly during the reduction process of the convex rolls with four transitional zone structures and then decreases slightly after reaching the maximum value and remains stable. This is because the alteration in equivalent strain encompasses both elastic and plastic strains. These strains recover after the reduction process ends, and as a result, a small decrease is exhibited. In addition, the equivalent effect of the bloom surface becomes much smaller during the reduction process of the CSC-Roll than that of the other three structures. After reduction with right angle convex roll, edge

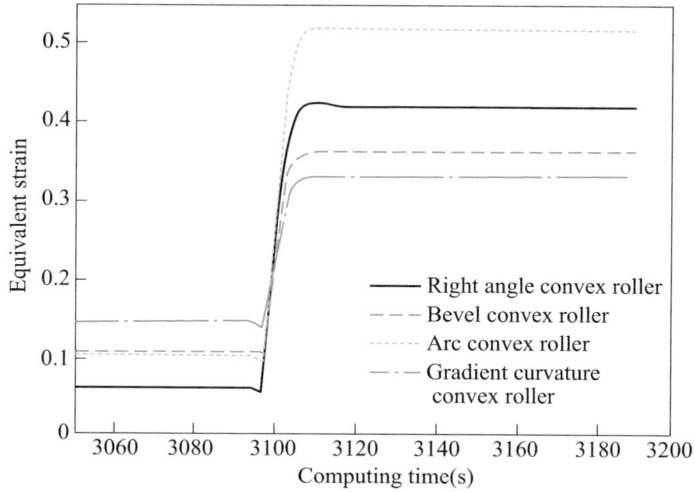

Fig. 7.43 Maximum equivalent strain history plot of the inner surface for convex rolls with different roll types

convex roll, arc convex roll, and CSC-Roll, the maximum equivalent strains on the surface of the bloom are 0.417, 0.359, 0.515, and 0.328, respectively. The minimum equivalent strain of the bloom surface with CSC-Roll means that the risk of surface crack defects is the lowest.

(B) The length of the gradual curvature region (Lgc)

The convex roll models with 50, 60, 70, 80, 90, and 100 mm transitional zone lengths were established to explore the optimal transitional zone lengths. The calculated contact reaction force and the maximum equivalent strain on the bloom surface are shown in Fig. 7.44a. Since the maximum output pressure in the plant is 1690 kN, the length of the gradient curvature zone should be less than 82.4 mm. At the same time, the longer the gradient curvature zone is, the smaller the maximum equivalent strain on the surface of the bloom will be. As shown in Fig. 7.44b, with the increase of the gradual curvature region, the maximum equivalent strain of the inner arc first decreases and then increases. The suitable Lgc is 80 mm considering comprehensively.

(C) The length of convex platform region (Lcp)

According to the size of the porosity zone, Lcp sets to 120 mm, 140 mm, …, 260 mm, respectively. Other conditions and parameters are the same, and the reduction amount of each example is 10 mm.

Figure 7.45 shows the distribution of reduction amount in thickness and width direction of porosity zone by different rolls with different Lcp. With the increase of Lcp, the amount of reduction transferred to the porosity zone gradually increases, the increase effect becomes smaller when Lcp exceeds 180 mm, and the amount of

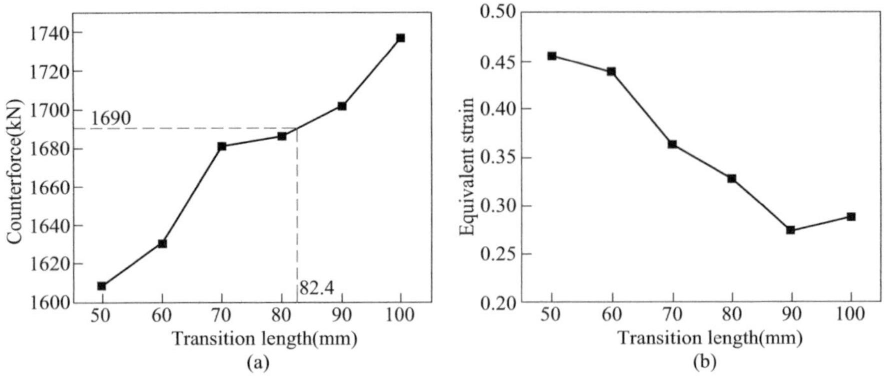

Fig. 7.44 Convex roller with different length of transition zone in the process of heavy reduction. **a** Contact reaction force; **b** maximum equivalent strain of slab surface

reduction is almost stable after exceeding 200 mm. When Lcp is longer than 200 m, the amount of reduction on porosity zone can be increased by more than 0.92 mm.

Figure 7.46a shows the equivalent strain diagram of the nodes on the centerline of the bloom during the reduction process of the convex rolls with different convex platform region lengths. The tendency of each curve is identical. The strain of the nodes in the middle of the bloom is relatively stable, while the two end positions fluctuate greatly due to the limited bloom length in the model. From the magnified view, it can be seen that the strain at the center node of the bloom increases with the increase of the length of the convex platform region. Average the different curves in the middle rectangle as shown in Fig. 7.46b. It shows that the equivalent strain of the bloom centerline node increases slowly with the increase of the length of the convex platform region, and gradually stabilizes after 200 mm. This is mainly because the longer the convex platform region is, the larger the reduction zone is, and the more

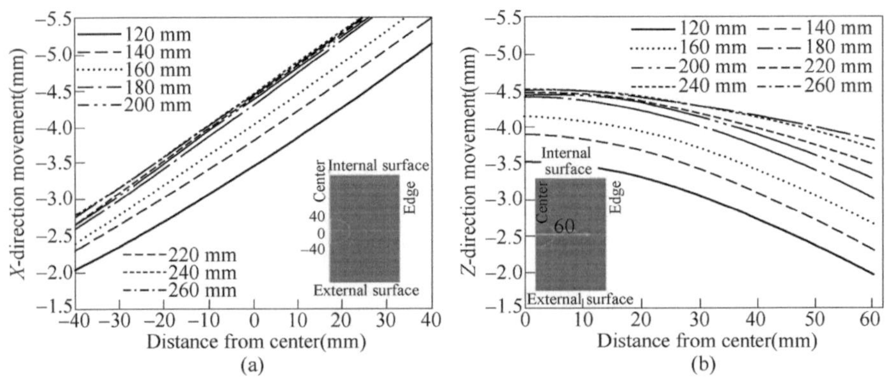

Fig. 7.45 Distribution of deformation in porosity zone of bloom after reduction of convex roll with different length of convex platform region. **a** Thickness direction; **b** width direction

pressure is transferred to the center. However, due to the constant reduction of the CSC-Roll, the central strain gradually reaches an extreme value with the increase of the length of the convex platform region.

As shown in Fig. 7.47, the counterforce increases with the increase of the length of the convex platform region and reaches the maximum unit pressure of 1690 kN when the length of the convex platform region is 201 mm. So, it is better to keep Lcp shorter than 201 mm.

Considering the limitation of hydraulic cylinder output pressure and the deformation in porosity zone of casting bloom after pressing, the optimum Lcp of CSC-Roll is 200 mm.

(D) The height of the convex platform (Hcp)

Under the same other conditions, the Hcp was set to 20, 25, 30, 35, and 40 mm to explore the optimal parameters. It can be seen from Fig. 7.48 that Hcp has little

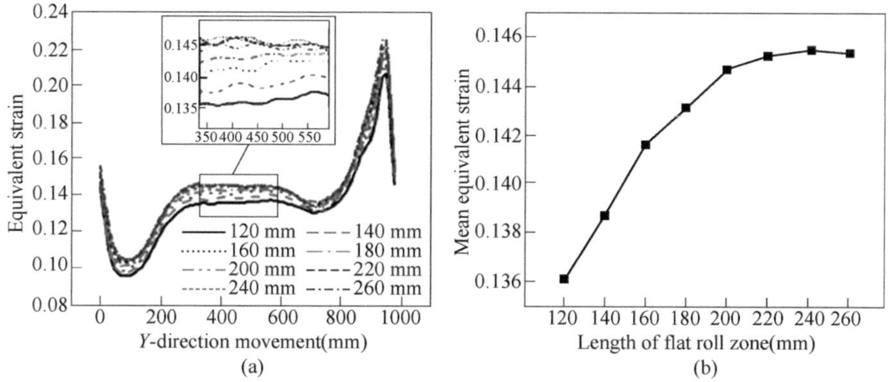

Fig. 7.46 **a** Equivalent strain and **b** equivalent strain of the bloom centerline node for CSC-Rolls of different convex platform region length

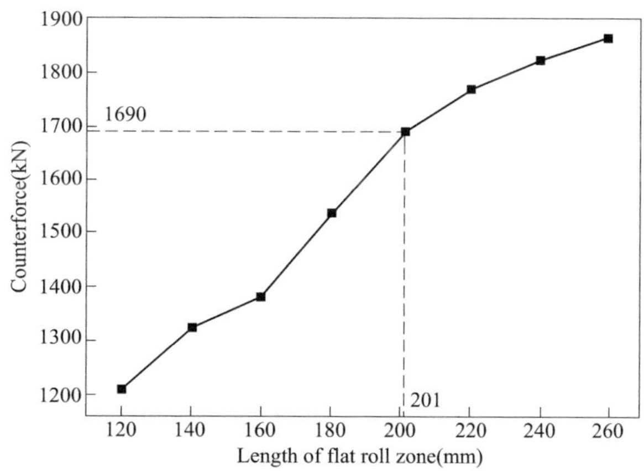

Fig. 7.47 Counterforce for CSC-Rolls of different convex platform region lengths

influence on the center quality of the bloom. In addition, Hcp has little effect on the amount of reduction in porosity zone. Therefore, it will not be discussed here.

As shown in Fig. 7.49, the counterforce decreases with the increase of Hcp and the maximum equivalent strain of the inner surface increases. When the raised height is 27.4 mm, the counterforce is 1690 kN, so the preferred raised height is 30 mm.

(E) Validation of applicability of calculation results

To verify the applicability of the above calculation of CSC-Roll, increase the reduction amount to 15 mm, and the distribution of the reduction in loose area under different Lcp is shown in Fig. 7.50. It can be seen that the reduction in porosity

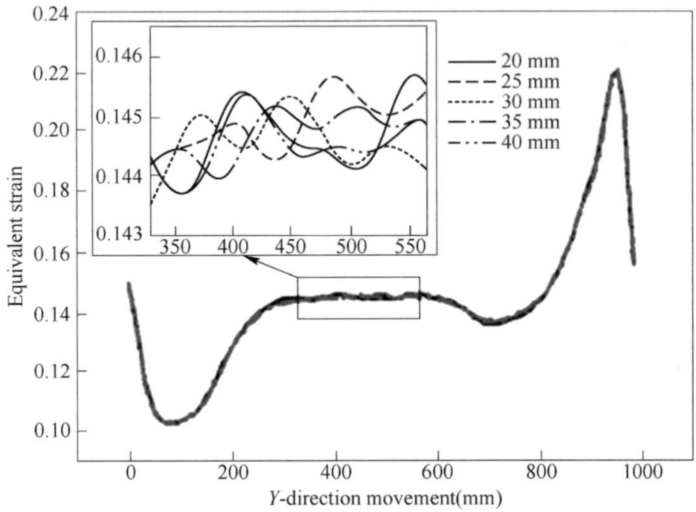

Fig. 7.48 Equivalent strain of the bloom centerline nodes for CSC-Rolls of different Hcp

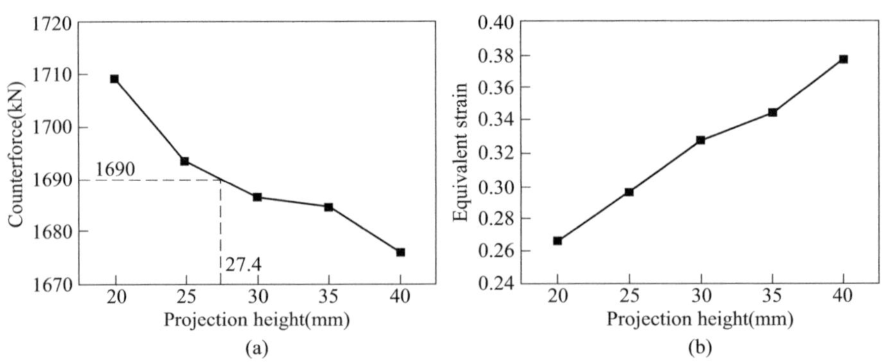

Fig. 7.49 **a** Counterforce and **b** maximum equivalent strain of the bloom inner surface for CSC-Rolls of different Hcp

zone is increasing with the Lcp, and the reduction in thickness and width direction decreases after the Lcp reaches 200 mm. It shows that the optimum structure of CSC-Roll calculated above has well applicability.

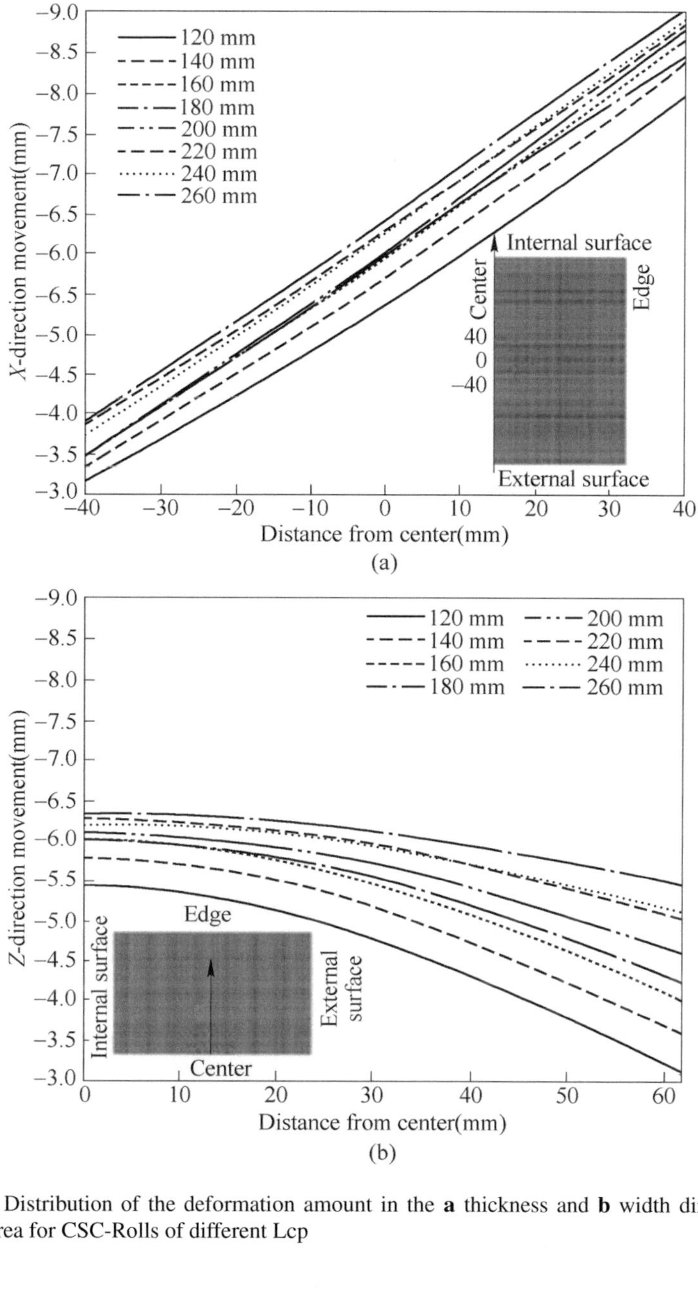

Fig. 7.50 Distribution of the deformation amount in the **a** thickness and **b** width directions of porosity area for CSC-Rolls of different Lcp

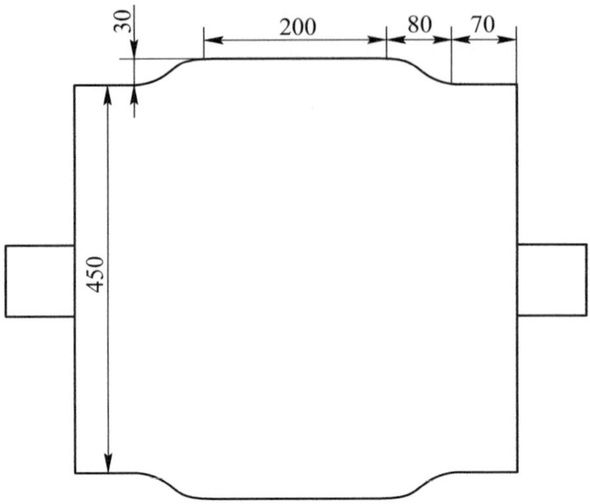

Fig. 7.51 Schematic of the optimal CSC-Roll structure

7.3.2.4 The Optimal CSC-Roll Structure

Based on the above simulation results, the key parameters of the CSC-Roll design for the 360 mm × 450 mm section No.45 steel bloom are as follows: the length of convex platform region is 200 mm, the height of the convex roll region is 30 mm, the length of the gradient curvature zone is 80 mm, the axial length is 500 mm, and the diameter of the edge zone is 450 mm. The schematic of the CSC-Roll is shown in is shown in Fig. 7.51.

7.4 Production Line Application of Solidification End Reduction Technology

Since Meishan steel introduced the first full streamline roll gap remote adjustable slab continuous casting machine in 2003, the solidification end soft reduction technology has been widely used in more than 100 production lines in China and has become the standard technology of advanced casting machines. In recent years, with the increasing demand for continuous casting production of high value-added steel products such as high carbon and high alloy, the ceaseless upgrading of the performance and specifications of steel products as well as the continuous clarification of solute transport law and crack initiation and propagation law at the end of solidification, the soft reduction technology has gradually broken through the limit of reduction $\leq 3\%$ slab thickness and central solid-phase rate fs ≤ 0.9 in the reduction interval. Furthermore, in view of the limitation that the conventional soft reduction deformation is difficult to fully penetrate into the core of the large section continuous casting billet and cannot meet the high homogeneity and high density requirements of the high-end large/thick section continuous casting billet, the heavy reduction technology at the solidification end has gradually become the preferred technology for thick slabs and blooms. In fact, whether it is light or heavy reduction, from the perspective of process control, it is necessary to apply a reasonable amount of reduction in a reasonable reduction range, so as to achieve the effect of improving segregation and porosity, which is the technical category of solidification end reduction. In this section, the application of solidification end reduction technology in four bloom and slab continuous casting production lines of Baosteel, Panzhihua Steel, and TangSteel in recent years is introduced.

7.4.1 Baosteel Meishan 230 mm Thick Continuous Casting Slab Production Line

No.3 slab continuous casting machine of Baosteel Meishan is a vertical bending slab continuous casting machine. Its section size is 230 mm × (900–1650) mm, and the length of the casting machine is 35.155 m, which is composed of 14 segments. In recent years, Baosteel Meishan has carried out product structure upgrading and adjustment, in which the content of C, Mn, and Cr in fine blanking steel for automobile parts is high, and the content of C can reach 0.65%. With the increase of the content of these easily segregated elements, the central segregation and porosity of the slab become more serious, resulting in a significant increase in the incidence of banded structure in hot rolled products. In regard to this, combined with the typical fine-blanking steel 16MnCr5 as an example, the solidification end reduction process of fine-blanking slab continuous casting was developed.

Figure 7.52 shows the change of solid fraction at the center point of 16MnCr5 continuous casting billet based on nail shooting experiment and thermal/mechanical simulation. It can be seen that in order to improve the solute segregation, Segment 8 and Segment 9 should be selected to complete the reduction. In this interval, the variation trend of reduction rate based on solute transport and reduction efficiency based on the deformation law of slab reduction process is shown in Fig. 7.53.

Table 7.9 shows the theoretical reduction calculated according to the reduction rate and reduction efficiency, in which the reduction of Segment 8 is 2.5 mm and the reduction of Segment 9 is 6.3 mm.

Additionally, thermal shrinkage deformation should also be taken into consideration. Figure 7.54 shows the thermal shrinkage deformation of the fine-blanking steel continuous casting slab along the casting flow. The self-thermal shrinkages of Segment 8 and Segment 9 are about 0.2 mm, so the reduction amount of Segment 8 and Segment 9 should be 2.7 mm and 6.5 mm. It can be seen that the total reduction amount of this process scheme is 8.2 mm, which is far more than the conventional light reduction range (the total reduction of the original process ≤ 3.5 mm).

The field test was carried out on the fine blanking steel produced by No.3 slab continuous casting machine of Meishan Iron and Steel Co., Ltd. The first-class adopts

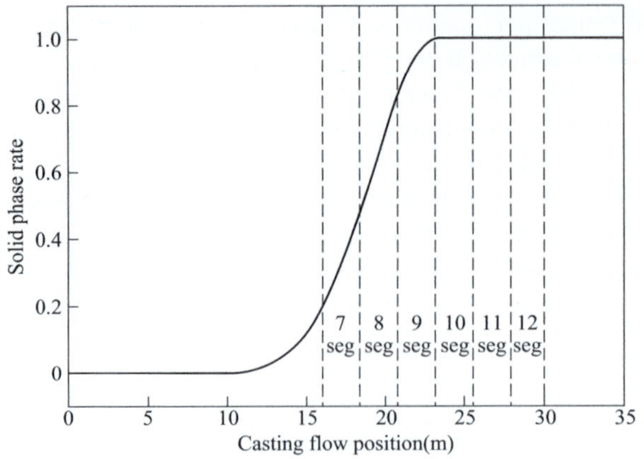

Fig. 7.52 Change trend of solid fraction at the center of slab

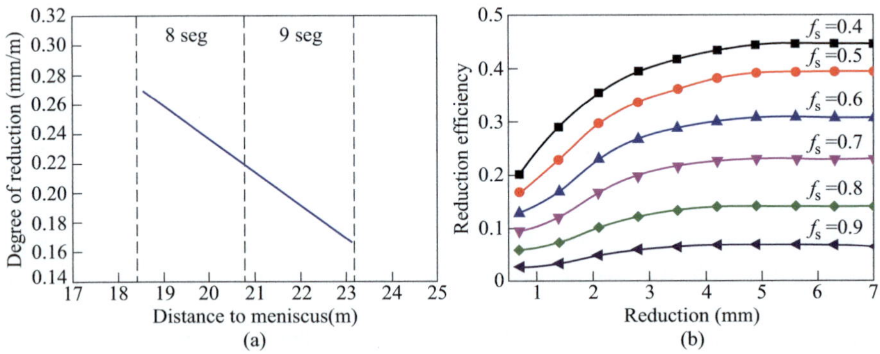

Fig. 7.53 Reduction range. **a** Reduction ratio and **b** reduction efficiency under different solid fraction conditions

the optimized reduction scheme, and the other first-class maintains the original reduction scheme. During the test, the two streams were sampled simultaneously and then compared. Figure 7.55 shows two sets of contrast samples of the acid etching samples of the slab and marks the results of the Manmark rating.

It can be seen that the macroquality of the slab produced after the optimization of the reduction scheme has been significantly improved. After large-scale popularization and application, the proportion of central segregation ≤ 2.2 grade of typical steel grades such as 40MnB and 16MnCr5 increased from 13 to 92.1%, and the rating of hot rolled banded structure was also significantly improved.

Table 7.9 Related parameters in the fan-shaped section under pressure

Segment	Solid-phase rate at the inlet	Solid-phase rate at the outlet	Average reduction rate	Interval length (m)	Liquid core reduction (mm)	Average solid-phase rate	Reduction efficiency	Theoretical reduction amount (mm)
8	0.5	0.8	0.25	2.4	0.60	0.65	0.27	2.5
9	0.8	1.0	0.20	2.4	0.48	0.90	0.07	6.3

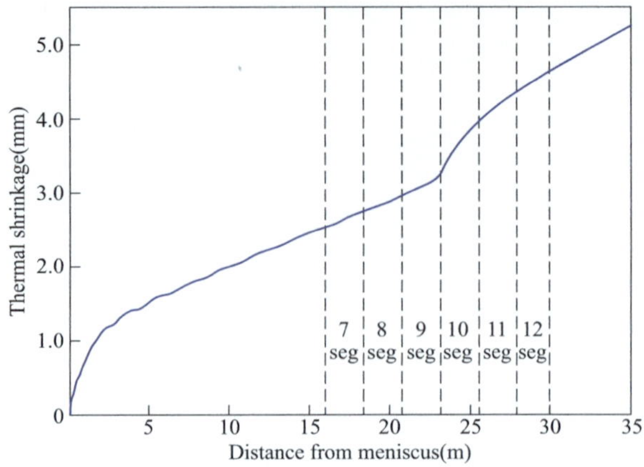

Fig. 7.54 Heat shrinkage law of fine blanking steel

7.4.2 Baosteel 320 mm × 425 mm Bloom Continuous Casting Production Line

The No.2 bloom caster in Baosteel is a typical vertical bending bloom caster. The section size is 320 mm × 425 mm, which is composed of 9 straightening machines. In order to improve its typical product (gear steel SAE5120H), the internal center segregation of gear steel is reduced. Combined with the production process of gear steel in Baosteel, the solidification end reduction process of typical gear steel bloom continuous casting was developed.

As shown in Fig. 7.56, the changes of the internal center solid of the gear steel under the typical casting speed of 0.65 m/min and different superheats are calculated. It can be seen that the solidification end point is located between the No.6 and No.7 straightening machines. Based on the mechanism that the solute elements enriched in the solid–liquid two-phase region are eliminated by the soft reduction extrusion at the solidification end of the continuous casting billet to reduce or eliminate the central segregation, the reduction roller and other factors are used as much as possible, and the reduction is selected at $fs = 0.3$–1.0.

To ensure the calculation accuracy, the infrared camera is used to randomly measure the narrow surface temperature or wide surface temperature distribution of the gear steel billet at different positions of the straightening machine in the actual production process and compared with the calculation results. It can be seen from Fig. 7.57 that the calculation results are more accurate.

The reduction efficiency of different rolls with solid fraction between 0.3 and 1.0 is given in Table 7.9. Based on the limitation of equipment conditions and the avoidance of reduction cracks, the reduction design is shown in Table 7.10.

Based on the above scheme and on-site implementation. As shown in Fig. 7.58, the internal quality of the gear steel slab is greatly improved after the implementation of the reduction scheme, the central shrinkage cavity on the longitudinal section of the slab is basically closed, and the V-shaped segregation on the cross-section

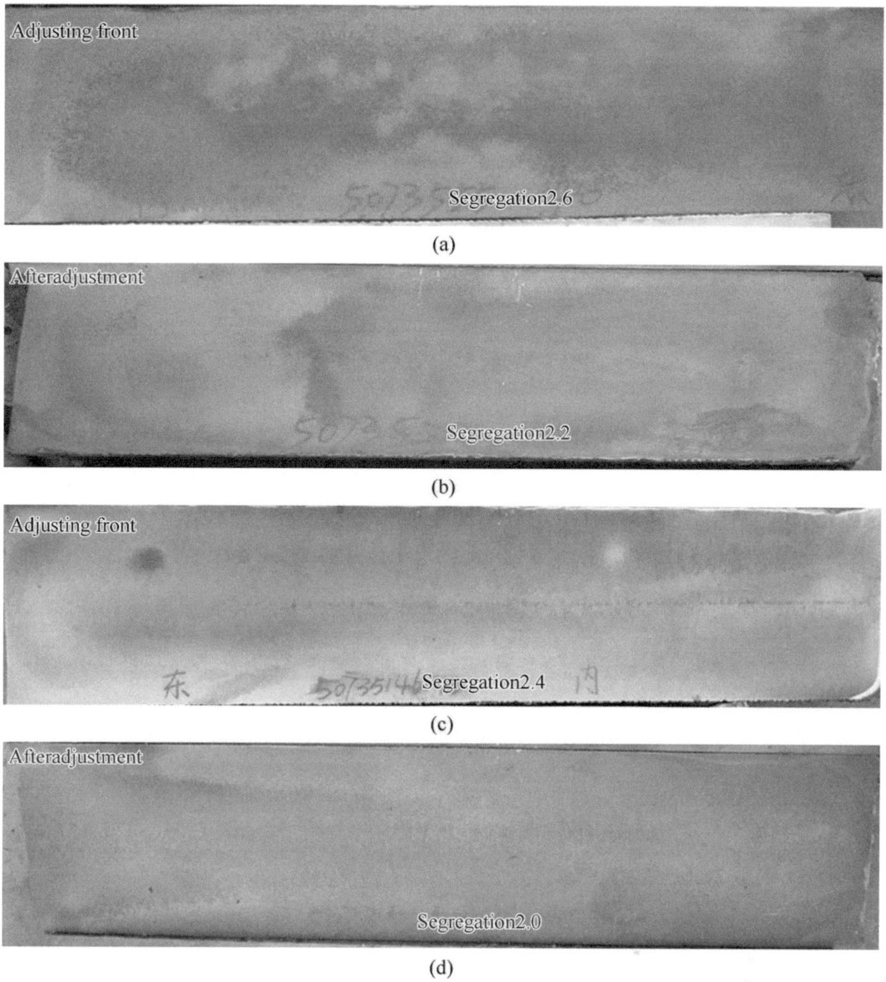

Fig. 7.55 Reduction scheme before and after the adjustment of low power contrast. **a** Before the adjustment of the scheme, the segregation level is 2.6; **b** after the adjustment of the scheme, the segregation level is 2.2; **c** after the adjustment of the scheme, the segregation level is 2.4; **d** after the adjustment of the scheme, the segregation level is 2.0

is obviously improved. According to the results of drilling cuttings sampling, the central carbon segregation of the slab decreased from 1.32 to 1.06.

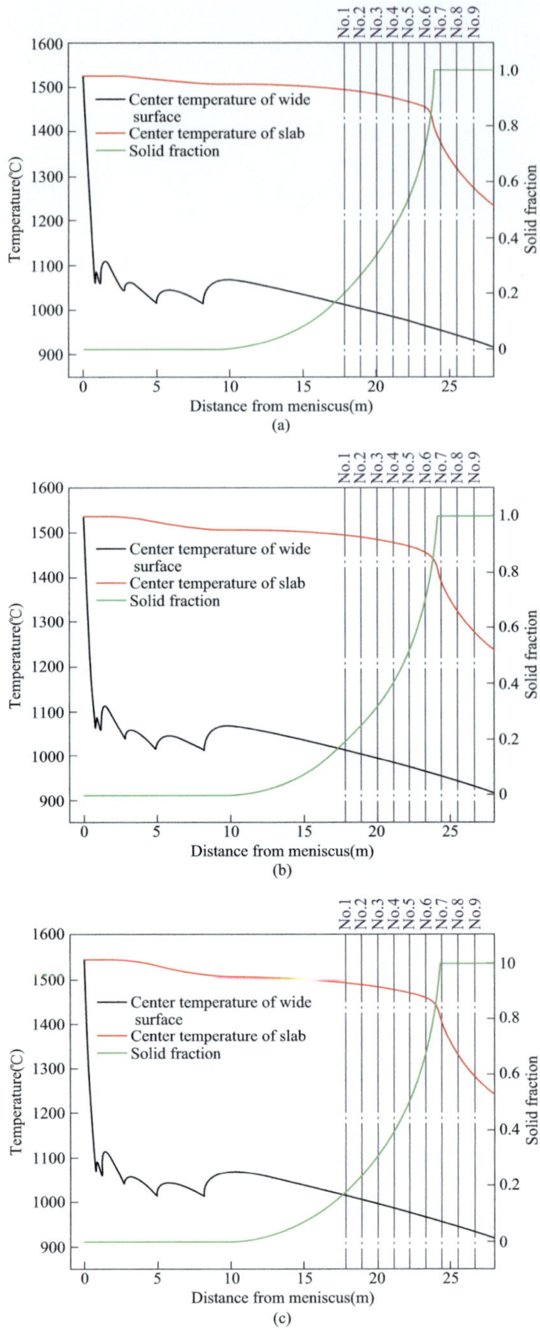

Fig. 7.56 0.65 m/min casting solidification end point calculation. **a** Superheat 20 °C; **b** superheat of 30 °C; **c** superheat of 40 °C

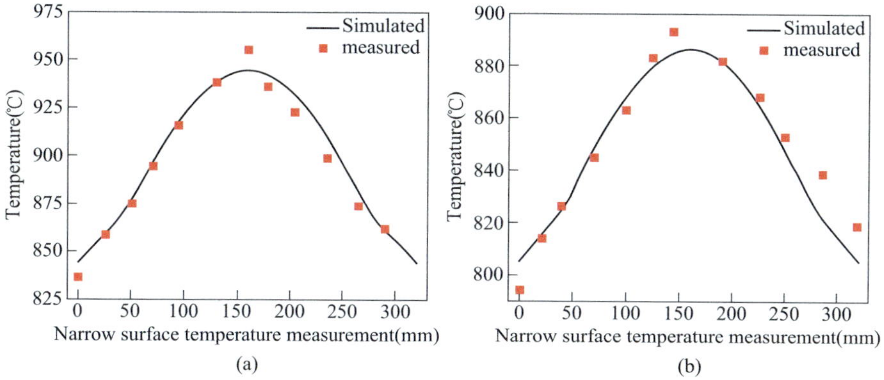

Fig. 7.57 **a** Comparison of the temperature measurement results of the front narrow surface of the No.5 straightening machine with the simulation results; **b** comparison of the temperature measurement results of the rear narrow surface of the No.9 straightening machine with the simulation results

Table 7.10 Reduction efficiency and reduction amount

Reduction roll	No.3	No.4	No.5	No.6	No.7
Reduction efficiency	0.425	0.368	0.315	0.172	0.163
Reduction amount (mm)	–	1	2	≥ 3	≥ 4

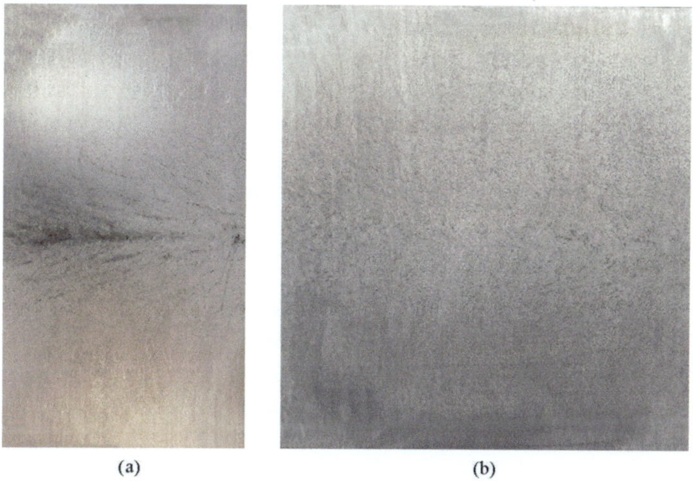

Fig. 7.58 Comparison of quality of longitudinal section slab before and after reduction of gear steel. **a** Before reduction; **b** after reduction

7.4.3 PanSteel 360 mm × 450 mm Bloom Continuous Casting Production Line

PanSteel No.2 bloom continuous casting machine has two sections, 360 mm × 450 mm section for the production of axle steel and other bar products, 320 mm × 410 mm section for the production of long-scale heavy-duty rail. Combined with the demand of section upgrading and transformation of No.2 billet continuous casting machine in PanSteel, the application of heavy reduction technology at the end of solidification was completed. This is also the first demonstration production line of heavy reduction of billet continuous casting in China, which can implement continuous and stable large reduction deformation after complete solidification of billet.

The meniscus of No.2 bloom caster in PanSteel is 750 mm from the bottom of mold. The secondary cooling zone includes five water spray cooling zones with a total length of 8.68 m. The specific secondary cooling zones are shown in Table 7.11.

Table 7.12 shows the position of the straightening machine of the No.2 bloom caster in PanSteel. It can be seen that there are a total of 7 straightening machines. The position distribution is from 20.351 m to 29.551 m, and the spacing between the straightening machines is about 1.5–1.6 m.

Taking the axle steel as an example, the calculation results of the solid-phase isotherm distribution in the thickness direction under different drawing speeds are shown in Fig. 7.59.

It can be seen from the figure that the distribution of solid-phase isotherms in the thickness direction is obviously different under different drawing speeds. The casting speed has a significant effect on the location of the solidification end point. When the casting speed is 0.5 m/min, the solidification end point is 20.67 m. When the casting speed is 0.55 m/min, the position of the solidification end point is 23.8 m; when the casting speed is 0.6 m/min, the position of the solidification end point is 26.38 m.

Table 7.11 Secondary cooling zone of bloom

Cooling zone	Length of each zone (m)	The distance from the end of cooling zone to the meniscus (m)
1	0.36	1.11
2	1.97	3.08
3	2.06	5.14
4	2.40	7.54
5	1.89	9.43

Table 7.12 Distance from each straightening machine to the meniscus

Number	No.1	No.2	No.3	No.4	No.5	No.6	No.7
Distance (m)	20.351	21.851	23.351	24.851	26.351	27.951	29.551

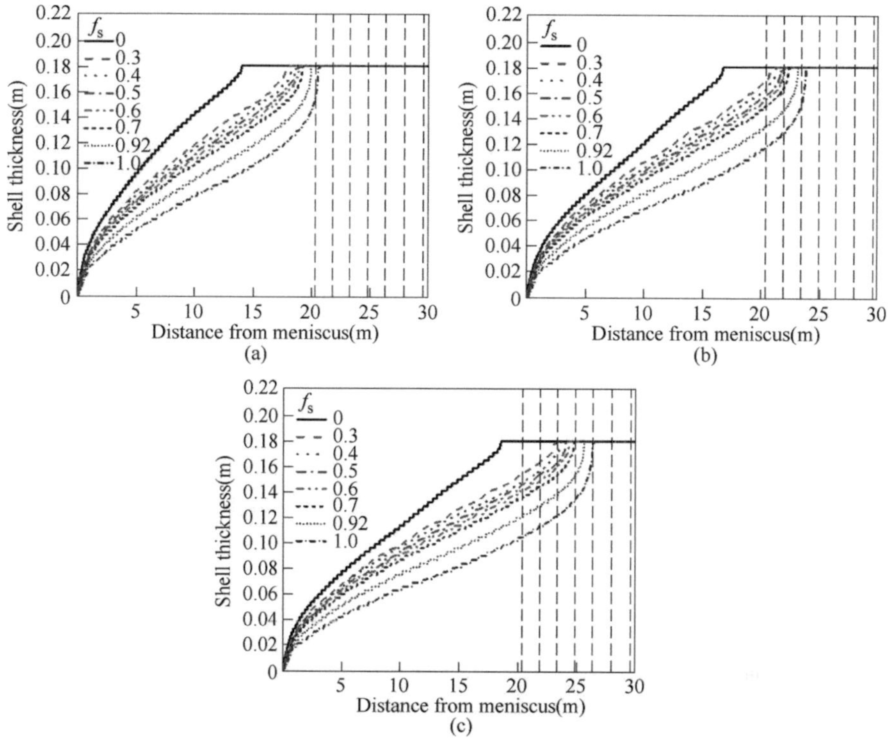

Fig. 7.59 Distribution of solid-phase isotherms in the thickness direction under different casting speed. **a** 0.50 m/min; **b** 0.55 m/min; **c** 0.60 m/min

When the casting speed increases by 0.05 m/min, the position of the solidification end point moves backward by 2.58 m.

In view of this, according to the above calculation results, the reduction interval is selected between No.2–No.6 straightening machines. Considering the influence of heat shrinkage on the reduction process, the constant casting speed mode is used for casting under heavy reduction. When the casting speed is 0.53 m/min, the process scheme is shown in Table 7.13.

Table 7.13 Axle steel reduction process

Parameter	Casting speed	No.1	No.2	No.3	No.4	No.5	No.6	No.7	Total
Pressure (MPa)	0.53 m/min	50	50	80	115	140	180	–	
Reduction amount (mm)		0	1.0	1.8	4.0	5.0	≥ 10	–	≥ 21.8

Figure 7.60 shows that the macroquality of the cross-section of the slab under different processes is compared. It can be seen from the figure that continuous shrinkage cavity defects are formed in the center of the slab without reduction, and V-shaped segregation is serious. The shrinkage cavity of the slab is improved during soft reduction, but there are still porosity and segregation bands in the center. After the implementation of heavy reduction, the central shrinkage cavity and segregation have almost entirely vanished.

Figure 7.61 shows the statistical results of macroquality under different reduction processes. With the gradual increase of reduction, the central segregation and porosity defects of slab have been improved simultaneously. When the heavy reduction with the convex roller is adopted, the proportion of the center porosity 0.5 grade of the slab reaches 100%.

With the increase of the density of the slab, the conventional detection method is not enough to characterize the improvement effect of the reduction process on the quality of the center of the bloom. In light of this, the absolute density of the slab under different reduction modes was measured and compared through the gas expansion replacement method. As shown in Fig. 7.62, as the reduction increases, the

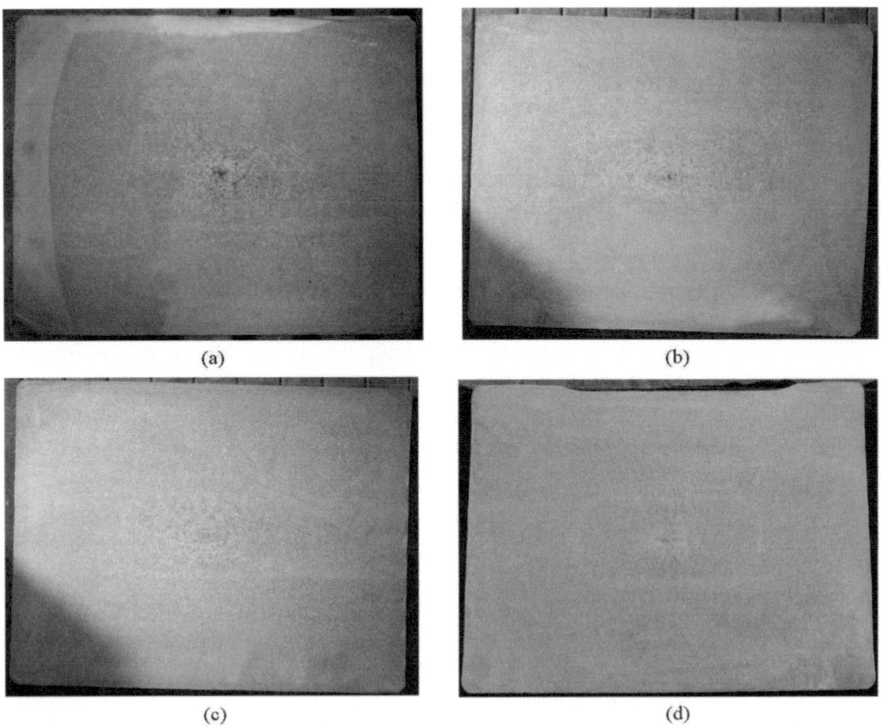

Fig. 7.60 Comparison of macroquality of axle steel rolled products with different reduction processes. **a** No reduction; **b** soft reduction; **c** heavy reduction with flat roller; **d** heavy reduction with convex roll

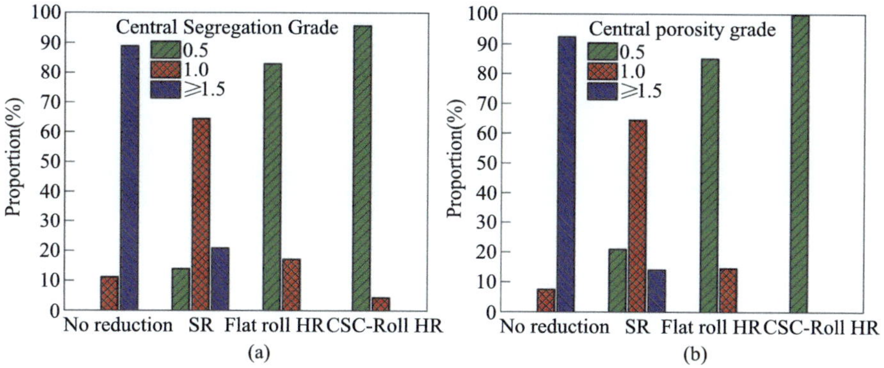

Fig. 7.61 Quality before and after the implementation of the technology. **a** Center segregation; **b** center porosity

true density of the center position of the slab gradually increases. After the CSC-Roll is put into use, the true density of the slab area is 0.21% higher than that of the flat roll heavy reduction and 0.86% higher than that of the soft reduction.

In order to quantitatively analyze the improvement effect of solidification end reduction process on the quality of rolled products. As shown in Fig. 7.63, after the implementation of heavy reduction, the density of the central area of the rolled axle with a diameter of 230 mm was increased by 13.7% (in situ analysis), and the corresponding qualified rate of axle flaw detection was increased from less than 70 to 100%.

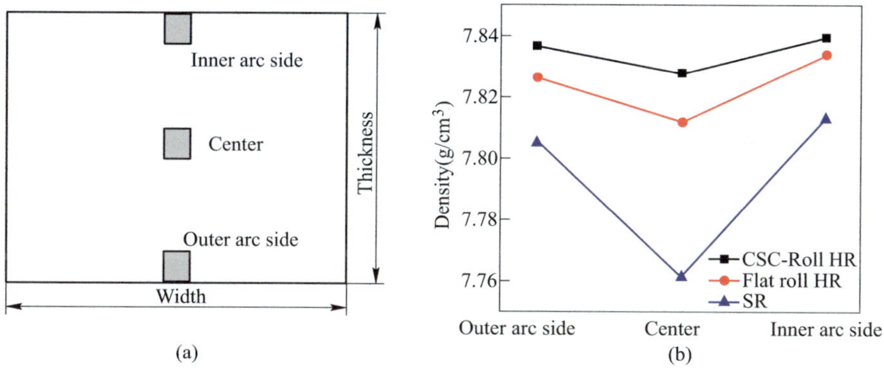

Fig. 7.62 Comparative analysis of the density at different positions of the slab. **a** Sampling position; **b** true density test results

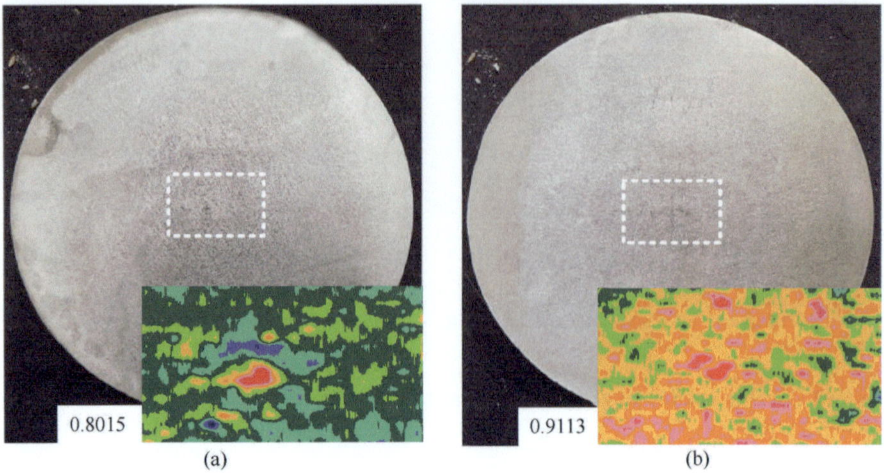

Fig. 7.63 Center relative density of rolling material of axle steel under different processes

7.4.4 TangSteel 280 mm Wide-Thick Slab Continuous Casting Production Line

In 2014, combined with the upgrading requirements of No.2 wide and heavy slab caster of TangSteel, the overall equipment, process, and control technology of heavy reduction at the solidification end of wide and heavy slab continuous casting were fully put into use, and the first wide-thick slab continuous casting production line with heavy reduction function in China was built and put into operation. The calculated steel grade is Q235, the casting speed is 0.83 m/min, and the parameters in the reduction process are shown in Table 7.14.

During the reduction process, the change trend of the reduction at different positions of the slab is shown in Fig. 7.64. During the reduction process, the reduction of the slab shows the trend of step-by-step, and near the corresponding position of the casting roller, due to the extrusion effect of the casting roller, the slab produces a certain bulging deformation, and the final reduction of the slab shows a sudden increase and then a rapid decrease trend. Due to the maximum reduction in the Segment 11 during the reduction process, the increase trend of the reduction of the slab in the segment is the fastest.

Table 7.14 Reduction process parameters

Segment	Roll diameter (mm)	Space between rolls (mm)	Relative reduction amount (mm)	Absolute reduction amount (mm)
10	390	410	5.38	5.38
11	390	410	16.27	21.65
12	390	410	6.40	28.05

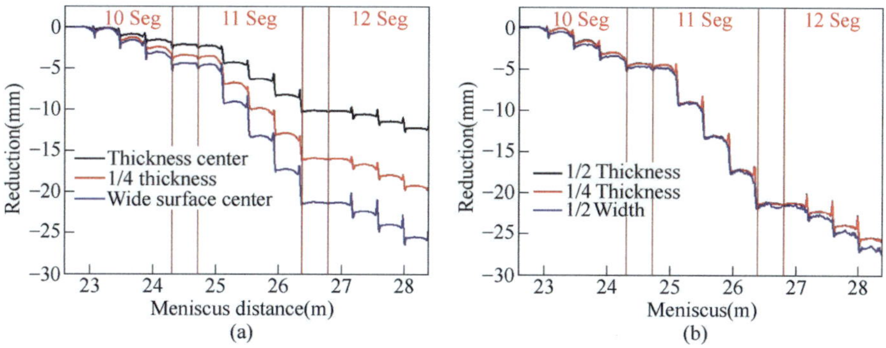

Fig. 7.64 Variation trend of reduction in different positions of slab. **a** Thickness direction and **b** width direction during the reduction process

From Fig. 7.64a, it can be seen that the closer the distance from the surface of the slab, the faster the reduction of the slab increases, and the more obvious the instantaneous bulging deformation formed near the casting roller position. At the end of the reduction process, the thickness center of the slab, the 1/4 position of the thickness, and the reduction of the slab surface are 12.1 mm, 19.5 mm, and 25.6 mm, respectively.

It can be seen from Fig. 7.64b that the temperature at the center of the wide surface of the slab is similar to that at the 1/4 position of the wide surface and the reduction effect of the casting roller is the same. Therefore, the change trend of the reduction of the above two points is basically the same. Because the temperature of the corner point of the slab is low, its resistance to deformation is strong, and during the reduction process, the resistance of the corner point of the slab along the width direction is smaller than that during the deformation process. Therefore, there is no instantaneous bulging deformation near the corresponding position of the casting roller.

As shown in Fig. 7.65, the wide 1/4–1/8 area with the most serious center segregation of the wide and thick slab was sampled by the designed reduction scheme. Figure 7.65 shows the comparison results of slab quality under different reductions (single segment) at the end of solidification. As the amount of reduction increases, the central segregation of the slab is gradually ameliorated. When the reduction reaches 11.5 mm, the center segregation of the slab basically disappears. At this time, the reduction deformation fully extrudes the molten steel rich in solute segregation elements in the liquid core.

As shown in Fig. 7.66, with the increase of reduction, the improvement effect of center segregation of slab is more obvious, and the C-grade rate of center segregation increases gradually. Based on the statistical outcomes of large-scale production after the implementation of heavy reduction, the center segregation C-grade rate of wide-thick slab reached 92.9%.

The in situ analysis method was used to measure the density of wide-thick slabs under different processes. The total reduction of the soft reduction process was

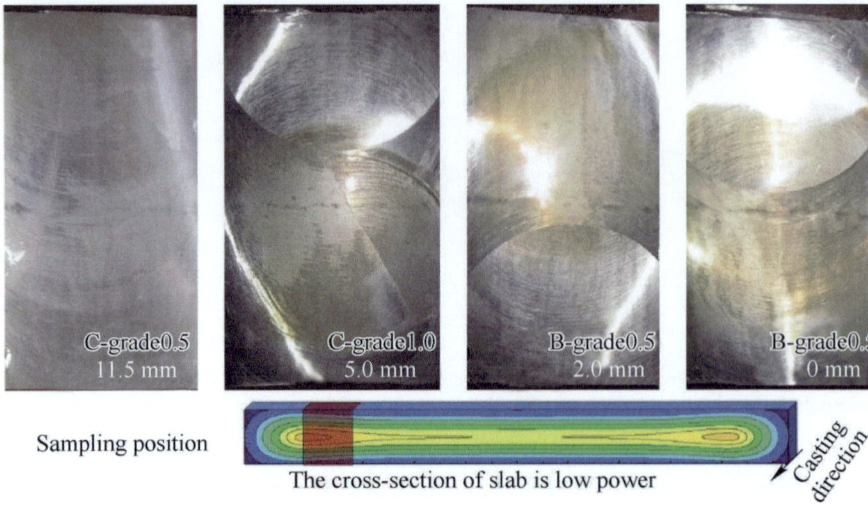

Fig. 7.65 Comparison of slab quality under different reduction amount

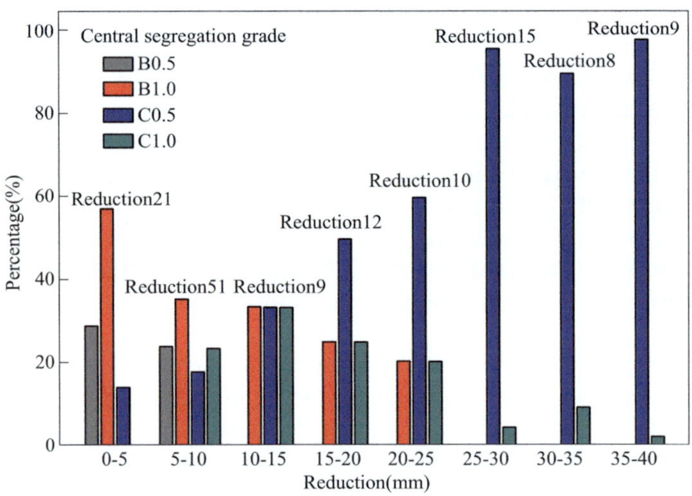

Fig. 7.66 Relationship between reduction and quality rating

8.9 mm, and the total reduction amount of the heavy reduction process was 28.9 mm. The sampling position and density test results of Q345GJ steel continuous casting billet are given in Fig. 7.67 and Table 7.15, respectively. It can be seen that the heavy reduction process has no obvious effect on the improvement of the inner and outer arc density of the slab, but the middle area is obviously improved. Compared with the soft reduction, the density of the central area of the slab is increased by 14.52% after the implementation of heavy reduction.

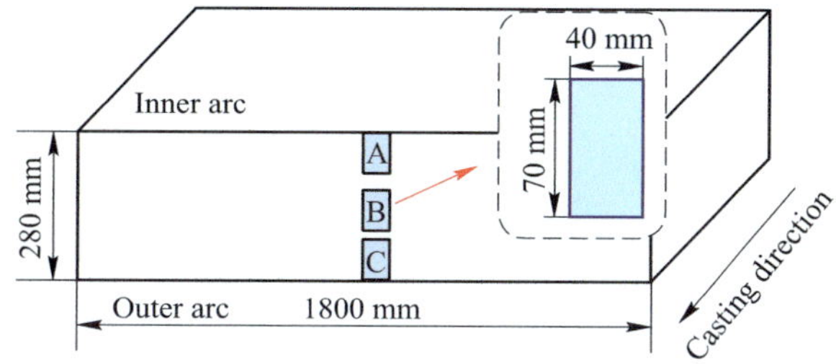

Fig. 7.67 Q345GJ steel continuous casting billet in situ analysis sampling position

Table 7.15 Q345GJ steel continuous casting billet density test results

	Inner arc A	Middle B	Outer arc C
Soft reduction	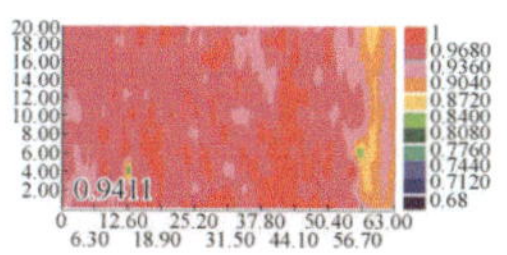	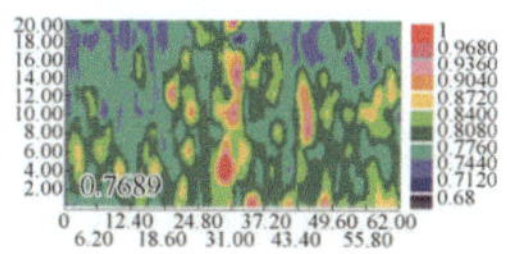	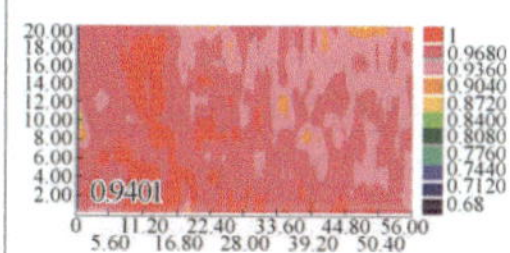
Heavy reduction	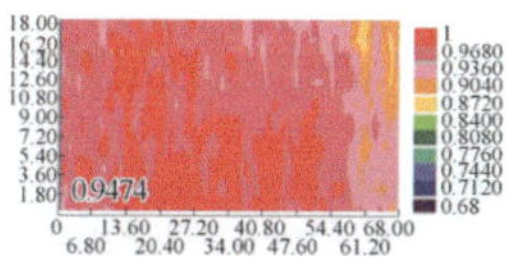	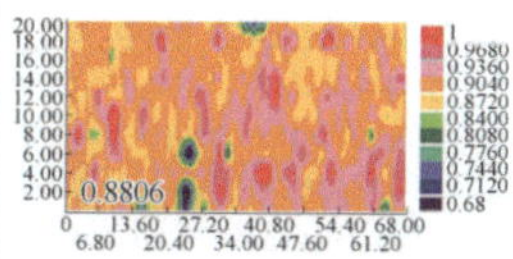	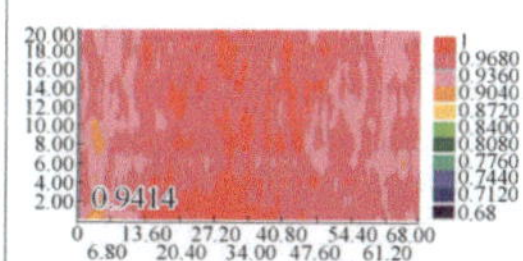

Table 7.16 Q345GJ steel plate density test results

	Inner arc A	Middle B	Outer arc C
Soft reduction	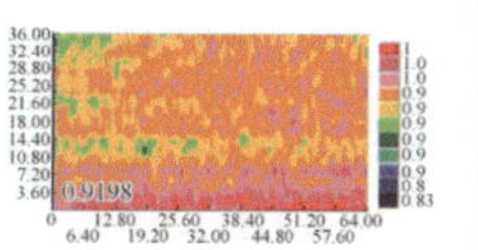	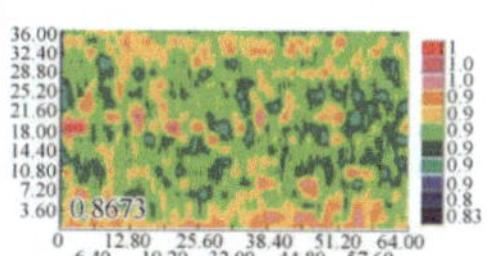	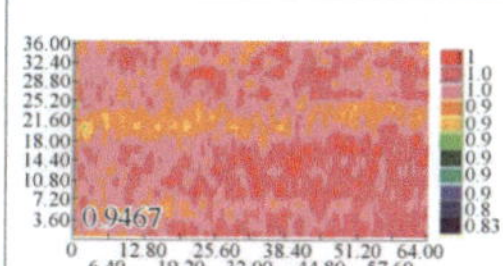
Heavy reduction	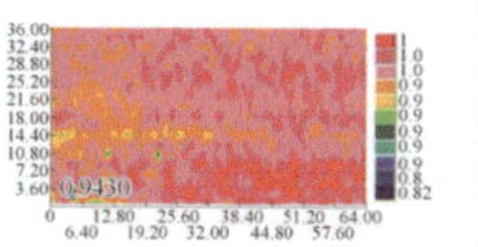	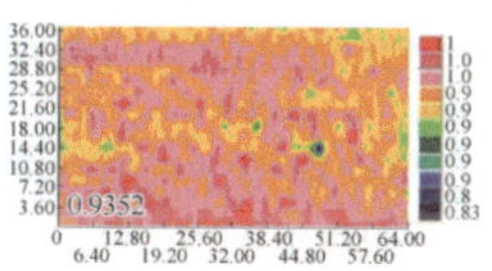	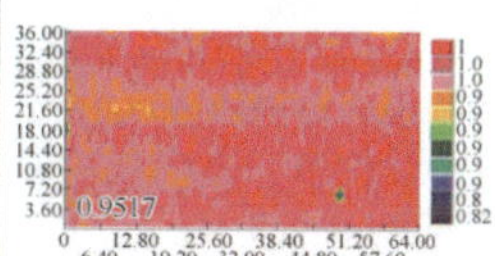

Further comparative analysis was conducted on the internal quality of thick plates obtained after rolling the wide and thick slabs produced by the above different processes, Fig. 7.68 and Table 7.16 respectively give the in situ analysis sampling position and density test results on the 120 mm thick Q345GJ extra-thick plate. Compared with the soft reduction, the density of the center area of the extra-thick plate rolled by the heavy reduction is increased by 7.83%.

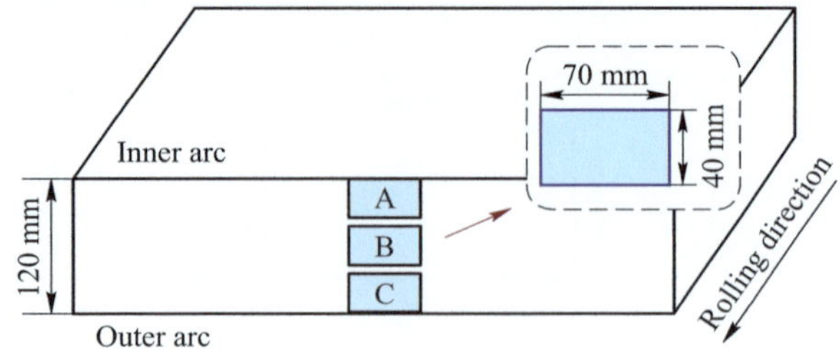

Fig. 7.68 Q345GJ steel plate in situ analysis sampling position

References

1. Ji C, Zhu M (2015) An online detection method of solidification end position in bloom continuous casting production process. China, CN104493121A[P]
2. Ji C, Wang C, Deng S et al (2017) A thermal tracking calculation method of continuous casting billet based on the distribution calculation of thermal physical parameters. China, CN201710004849.X [P]
3. Wu C, Ji C, Wang L et al (2015) A method for formulating the basic roll gap of wide and thick slab continuous casting machine, China, CN105033214B[P]
4. Ji C, Luo S, Zhu M et al (2014) Uneven solidification during wide-thick slab continuous casting process and its influence on soft reduction zone. ISIJ Int 54(1):103–111
5. Cheng J, Yi J, Wenzhong X et al (2012) Research and application of roll gap on-line calibration method for withdrawal machines. China Metall 22(2):10–13
6. Ji C, Zhu M (2015) A torque control method of stretching machine to improve the reduction effect of solidification end of continuous casting slab. China, CN104889354B[P]
7. Ji C, Zhu M (2016) A two-stage continuous dynamic heavy reduction method for continuous casting billets. China, CN106001476B[P]
8. Ji C, Zhu M, Zhang H et al (2017) A single point and continuous heavy reduction process for solidification end of continuous casting slab. China, CN106735026A[P]
9. Ji C, Zhu M, Li G (2020) Critical strain measurement and crack propagation prediction method for surface crack propagation of continuous casting slab. China, CN202010088411.6[P]
10. Ji C, Zhang J, Li G et al (2020) A method for predicting the crack risk of continuous casting slab and its application. China, CN201911315780.8[P]
11. Li G, Ji C, Zhu M (2021) Prediction of internal crack initiation in continuously cast blooms. Metall Mater Trans B 52(2):1164–1178
12. Ji C, Zhu M, Li D (2009) Process simulation and reproduction system of bloom continuous casting production process. China, CN200910011836.0 [P]
13. Wanguo L, Lianyu R, Kai T (2015) Some basic questions of roller layout and related equipment design of slab caster. Contin Cast 40(6):48–56
14. Yeping Y, Lishi D,Xuan C (2016) Brief introduction of slab caster segment rollers. Metall Equip (6):45–48
15. Zhou B (2010) Selection & design of the segment rollers for slab continuous caster. Heavy Mach (s1):208–211
16. Jun J (2008) Research for segment rollers of slab caster. Heavy Mach 5:29–33
17. Yin Z (1999) The deflection of statically determinate beam and statically indeterminate beam is solved by finite difference method. J Mach Des 16(8):34–36
18. Kang F, Zhongci S (2010) Mathematicaltheory of elastic structure. Science Press, Beijing
19. Xucheng W (2003) Finite element method. Tsinghua University Press, Beijing
20. Mingyan L, Ping L, Xingjia L et al (1990) Design and calculation of slab continuous casting machine (I). China Machine Press, Beijing
21. Daxian C (2017) Mechanical design manual (6th edn). Chemical Industry Press, Beijing

22. Jinli W, Liquan L, Shaogang L (2014) The research on oil-air lubrication in grooved sliding bearing. Key Eng Mater 572:384–387
23. Liu J-S, Parks GT, Clarkson PJ et al (2005) Topology/shape optimization of axisymmetric continuum structures metamorphic development approach. Struct Multidiscip Optim 29:73–83
24. Okamoto M, Okimori M, Kaneko N et al (1990) Roll shape of soft reduction for bloom. CAMP—ISIJ 3(4):1174
25. Masaya T, Yukihiro M, Yasuaki M et al (2015) NSENGI's new developed bloom continuous casting technology for improving internal quality of special bar quality. METEC and 2nd ESTAD 307–318
26. Ogibayashi S, Uchimura M, Isobe K et al (1990) Improvement of center segregation in continuously cast blooms by soft reduction in the final stage of solidification. In: Proceedings of the 6th International Iron and Steel Congress, Nagoya Math J 271–278
27. Moon CH, OH KS, Lee JD (2012) Effect of the roll surface profile on centerline segregation in soft reduction process. ISIJ Int 52(7):1266–1272
28. Ji C, Zhu M (2015) A tapered curvature convex roller for bloom continuous casting and its use method. China, CN201410666353.5[P]

MIX
Papier aus verantwortungsvollen Quellen
Paper from responsible sources
FSC® C105338

If you have any concerns about our products,
you can contact us on
ProductSafety@springernature.com

In case Publisher is established outside the EU,
the EU authorized representative is:
Springer Nature Customer Service Center GmbH
Europaplatz 3, 69115 Heidelberg, Germany

Printed by Libri Plureos GmbH
in Hamburg, Germany